Sustainability Sciences in Asia and Africa

Agroecosystem Dynamics and Sustainable Practices

Sustainability as a scientific endeavor requires broad understanding of the interconnections in our global environment. It is therefore positioned at the crossroads of research in natural sciences, technology and human sciences. This book series supports the global efforts by providing timely coverage of the progress, opportunities, and challenges towards achieving sustainable development goals (SDGs) in Asia and Africa. It fosters the research in transdisciplinary academic fields spanning across agriculture and global food security, health and diseases, access to clean water and affordable energy, climate resilience and effects of global climate change on biodiversity, water and land. This series is planned to narrate the stories of scientific development and progression towards SDGs from the regions of Asia and Africa. The overall aim of the series is to provide a comprehensive publication platform and act as a knowledge engine in the growth of sustainability sciences with a special focus on developing nations. The series constitute of four sub-series each focusing on a specific theme closely linked to a set of SDGs. Each sub-series is leaders and experts in agricultural sciences, environmental sciences, biomedicine and medical sciences respectively. Books in this series are useful for environmentalists, biotechnologists, plant scientists, medical practitioners, conservation biologists, policy makers, NGOs and scientific bodies working in the areas of sustainability sciences in the developing regions.

This book series publishes research associated with the prevailing opportunities and challenges faced by agroecosystems in Asia and Africa. It also explores the future scope of agroecological techniques and offers guidance on sustainable management of these systems. The series discusses important areas such as sustainable agricultural production, soil health, water resource management, biodiversity conservation, animal health and productivity, insect management strategies, genome editing and climate change mitigation. Further, it includes the challenges faced by African agriculture in arid and semi-arid regions, particularly in response to climate change. It highlights issues such as water shortages, soil degradation, desertification, and the need for sustainable farming techniques.

Sustainable agroecosystem maintenance entails strategies that improve overall environmental health and economic viability while limiting negative environmental impacts. The series takes a multidisciplinary approach to investigate creative ways for boosting sustainable agriculture, increasing food security, and improving lives for people in global south. This book series acts as a global publishing platform to collate and disseminate knowledge on fostering resilient and productive agroecosystems, thus enhancing accessibility, and fostering informed decision-making.

Kamel A. Abd-Elsalam • Heba I. Mohamed
Editors

Elicitors for Sustainable Crop Production

Overcoming Biotic Stress Challenges in Plants

Springer

ARC

Editors
Kamel A. Abd-Elsalam
Plant Pathology Research Institute
Agricultural Research Center
Giza, Egypt

Heba I. Mohamed
Biological and Geological Sciences
Department
Ain Shams University
Cairo, Egypt

ISSN 2730-6771 ISSN 2730-678X (electronic)
Sustainability Sciences in Asia and Africa
ISSN 3059-247X ISSN 3059-2488 (electronic)
Agroecosystem Dynamics and Sustainable Practices
ISBN 978-981-96-6571-6 ISBN 978-981-96-6569-3 (eBook)
https://doi.org/10.1007/978-981-96-6569-3

This Springer imprint is published by the registered company Springer Nature Singapore Pte Ltd.
The registered company address is: 152 Beach Road, #21-01/04 Gateway East, Singapore 189721, Singapore

Preface

Climate change poses a threat to agriculture, and its effects on the sustainability and financial success of agricultural production systems are significant. The impact of biotic and abiotic stresses on crops is expected to grow due to climate change. Biotic stress is stress caused by other organisms, including bacteria, viruses, fungi, parasites, advantageous and detrimental insects, weeds, and native or cultivated plants. Since cash crops suffer enormous financial losses, biotic stresses are a primary area of study in agriculture. Both practical and economic considerations are impacted by the connection between biotic stress and plant production. Population dynamics, plant-stressor coevolution, and ecosystem nutrient cycling are all impacted by biotic injury's effect on crop yield.

Agriculture and food security are still seriously threatened by crop plant losses due to pests and disease. In order to control pests and diseases, agriculture became more dependent on synthetic chemical pesticides in the second half of the twentieth century, particularly in the intensive agricultural methods that are typical in the industrialized world. However, this reliance on chemical control is becoming unsustainable in the twenty-first century. Due to the development of resistance in the target pests, pesticides often have a short shelf life. They are also increasingly recognized to have detrimental effects on biodiversity, the health of agricultural workers, and even customers.

The majority of disease control is accomplished by chemical substances like herbicides, insecticides, and fungicides. However, because these pesticides are harmful to the environment and living things, new, safe ways to control disease must be found. Inducing the expression of genes linked to pathogenesis and shielding plants from illness are two ways that elicitors start systemic acquired resistance or induced systemic resistance in the host.

Elicitors are substances derived from both biotic (living) and abiotic (non-living) sources that can cause plants to react to stress by intensifying the manufacture of secondary plant metabolites; this process is known as "elicitation." Elicitors are divided into two categories based on their origins: biotic and abiotic elicitors. In addition to heavy metals and intracellular signaling molecules (such as methyl jasmonate, salicylic acid, brassinosteroids, polyamines, abscisic acid, gibberellic acid, and gaseous molecules like ethylene and nitric oxide), abiotic elicitors include physical, chemical, and hormonal factors such as UV radiation, osmotic stress, drought, salinity, and temperature stress (both high and low). On the other hand, biotic elicitors are biologically derived compounds, such as polysaccharides and pathogens (extracted from bacteria, fungi, and yeast).

Elicitors are natural or synthetic compounds that cause systemic resistance in plants. They can be secreted by bacteria or by a chemical combination of components. The infections raise the intracellular concentration of reactive oxygen species (ROS) and calcium in the plant cell, which results in programmed cell death (PCD) or apoptosis.

In general, seven categories of substances that act as biostimulants can be defined: (i) chitin and chitosan derivatives; (ii) free amino acids and N-containing substances (peptides, polyamines, and betaines); (iii) antitranspirants (polyacrylamide); (iv) plant growth-promoting rhizobacteria (PGPR); (v) arbuscular mycorrhizal fungi (AMF); (vi) phytohormones (jasmonic acid, melatonin); and (vii) *Trichoderma* spp. The use of biostimulants as biopesticides is a promising alternative for the control and prevention of fungal and fungal-like diseases. They are currently considered an emerging class of crop management products that aim to moderate crop stress to increase crop productivity.

This book explores the biosynthesis of biostimulants (elicitors) and their role in reducing biotic stress and increasing plant growth and productivity. It provides a summary of recent research in biostimulants, molecular and physiological interactions, and field monitoring of biotic stress in plants. This will be the first comprehensive book to provide a summary of the latest research on elicitors assisting in biotic stress management in crops, with a focus on sustainability in agriculture. The current volume is interdisciplinary and will be very useful for students, teachers, researchers, plant physiologists, environmental scientists, and agrochemical companies. It will also be of immense value to graduate students in agriculture, biotechnology, and environmental university programs. Professionals or practitioners in research institutes involved in plant science, material science, and crop production research will also benefit from this book.

Giza, Egypt — Kamel A. Abd-Elsalam
Cairo, Egypt — Heba I. Mohamed

Contents

Editors and Contributors

About the Editors

Kamel A. Abd-Elsalam Ph.D., is currently a Research Professor at the Plant Pathology Research Institute, Agricultural Research Center, Giza, Egypt. Dr. Kamel's research interests include developing and improving plant biosecurity diagnostic tools, understanding fungal pathogen genomes, and creating eco-friendly hybrid nanomaterials for controlling toxigenic fungi and plant diseases as well as applications in agro-ecosystems. Kamel earned his Ph.D. in Molecular Plant Pathology from Christian Albrechts University of Kiel (Germany) and Suez Canal University (Egypt). In 2008, he was awarded a postdoctoral fellowship from the same institution. Dr. Kamel served as a visiting associate professor at Mae Fah Luang University in Thailand, the Institute of Microbiology at Technical University of Munich (TUM) in Germany, the Laboratory of Phytopathology at Wageningen University in the Netherlands, and the Plant Protection Department at Sassari University in Italy. He has published 26 books related to nanobiotechnology applications in agriculture and plant protection with major publishing houses such as Springer, Taylor & Francis, and Elsevier. Since 2019, he has served as the Editor-in-Chief of the Elsevier book series Nanobiotechnology for Plant Protection and as the Series Editor for Applications of Genome Modified Plants and Microbes in Food and Agriculture. Additionally, he has been an active member of the

Elsevier Advisory Panel since 2020, providing feedback and suggestions to improve Elsevier's products and services. He has published more than 252 scientific research articles in international and regional specialized journals with high impact factors, and has an h-index of 44, an i-10 index of 137, and over 7421 citations. He served as a Guest Editor for the *Journal of Fungi, Plants, and Microorganisms*, and as a Review Editor for *Frontiers in Genomic Assay Technology*, *Agrochemicals*, and *Plant Nano Biology*. He was ranked in the top 2% of the most influential scientists in the world in nanobiotechnology for 2020, 2021, and 2022 by Elsevier and Stanford University. In 2014, he was awarded the Federation of Arab Scientific Research Councils Prize for excellent scientific research in biotechnology (fungal genomics), ranking first. Additionally, he received the Egyptian State Award for Agricultural Scientific Excellence from The Academy of Scientific Research and Technology (ASRT), Egypt, in 2022 with a silver medal.

Heba I. Mohamed is a Professor of Plant Physiology, at the Faculty of Education, Biological and Geological Sciences Department, Ain Shams University. Dr. Heba completed her M.Sc. and Ph.D. in Plant Physiology at the Faculty of Education, Ain Shams University. Her research interests include biotic and abiotic stresses, plant biochemistry, use of eco-friendly compounds to alleviate plant stress, plant secondary metabolites, and genetic differences between different genotypes. She has edited 6 books in Springer Nature. She has published 40 book chapters, 10 review articles, and more than 120 research articles in international peer-reviewed journals. She is a reviewer of international journals. She also is an editor of the Phyton journal. She has obtained a certificate of recognition in honor of achievement in an international publication that supports Ain Shams University World Ranking of 25/6/2012, 2017, 2018, 2019–2025. She obtained the Country Encouragement Prize in Cairo in 2018. Also, Dr. Heba was listed among the world's top 2% scientists by Stanford University for the years 2021, 2022, 2023, and 2024. This recognition is based on a global ranking that identifies the most influential researchers worldwide.

Contributors

Toka E. Abdelkhalek Biotechnology English Program, Faculty of Agriculture, Cairo University, Giza, Egypt

Kamel A. Abd-Elsalam Plant Pathology Research Institute, Agricultural Research Center, Giza, Egypt

Haseeb Ahmad Guangxi Key Laboratory of Agro-Environment and Agro-Products Safety, Key Laboratory of Crop Cultivation and Physiology, College of Agriculture, Guangxi University, Nanning, China

Sana AlHajj Ahmad Faculty of Agricultural and Food Sciences, American University of Beirut, Beirut, Lebanon

Jannat Akram Department of Plant Breeding and Genetics, University of the Punjab, Lahore, Pakistan

Lourd Abd Al Ahad Faculty of Agricultural and Food Sciences, American University of Beirut, Beirut, Lebanon

Amjad Ali Department of Plant Protection, Faculty of Agricultural Sciences and Technologies, Sivas University of Science and Technology, Sivas, Turkey

Hassan Almoammar Wellness and Preventive Medicine Institute, Health Sector, King Abdulaziz City for Science and Technology (KACST), Riyadh, Saudi Arabia

Muhammad Tanveer Altaf Department of Field Crops, Faculty of Agriculture and Natural Sciences, Recep Tayyip Erdoğan University, Rize, Turkey

Mohd Anas Department of Botany, Faculty of Life Sciences, Aligarh Muslim University, Aligarh, India

Moh Sajid Ansari Section of Environmental Botany and Plant Pathology, Department of Botany, Faculty of Life Sciences, Aligarh Muslim University, Aligarh, UP, India

Ramadan A. Arafa Plant Pathology Research Institute, Agricultural Research Center, Giza, Egypt

Debdulal Banerjee Department of Botany and Forestry, Microbiology and Microbial Biotechnology Laboratory, Vidyasagar University, Midnapore, West Bengal, India

Centre for Life Sciences, Vidyasagar University, Midnapore, West Bengal, India

Abdul Basit Department of Horticulture, Graduate School, Kyungpook National University, Daegu, South Korea

Zakaria Boual Laboratory of Preservation of Ecosystems in Arid and Semi-Aride areas, Faculty of Nature and Life Sciences, University of KASDI Merbah-Ouargla, Ouargla, Algeria

Fatima Zahra Chakou Laboratory of Preservation of Ecosystems in Arid and Semi-Aride Areas, Faculty of Nature and Life Sciences, University of KASDI Merbah-Ouargla, Ouargla, Algeria

Mohammad Danish Institute of Genetics and Biotechnology, Hungarian University of Agricultural and Life Sciences, Gödöllő, Hungary

Mona F. A. Dawood Botany and Microbiology Department, Faculty of Science, Assiut University, Assiut, Egypt

Israa El Ghandour Faculty of Agricultural and Food Sciences, American University of Beirut, Beirut, Lebanon

Marwa T. El-Mahdy Department of Pomology, Faculty of Agriculture, Assiut University, Assiut, Egypt

Rania Francis Department of Agriculture and Food Engineering, School of Engineering, Holy Spirit University of Kaslik, Jounieh, BP, Lebanon

School of Technology, Al-Kafaàt University, Ain Saadeh, Lebanon

Karima Guehaz Laboratory of Preservation of Ecosystems in Arid and Semi-Aride Areas, Faculty of Nature and Life Sciences, University of KASDI Merbah-Ouargla, Ouargla, Algeria

Rawan K. Hassan Biotechnology English Program, Faculty of Agriculture, Cairo University, Giza, Egypt

Muhammad Faheem Jan College of Agriculture, Northeast Agricultural University, Harbin, China

Yachana Jha N. V. Patel College of Pure and Applied Sciences, Anand, Gujarat, India

Amin Jurdi Faculty of Agricultural and Food Sciences, American University of Beirut, Beirut, Lebanon

Anu Kalia Electron Microscopy and Nanoscience Laboratory, Punjab Agricultural University, Ludhiana, Punjab, India

Abrar Ahmad Khan Section of Environmental Botany and Plant Pathology, Department of Botany, Faculty of Life Sciences, Aligarh Muslim University, Aligarh, UP, India

Adnan Khan Section of Environmental Botany and Plant Pathology, Department of Botany, Faculty of Life Sciences, Aligarh Muslim University, Aligarh, UP, India

Ayesha Khan Laboratory of Plant Physiology, Lithuanian Research Centre for Agriculture and Forestry, Kėdainiai, Lithuania

Ehtisham Hassan Khan National Key Laboratory for Germplasm Innovation and Utilization of Horticultural Crops, The Center of Crop Nanobiotechnology, College of Plant Science and Technology, Huazhong Agricultural University, Wuhan, China

Waqas Liaqat Department of Field Crops, Faculty of Agriculture, Institute of Natural and Applied Sciences, Cukurova University, Adana, Turkey

Aquino Macwan N. V. Patel College of Pure and Applied Sciences, Anand, Gujarat, India

Sadia Majeed Department of Botany, Lahore College for Women University, Lahore, Pakistan

Mohammad Imran Mir Department of Plant Sciences, University of Hyderabad, Hyderabad, Telangana, India

Debasis Mitra Department of Microbiology, Graphic Era (Deemed to be University), Dehradun, Uttarakhand, India

Asmaa M. Mogazy Biological and Geological Sciences Department, Faculty of Education, Ain Shams University, Cairo, Egypt

Heba I. Mohamed Biological and Geological Sciences, Faculty of Education, Ain Shams University, Cairo, Egypt

Murad Muhammed State Key Laboratory of Desert and Oasis Ecology, Key Laboratory of Ecological Safety and Sustainable Development in Arid Lands, Xinjiang Institute of Ecology and Geography, Chinese Academy of Sciences, Urumqi, China

University of Chinese Academy of Sciences, Beijing, People's Republic of China

Bader Un Nisa Department of Botany, Lahore College for Women University, Lahore, Pakistan

Md Affan Parwez Microbial Biotechnology and Microbiomics Group, Institute of Genetics and Biotechnology, Hungarian University of Agricultural and Life Sciences, Gödöllő, Hungary

Keyanna Radi Faculty of Agricultural and Food Sciences, American University of Beirut, Beirut, Lebanon

Abdul Razzak Institute of Forest Ecology, Slovak Academy of Sciences, Technical University in Zvolen, Zvolen, Slovakia

Tony Kevork Sajyan Faculty of Agricultural and Food Sciences, American University of Beirut, Beirut, Lebanon

Department of Agriculture and Food Engineering, School of Engineering, Holy Spirit University of Kaslik, Jounieh, BP, Lebanon

Hiran Kanti Santra Department of Botany and Forestry, Microbiology and Microbial Biotechnology Laboratory, Vidyasagar University, Midnapore, West Bengal, India

Muhammad Shafiq Department of Horticulture, Faculty of Agricultural Sciences, University of the Punjab, Lahore, Pakistan

Syed Tanveer Shah Department of Agriculture, Hazara University Mansehra, Khyber Pakhtunkhwa, Dhodial, Pakistan

M. V. Sreelakshmi Department of Microbiology, College of Basic Sciences and Humanities, Punjab Agricultural University, Ludhiana, Punjab, India

Alia Telli Laboratory of Preservation of Ecosystems in Arid and Semi-Arid Areas, Faculty of Nature and Life Sciences, University of KASDI Merbah-Ouargla, Ouargla, Algeria

Samar G. Thabet Department of Botany, Faculty of Science, Fayoum University, Fayoum, Egypt

Mohammad Umar Institute of Plant Genetics and Biotechnology, Plant Science and Biodiversity Center, Slovak Academy of Sciences, Nitra, Slovakia

Muhammad Adil Younas Department of Agriculture, Hazara University Mansehra, Khyber Pakhtunkhwa, Dhodial, Pakistan

Chapter 1
Plant Defense Mechanisms Against Biotic Stress

Mohammad Umar, Abdul Razzak, Mohd Anas, Md Affan Parwez, and Mohammad Danish

Abstract Plants face severe biotic stress, including pathogen invasion and herbivory, which significantly impacts plant growth and survival. To alleviate these constraints, plants have evolved a range of defense systems at the structure, biochemistry, and molecular biology levels. Morphological defenses, trichomes, wax layers, and rigidified walls can defend herbivores against herbivores and deflect pathogens from adhering. Biochemical defense (secondary metabolites, including alkaloids, phenolics, and terpenoids) disrupts herbivore feeding and pathogen growth while phytohormones jasmonic acid (JA) and salicylic acid (SA) form the overall defense mechanisms. Pattern-triggered immunity (PTI) and effector-triggered immunity (ETI), respectively, triggered by pathogen-associated molecular patterns (PAMPs) and effector molecules are established at the basis of plant–pathogen interactions as represented by a zigzag model. The modern development in biotechnology has overcome crop resistance to biotic stresses. Transgenic crops, such as Bt cotton and virus-resistant papaya, have reduced pesticide use, improved production, and contributed to sustainable cropping. Tools, including CRISPR-Cas9 and RNA interference (RNAi), provide molecularly precisely defined genetic changes—in terms of

M. Umar (✉)
Institute of Plant Genetics and Biotechnology, Plant Science and Biodiversity Center, Slovak Academy of Sciences, Nitra, Slovakia

A. Razzak
Institute of Forest Ecology, Slovak Academy of Sciences, Technical University in Zvolen, Zvolen, Slovakia

M. Anas
Department of Botany, Faculty of Life Sciences, Aligarh Muslim University, Aligarh, India

M. A. Parwez
Microbial Biotechnology and Microbiomics Group, Institute of Genetics and Biotechnology, Hungarian University of Agricultural and Life Sciences, Gödöllő, Hungary

M. Danish
Institute of Genetics and Biotechnology, Hungarian University of Agricultural and Life Sciences, Gödöllő, Hungary

K. A. Abd-Elsalam, H. I. Mohamed (eds.), *Elicitors for Sustainable Crop Production*, Sustainability Sciences in Asia and Africa,
https://doi.org/10.1007/978-981-96-6569-3_1

genetic modulation of susceptibility genes or repression of pest-specific genes. These methods are cofactors to conventional breeding techniques using molecular markers for the intelligent production of adaptive crop genotypes. This chapter reviews and explains some deep processes of plant resistance to biotic threats and their functions of phytohormonal signaling, transcriptional regulation, and secondary metabolite biosynthesis. It also explains its innovative role in present biotechnological intervention, such as its contribution to the development of ecologically sound agricultural systems. Management of biotic stress using integrative strategies ensures food security, ecological stability, and resistance to climate change and the demands of the human world.

Keywords Biotic stress · Plant defense mechanisms · Transgenic crop · CRISPR-Cas9 · Sustainable agriculture

1.1 Introduction

Plants have been constantly challenged with many biotic stresses, such as pathogen and herbivores, which can efficiently reduce the plant growth and yield (Ali et al. 2024a, b). To counteract these threats, plants have developed intricate defense systems acting at the level of structure, biochemistry, and molecular levels. These alterations play important functions in improving agricultural resilience and global food security (Waterman et al. 2021). Plant defense mechanisms can be categorized as defense against the pathogen and defense against the herbivores. Defense mechanisms of pathogens are triggered in response to plants' recognition of pathogen-associated molecular patterns (PAMPs) via pattern recognition receptors (PRRs), resulting in PAMP-triggered immunity (PTI) (Guerreiro and Marhavý 2023). Nevertheless, most pathogens have developed effector(s) capable of inhibiting immune responses, resulting in effector-induced susceptibility (ETS) (Peng et al. 2023). Parallelly plants express functional resistance (R) genes that are effective in the turnover of effector determinants to reinstate the activities of their immune system through effector-triggered immunity (ETI) (Thompson et al. 2022). This arms race of plant defenses versus pathogen tactics is summarized by the "zigzag" theory of plant defense, that is, how the process of this interaction evolves. Herbivore defense mechanisms encompass both structural and chemical responses. Structural defenses, including trichomes and toughened leaves, mechanically repel herbivores, and it is difficult for herbivores to feed (Nishad et al. 2020; Zhou et al. 2022). From a biochemical, however, secondary metabolites like alkaloids and terpenoids produced by plants decrease the nutritive value of the cells and allure natural enemies of herbivores are thus built-up (Huang et al. 2022; Kaur et al. 2022). Phytohormones, especially jasmonic acid (JA) and salicylic acid (SA), have critical functions in the regulation of these defenses and enable the plant to adjust defenses in response to the specific aggressors it is trying to protect against (Kanda et al. 2023; Xu et al. 2021). Recent advances in biotechnology have allowed the development of crops with enhanced resistance to biotic stresses. Genetically modified (GMO) crops, for example, *Bacillus thuringiensis* (Bt) cotton and Bt maize, contain Bt toxins that are only toxic to the insect pest, resulting in decreased pesticide use and ecological sustainability (Magara et al. 2020; Pearse et al. 2020). Additionally, CRISPR-Cas9 technology has developed to be an effective single-engineering tool for the targeted modification of susceptibility genes and resistance of pathogens of the bacterium and fungi (Zhang et al. 2024a, b). RNA interference (RNAi) technology presents novel approaches to target functionally relevant genes in insects or pathogens, which reduce their damaging effects on plants (Chacón-Fuentes et al. 2023; Xu et al. 2021). Combining classical breeding practices with the use of these modern biotechnological approaches opens a promising pathway to developing maize varieties able to withstand biotic stresses. This integration design not only can support more sustainable agricultural practices, but it also addresses the current highest priority of food security on a global level with relevance to climate change and a growing population food demand (Waterman et al. 2021). The purpose of this chapter is to discuss, in detail, plant defense against biotic stress and the importance of biotechnological applications, such as CRISPR-Cas9 and transgenic plants, in the creation of stress-resistant, high-producing crops as part of sustainable agriculture and global food security.

1.2 Types of Stress Responses

Biotic stress is damage caused to plants by living organisms, including pathogens like bacteria, fungi, viruses, and herbivores. In response to these enemies, plants have evolved elaborate defense mechanisms. This module examines ways in which plants defend against both pathogens and herbivores.

1.2.1 Pathogen Defense

Plant–biotic stress interaction is one of the fastest developing fields concerned with the interaction of plants and living organisms. Until lately, much research was focused on the relationship between plants and pathogens, although recently the concept of plant–arthropod pest interaction has also received much attention. Because plants cannot move, they have evolved sophisticated mechanisms for self-defense against biotic stresses caused by pathogens or pests. Interactions involving multilevel molecular networks, signaling pathways, and defense strategies—where both the plant and its attacker engage in an endless game of adapts-to-overcome—continue between them (Zehra et al. 2021). Receptors that recognize specific pathogens (including fungi, bacteria, and viruses) or pests (like herbivorous insects, mites, and nematodes) initiate plant defense responses. Even volatile compounds emitted by neighboring plants are under attack act as signals for plant defense mechanisms. With the advancement of molecular research, several regulatory elements, key molecules, and pathways in plant defense mechanisms have come to light against biotic stresses. For plants, the rapid identification of a pathogen is a must to switch on their defense enzymes (Zehra et al. 2021). They recognize pathogens by some molecules that they release that act like biological elicitors to initiate defensive responses. These involve the activation of various genes responsible for producing proteins, secondary metabolites, and pathogen-related (PR) proteins. Plants rely on both pre-formed and induced mechanisms to protect themselves from pathogens. Structural barriers and immune responses against specific pathogen-derived molecules known as PAMPs are a way for plants to recognize certain pathogens. Such PAMPs include microbial cell wall components such as chitin and glucan, which elicit sequential defense responses in many plant species. Plants perceive general microbial structures with PAMPs, including various proteins from bacteria and fungi. Once these elicitors bind to plant receptors, they activate signaling pathways that result in the production of phytoalexins, antimicrobial proteins, and the strengthening of the plant cell wall (Zehra et al. 2021). Plants have disease tolerance, which runs a spectrum from vulnerability through to resistance. The plant is assisted in either tolerating or resisting biotic stress through systemic and induced responses. In response to an attack, plants synthesize defense proteins, which are generally effective in protecting them against pathogens by deploying a multitude of defense mechanisms. The presence of an elicitor has been known to activate the expression of pathogenesis-related genes for a more effective defense by the plant against invaders (Collinge and Sarrocco 2022). A more holistic approach to disease management is the combined application of beneficial microbes, their metabolites, synthetic chemicals, and plant extracts, which can aid farmers in better managing plant diseases and pests, increasing crop yields, and thereby sustaining food security (Collinge and Sarrocco 2022).

Plants accumulate a wide variety of sRNAs in their tissues, which, despite their variation in size, sequence, and biogenesis, all act in repressing gene activity via RNA silencing. These sRNAs, important regulators of development and genome stability, are crucial players during the plant's responses to biotic stress. Initially, RNA silencing was largely viewed as an antiviral defense mechanism in which virus-derived sRNAs suppress viral genes, and viruses counter the production of proteins that inhibit this silencing. However, it has now become clear that plant-derived sRNAs also regulate responses to a wide variety of microbes, and, in turn, microbes have evolved ways to circumvent or manipulate the plant's silencing mechanisms. This interaction exemplifies the ongoing molecular arms race between plants and their pathogens (Appu et al. 2021).

1.2.2 Plant Immune System

A complex network of intracellular signaling processes is used in the immune system of plants; it cannot be generalized to all plants due to species-to-species differences. A simple model that explains the concept of plant immunity is the "zig-zag model" as it explains the dynamic battle between hosts and pathogens. Pathogens possess distinct molecular patterns, also called microbe-associated or pathogen-associated molecular patterns (MAMPs/PAMPs), like chitin found in fungal cell walls or flg22 of bacterial pathogens. Plants contain pattern recognition receptors (PRRs) that recognize such MAMPs/PAMPs, thereby initiating a first line of defense called PAMP-triggered immunity (PTI) (Ngou et al. 2022). The first line of defense includes an influx of Ca2+, followed by the production of ROS, which can further lead to an oxidative burst. These bursts may result in hypersensitive response (HR)-mediated cell death, although ROS may also serve as signals for further defensive actions (Ngou et al. 2022).

While PTI initially restricts pathogen invasion, some pathogens can evade this defense by producing toxins or effectors, leading to a second phase called effector-triggered susceptibility (ETS). Effectors, produced by pathogens' virulence genes, target specific areas within the host. These toxins can either be specific to a host or affect multiple hosts. For example, several *Alternaria* species produce host-specific toxins such as AM, AF, AK, ACR, and AAL toxins. Likewise, some *Fusarium species* produce mycotoxins such as fusaric acid and beauvericin during disease. In response to these toxins, plants have developed resistance (R) genes that allow them to produce antitoxins, which neutralize the toxic actions of toxins produced by the pathogen (Tzipilevich et al. 2021). Systemic acquired resistance and induced systemic resistance are two ways to resist attacks by pathogens or parasites that are activated by pre-existing infection or treatment. SAR occurs throughout the plant in response to specific inoculate, be they virulent, avirulent, or even non-pathogenic microbes, or artificial chemical agents like salicylic acid (SA) and chitosan. ISR represents the arsenal of defense mechanisms activated against invading pathogens; these mechanisms do not, however, directly kill or inhibit the invader (Kamle et al. 2020). SAR fires up when a local infection stimulates an SA-dependent signal transduction pathway that translates to broad-spectrum and long-term resistance against a whole bust of pathogens, such as bacteria, fungi, and viruses. Unlike direct antimicrobial responses, ISR boosts the plant's overall defense readiness, making it more resistant to future challenges. This section outlines the roles and mechanisms of SAR and ISR in protecting plants from phytopathogens (Kamle et al. 2020).

1.2.3 Herbivore Defense

The SAR response in plants is initiated through a local infection that activates a salicylic acid-dependent signaling pathway in the plant, which activates broad-spectrum durable resistance against a wide array of pathogens, including fungi, bacteria, and viruses. Contrary to the direct antimicrobial-type responses, ISR primes general defense preparedness in the plant, thereby rendering it more resistant to further attacks. This section covers the roles and mechanisms of SAR and ISR that protect plants against phytopathogens (Kamle et al. 2020).

Plant–insect interactions may be grouped into two: mutualistic associations with pollinators and antagonistic interactions with herbivores. Many research works have focused on the latter antagonistic plant–herbivore interactions to understand the defensive mechanisms of plants against feeding insects. Plants utilize a wide range of defensive mechanisms against sudden biotic attacks using physical and chemical barriers. The main physical barriers are the structures such as epidermal layers, hairs, thorns, and trichomes. Upon contact with an insect, chemical alterations take place in the plant (Balaji and Jambagi 2024). The first step involves changes in the plant's cell walls, while signal communication occurs, with receptors perceiving signals from the insect, thus continuing the activation of the plant's immune system. The majority of plant infections from such pathogens, bacteria, fungi, or viruses are due to an attack from herbivores. In the site of infection, the receptors present on the cell surface recognize these molecular signals of the attack and activate PTI. The herbivores produce effector proteins that activate ETI. Thus, mobile signals are generated, which move to different tissues, activating the secondary immune system (SAR) in plants. Several changes in the plasma membrane occur upon herbivore attack. Initiation of the process begins when the plant recognizes the molecular triggering signals and effector proteins leading to increased cytosolic calcium concentrations, depolarization of the plasma membrane potential, and the activation of mitogen-activated protein kinases (MAPKs). These MAPKs are involved in protein phosphorylation, the activation of NADPH oxidase, and the production of reactive oxygen species (ROS) and reactive nitrogen species (RNS) (Balaji and Jambagi 2024). It thus follows with the generation and quick induction of phytohormones related to stress, such as endogenous jasmonic acid (JA), salicylic acid (SA), and ethylene in quantity. These phytohormones are the significant mediators in the pathways of the stress signal. Plants also accumulate secondary metabolites as a late-stage defense response. This includes the production of VOCs, which are toxic or feeding deterrents that reduce the nutritional value of the plant for herbivores (Balaji and Jambagi 2024).

Also, the ATPases, which have a critical role in plant cell energy supply, exert an important extracellular function under stress conditions. Extracellular ATP signaling is closely linked to the action of secondary messengers such as Ca^{2+}, ROS, and NO. Ca^{2+}-ATPases transport ions across cellular membranes and are members of the P-type ATPase superfamily, further subgrouped into two main categories based on their subcellular localization in plant cells. These include P-IIA ER-type Ca2+-ATPases (ECAs) and P-IIB autoinhibited Ca^{2+}-ATPases (ACAs). While ECAs are like animal sarcoplasmic-endoplasmic reticulum Ca^{2+}-ATPases, ACAs resemble animal PM-type ATPases. ACAs are more selective, transporting only Ca^{2+}, whereas ECAs can transport other ions such as Cd^{2+}, Mn^{2+}, and Zn^{2+}. The calcium signaling pathways activated by herbivore attacks involve these key processes as shown in Fig. 1.1 (Kamle et al. 2020).

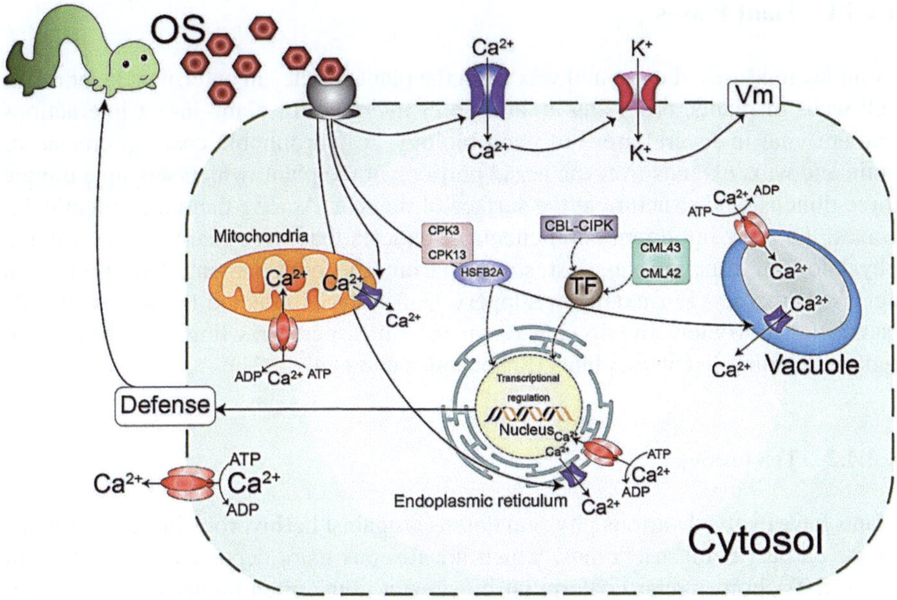

Fig 1.1 A molecular signal model for the early signal network of plant defense response to herbivorous insects. (Kamle et al. 2020)

1.2.4 Structural Defenses

Structural defenses are the plant's prerogative in the arsenal of weaponry against herbivores, which provides it with its first line of defense in interactions with herbivores. These defenses imply physical barriers that help deter most herbivores from feeding on a host plant and make feeding on it difficult for them (Balaji and Jambagi 2024). First are structures located on leaves and stems that encourage herbivores to have a difficult time holding and feeding upon the plant. Also, these structures assist the plant in overcoming other environmental challenges such as drought, evaporation, and too much sunlight. They explain herbivores' pressures, which would lead to the need for plants to evolve arrays of specialized defenses, finely tuned to mitigate herbivory impacts on growth, development, and fitness. The physical defenses available to plants include waxy coatings, hair-like structures (trichomes), toughened leaves, thorns, spines, and sharp crystalline structures (raphides) of calcium oxalate or calcium carbonate. Within the vegetative materials, some do make the ingestion of the vegetative tissue difficult, thus decreasing the optimal food value for the herbivores and, in some cases, acting poisonously through the mechanical destruction of the physical structure of the herbivore or interference with biochemical processes (Verma et al. 2021).

1.2.4.1 Plant Waxes

Complex mixtures of cutin and wax form the plant cuticle, an outermost layer of the cell walls of plants, providing an avenue or interface for plant–insect interactions that are vital in several areas of plant biology. A hydrophobic coating, created of cutin and wax, extends over the aerial portions of the plant, which sets up a unique three-dimensional structure at the surface of the leaf. Among them are epicuticular waxes, the most important components of cuticles that serve many ecological and physiological functions against stresses from the environment, like protection against herbivory. For example, slippery leaf surfaces affected by waxes impede caterpillar movement by slowing down the time spent crawling, which leads to reduced feeding in canola plants (Shaheenuzzamn et al. 2021).

1.2.4.2 Trichomes

Plants have evolved various physical defenses against herbivorous insects, with one of the earliest being trichomes, which are fuscous hairs deposited along the leaf surface. Trichomes retard caterpillar movement, thus, often hindering their movement or contact to the leaf surface. They may take various shapes, be multicellular or unicellular, and possess important roles in tolerance to biotic and abiotic stresses (Nandi et al. 2022). In the case of their non-glandular variety, the trichomes, which do not secrete any substances, are physical barriers against insects, disrupting or blocking movement patterns. Trichomes can ensnare soft-bodied insects, hence reducing feeding activity and in some cases injuring the insects. Young tobacco hornworm caterpillars, for example, often experience difficulties in feeding because of the abundance of trichomes on a plant in question, starving them to eventual desiccation or injury (Nandi et al. 2022).

Spinescence encompasses equipping structures like spines, thorns, and prickles that emerge from various parts of the plant vegetable, such as petioles, veins, or the epidermis. These structures inflict mechanical injuries on herbivores; once inside, they may introduce pathogens, thus discouraging further feeding. Some plants use striking sunlit colors as warning signals for potential danger. Or, to reiterate, research of European holly indicates that spiny leaves are less attacked by herbivores than those having fewer spines (Li et al. 2023a, b).

1.2.4.3 Sclerophylly

Sclerophylly, hard and thick leaves, serve many purposes ranging from support to moisture conservation and tolerance of environmental oscillation. They are tougher, lower in nitrogen, higher in secondary compounds, and are unpalatable due to being tougher for herbivores to digest. Thickening involves some layers, such as the cuticle and the outer epidermal walls, and less frequently a different hypodermal layer, leading to the development of a tough leaf structure that resists herbivores and lowers their fitness if consumed (Sereneski De Lima et al. 2022).

1.2.4.4 Raphides

Calcium oxalate crystals or raphides are ubiquitous in the plant kingdom and exist in various forms, such as crystal sand and styloids. Needle-like structures densely packed in specialized cells are called idioblasts, which act as a physical and hence a passive means of defense, deterring herbivores. Found in all parts of plant tissues, raphides are common among various plant taxa and presumably quite effective against herbivory through their sharpness and irritation or physical damage to soft tissues when ingested (Lawrie et al. 2023).

1.2.5 Chemical Defenses (Secondary Metabolites Like Alkaloids and Terpenoids)

There are over 400,000 vascular plant species on Earth that serve as sources for thousands of metabolites, yet the function, structure, and utility of these metabolites remain largely underexplored. As sessile organisms exposed to constantly changing environments, plants exhibit far more metabolic diversity compared to other organisms. Biotic stress caused by bacteria, viruses, fungi, and nematodes leads to significant losses in crop yield and productivity (Al-Khayri et al. 2023). The response of a plant to biotic stress is very complicated and requires interactions between two living organisms. Pathogens kill plant cells and, therefore, can weaken the host plant to obtain various nutrients. However, plants have evolved multilayer protective and defense mechanisms to avert a given invasion (Al-Khayri et al. 2023). The interaction between the plants and pathogens is a dynamic process whereby the pathogens secrete biomolecules, such as lipopolysaccharides, sugars, and proteins into the plant tissues, beginning the infection. These components derived from plant pathogens are central to the infection since they confer pathogen fitness in colonizing plant cells. In response, plants secrete effector molecules that either prevent infection by negative-feedback regulation or potentiate an active plant defense response (Al-Khayri et al. 2023).

Several strategies can be taken by pathogens to acquire nutrients. These pathogens belong to the groups necrotrophic, biotrophic, and hemibiotrophic. The biotrophic pathogens obtain nutrients from living host tissues continuously while necrotrophic pathogens kill the host to exploit the nutrients released from dead cells (Divekar et al. 2022). Hemibiotrophic pathogens start their infection in a biotrophic manner, but afterward turn into a necrotrophic one. Plant health depends on its natural immunity, where sometimes performed defense responses, like structural barriers, are present alongside antimicrobial compounds. However, many factors contribute to the infection's success, such as the specific surface receptors neighboring plant cells, the ability of the pathogen to secrete toxins, and the ability to preclude plant defense responses, along with the whole biochemical, physiological, and genetic makeup of the plant. Plants employ mechanisms in parallel to restrict infection (Divekar et al. 2022).

1.2.5.1 Phenolic Metabolites

Phenolic metabolites are compounds having varying numbers of hydroxyl groups and one or more phenolic structures, in free and conjugated forms. These metabolites are found in chloroplasts in young plant cells and accumulate in vacuoles in mature cells. They are synthesized and accumulate from precursors such as shikimic acid and phenylalanine, depending on chloroplast function and structure. This group of compounds includes simple phenols, phenolic acids, quinones, and flavonoids; all of these are important in the metabolism of plants and have antibacterial, allelopathic, and antifungal activities toward pathogen-related elicitor compounds. The prominent secondary metabolites are phenolic compounds, falling mostly as intrinsically intermediary metabolites for underlying metabolic pathways (Abu-Shahba et al. 2022). Besides belonging to different structures and biosynthetic pathways, alkaloids form a class of secondary metabolites and act here to bootstrap the plant's defenses under stress. Found across vascular plants, there are more than 20,000 types of alkaloids that add to the levels of resistance in the plant. These nitrogen-based compounds, including substances such as quinolones and metronidazole, disrupt microbial enzyme activity by targeting important pathways. For instance, squalamine acts via a detergent-like action to disrupt type Gram-negative membranes and cause depolarization in Gram-positive membranes (Gaur and Gänzle 2023).

1.2.5.2 Saponin Metabolites

Saponins are a class of special secondary metabolites containing a triterpene or steroidal aglycone moiety combined with sugar chains, playing an important role in the mechanism of plant defense. These well-known phytochemical amphiphilic compounds possessing foaming and emulsifying activities are widely distributed in higher plants. The content of these substances in plants is very diverse and depends on many factors, among which the action of some biotic stressors is included. In this regard, the quantity of saponins inside the infected cells demonstrates strong variation, therefore playing a critical role in the mechanisms of plant defense (Yu et al. 2022).

Terpenes are volatile secondary metabolites responsible for the color, odor, and flavor of plant parts. Terpenes are considered one of the most chemically diverse groups; they are hydrocarbons, while terpenoids are their modified forms. These compounds help protect plants against herbivores by attracting their natural predators. In addition, terpenoids assist plant interaction with other organisms, including neighboring plants and insects, and act as precursors for the synthesis of other classes of compounds such as carotenoids, sterols, chlorophyll components, and some hormones (Zhang et al. 2024a, b).

1.2.5.3 Transcription Factors and Genes Linked to Secondary Metabolites in Biotic Stress Response

Biotic stress in plants arises due to interactions with fungi, bacteria, viruses, parasites, and weeds. Transcription factors (TFs) are DNA-binding proteins that interact with the promoter regions of target genes, thus controlling transcription by regulating RNA polymerase activity. TFs have a key role in combining internal and external signals to regulate the expression of defense-related genes and the accumulation of secondary metabolites needed for resistance to biotic stress (Anjali et al. 2023). Multiple TFs operate at one of several regulatory levels to regulate genes involved in secondary metabolite biosynthetic pathways. More recently, research began focusing on the identification and interaction processes of TFs involved in secondary metabolite biosynthesis.

Certain TFs are particularly important in modulating pathways for defense-related secondary metabolites, such as systemic acquired resistance (SAR), which promotes a salicylic acid-mediated response (Kumar et al. 2023). For instance, the pathogen *Plasmopara viticola* triggers genes linked to the jasmonic acid (JA) pathway, enhancing tolerance to downy mildew through the VvWRKY1 transcription factor. In addition, OsWRKY31 and OsWRKY45 from rice have been identified to induce resistance against secondary metabolite synthesis in the field of diseases, targeting *Magnaporthe oryzae* (Anjali et al. 2023) (Table 1.1).

1.3 Plant Response via Signaling Pathways Against Biotic Stress

1.3.1 Salicylic Acid (SA)

Salicylic acid (SA) is a key phytohormone in plant defense signaling, especially when responding to biotrophic pathogens. It has been well established that SA is a key signal for the activation of systemic acquired resistance (SAR), with subsequent induction of genes and proteins associated with defense that prepares the plant for enhanced resistance following a challenge by pathogens. For example, the exogenous application of SA can trigger the synthesis of pathogenesis-related (PR) proteins that play an important role in plant defense against fungal and bacterial pathogens (Silva et al. 2022). Not only does this induction take place locally at the site of application, but also systematically throughout the plant, indicating that SA functions as a signaling molecule to inform surrounding tissues about pathogen attack (Khotimah 2022; Silva et al. 2022). When biotrophic pathogens invade a

Table 1.1 Types of stress responses

Defense mechanism	Category	Description	Example	Impact on biotic stress	Citations
Structural defenses	Morphological	Physical barriers to pathogen and herbivores	Trichromes, waxy cuticle	Prevents feedings and pathogen adherence	Zehra et al. (2021)
	Anatomical	Internal reinforcement to stress	Suberized tissue, sclerenchyma cells	Limits pathogen invasion	Abu-Shahba et al. (2022)
Biochemical defenses	Secondary metabolites	Toxic compounds repel herbivores and pathogen	Alkaloids, phenolics, terpenoids	Disrupts feeding and growth of invaders	Al-Khayri et al. (2023)
	Antimicrobial compound	Inhibits pathogen growth directly	Phytoalexins, chitinases	Neutralize pathogen	Taylor et al. (2021)
Phytohormonal regulation	Hormonal signaling	Activities systematic and localized defenses	Jasmonic acid, salicylic acids	Modulates responses to stressors	Kanda et al. (2023) and Xu et al. (2021)
Immune response	Molecular recognition	Detects and counter pathogens	Pattern-triggered immunity (PTI)	Prevents pathogen establishment	Guerreiro and Marhavý (2023)
	Effector-based response	Counteracts pathogen effector proteins	Effector-triggered immunity (ETI)	Restores immune functionality	Thompson et al. (2022)
Systemic defenses	Systemic acquired resistance (SAR)	Broad-spectrum resistance induced by local infections	Salicylic acid signaling	Prepare plants for future attacks	Kumar et al. (2023)
	Induced systemic resistance (ISR)	Beneficial microbes enhance resistance	Rhizobacteria-triggered defense	Improve plant resistance and growth	Kamle et al. (2020)

host, they will keep the host tissue alive and thus there is a remarkable interaction between SA and pathogen resistance. Previous studies showed that when pathogens infect the plant accumulation of SA is induced and several steps of resistance response such as expression of PR genes like PR1 or PR2 that give a good response against *Phytophthora sojae* are activated (Gülser and Sönmez 2022; Han et al. 2022). In addition, the implication of SA in controlling the expression of 15 of these genes holds since some studies demonstrate that SA mediates transcription factor phosphorylation to further enhance defense-related gene expressions (Han et al. 2022). This regulatory system does highlight a key role that plays in engaging SA to activate the highest possible and protective level of immune response in plants. SA not only activates numerous defense genes, but it is also implicated in the lactation of other phytohormones that are crucial for plant defenses. The interaction between salicylic acid (SA) and jasmonic acid (JA) has been well established. SA is mainly involved in biotrophic pathogen responses while JA plays a role in necrotrophic pathogens responses (Silva et al. 2022). This differential signaling route enables the plants with a pathway to adapt their defenses depending on what kind of pathogen infected them so that they can survive better. The relative balance between SA and JA signaling is important in determining the overall plant strategy for defense and can affect outcomes of pathogen interactions (Silva et al. 2022). The biosynthesis of SA is also closely related to the protective response capability of plants. Upon pathogen attack, pathways for SA production such as isochromatic synthase (ICS) and phenylalanine ammonia-lyase (PAL) are induced (Silva et al. 2020). Such a biosynthetic response is essential because it allows the production of a sufficient amount of SA to trigger and support the defense response. The high SA has been found in the infected tissues and could be linked to the induction of a whole series of defense responses, which involve, for example, reactive oxygen species (ROS) production or cell wall lignification (Nurcahyani 2024). In addition, SA is also a signaling molecule but also regulates some physiological activities that are of critical value for plant growth and development. In addition, SA has been shown to play roles in the control of photosynthesis, respiration, and flower formation; at the same time, major roles are played in plant survival under stress conditions (Khotimah 2022). SA's multifunctional role encompasses not only the defense pathway but also the whole plant level of physiology thus emphasizing its role as a key player in developing plants' tolerance to biotic stressors. The treatment of SA, particularly, to the plants to increase their resistance to some diseases is widely practiced in agriculture. The application of this hormone was shown in field studies to increase the resistance of plants against a wide range of pathogens such as *Colletotrichum capsici* and *Fusarium oxysporum* (Khotimah 2022; Kumawat et al. 2023). Thus, such results imply that SA can be a sustainable, eco-friendly, and feasible plant regulator to enhance plant health and yield in agronomic systems with crops vulnerable to pathogens. Furthermore, the capability of SA to elicit systemic resistance reinforces the potential of SA as a biopesticide or plant activator in sustainable agriculture (Kumawat et al. 2023).

1.3.2 Jasmonic Acid (JA)

Jasmonic acid (JA) is an important phytohormone that mediates plant defense responses against necrotrophic pathogens and herbivores. The signaling pathway, as the name suggests, is the main player in the process of mutual protection as it is responsible for the activation of the genes responsible for secondary metabolites and pathogenesis-related proteins. This is, however, more clearly seen in *Arabidopsis* and tomato species, wherein JA is the mediator of resistance to a wide range of pathogens, including *Botrytis cinerea* and *Alternaria brassicicola* (Genva et al. 2020; Muñoz et al. 2021; Macioszek et al. 2023). However, the crosstalk of JA with other phytohormones, especially salicylic acid (SA) and ethylene (ET), is quite complicated and plays a key role in defense signaling (Chen et al. 2020). JA signaling is triggered to induce the production of defensive compounds upon wounding or herbivory to deter herbivores and restrict pathogen growth. Further evidence for the importance of JA in mediating plant responses to herbivore damage is gained from its ability to induce the expression of these genes that specify protease inhibitors and other defensive proteins (Liu et al. 2024a, b). Recent studies have implicated JA as a linchpin of plant innate immunity. For instance, exposure to JA signaling has also been shown to elicit an increased production of reactive oxygen species, which are important for cell death regulation, localized, and in turn, spatially reinforce cell wall formation to confine pathogen spread (Zhang et al. 2021; Ryalls et al. 2023). Further crosstalk between the jasmonate and a diverse group of other signaling pathways, including auxins and ABA, shows further proof of the multifarious function of jasmonate in plant defense (Han and Luthe 2022). Besides, the crosstalk between these hormones is extremely important since an SA excess causes the plant to become susceptible to necrotrophic pathogens, and jasmonate takes charge of defense in this case (Ghozlan et al. 2020). In addition, some studies focus on how pathogens can manipulate JA signaling as many necrotrophic fungi have developed the ability to suppress plant defenses that are based on the JA pathway and therefore thrive at host plants (Fu et al. 2023; Waterman et al. 2021).

1.3.3 Crosstalk Between Signaling Pathways

Salicylic acid (SA), jasmonic acid (JA), and ethylene (ET) signaling pathways are frequently engaged in deep crosstalk, which plays a vital role in plant defense against biotic stress. These phytohormones have a significant effect on the plant's accommodation to a multitude of biotic stresses (i.e., pathogens and herbivores). Interactions among these signaling pathways are mainly synergistic and antagonistic, the outcome of which is orchestrated to decide the plant's defense strategy. SA is mostly related to biotrophic pathogens that feed on living plant tissues. Activation of SA signaling is associated with the expression of pathogenesis-related (PR) genes, including PR1, that are important for systemic acquired resistance establishment (SAR) (Agarwal et al. 2020; Ali et al. 2024a, b). Plant NPR1 (non-expression of PR genes 1) is a prominent regulatory protein within the SA signaling pathway that serves as a molecular switch to facilitate the expression of Pronounced Response (PH) gene induction upon accumulation of SA (Chen et al. 2021; Guan et al. 2023). On the other hand, jasmonic acid largely contributes during attacks by necrotrophic pathogens and herbivores, triggering a distinctive repertoire of defense genes that potentialize resistance to these adversaries (Jiao et al. 2022). JAZ (jasmonate

ZIM-domain) proteins often mediate this JA signaling pathway by inhibiting the expression of JA-responsive genes until a high enough level of JA is achieved to release their repression (Wang et al. 2020). Another class of important phytohormone, ethylene crossover interacts with both salicylic acid (SA) and JA signaling pathways, as well as making the regulatory network rather complex in plant defense. Ethylene is recognized by the stronger response of the plant to the JA induced by necrotrophic pathogens; thus, it helps the plant to have a higher level of defense (Wang et al. 2021a, b). The ET-Sorti relationship, however, is more complicated; SA is the inhibitor of the JA response, while ET makes the JA response less sensitive to the SA-induced inhibition, so there may be a sophisticated coordination among these hormones (Seo et al. 2020). This interconnectivity mainly regulates the local defense of the plant that allows the plant to develop its own unique defense response depending on the type of biotic stress experienced. Studies have indicated that the crosstalk of these two pathways is not just an antagonistic one, but it also can be a cooperative one. For example, the combined actions of JA and SA on tea plants have shown increased volatile emission, which in turn reduces the herbivore attack, thus reflecting a successful interaction between those two pathways (Jiao et al. 2022). Along the same lines, the inclusion of nitric oxide (NO) into the signal transduction routes of SA and JA has been hypothesized as a pathway to plant defense against pathogens, thus showcasing the complexity of these interactions (Agarwal et al. 2020). Besides, the regulatory function of abscisic acid (ABA) in this interactome cannot be overestimated. ADHD acts as a negative regulator, but recent studies confirm that ABA can work with the SA and JA along with some stressors, thereby causing the plant's overall longer-lasting defense against abiotic and biotic stresses (Yuan and Poovaiah 2022). This means that the signaling networks are not isolated but interlinked, with many hormones being the determinants of stress responses. The transcription control of the pathways is achieved by different transcription factors, for example, WRKY and MYB proteins, which receive signals from SA, JA, and ET and thus either down- or upregulate genes in stress conditions (Ali et al. 2024a, b; Han et al. 2022). Moreover, the JAZ proteins that are the repressors of the JA signaling pathway can also interact with other transcription factors in the SA signaling, which can then give a deeper sense of the interdependence of these regulatory networks as shown in Fig. 1.2 (Nomoto et al. 2021).

1.4 Biotechnology and Crop Improvement in Response to Biotic Stress

1.4.1 Introduction to Agricultural Biotechnology

Modern agriculture has been greatly changed by biotechnology for improving crop resistance to different biotic and abiotic stresses, leading to better productivity and sustainability (Liu et al. 2022). An important advance in this field has been the

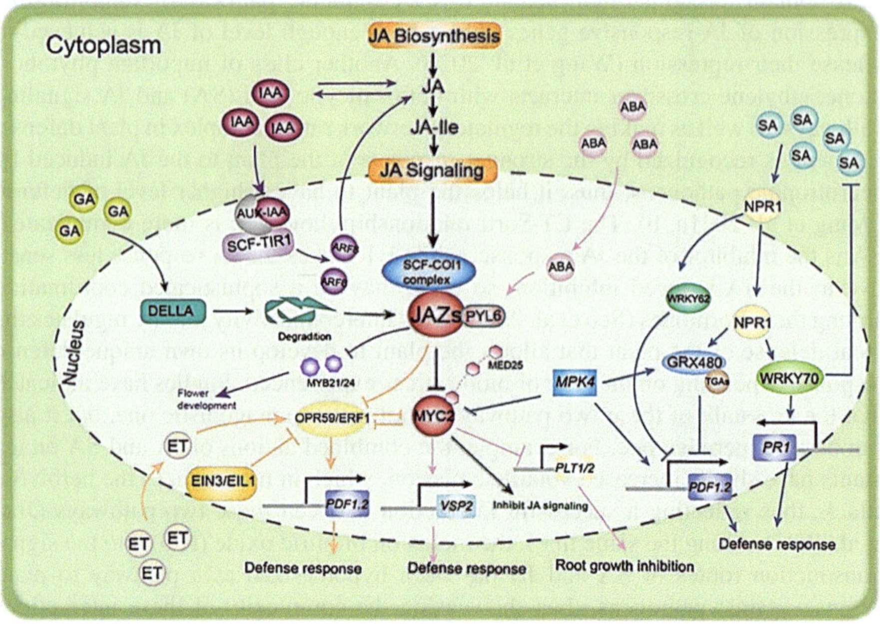

Fig. 1.2 Crosstalk between signaling pathways

development of genetically modified (GM) crops that resist biotic stressors, like pests and pathogens. One of the best examples of biotechnology is the introduction of *Bacillus thuringiensis* (Bt) genes into crops. Cry proteins, toxic to certain insect pests, that can be produced in commercial Bt crops (cotton and maize) have reduced the need for chemical insecticides by a factor of approximately 100× (Chang et al. 2020). The economic benefit of this advancement is that it cuts down on production costs as well as pollution from pesticide use, the latter two not being the least concerning issues (Sheng et al. 2020). Bt crop adoption has been very often credited with increasing yields and reducing losses from pest damage. Not only has biotechnology allowed the development of crops with biotic stress resistance but crops also have biotechnological development to resist various abiotic stresses, including drought, salinity, and extreme temperatures. For example, genes that help maize tolerate drought have been introduced into those crops, so that they can continue to yield even under conditions of water limitation (Kumari et al. 2022). The expression of salt-tolerant genes in crops such as rice and wheat can grow in saline soils, normally unsuited for regular farming (Xuebo et al. 2023). Genetically modified crops help to keep up food production in areas wherein there are disputes of bad climatic conditions, thereby safeguarding ways to obtain food and the sustained agricultural economy. The advent of gene-editing technologies, including CRISPR-Cas9, has made very precise genome modification of crops possible in a way not previously possible. Targeted gene knockout using CRISPR-Cas9 creates the opportunity for rapid crop varietal improvement by enabling the editing of genes conferring stress resistance (Wang et al. 2021a, b). This technology has been applied to various crops for biotic as well as abiotic stress resistance. CRISPR-Cas9 has been used to produce rice varieties with increased resistance to bacterial light and tolerance to drought (Hamdan et al. 2022; Muha et al. 2023) and is relevant to the potential for addressing global food security issues. Still, another biotechnological tool to facilitate very efficient crop breeding programs is marker-assisted selection (MAS). The basic idea of MAS is that molecular markers linked to desirable traits can be used to select stress-resistant plants efficiently. The use of this technique is an advantage because it has accelerated breeding and increases the precision of designing new crop varieties with superior stress resistance (Biswas et al. 2023). However, breeders can more efficiently develop crop plants that are resilient to environmental stresses by identifying and selecting plants that carry specific genetic markers that are correlated with desirable traits. Biotechnology for agriculture is an enormous environmental and economic benefit. Biotech crops favor biodiversity, reducing reliance on chemical pesticides and fertilizers, therefore, decreasing environmental contamination (Rasheed et al. 2021). Increase yields or reduce production costs in the farm level as well as at the level of farmers and improve food security and economic stability (Shu et al. 2020). Less pesticide use also means that runoff of chemicals into water bodies is also lessened, which reduces the risk of chemical runoff into water bodies, resulting in the protection of aquatic ecosystems and soil health. The potential of biotechnology in agriculture is looking huge in the future. Continued advances in the areas of genomics, transcriptomics, and metabolomics will help us better understand plant stress responses, thus compelling the improvement of more resilient crops. However, many of the challenges posed by climate change and the need for sustainable agriculture will continue to be addressed through the integration of biotechnological approaches with currently used breeding methods (Zhao et al. 2023). Biotechnology is demonstrated to profoundly affect stress resistance in crops and has provided ingenious ways of increasing crops' growth and productivity and furthering agricultural sustainability. Beginning with genetic engineering, gene editing, and marker-assisted selection, scientists have created crops resistant to many biotic and abiotic stresses, keeping food security and environmental conservation. Pests and pathogens cause biotic stress on crop productivity throughout the globe. Primary agents of biotic stress are insects, nematodes, fungi, bacteria, and viruses capable of doing extensive crop damage. Direct yield losses incurred by insect feeding (e.g., cotton bollworm and European corn borer) and a physiologically compromised crop are twice as susceptible to secondary infections (Kivrak et al. 2021; Yan et al. 2022). Root-knot nematodes, and other nematodes, attack plant roots to disrupt nutrient and water uptake stunted growth and reduce crop yield (Ferrari 2022). Crops are infected by fungal pathogens like rust, blight, and mildew that negatively affect photosynthesis and plant health. Bacterial pathogens of the *Xanthomonas* and *Pseudomonas* genera cause leaf spots, wilting, and blight diseases that result in productivity losses through the reduced photosynthetic surface area and poor plant health (Sun et al. 2024). Crop yields are reduced by conditions induced by viral pathogens, often spread by insect vectors, such as stunting, chlorosis, and malformed growth. These biotic stressors have a large economic impact, causing annual agricultural output losses of billions of dollars plus the costs of addressing these threats (Adhikari and Poudel 2020). Biotic stress needs to be managed effectively for crop productivity, and it is a key ingredient for agriculture and food security. Pest management strategies, including biological, cultural, mechanical, and chemical controls, such as integrated pest management (IPM), were effective at mitigating pest populations and minimizing the environmental impact of pest management (Zegeye et al. 2022). Additionally, the growing development of biotechnological advancements, including the marketing of genetically engineered crops that produce insecticidal proteins, has reduced crop losses attributable to pests and pathogens (Chang et al. 2020). Such approaches do more than strengthen crop resiliency to biotic stress; they are also a means by which to reduce the need for chemical pesticides and environmental contamination.

1.4.2 Role of Transgenic Crops in Biotic Stress Management

Biotic stress management is highly reliant on transgenic crops, which is a powerful tool for combating pests and pathogens that threaten world agricultural productivity. Genetically engineered to express the desired traits conferring resistance to a variety of biotic stressors, these crops decrease the need for chemical pesticides and increase crop yields. Bt cotton and Bt maize are examples of such success through Bt crops having been engineered to produce Cry proteins from *Bacillus thuringiensis* (Bt). But these proteins are toxic to a whole range of insect pests, including the cotton bollworm and the European corn borer, and are safe for humans and most beneficial insects (Abebe and Tafa 2022; Bolaji Umar et al. 2022). The increasing use of Bt crops has made it possible to abandon the extensive use of pesticides, which has reduced the production costs for farmers and minimized chemical pollution from runoff (Abebe and Tafa 2022). In addition, transgenic crops resistant to viral, bacterial, and fungal infections have been created. For instance, papaya has been altered due to genetic engineering to be resistant to papaya ringspot virus (PRSV), which was once seen as a great threat to the papaya industry of Hawaii (Yang et al. 2023). Likewise, genetically engineered potatoes have been developed with late blight, caused by *Phytophthora infestans*, resistance genes that reduce fungicide treatment frequency (Sharma et al. 2021). Besides pest and pathogen interactions, transgenic crops support sustainable agriculture principles through the enhancement of biodiversity and ecological balance as fewer pesticides contribute to the health of beneficial insects and other organisms in the soil. Further developing and applying transgenic crops to improve food security and agricultural sustainability is critical in tackling greater pressure from biotic stress factors (Lee et al. 2020). The creation of crops resistant to the attacks of insect pests and pathogens is a breakthrough in the development of modern agricultural biotechnology and comes in handy in combating challenges that adversely impact the security and yield of crops. This process employs the introduction of appropriate genes into the genomes of crops to endow the crops with specific resistance traits. The introduction of *Bacillus thuringiensis*

(Bt) genes to crops is one of the key developments in the biotechnology industry. Bt genes produce Cry proteins that are toxic to certain insect pests but are safe to humans and most beneficial insects. For example, Bt cotton and Bt maize are types of Bt crops that possess these proteins, which target the gut cells of cotton bollworm and European corn borer pests, stopping them from feeding and eventually leading to their death (Muthu et al. 2020). The growing popularity of Bt crops has resulted in reduced application of chemical insecticides, which in return brought down the production costs and environmental and health risks associated with the use of pesticides (Verma et al. 2022).

In addition to Bt technology, genetic modifications have also been made to provide resistance against other fungal, bacterial, and viral pathogens. It has been established, for instance, that plants' defenses can be improved by the introduction of genes coding for pathogenesis-related (PR) proteins. Cell wall degrading enzymes such as chitinases and glucanases that are classified as PR proteins stop the establishment of fungal infections in plants and their spread in the tissues (Taylor et al. 2021). In addition, the evolution of RNA interference (RNAi) technology has resulted in the creation of crops that silence certain genes in pests or pathogens, thereby impairing their ability to infect and attack crops. The principle of RNAi is that double-stranded RNA, which is complementary to the target gene, is introduced and results in the degradation of the corresponding mRNA, leading to the inhibition of gene expression (Qin et al. 2021). Genetic engineering has been used in producing improved crops that are virus-resistant. Among the very best is the transformation of the papaya genome into having the PRSV coat protein gene, which makes it resist the PRSV, leading to resistance through disruption in the ability of the virus to replicate and spread within the plant (Suresh Govindarao Borkar et al. 2023). For example, blight-resistant genetically modified potatoes, for instance, is such an example where biotechnology intervention can impact major diseases of agriculture significantly caused by the fungus-like pathogen *Phytophthora infestans* (Yang et al. 2022). Such developments of resistant crops against insects and pathogens will bring far more advantages than merely crop yields with reduced pesticide applications. It is handy for sustainability in agriculture as it develops ecological balance with less environmental impact. This saves useful insects and prevents the entry of pesticides into water bodies, thus conserving aquatic ecosystems and soil health due to reduced chemical inputs in biotech crops (Liu et al. 2024a, b). To come will be the yield and visibility of genetically modified organisms into the future, which will bring about vital, missing global food security as well as the concerns associated with climate change and growing human populations. The decoupling of chemical pesticide use with biotech and transgenic crops has been a transformative achievement in agriculture presenting eco-friendly agricultural systems and improving humankind and environmental health. Engineered crops, which carry a single gene conferring resistance against pests and pathogens, have already significantly reduced the use of artificial chemical pesticides. A seminal illustration of such an innovation is the creation of *Bacillus thuringiensis* (Bt) crop. These genetically engineered plants are intended to synthesize Cry proteins toxic to some insect pests, but non-toxic to humans, animals, and beneficial insects. However, for instance, Bt cotton and Bt maize have been genetically modified to withstand the pests of the cotton bollworm and the European corn borer, respectively (Yang et al. 2021a, b). The widespread planting of Bt crops has been accompanied by an extraordinary decrease in the use of chemical insecticides, with a significant financial saving for farmers who no longer need to purchase and spray as much pesticide as possible. This decrease in pesticide application also reduces the environmental contamination generally associated with chemical leaching that negatively impacts aquatic environments and soil quality (Li et al. 2023a, b).

Specifically, the application of Bt crops has also been proven to have beneficial effects on nontarget species. Through the reduction of the average level of chemical load in agricultural systems, these crops can foster beneficial insects (pollinators and natural predators of pests) and to preserve soil microbiota, essential for plants' ability to perform nutrient cycling and manage soil fertility (Wang et al. 2022). Besides Bt technology, other biotechnological advances have also contributed to reducing chemical pesticide usage [e.g., development of virus-resistant crops, for example, papaya virus resistant to the papaya ringspot virus (PRSV)] disease outbreaks without the use of chemical controls (Li and Zhu 2023). Also, plants engineered to produce antimicrobial peptides can withstand destructive bacterial and fungal infections and, therefore, avoid the use of fungicidal and bactericidal agents (Hai and Leng 2022). These biotechnological interventions do not only enhance the drought resistance of the crops and their productivity but, in turn, serve to achieve the long-term ecological sustainability of the agricultural practices by re-establishing ecological equilibrium. Biotechnology contributes to reducing the use of chemical pesticides, thus alleviating the detrimental consequences of agricultural pesticide application on the environment, including contamination of groundwater, increased physiological resistance of pest populations to pesticides, and damage to nontarget organisms including humans, as these organisms can be exposed to pesticide residues. There are huge environmental and public health benefits of reduction in the use of chemical pesticides by biotechnology and in the direction of a more sustainable and safe food production process. Continued research and development and application of transgenic plants remain to hold much potential for future agricultural success. Ongoing research into novel genomic manipulations and biotechnological strategies is sure to result in more efficient and precise pest and pathogen control and thus requires less intensive chemical applications. As biotechnological approaches and technologies evolve, such approaches and technologies will play an increasingly important role in their integration into crop management systems to improve global food security and sustainable agriculture. The past decade has seen the development of crops that are resistant to insect pests and pathogens, enabling better crop yields, reduced pesticide use, and enhanced sustainability. Bt crops are one of the most significant advances in this area. Genetically engineered, these crops produce proteins from *Bacillus thuringiensis* (Bt), a bacterium natural to the soil that produces toxins that kill insects. Bt cotton and Bt maize are high-volume genetically engineered crops. Cry proteins that target the cotton bollworm, a devastating pest in cotton production, have been engineered in Bt cotton. The result has been big reductions in the need for chemical insecticides, which have reduced production costs and are less likely to cause environmental contamination (Tudi et al. 2021). Farmers using Bt cotton report benefits in the form of increased yields and increased profitability resulting from decreased damage to the crop and reduced costs of pest management. Like Bt maize, Bt cotton has also transformed cotton cultivation resisting the European corn borer and corn rootworm, two of the most damaging pests in cotton production. In addition to raising maize yields, Bt maize introduction has also led to lower chemical insecticide needs, a move toward more sustainable farming practice (Yang et al. 2022). In addition to providing greater yield, it has long-term economic and environmental benefits beyond the increased productivity: the reduction in chemical pesticide use preserves beneficial insect populations and reduces the potential for pesticide runoff into water, protecting aquatic ecosystems and maintaining soil health. The successful transgenic crop Bt soybean has been engineered to resist the lepidopteran pests. The adoption of Bt soybean in countries like Brazil and Argentina has been associated with increases in yields, relative to yields with no Bt, and corresponding reductions in pesticide applications (Meissner Filho et al. 2021). Not only does this benefit the environment by reducing the chemical load in agricultural ecosystems or decreasing pesticide use, it also helps the farmer's bottom line by reducing pest management costs and increasing crop resilience (Saleem et al. 2020). Crops resistant to pathogens have been developed significantly. Genetically modified papaya resistant to the papaya ringspot virus (PRSV) has saved the papaya industry in Hawaii from collapse. Scientists introduced the gene encoding the PRSV coat protein into the papaya genome to create a transgenic papaya that is immune to the virus, controlling the disease and allowing farmers to continue to grow papaya without resistance to the disease. Another well-publicized example is the creation of potato resistant to *Phytophthora infestans*, the pathogen that causes late blight (Mo et al. 2020). By genetic engineering of resistance genes from wild relatives of potatoes into commercial varieties, transgenic potatoes have been produced that are very resistant to late blight. This advancement shows much promise to significantly decrease fungicide use, which is common to control late blight, decreasing production costs and lowering environmental impact (An et al. 2022). These successful transgenic crops—Bt cotton, Bt maize, Bt soybean, virus-resistant papaya, and late blight-resistant potatoes—show how biotechnology has gone from being a small but very effective part of modern agriculture. These innovations help reduce farms' dependency on chemical pesticides, increase crop resilience, promote more sustainable farming practices, and provide food security and economic viability of agriculture.

1.4.3 Integration of Biotechnological Approaches with Traditional Breeding

Biotechnological approaches, integrated with traditional breeding, are recognized as a powerful strategy for improving crop resilience, productivity, and sustainability. For centuries, traditional breeding methods, in which plants with desirable traits are repeatedly selected and crossed, have been the standard agricultural improvement method. However, these methods are often time-consuming and often limited by the amount of genetic diversity available in the breeding pool. Precision and efficiency that come along with genetic modification and gene editing represent new biotechnological approaches to complement traditional breeding techniques and introduce crops' desired traits (Tabe-Ojong et al. 2023). Molecular markers can be employed to screen plants at an early stage of the breeding process to identify plant selections with desirable genetic traits that can prevent long breeding cycles to develop improved varieties (Martínez-Fortún et al. 2022). Advanced gene-editing tools like CRISPR-Cas9 also enable precise genome modification in plants with traits like pest resistance, drought tolerance, and increased nutritional content (Gustafson et al. 2021). The integration of these biotechnological tools with conventional breeding should allow scientists to develop crop varieties that exhibit better agronomic traits, as well as adapt more rapidly to environmental change. Based on such a synergistic approach, both methods' strengths are used to produce robust, high-yielding crop varieties, resistant to biotic and abiotic stresses. In the end, the combination of biotechnology with traditional breeding has proven to be essential to solving the world's problem of food security and sustainable agriculture (Martínez-Fortún et al. 2022). Strategic combination of conventional breeding and genetic modification is a way to extend crop resilience to biotic stresses (pests and pathogens) by capitalizing on different strengths of the two methodologies. Conventional plant breeding involves using the traits we want and choosing the plants to bring them together by crossing them over many generations to create new varieties with desirable characteristics. The use of this method has proved fundamental to crop improvement in centuries, yet it is often tedious and less accurate. Genetic modification, however, offers the opportunity to specifically insert a certain gene into the plant genome, thereby enabling much faster and more targeted insertion of genes encoding resistance to biotic stresses (Razzaq et al. 2021). Combining these two approaches allows scientists to develop robust crop varieties that give resistance, and the agronomic characteristics desired for farmers on faster time scales. A good example of this is the use of molecular markers to follow beneficial genes through the breeding process, selectively choosing plants that perform at the highest level until they reach the final stage of development. Gene-editing tools like CRISPR-Cas9 also allow for the precise modification of plant genomes, including the introduction of traits like pest resistance or disease tolerance without lengthy breeding cycles (Acevedo et al. 2020). The benefits of this combined approach go beyond improving crop variety development process efficiency; it also ensures that the developed varieties continue to have high yields and can adapt to a range of growing conditions. But the combination of conventional breeding and genetic modification offers a powerful twist on this problem and will help generate more sustainable and resilient agricultural systems. Advancements in molecular marker integration and gene-editing tools such as CRISPR-Cas9 have changed the rates of modern agriculture by facilitating precision breeding and of the speed of developing crop varieties resistant to biotic and abiotic stresses (Nales and Fischer 2023). Specific DNA sequences associated with specific traits, used to follow genes as they are passed from generation to generation, molecular markers are. This marker-assisted selection (MAS) technique enables breeders to weed out plants with desirable traits in the early stages of the breeding process so that improved varieties can be developed much more quickly (Liu et al. 2020). In addition, MAS has been vital in developing crops with resistance to diseases, pests, and environmental stress because it is a much more accurate and efficient method for choosing the best-performing plants (Mehra et al. 2024). For example, MAS has generated rice varieties with enhanced resistance to bacterial blights and blasts, resulting in increased yield stability in areas that are vulnerable to these diseases (Kalaitzandonakes et al. 2023). Plant breeding with unprecedented precision made possible by gene-editing tools such as CRISPR-Cas9 has made genome modification by genetic engineering based on rational genetic introgression a reality. At its core, CRISPR-Cas9 works by creating a double-stranded break at a particular location in the DNA, facilitated by a short RNA molecule matching the target sequence. The cell's natural mechanisms can also use this break to repair it, without introducing new genetic material if desired, or with the introduction of new genetic material to enable precise edits that increase desirable traits (Mehra et al. 2024; Shahriar et al. 2021). CRISPR Cas9 is revolutionary because it is easy to use, efficient, highly versatile, and simple. Genes such as these have been knocked out or modified in wheat to enhance resistance to a variety of biotic stresses, including fungal pathogens, and in tomatoes to increase resistance to powdery mildew. Molecular markers and gene-editing combination provide a powerful approach of crop improvement. Candidate genes for resistance traits can be precisely edited using CRISPR-Cas9 to increase or reintroduce them into new varieties using molecular markers that can identify them. Not only does this approach speed up the process of breeding crops, but it also opens the door to a much greater number of traits that can be added or improved in the crop. For example, rice varieties that express increased resistance to bacterial blight were developed by editing the promoter regions of susceptibility genes in rice using CRISPR/Cas-9, leading to decreased expression of the genes and enhanced disease resistance (Kalaitzandonakes et al. 2023). Also, molecular markers have been applied to track the inheritance of edited genes in the next generations to guarantee stability and efficiency with which the introduced traits are transmitted. The use of molecular markers and gene engineering combined with disease resistance has been applied in addition to other important traits, including drought tolerance, nutrient efficiency, and crop yield. For instance, CRISPR–Cas9 has been used to edit genes in plants that regulate stomata and water use efficiency in rice and wheat varieties that tolerate such conditions and still yield (Yadav et al. 2021). Traits such as high-yielding genotypes with improved beneficial traits of high nutrient uptake and utilization have been molecularly selected for, resulting in improved overall crop performance. The integration of traditional breeding with these new biotechnological tools should represent real progress in crop improvement. This allows for the development of robust, high-yielding crop varieties that can cope with the changes driven by a changing climate and pest and disease pressures. The availability of strong molecular markers and genome-editing tools enables breeders to design crops that not only meet the requirements of modern agriculture but also provide for sustainability and resilience in farming systems. Biotechnology is critical to reduce dependence on dangerous pesticides and help for ecological balance with breeds that are naturally resistant to pests and disease. Using this creative solution, agricultural productivity is improved, while at the same time reducing the environmental impact of such extensive pesticide use. The most important contribution of biotechnology is to create genetically modified (GM) crops like Bt crops. By engineering to express *B. thuringiensis* (Bt) toxins, these crops eliminate key agricultural pests, including cotton bollworm and European corn borer (Afolabi et al. 2023; Matsumoto et al. 2022). Bt crops are reducing the need for chemical insecticides and have had large reductions in pesticide uses, leading to large money savings for farmers and less pesticide runoff contaminating the environment (Afolabi et al. 2023). The reduction in pesticide use also reduces the use of pesticides on plants that do not produce harvestable crop seeds, acting to protect nontarget organisms, including beneficial insects such as pollinators and natural pest predators, that help maintain ecological balance (Tyagi et al. 2020). Biotechnology, however, has expanded beyond Bt technology—these crops are resistant to viral, bacterial, and fungal pathogens. For example, virus-resistant papaya (genetically modified so it cannot get the papaya ringspot virus) has been made to prevent disease outbreaks and reduce the need for chemical treatments (Suresh Govindarao Borkar et al. 2023). In addition, biotechnology has promoted the production of plants synthesizing antimicrobial peptides, which confer resistance to bacterial and fungal infection and decrease the need for chemically produced fungicides and bactericides (Hamdan et al. 2022). The result is more resilient and higher-yielding crops, with the bonus that they also facilitate long-term sustainability by promoting healthier ecosystems. In addition, the adoption of biotech crops has helped to preserve soil health and keep water pollution at bay. Reduced use of chemical pesticides reduces the risk of chemical runoff into water bodies, which can do major harm and affect the lives of aquatic creatures. Biotechnology helps to reduce pesticide applications and maintain soil biodiversity and fertility, which are essential for sustainable agricultural practices. Its use also diminishes pesticide dependency and lowers the risk of pests developing investment in chemical treatments due to its great utilization in conventional agriculture. Thus, we may say that the major role of biotechnology is not limited to immediate control of the pest and pathogen. Integrated pest management (IPM) is supported, and practices are a combination of biological, cultural, mechanical, and chemical controls that help in maintaining pest populations in an effective and sustainable manner (Tyagi et al. 2020). IPM strategies, using biotechnological tools, are more efficient and less reliant on chemical pesticides, and promote ecological balance and sustainable farming practice. In short, biotechnology helps to decrease the dependence on harmful pesticides and maintain ecological balance by developing the crops having intrinsic resistance to treat the pests and diseases. This advance also brings economic benefits to farmers and contributes to protecting and enhancing biodiversity through environmental conservation and sustainable agricultural practices (Hamdan et al. 2022).

1.4.4 Biotechnological Innovations in Pesticide Reduction

Biotechnology's influence on environmental contamination reduction and green agriculture is deep and wide range. The development of genetically modified (GM) crops has dramatically lessened the impact chemical pesticides and fertilizers impose on the environment. The adoption of Bt crops, such as Bt cotton and Bt maize engineered to produce the *B. thuringiensis* (Bt) toxins, is one of the most obvious examples. Specifically aimed at and killing major agricultural pests such as the cotton bollworm and the European corn borer, these toxins replace the use of chemical insecticides (Brookes 2022). By reducing pesticide use, this also reduces production costs for farmers and helps to lessen the risk of pesticides running off to water bodies, contaminating water bodies and harm nontarget species, such as beneficial insects and soil microorganisms (Brookes and Barfoot 2020). Also, GM crops intended to be tolerant of herbicides, such as glyphosate-tolerant soybeans, result in a reduction in the number of herbicides used and decrease in the chemical inputs related to it (Brookes and Barfoot 2020). Biotechnology broadens sustainable agriculture by reducing reliance on chemical pesticides and fertilizers, thus improving healthier soil ecology and water quality. Reduced use of chemical pesticides assures soil biodiversity; this is the sine qua non condition for preserving soil structure, fertility, and the general health of the agroecosystems. In addition, reducing the use of herbicides and pesticides also helps grow the number of beneficial insects that eat the pollinators of our food sources and eat the pests we feed (Jose et al. 2020). Additionally, the GM crops that have been developed to be pest-resistant or stress-tolerant increase crop yields and stability and subsequently increase food security and reduce the requirement for terrestrial habitat for agricultural land conversion. Natural habitats are very important for natural biodiversity and ecological balance; therefore, through the preservation of these natural habitats, whatever harms them must be avoided. Papaya ringspot virus (PRSV), for example, in Hawaii, the cultivation of virus-resistant papaya has not only saved papaya industry from collapse but reduced the need for chemical control damaging papaya ringspot virus, a case in itself that minimizes environmental contamination (Seixas et al. 2022). Biotechnology has many other benefits for integrated pest management (IPM) strategies that undertake biological, cultural, mechanical, and chemical controls over populations to control pests effectively and sustainably. IPM can be improved by biotechnological tools to new modes of pest control, with minimal environmental impact (Brookes 2021). For instance, the integration of Bt crops into IPM programs leads to less frequent and a smaller number of pesticides applied in IPM programs that are more sustainable for pest management. Biotechnology in general reduces environmental contamination, but it also promotes sustainable agriculture through intrinsic pest and disease resistance in the crop. Its benefits are it brings food, economic benefits for farmers, environmental conservation, and sustainable agricultural practices protecting and enhancing biodiversity.

1.4.5 Advancements and Emerging Trends in Biotechnology

Biotechnology holds bright prospects for transforming the practice of agriculture, healthcare, and environmental management, delivering new and creative ideas for some of the challenges facing the world today. Gene-editing technologies such as CRISPR-Cas9 are poised to reap major advances in the agronomic use of these technologies through the potential for improved crop resistance to pests, diseases, and environmental stresses (Abdul Aziz et al. 2022; Brookes and Barfoot 2020). With this precision breeding, crops become better for nutrition, bigger and better able to stand up to climate change. Biotechnology is leading the way in healthcare, to personalized medicine, making treatments and therapies personalized to individual patient genetics, which increases efficacy and reduces side effects (Anyanwu et al. 2024). Techniques like gene therapy or regenerative medicine can cure genetic disorders and repair damaged tissues. Furthermore, biotechnology has important uses in terms of environmental sustainability in developing bio-based products and processes. For example, genetically engineered microorganisms are being employed for cleaning oil spillages and plastic waste degradation, ecofriendly approaches as opposed to the conventional approach (Anyanwu et al. 2024). Also, with the combination of biotechnology, artificial intelligence, and data analytics, research and innovation will accelerate and new frontiers will be opened in the area of synthetic biology and bioengineering. While the ethical and regulatory frameworks will continue to evolve, society should hold on tightly to them, to develop and implement the biotechnological advancement on a responsible basis (Clemensen et al. 2020). Overall, biotechnology has a bright future, with the capacity to lead us to dramatic progress within a wide variety of sectors, resulting in a healthier, more sustainable world. Biofortification is a revolutionary agricultural biotechnology approach for improving the nutritional quality of staple food crops, particularly in the form of micronutrients such as iron and zinc that are crucial for human health but suffer from widely prevalent deficiencies among the world's poor. The process here is a means to get either via traditional breeding methods, genetic modification, or agronomic practices a higher level of essential vitamins and minerals in crops to raise the nutritional quality of the food supply. Scientists can increase the nutrient content of crops through biofortification: breeding a crop that is higher in nutrients like iron, zinc, vitamin A, and folate, all of which are important for human health (Raspor and Cingel 2021). The development of golden rice is a huge milestone in biofortification. Golden rice has been genetically modified to produce beta-carotene, a precursor of vitamin A, a deficiency of which is a major cause of preventable blindness and increased deaths in children from developing areas (Raspor and Cingel 2021). Similarly, other staple crops, including beans, wheat, and maize, have developed biofortified varieties containing higher levels of iron and zinc. They also are micronutrients with an important role in preventing anemia and supporting immune function. For example, HarvestPlus, a global biofortification initiative, has successfully released iron biofortified beans and pearl millet; zinc biofortified wheat and rice; and provitamin A biofortified maize and cassava in various developing countries (Manganda et al. 2024). In addition to being nutritionally enhanced, these biofortified crops are also bred to retain or enhance their yields, so that farmers can adopt them without sacrificing production. One advantage of biofortification over other types of nutritional interventions, such as supplementation and fortification of food, is that this intervention offers long-term, sustainable solutions to micronutrient deficiency. It ensures that essential nutrients are not missing from the crops consumed by individuals; this directly accesses populations that may not have accessed fortified foods and supplements. It is highly advantageous for people staying in rural areas of developing countries where there is relatively little dietary diversity and health care services are not readily available. Biofortification can be combined with other agricultural practices for improved efficacy. For example, agronomic biofortification refers to the application of mineral fertilizers directly to soils or crops to increase the uptake of nutrients such as zinc and selenium. This complements the genetic biofortification efforts (Manganda et al. 2024; Raspor and Cingel 2021). The integrated approach then allows for biofortified crops to be grown under diverse environmental conditions, and thus they are resilient to climate change and other different agricultural challenges. New technologies and breeding techniques ensure that there is continuous evaluation to be able to further enhance the nutritional profiles of crops. The genome-editing tools like CRISPR/Cas9 offer unprecedented accuracy in modifying plant genomes to allow for targeted enhancement of nutrient content without affecting other desirable traits (Anusha and Amir 2024; Yang et al. 2021a, b). Biofortification therefore is one of the approaches critical to solving malnutrition problems globally through staple crops whose nutritional value can be enhanced. The process offers an effective sustainable means of tackling micronutrient deficiencies but also aids agriculture in improving productivity and resilience for enhanced health benefits and food security. Bioremediation and phytoremediation are two promising approaches for sustainable farm systems, using natural processes to remediate contaminated environments, which will be useful in enhancing soil health and agricultural productivity. Bioremediation utilizes microorganisms, like bacteria, fungi, and algae, taking advantage of their natural metabolic capabilities to degrade or detoxify pollutants in soil and water into less toxic substances. This is very effective for organic pollutants such as pesticides, petroleum hydrocarbons, and industrial solvents, which are often present in agricultural sites as common contaminants (Mustafa et al. 2022). Therefore, bioremediation is the least-pollutive and lowest-cost method compared to conventional methods of soil remediation that have relied on the use of toxic chemicals or excavation of contaminated soils. Phytoremediation, in turn, is the process by which plants absorb, accumulate, and detoxify contaminants in the environment. Some plant species, called hyperaccumulators, take up large amounts of heavy metals and certain radionuclides and organic pollutants from the soil through their roots. These plants then transfer the contaminants to their shoots, which can be harvested and disposed of safely and minimize the toxicity of the toxins in the soil (Beillouin et al. 2021). The major role of phytoremediation is remediation through heavy metals since they permanently exist in the environment and severely constitute hazards to the human health system as well as the ecosystem. This method not only purifies the soil, but it also improves its physical and chemical properties, hence, becoming rich and structured. There are significant advantages to integrating bioremediation and phytoremediation into sustainable agriculture practices. These biological treatments have an environmental advantage; thus, they can be performed with very minimal structural damage to the soil system and its surrounding ecosystems; consequently, they support restoration of contaminated land that might once again be used properly for agricultural purposes. For instance, the use of plant species in the processes may help restore the fertility of soils polluted with heavy metals, thus enabling cultivation on lands that have earlier become unproductive (Choudhary et al. 2020). Bioremediation and phytoremediation can also complement sustainable farming practices such as crop rotation and organic farming by achieving maximum effectiveness and developing healthy soil in general. Adoption will help in controlling the environmental hazards related to agro activities through bioremediation and phytoremediation technologies. These include prevention and remediation of contaminants in soils and waters, which would protect biodiversity and enhance ecosystem resilience. It would, therefore, promote long-term sustainability in farming systems by reducing dependency on chemical inputs and natural recovery of degraded land areas over time. Continuous research and development in this field have expanded its applications and effectiveness of bioremediation and phytoremediation, which are both contributing to sustainable agriculture (Abdulkadir 2023). In conclusion, bioremediation and phytoremediation provide innovative and sustainable methods of managing environmental contamination in agricultural systems. That these technologies improve soil health, crop productivity, and ecological balance emphasizes their significance in promoting sustainable farming practices. New frontiers in genetic engineering for biotic stress management appear at rapid speeds and are now promising new solutions in crop resilience against pests and pathogens. Among the most revolutionary tools for that is gene-editing CRISPR-Cas9 technology, which enables the precise modification of plant genomes. Through the application of CRISPR-Cas9, scientists can precisely target genes specifically responsible for the susceptibility to stress and modify it to confer resistance (Adhikari and Menalled 2020). This precision not only ensures the speed of the development of resistant crop varieties but also ensures that mutations do not come at other desirable costs. There are even already real-life examples: for example, CRISPR-edited rice genes that can enhance resistance to bacterial blight by disrupting the susceptibility gene OsSWEET14 in such a way that a pathogen cannot hijack the cellular machinery of the plant (Ferreira et al. 2023; Singh et al. 2021). Another frontier that is just emerging is the use of RNAi technology, whereby gene expression is specifically switched off in pests or pathogens to make them innocuous to the crop. This has promised to manage viral diseases in crops like tomatoes and cassava targeting the genes that replicate the viruses in the plant (Ramamoorthy and Natarajan 2020). Synthetic biology also opens new pathways in the management of biotic stresses by inventing novel genetic circuits and pathways that can be introduced into plants to enhance their defense mechanisms. These developments do not only make biotic stress management more efficient but also reduce the use of chemical pesticides, thus promoting sustainable agricultural practices. As these technologies in genetic engineering advance, they promise to revolutionize agriculture through the provision of robust and environmentally friendly solutions to biotic stress challenges. The near future of biotechnology in biotic stress management offers much promise for sustainable means of improving global food security without much negative impact on the environment (Zainal-Abidin et al. 2021) (Table 1.2).

1.5 Conclusion and Future Plans

Physiological plant defense mechanisms elicited by biotic stress represent the complex interaction between biology and environmental stress. Plant reactive mechanisms, ranging from molecular signaling pathways involving salicylic acid and jasmonic acid to a physical level of trichome emergence or wax sheaths, show an impressive array of tactics to protect against pathogens and herbivores. Recent biotechnological achievements, including genetic engineering, RNA interference, and CRISPR-Cas9, have also played a role in increasing our ability to improve plant immunity to biotic stress. Illustrative examples of such a technology, for example, Bt crops and virus-resistant papaya, are shown to illustrate the change-inducing abilities of biotechnology in reducing the use of chemical pesticides and in creating sustainable agriculture. Furthermore, the combination of conventional breeding methods and modern biotechnological approaches is a complementary strategy to ameliorate crop tolerance, yield, and sustainability. Transgenic crop and biofortification crop breeding not only alleviate food hunger, but also adaptive measures for the causes of environmental problems such as pesticide drift, loss of biodiversity, and environmental damage. In the future, the further development of the gene-editing technology, together with a better understanding of plant stress responses will promise sustainable and efficient agriculture. Enabled by this progress, we can not only overcome the conflate problems of climate change and population growth but also provide the world with a permanent and sustainable food supply from time immemorial.

Table 1.2 Biotechnological advancements in agriculture

Aspect	Key focus	Examples	Environmental benefits	References
Impact of biotechnology in agriculture	Improved productivity and sustainability	GM crops enhancing resistance to biotic/abiotic stresses	Reduced reliance on chemical inputs	Liu et al. (2022)
Development of GM crops	Reduction of pesticide use	Bt cotton, Bt maize	Less pesticide runoff reduced environmental contamination	Chang et al. (2020) and Sheng et al. (2020)
Bt technology in crop improvement	Resistance to pests	Cry proteins targeting bollworm, European corn borer	Preservation of beneficial insects	Abebe and Tafa (2022) and Muthu et al. (2020)
Drought and salinity tolerance	Resilience to abiotic stresses	Drought-tolerant maize, salt-tolerant rice and wheat	Sustains productivity in stressed environments	Kumari et al. (2022) and Xuebo et al. (2023)
CRISPR-Cas9 in crop breeding	Precision genome editing	Stress-resistant rice varieties	Reduces the need for chemical treatments	Wang et al. (2021a, b) and Muha et al. (2023)
Marker-assisted selection (MAS)	Faster breeding cycles	Rice varieties with bacterial blight resistance	Reduced development time for resilient crops	Biswas et al. (2023) and Kalaitzandonakes et al. (2023)
RNA interference (RNAi)	Targeted gene silencing	Pathogen-resistant crops	Reduces fungicide and bactericide use	Qin et al. (2021)
Pathogen resistance	Resistance to diseases	PRSV-resistant papaya, late blight-resistant potatoes	Reduces chemical fungicide usage	Suresh Govindarao Borkar et al. (2023) and Sharma et al. (2021)
Bt crops and environmental impact	Reduced pesticide application	Bt soybean in Brazil, Argentina	Protects soil microbiota and water bodies	Yang et al. (2022) and Li et al. (2023a, b)
Economic benefits of GM crops	Reduced costs and increased yields	Bt crops with lower pest management costs	Improved farmer profitability	Meissner Filho et al. (2021) and Saleem et al. (2020)
Integrated pest management (IPM)	Sustainable pest control	Bt crops as part of IPM	Reduces reliance on chemical pesticides	Zegeye et al. (2022) and Tyagi et al. (2020)
Sustainable agriculture	Biodiversity and habitat conservation	Reduced habitat conversion for agriculture	Promotes ecological balance	Seixas et al. (2022)

(continued)

Table 1.2 (continued)

Aspect	Key focus	Examples	Environmental benefits	References
Transgenic innovations	Resistance to biotic stresses	Late blight-resistant potatoes, PRSV-resistant papaya	Minimizes fungicide use and environmental contamination	Mo et al. (2020) and An et al. (2022)
Future potential of biotechnology	Genomics and metabolomics advancements	Biotic/abiotic stress-resistant crops	Facilitates sustainable agricultural practices	Zhao et al. (2023) and Martínez-Fortún et al. (2022)
Environmental benefits of reduced pesticides	Reduction of pesticide runoff	GM crops reducing environmental contamination	Protects aquatic ecosystems	Brookes (2022) and Hamdan et al. (2022)
Gene-editing and molecular markers	Precision breeding	CRISPR-Cas9 for drought and pest resistance	Develop crops with better environmental adaptability	Shahriar et al. (2021) and Acevedo et al. (2020)
Biotechnology in food security	Enhanced resilience and yield	Disease-resistant rice, high-yield soybean	Ensures global food security	Yadav et al. (2021) and Matsumoto et al. (2022)

References

Abdul Aziz M, Brini F, Rouached H, Masmoudi K (2022) Genetically engineered crops for sustainably enhanced food production systems. Front Plant Sci 13:1027828. https://doi.org/10.3389/fpls.2022.1027828

Abdulkadir KN (2023) Cropping systems diversification as an approach to enhancing crop productivity. https://doi.org/10.20944/preprints202312.2115.v1

Abebe A, Tafa Z (2022) Review on: impact of genetic engineering in biotic stresses resistance crop breeding. Int J Resin Appl Sci Biotechnol 9(2):243–250. https://doi.org/10.31033/ijrasb.9.2.18

Abu-Shahba MS, Mansour MM, Mohamed HI, Sofy MR (2022) Effect of biosorptive removal of cadmium ions from hydroponic solution containing indigenous garlic peel and mercerized garlic peel on lettuce productivity. Sci Hortic 293:110727. https://doi.org/10.1016/j.scienta.2021.110727

Acevedo M, Pixley K, Zinyengere N, Meng S, Tufan H, Cichy K, Bizikova L, Isaacs K, Ghezzi-Kopel K, Porciello J (2020) A scoping review of adoption of climate-resilient crops by small-scale producers in low- and middle-income countries. Nat Plants 6(10):1231–1241. https://doi.org/10.1038/s41477-020-00783-z

Adhikari S, Menalled FD (2020) Supporting beneficial insects for agricultural sustainability: the role of livestock-integrated organic and cover cropping to enhance ground beetle (Carabidae) communities. Agronomy 10(8):1210. https://doi.org/10.3390/agronomy10081210

Adhikari P, Poudel M (2020) CRISPR-Cas9 in agriculture: approaches, applications, future perspectives, and associated challenges. Malays J Halal Res 3(1):6–16. https://doi.org/10.2478/mjhr-2020-0002

Afolabi F, Matsumoto R, Akinwande B, Otegbayo B, Asfaw A (2023) Effect of mineral fertilisation on tuber yield and quality in yams (Dioscorea alata and Dioscorea rotundata). Agriculture 13(12):2240. https://doi.org/10.3390/agriculture13122240

Agarwal N, Srivastava R, Verma A, Rai KM, Singh B, Verma PC (2020) Unravelling cotton nonexpressor of pathogenesis-related 1(NPR1)-like genes family: evolutionary analysis and putative role in fiber development and defense pathway. Plan Theory 9(8):999. https://doi.org/10.3390/plants908099

Al-Khayri JM, Rashmi R, Topp V, Chole PB, Banadka A, Sudheer WN, Nagella P, Shehata WF, Al-Mssallem MQ, Alessa FM, Almaghasla MI, Rezk AA-S (2023) Plant secondary metabolites: the weapons for biotic stress management. Metabolites 13(6):716. https://doi.org/10.3390/metabo13060716

Ali J, Mukarram M, Abbas A, Umar M, Fleischer P, Mohamed HI (2024a) Wound to survive: mechanical damage suppresses aphid performance on brassica. J Plant Dis Prot 131(3):781–792. https://doi.org/10.1007/s41348-024-00871-8

Ali M, Shafiq M, Haider MZ, Sami A, Alam P, Albalawi T, Kamran Z, Sadiq S, Hussain M, Shahid MA, Jeridi M, Ashraf GA, Manzoor MA, Sabir IA (2024b) Genome-wide analysis of NPR1-like genes in citrus species and expression analysis in response to citrus canker (Xanthomonas axonopodis pv citri). Front Plant Sci 15. https://doi.org/10.3389/fpls.2024.1333286

An C, Sun C, Li N, Huang B, Jiang J, Shen Y, Wang C, Zhao X, Cui B, Wang C, Li X, Zhan S, Gao F, Zeng Z, Cui H, Wang Y (2022) Nanomaterials and nanotechnology for the delivery of agrochemicals: strategies towards sustainable agriculture. J Nanobiotechnol 20(1):11. https://doi.org/10.1186/s12951-021-01214-7

Anjali KS, Korra T, Thakur R, Arutselvan R, Kashyap AS, Nehela Y, Chaplygin V, Minkina T, Keswani C (2023) Role of plant secondary metabolites in defence and transcriptional regulation in response to biotic stress. Plant Stress 8:100154. https://doi.org/10.1016/j.stress.2023.100154

Anusha MR, Amir M (2024) Integrating omics approaches for climate-resilient crops: a comprehensive review. J Adv Biol Biotechnol 27(6):351–363. https://doi.org/10.9734/jabb/2024/v27i6895

Anyanwu EC, Arowoogun JO, Odilibe IP, Akomolafe O, Onwumere C, Ogugua JO (2024) The role of biotechnology in healthcare: a review of global trends. World J Adv Res Rev 21(1):2740–2752. https://doi.org/10.30574/wjarr.2024.21.1.0382

Appu M, Ramalingam P, Sathiyanarayanan A, Huang J (2021) An overview of plant defense-related enzymes responses to biotic stresses. Plant Gene 27:100302. https://doi.org/10.1016/j.plgene.2021.100302

Balaji BN, Jambagi SR (2024) Structural defences in plants against herbivores – a review. Indian J Entomol:1–9. https://doi.org/10.55446/IJE.2024.2031

Beillouin D, Ben-Ari T, Malézieux E, Seufert V, Makowski D (2021) Positive but variable effects of crop diversification on biodiversity and ecosystem services. Glob Chang Biol 27(19):4697–4710. https://doi.org/10.1111/gcb.15747

Biswas S, Das R, Lay Nwe L (2023) Organic farming to mitigate abiotic stresses under climate change scenario. https://doi.org/10.5772/intechopen.111620

Bolaji Umar O, Amudalat Ranti L, Shehu Abdulbaki A, Lukman Bola A, Khadijat Abdulhamid A, Ramat Biola M, Oluwagbenga Victor K (2022) Stresses in plants: biotic and abiotic. In: Current trends in wheat research. IntechOpen. https://doi.org/10.5772/intechopen.100501

Brookes G (2021) Environmental impacts of genetically modified (GM) crop use: impacts on pesticide use and carbon emissions. In: Plant biotechnology. Springer, pp 87–101. https://doi.org/10.1007/978-3-030-68345-0_7

Brookes G (2022) Genetically modified (GM) crop use 1996–2020: environmental impacts associated with pesticide use change. GM Crops Food 13(1):262–289. https://doi.org/10.1080/21645698202222118497

Brookes G, Barfoot P (2020) Environmental impacts of genetically modified (GM) crop use 1996–2018: impacts on pesticide use and carbon emissions. GM Crops Food 11(4):215–241. https://doi.org/10.1080/21645698.2020.1773198

Chacón-Fuentes M, Bardehle L, Seguel I, Espinoza J, Lizama M, Quiroz A (2023) Herbivory damage increased VOCs in wild relatives of Murtilla plants compared to their first offspring. Metabolites 13(5):616. https://doi.org/10.3390/metabo13050616

Chang Y, Zhu C, Jiang J, Zhang H, Zhu J, Duan C (2020) Epigenetic regulation in plant abiotic stress responses. J Integr Plant Biol 62(5):563–580. https://doi.org/10.1111/jipb.12901

Chen L, Kuai P, Ye M, Zhou S, Lu J, Lou Y (2020) Overexpression of a cytosolic 6-phosphogluconate dehydrogenase gene enhances the resistance of rice to Nilaparvata lugens. Plan Theory 9(11):1529. https://doi.org/10.3390/plants9111529

Chen H, Li M, Qi G, Zhao M, Liu L, Zhang J, Chen G, Wang D, Liu F, Fu ZQ (2021) Two interacting transcriptional coactivators cooperatively control plant immune responses. https://doi.org/10.1101/2021.03.21.436112

Choudhary AK, Varatharajan T, Rohullah R, Bana RS, Pooniya V, Dass A et al (2020) Integrated crop management technology for enhanced productivity, resource-use efficiency and soil health in legumes – a review. Indian J Agric Sci 90(10):1839–1849. https://doi.org/10.56093/ijas.v90i10.107882

Clemensen AK, Provenza FD, Hendrickson JR, Grusak MA (2020) Ecological implications of plant secondary metabolites – phytochemical diversity can enhance agricultural sustainability. Front Sustain Food Syst 4. https://doi.org/10.3389/fsufs.2020.547826

Collinge DB, Sarrocco S (2022) Transgenic approaches for plant disease control: status and prospects 2021. Plant Pathol 71(1):207–225. https://doi.org/10.1111/ppa.13443

Divekar PA, Narayana S, Divekar BA, Kumar R, Gadratagi BG, Ray A, Singh AK, Rani V, Singh V, Singh AK, Kumar A, Singh RP, Meena RS, Behera TK (2022) Plant secondary metabolites as defense tools against herbivores for sustainable crop protection. Int J Mol Sci 23(5):2690. https://doi.org/10.3390/ijms23052690

Ferrari L (2022) Farmers' attitude toward CRISPR/Cas9: the case of blast resistant rice. Agribusiness 38(1):175–194. https://doi.org/10.1002/agr.21717

Ferreira L d S, Oliveira V d S, Marchiori JJ d P, Gomes M d P, Ferreira TC, Vargens FN, Bolsoni EZ, Carriço E, Holtz AM (2023) Soil fertility indicators in crop-livestock-Forest integration systems. Int J Plant Soil Sci 35(20):1093–1104. https://doi.org/10.9734/ijpss/2023/v35i203906

Fu Y, Liu T, Wang X, Wang Y, Gong Q, Li G, Lin Q, Zhu S (2023) Untargeted metabolomics reveal rhizosphere metabolites mechanisms on continuous ramie cropping. Front Plant Sci 14. https://doi.org/10.3389/fpls.2023.1217956

Gaur G, Gänzle MG (2023) Conversion of (poly)phenolic compounds in food fermentations by lactic acid bacteria: novel insights into metabolic pathways and functional metabolites. Curr Res Food Sci 6:100448. https://doi.org/10.1016/j.crfs.2023.100448

Genva M, Andersson MX, Fauconnier M-L (2020) Simple liquid chromatography-electrospray ionization ion trap mass spectrometry method for the quantification of galacto-oxylipin arabidopsides in plant samples. Sci Rep 10(1):11957. https://doi.org/10.1038/s41598-020-68757-x

Ghozlan MH, Eman EA, Tokgöz S, Lakshman DK, Mitra A (2020) Plant defense against necrotrophic pathogens. American Journal of Plant Sciences 11(12):2122–2138. https://doi.org/10.4236/ajps.2020.1112149

Guan Q, David L, Moran R, Grela I, Ortega A, Scott P, Warnock L, Chen S (2023) Role of NPR1 in systemic acquired stomatal immunity. Plan Theory 12(11):2137. https://doi.org/10.3390/plants12112137

Guerreiro J, Marhavý P (2023) Unveiling the intricate mechanisms of plant defense. Front Plant Physiol 1. https://doi.org/10.3389/fphgy.2023.1285373

Gülser F, Sönmez F (2022) Effects of mycorrhizae and salicylic acid on growth, cadmium content and uptake of maize (Zea mays L) seedlings in cadmium contaminated media. Uluslar Tarım Yaban Hayatı Bilimleri Dergisi 8(1):133–141. https://doi.org/10.24180/ijaws.1011361

Gustafson D, Asseng S, Kruse J, Thoma G, Guan K, Hoogenboom G, Matlock M, McLean M, Parajuli R, Rajagopalan K, Stöckle C, Sulser TB, Tarar L, Wiebe K, Zhao C, Fraisse C, Gimenez C, Intarapapong P, Karimi T et al (2021) Supply chains for processed potato and tomato products in the United States will have enhanced resilience with planting adaptation strategies. Nat Food 2(11):862–872. https://doi.org/10.1038/s43016-021-00383-w

Hai Y, Leng G (2022) A more than four-fold sex-specific difference of autism spectrum disorders and the possible contribution of pesticide usage in China 1990–2030. Front Public Health 10. https://doi.org/10.3389/fpubh.2022.945172

Hamdan MF, Karlson CKS, Teoh EY, Lau S-E, Tan BC (2022) Genome editing for sustainable crop improvement and mitigation of biotic and abiotic stresses. https://doi.org/10.20944/preprints202207.0404.v1

Han Y, Luthe D (2022) Key genes in the JAZ signaling pathway are up-regulated faster and more abundantly in caterpillar-resistant maize. J Chem Ecol 48(2):179–195. https://doi.org/10.1007/s10886-021-01342-2

Han Q, Tan W, Zhao Y, Yang F, Yao X, Lin H, Zhang D (2022) Salicylic acid-activated BIN2 phosphorylation of TGA3 promotes Arabidopsis PR gene expression and disease resistance. EMBO J 41(19):e110682. https://doi.org/10.15252/embj.2022110682

Huang W, Wang Y, Tian W, Cui X, Tu P, Li J, Shi S, Liu X (2022) Biosynthesis investigations of terpenoid, alkaloid, and flavonoid antimicrobial agents derived from medicinal plants. Antibiotics 11(10):1380. https://doi.org/10.3390/antibiotics11101380

Jiao L, Bian L, Luo Z, Li Z, Xiu C, Fu N, Chen Z (2022) Enhanced volatile emissions and antiherbivore functions mediated by the synergism between jasmonic acid and salicylic acid pathways in tea plants. Horticulture Research 9 uhac144. https://doi.org/10.1093/hr/uhac144

Jose M, Vertuan H, Soares D, Sordi D, Bellini LF, Kotsubo R, Berger GU (2020) Comparing agronomic and phenotypic plant characteristics between single and stacked events in soybean, maize, and cotton. PLoS One 15(4):e0231733. https://doi.org/10.1371/journal.pone.0231733

Kalaitzandonakes N, Willig C, Zahringer K (2023) The economics and policy of genome editing in crop improvement. Plant Genome 16(2). https://doi.org/10.1002/tpg2.20248

Kamle M, Borah R, Bora H, Jaiswal AK, Singh RK, Kumar P (2020) Systemic acquired resistance (SAR) and induced systemic resistance (ISR): role and mechanism of action against phytopathogens, pp 457–470. https://doi.org/10.1007/978-3-030-41870-0_20

Kanda Y, Shinya T, Maeda S, Mujiono K, Hojo Y, Tomita K, Okada K, Kamakura T, Galis I, Mori M (2023) BSR1, a rice receptor-like cytoplasmic kinase, positively regulates defense responses to herbivory. Int J Mol Sci 24(12):10395. https://doi.org/10.3390/ijms241210395

Kaur S, Samota MK, Choudhary M, Choudhary M, Pandey AK, Sharma A, Thakur J (2022) How do plants defend themselves against pathogens-biochemical mechanisms and genetic interventions. Physiol Mol Biol Plants 28(2):485–504. https://doi.org/10.1007/s12298-022-01146-y

Khotimah K (2022) Agronomy Performance and Resistance of Shallots against Fusarium Wilt Disease under Various Salicylic Acid Treatments. Jurnal Online Pertanian Tropik 9(2):164–169. https://doi.org/10.32734/jopt.v9i2.9500

Kivrak E, Pauzaite T, Copeland N, Hardy J, Kara P, Firlak M, Yardimci A, Yilmaz S, Palaz F, Ozsoz M (2021) Detection of CRISPR-Cas9-mediated mutations using a carbon nanotube-modified electrochemical genosensor. Biosensors 11(1):17. https://doi.org/10.3390/bios11010017

Kumar S, Korra T, Thakur R, Arutselvan R, Kashyap AS, Nehela Y, Keswani C (2023) Role of plant secondary metabolites in defence and transcriptional regulation in response to biotic stress. Plant stress 8:100154. https://doi.org/10.1016/j.stress.2023.100154

Kumari D, Prasad B, Dwivedi P (2022) Genome editing platforms in rice (Oryza sativa L): basic methodology and troubleshooting. https://doi.org/10.21203/rs.3rs-1961833/v1

Kumawat RK, Tiwari G, Ramakrishnan RS, Debnath S, Thakur S, Bhayal D, Samaiya RK, Upadhyay A, Bhayal L (2023) Sowing time, plant growth regulators and their application scheduling affect quality traits of wheat (Triticum aestivum L). Int J Plant Soil Sci 35(18):121–126. https://doi.org/10.9734/ijpss/2023/v35i183274

Lawrie NS, Cuetos NM, Sini F, Salam GA, Ding H, Vancolen A, Nelson JM, Erkens RHJ, Perversi G (2023) Systematic review on raphide morphotype calcium oxalate crystals in angiosperms. AoB Plants 15(4). https://doi.org/10.1093/aobpla/plad031

Lee S, Lee T, Yang S, Lee I (2020) BarleyNet: a network-based functional omics analysis server for cultivated barley, Hordeum vulgare L. Front Plant Sci 11. https://doi.org/10.3389/fpls.2020.00098

Li X, Zhu M (2023) The role of agricultural mechanization services in reducing pesticide input: promoting sustainable agriculture and public health. Front Public Health 11. https://doi.org/10.3389/fpubh.2023.1242346

Li C, Mo Y, Wang N, Xing L, Qu Y, Chen Y, Yuan Z, Ali A, Qi J, Fernández V, Wang Y, Kopittke PM (2023a) The overlooked functions of trichomes: water absorption and metal detoxication. Plant Cell Environ 46(3):669–687. https://doi.org/10.1111/pce.14530

Li C, Wang J, Ling F, You A (2023b) Application and development of Bt insect resistance genes in rice breeding. Sustainability 15(12):9779. https://doi.org/10.3390/su15129779

Liu Q, Yang X, Tzin V, Peng Y, Romeis J, Li Y (2020) Plant breeding involving genetic engineering does not result in unacceptable unintended effects in rice relative to conventional cross-breeding. Plant J 103(6):2236–2249. https://doi.org/10.1111/tpj.14895

Liu H, Chen W, Li Y, Sun L, Chai Y, Chen H, Nie H, Huang C (2022) CRISPR/Cas9 technology and its utility for crop improvement. Int J Mol Sci 23(18):10442. https://doi.org/10.3390/ijms231810442

Liu J, Li L, Xiong Z, Robert CAM, Li B, He S, Chen W, Bi J, Zhai G, Guo S, Zhang H, Li J, Zhou S, Zhang X, Song C (2024a) Trade-offs between the accumulation of cuticular wax and jasmonic acid-mediated herbivory resistance in maize. J Integr Plant Biol 66(1):143–159. https://doi.org/10.1111/jipb.13586

Liu Z, Liao C, Zou L, Jin M, Shan Y, Quan Y, Yao H, Zhang L, Wang P, Liu Z, Wang N, Li A, Liu K, Tabashnik BE, Heckel DG, Wu K, Xiao Y (2024b) Retrotransposon-mediated disruption of a chitin synthase gene confers insect resistance to Bacillus thuringiensis Vip3Aa toxin. PLoS Biol 22(7):e3002704. https://doi.org/10.1371/journal.pbio.3002704

Macioszek VK, Jęcz T, Ciereszko I, Kononowicz AK (2023) Jasmonic acid as a mediator in plant response to necrotrophic fungi. Cells 12(7):1027. https://doi.org/10.3390/cells12071027

Magara HJO, Mutyambai DM, Charles MAO, Otieno SA, Nyaga TM, Niassy S, Khan ZR (2020) Responses of stemborer Chilo partellus to volatiles emitted by maize landraces exposed to signal grass (Brachiaria brizantha). J Plant Interact 15(1):345–357. https://doi.org/10.1080/17429145.2020.1827056

Manganda AS, Sehnem S, Lara AC (2024) Transition to the circular economy: innovative and disruptive production technologies adopted by agribusiness startups. Environ Qual Manag 34(1). https://doi.org/10.1002/tqem.22293

Martínez-Fortún J, Phillips DW, Jones HD (2022) Natural and artificial sources of genetic variation used in crop breeding: a baseline comparator for genome editing. Front Genome Ed 4. https://doi.org/10.3389/fgeed.2022.937853

Matsumoto R, Asfaw A, Ishikawa H, Takada K, Shiwachi H, Asiedu R (2022) Biomass production and nutrient use efficiency in white Guinea yam (Dioscorea rotundata Poir) genotypes grown under contrasting soil mineral nutrient availability. Front Plant Sci 13. https://doi.org/10.3389/fpls.2022.973388

Mehra P, Patel SS, Pansare UD (2024) Application of molecular markers in vegetable improvement: a review. Plant Arch 24(1). https://doi.org/10.51470/PLANTARCHIVES.2024.v24.no.1.050

Meissner Filho PE, Vilarinhos AD, Jesus V, Oliveira VJ d S d, Silva D d CS d, Santos V d S, Dantas JLL (2021) Resistance of transgenic papaya trees to papaya ringspot in Brazil. Pesq Agrop Brasileira 56. https://doi.org/10.1590/s1678-3921.pab2021.v56.01954

Mo C, Wu Z, Xie H, Zhang S, Li H (2020) Genetic diversity analysis of papaya leaf distortion mosaic virus isolates infecting transgenic papaya "Huanong No 1" in South China. Ecol Evol 10(20):11671–11683. https://doi.org/10.1002/ece3.6800

Muñoz A, Santamaria ME, Fernández-Bautista N, Mangano S, Toribio R, Martínez M, Berrocal-Lobo M, Diaz I, Castellano MM (2021) The co-chaperone HOP3 participates in jasmonic acid signaling by regulating CORONATINE-INSENSITIVE 1 activity. Plant Physiol 187(3):1679–1689. https://doi.org/10.1093/plphys/kiab334

Mustafa A, Brtnicky M, Hammerschmiedt T, Kucerik J, Kintl A, Chorazy T, Naveed M, Skarpa P, Baltazar T, Malicek O, Holatko J (2022) Food and agricultural wastes-derived biochars in combination with mineral fertilizer as sustainable soil amendments to enhance soil microbiological activity, nutrient cycling and crop production. Front Plant Sci 13. https://doi.org/10.3389/fpls.2022.1028101

Muthu V, Abbai R, Nallathambi J, Rahman H, Ramasamy S, Kambale R, Thulasinathan T, Ayyenar B, Muthurajan R (2020) Pyramiding QTLs controlling tolerance against drought, salinity, and submergence in rice through marker assisted breeding. PLoS One 15(1):e0227421. https://doi.org/10.1371/journal.pone.0227421

Muha-Ud-Din G, Ali F, Hameed A, Naqvi SAH, Nizamani MM, Jabran M, Yong W (2024) CRISPR/Cas9-based genome editing: A revolutionary approach for crop improvement and global food security. Physiological and Molecular Plant Pathology, 129:102191. https://doi.org/10.1016/j.pmpp.2023.102191

Nales P, Fischer ARH (2023) Breeding by intervening: exploring the role of associations and deliberation in consumer acceptance of different breeding techniques. Public Underst Sci 32(7):889–906. https://doi.org/10.1177/09636625231168087

Nandi D, Chakraborty A, Biswas T, Meher D, Singh AP (2022) Role of trichomes in plant defence—a crop specific review. Crop Res 57(5 and 6). https://doi.org/10.31830/2454-1761.2022.895

Ngou BPM, Ding P, Jones JDG (2022) Thirty years of resistance: zig-zag through the plant immune system. Plant Cell 34(5):1447–1478. https://doi.org/10.1093/plcell/koac041

Nishad R, Ahmed T, Rahman VJ, Kareem A (2020) Modulation of plant defense system in response to microbial interactions. Front Microbiol 11. https://doi.org/10.3389/fmicb.2020.01298

Nomoto M, Skelly MJ, Itaya T, Mori T, Suzuki T, Matsushita T, Tokizawa M, Kuwata K, Mori H, Yamamoto YY, Higashiyama T, Tsukagoshi H, Spoel SH, Tada Y (2021) Suppression of MYC transcription activators by the immune cofactor NPR1 fine-tunes plant immune responses. Cell Rep 37(11):110125. https://doi.org/10.1016/j.celrep.2021.110125

Nurcahyani E, Afifah SN, Astary A, Irawan B, Wahyuningsih S (2024) Analysis of protein profile and reducing sugar content of cassava plants (Manihot esculenta Crantz) results of induced resistance with salicylic acid. https://doi.org/10.30574/wjarr.2024.22.2.1372

Pearse IS, LoPresti E, Schaeffer RN, Wetzel WC, Mooney KA, Ali JG, Ode PJ, Eubanks D, Bronstein JL, Weber MG (2020) Generalising indirect defence and resistance of plants. Ecol Lett 23(7):1137–1152. https://doi.org/10.1111/ele.13512

Peng H, Feng H, Zhang T, Wang Q (2023) Plant defense mechanisms in plant-pathogen interactions. Front Plant Sci 14:1292294. https://doi.org/10.3389/fpls.2023.1292294

Qin J, Guo L, Ye F, Kang S, Sun D, Zhu L, Bai Y, Cheng Z, Xu L, Ouyang C, Xiao L, Wang S, Wu Q, Zhou X, Crickmore N, Zhou X, Guo Z, Zhang Y (2021) MAPK-activated transcription factor PxJun suppresses PxABCB1 expression and confers resistance to bacillus thuringiensis Cry1Ac toxin in Plutella xylostella (L). Appl Environ Microbiol 87(13):e0046621. https://doi.org/10.1128/AEM.00466-21

Ramamoorthy K, Natarajan S (2020) Sustainable crop diversification to stabilize system productivity and profitability. Int J Chem Stud 8(1):1390–1397. https://doi.org/10.22271/chemi.2020.v8.i1t.8450

Rasheed A, Gill RA, Hassan MU, Mahmood A, Qari S, Zaman QU, Ilyas M, Aamer M, Batool M, Li H, Wu Z (2021) A critical review: recent advancements in the use of CRISPR/Cas9 technology to enhance crops and alleviate global food crises. Curr Issues Mol Biol 43(3):1950–1976. https://doi.org/10.3390/cimb43030135

Raspor M, Cingel A (2021) Genetically modified potato for pest resistance: thrift or threat? In: Solanum tuberosum – a promising crop for starvation problem. IntechOpen. https://doi.org/10.5772/intechopen.98748

Razzaq A, Zafar MM, Li P, Qun G, Deng X, Ali A, Hafeez A, Irfan M, Liu A, Ren M, Shang H, Shi Y, Gong W, Yuan Y (2021) Transformation and overexpression of primary cell wall synthesis-related zinc finger gene Gh_A07G1537 to improve fiber length in cotton. Front Plant Sci 12. https://doi.org/10.3389/fpls.2021.777794

Ryalls JMW, Gherlenda AN, Rowe RC, Moore BD, Johnson SN (2023) Silicon supplementation and jasmonate activation synergistically increase phenolic defences against a legume herbivore. J Ecol 111(10):2208–2217. https://doi.org/10.1111/1365-2745.14170

Saleem A, Ali Z, Yeh S-D, Saeed W, Imdad A, Akbar M, Naseem S (2020) Genetic variability and evolutionary dynamics of atypical papaya ringspot virus infecting papaya. https://doi.org/10.21203/rs.3rs-118033/v1

Seixas RL, da Silveira JMFJ, Ferrari VE (2022) Assessing environmental impact of genetically modified seeds in Brazilian agriculture. Front Bioeng Biotechnol 10. https://doi.org/10.3389/fbioe.2022.977793

Seo SY, Wi SJ, Park KY (2020) Functional switching of NPR1 between chloroplast and nucleus for adaptive response to salt stress. Sci Rep 10(1):4339. https://doi.org/10.1038/s41598-020-61379-3

Sereneski De Lima C, Torres Boeger MR, Larcher De Carvalho L (2022) The influence of edaphic factors on the sclerophylly of mangrove tree species in southern Brazil. https://doi.org/10.5380/rf.v52i2.73959

Shaheenuzzamn M, Shi S, Sohail K, Wu H, Liu T, An P, Wang Z, Hasanuzzaman M (2021) Regulation of cuticular wax biosynthesis in plants under abiotic stress. Plant Biotechnol Rep 15(1):1–12. https://doi.org/10.1007/s11816-020-00656-z

Shahriar SA, Islam MN, Chun CNW, Rahim MA, Paul NC, Uddain J, Siddiquee S (2021) Control of plant viral diseases by CRISPR/Cas9: resistance mechanisms, strategies and challenges in food crops. Plan Theory 10(7):1264. https://doi.org/10.3390/plants10071264

Sharma S, Sharma R, Govindaraj M, Mahala RS, Satyavathi CT, Srivastava RK, Gumma MK, Kilian B (2021) Harnessing wild relatives of pearl millet for germplasm enhancement: challenges and opportunities. Crop Sci 61(1):177–200. https://doi.org/10.1002/csc2.20343

Sheng X, Sun Z, Wang X, Tan Y, Yu D, Yuan G, Yuan D, Duan M (2020) Improvement of the rice "easy-to-shatter" trait via CRISPR/Cas9-mediated mutagenesis of the qSH1 gene. Front Plant Sci 11. https://doi.org/10.3389/fpls.2020.00619

Shu H, Luo Z, Peng Z, Wang J (2020) The application of CRISPR/Cas9 in hairy roots to explore the functions of AhNFR1 and AhNFR5 genes during peanut nodulation. BMC Plant Biol 20(1):417. https://doi.org/10.1186/s12870-020-02614-x

Silva AAR, Lima GS, Azevedo CAV, Veloso LLSA, Lacerda CN, Gheyi HR (2020) Salicylic acid as an attenuator of salt stress in soursop. Revista Caatinga 33:1092–1101. https://doi.org/10.1590/1983-21252020v33n424rc

Silva AAR, Lima GS, Azevedo CAV, Veloso LLSA, Lacerda CN, Gheyi HR, Pereira WE, Silva VR, Soares LAA (2022) Methods of application of salicylic acid as attenuator of salt stress in cherry tomato. Braz J Biol 82:e265069. https://doi.org/10.1590/1519-6984.265069

Singh RK, Sood P, Prasad A, Prasad M (2021) Advances in omics technology for improving crop yield and stress resilience. Plant Breed 140(5):719–731. https://doi.org/10.1111/pbr.12963

Sun J, Wang Y, Guo C, Ge R, Naren T, Jiang L (2024) CRISPR/Cas9 effectively generate chromosome structural variations in rice protoplasts. Crop Sci 64(6):3430–3437. https://doi.org/10.1002/csc2.21334

Suresh Govindarao Borkar, Bhosale AN, Ajayasree TS (2023) Variability in bacterial endophytes in leaves of transgenic Bt and non-Bt cotton crop varieties and their role in suppression of bacterial leaf blight pathogen X a pv malvacearum and the incited disease reaction. World J Adv Res Rev 20(2):626–636. https://doi.org/10.30574/wjarr.2023.20.2.2216

Tabe-Ojong MPJ, Lokossou JC, Gebrekidan B, Affognon HD (2023) Adoption of climate-resilient groundnut varieties increases agricultural production, consumption, and smallholder commercialization in West Africa. Nat Commun 14(1):5175. https://doi.org/10.1038/s41467-023-40781-1

Taylor KL, Hamby KA, DeYonke AM, Gould F, Fritz ML (2021) Genome evolution in an agricultural pest following adoption of transgenic crops. Proc Natl Acad Sci 118(52). https://doi.org/10.1073/pnas.2020853118

Thompson MN, Grunseich JM, Marmolejo LO, Aguirre NM, Bradicich PA, Behmer ST, Suh CP-C, Helms AM (2022) Undercover operation: belowground insect herbivory modifies systemic plant defense and repels aboveground foraging insect herbivores. Front Ecol Evol 10. https://doi.org/10.3389/fevo.2022.1033730

Tudi M, Daniel Ruan H, Wang L, Lyu J, Sadler R, Connell D, Chu C, Phung DT (2021) Agriculture development, pesticide application and its impact on the environment. Int J Environ Res Public Health 18(3):1112. https://doi.org/10.3390/ijerph18031112

Tyagi S, Kesiraju K, Saakre M, Rathinam M, Raman V, Pattanayak D, Sreevathsa R (2020) Genome editing for resistance to insect pests: an emerging tool for crop improvement. ACS Omega 5(33):20674–20683. https://doi.org/10.1021/acsomega.0c01435

Tzipilevich E, Russ D, Dangl JL, Benfey PN (2021) Plant immune system activation is necessary for efficient root colonization by auxin-secreting beneficial bacteria. Cell Host Microbe 29(10):1507–1520e4. https://doi.org/10.1016/j.chom.2021.09.005

Verma KK, Song X-P, Tian D-D, Guo D-J, Chen Z-L, Zhong C-S, Nikpay A, Singh M, Rajput VD, Singh RK, Minkina T, Li Y-R (2021) Influence of silicon on biocontrol strategies to manage biotic stress for crop protection, performance, and improvement. Plan Theory 10(10):2163. https://doi.org/10.3390/plants10102163

Verma KK, Song X-P, Budeguer F, Nikpay A, Enrique R, Singh M, Zhang B-Q, Wu J-M, Li Y-R (2022) Genetic engineering: an efficient approach to mitigating biotic and abiotic stresses in sugarcane cultivation. Plant Signal Behav 17(1). https://doi.org/10.1080/15592324.2022.2108253

Wang C, Dai S, Zhang Z, Lao W, Wang R, Meng X, Zhou X (2021a) Ethylene and salicylic acid synergistically accelerate leaf senescence in Arabidopsis. J Integr Plant Biol 63(5):828–833. https://doi.org/10.1111/jipb.13075

Wang W, Wang H, Liu J, Li T, Zhao H (2021b) A highly efficient protocol to establish regeneration and genetic transformation systems and create targeted mutations using CRISPR/Cas9 in Lycium Ruthenicum. https://doi.org/10.21203/rs.3.rs-398418/v1

Wang X, Chi Y, Li F (2022) Exploring China stepping into the dawn of chemical pesticide-free agriculture in 2050. Front Plant Sci 13. https://doi.org/10.3389/fpls.2022.942117

Wang X, Zhang J, Song J, Huang M, Cai J, Zhou Q, Jiang D (2020) Abscisic acid and hydrogen peroxide are involved in drought priming-induced drought tolerance in wheat (Triticum aestivum L.). Plant Biology 22(6):1113–1122. https://doi.org/10.1111/plb.13143

Waterman JM, Hall CR, Mikhael M, Cazzonelli CI, Hartley SE, Johnson SN (2021) Short-term resistance that persists: rapidly induced silicon anti-herbivore defence affects carbon-based plant defences. Funct Ecol 35(1):82–92. https://doi.org/10.1111/1365-2435.13702

Xu M, Jiang Y, Chen S, Chen F, Chen F (2021) Herbivory-induced emission of volatile terpenes in Chrysanthemum morifolium functions as an indirect defense against Spodoptera litura larvae by attracting natural enemies. J Agric Food Chem 69(34):9743–9753. https://doi.org/10.1021/acs.jafc.1c02637

Xuebo W, Dandan L, Xiaolei T, Changchun C, Xinyao Z, Zhan S, Aiguo Y, Xiankui F, Dan L (2023) CRISPR/Cas9-mediated targeted mutagenesis of two homoeoalleles in tobacco confers resistance to powdery mildew. Euphytica 219(6):67. https://doi.org/10.1007/s10681-023-03196-z

Yadav G, Jayaswal D, Jayaswall K, Bhandawat A, Singh A, Tilgam J, Rai AK, Chaturvedi R, Kumar A, Kumar S, Kumar S P (2021) Identification and characterization of chickpea genotypes for early flowering and high seed germination through molecular markers. https://doi.org/10.21203/rs.3.rs-1033758/v1

Yan P, Zhu Y, Wang Y, Ma F, Lan D, Niu F, Dong S, Zhang X, Hu J, Liu S, Guo T, Xin X, Zhang S, Yang J, Cao L, Luo X (2022) A new RING finger protein, PLANT ARCHITECTURE and GRAIN NUMBER 1, affects plant Architecture and Grain yield in Rice. Int J Mol Sci 23(2):824. https://doi.org/10.3390/ijms23020824

Yang F, Kerns DL, Little NS, Santiago González JC, Tabashnik BE (2021a) Early warning of resistance to Bt toxin Vip3Aa in Helicoverpa zea. Toxins 13(9):618. https://doi.org/10.3390/toxins13090618

Yang Y, Saand MA, Huang L, Abdelaal WB, Zhang J, Wu Y, Li J, Sirohi MH, Wang F (2021b) Applications of multi-omics technologies for crop improvement. Front Plant Sci 12. https://doi.org/10.3389/fpls.2021.563953

Yang F, Kerns DL, Little N, Brown SA, Stewart SD, Catchot AL, Cook DR, Gore J, Crow WD, Lorenz GM, Towles T, Tabashnik BE (2022) Practical resistance to cry toxins and efficacy of Vip3Aa in Bt cotton against Helicoverpa zea. Pest Manag Sci 78(12):5234–5242. https://doi.org/10.1002/ps.7142

Yang Y, Chen Y, Bo Y, Liu Q, Zhai H (2023) Research progress in the mechanisms of resistance to biotic stress in sweet potato. Genes 14(11):2106. https://doi.org/10.3390/genes14112106

Yu B, Patterson N, Zaharia LI (2022) Saponin biosynthesis in pulses. Plants 11(24):3505. https://doi.org/10.3390/plants11243505

Yuan P, Poovaiah BW (2022) Interplay between Ca2+/calmodulin-mediated signaling and AtSR1/CAMTA3 during increased temperature resulting in compromised immune response in plants. Int J Mol Sci 23(4):2175. https://doi.org/10.3390/ijms23042175

Zainal-Abidin R-A, Ruhaizat-Ooi I-H, Harun S (2021) A review of omics technologies and bioinformatics to accelerate improvement of papaya traits. Agronomy 11(7):1356. https://doi.org/10.3390/agronomy11071356

Zegeye WA, Tsegaw M, Zhang Y, Cao L (2022) CRISPR-based genome editing: advancements and opportunities for rice improvement. Int J Mol Sci 23(8):4454. https://doi.org/10.3390/ijms23084454

Zehra A, Raytekar NA, Meena M, Swapnil P (2021) Efficiency of microbial bio-agents as elicitors in plant defense mechanism under biotic stress: a review. Curr Res Microb Sci 2:100054. https://doi.org/10.1016/j.crmicr.2021.100054

Zhang L, Liu Z, Song Y, Sui J, Hua X (2024a) Advances in the involvement of metals and metalloids in plant defense response to external stress. Plan Theory 13(2):313. https://doi.org/10.3390/plants13020313

Zhang Y, Hao R, Chen J, Li S, Huang K, Cao H, Farag MA, Battino M, Daglia M, Capanoglu E, Zhang F, Sun Q, Xiao J, Sun Z, Guan X (2024b) Health benefits of saponins and its mechanisms: perspectives from absorption, metabolism, and interaction with gut. Crit Rev Food Sci Nutr 64(25):9311–9332. https://doi.org/10.1080/10408398.2023.2212063

Zhang Y, Yu J, Wang X, Durachko DM, Zhang S, Cosgrove DJ (2021) Molecular insights into the complex mechanics of plant epidermal cell walls. Science 372(6543):706–711. https://doi.org/10.1126/science.abf2824

Zhao T, Wu X, Jiang M (2023) Editorial: advanced breeding for abiotic stress tolerance in crops. Front Plant Sci 14. https://doi.org/10.3389/fpls.2023.1265339

Zhou H, Hua J, Zhang J, Luo S (2022) Negative interactions balance growth and defense in plants confronted with herbivores or pathogens. J Agric Food Chem 70(40):12723–12732. https://doi.org/10.1021/acs.jafc.2c04218

[illegible]

Chapter 2
Elicitors in Plant Biology: Types and Defense Mechanisms

Ehtisham Hassan Khan, Waqas Liaqat, Muhammad Faheem Jan, Muhammad Tanveer Altaf, Amjad Ali, and Haseeb Ahmad

Abstract The management of diseases predominantly depends on the utilization of insecticides, bactericides, and fungicides, which are chemical compounds that exhibit toxicity toward plant pathogens, vectors, or causative agents associated with plant diseases. Nonetheless, the detrimental impacts of such chemical substances or the products of their degradation on the ecosystem and human health highlight the critical need for the exploration of novel, nonharmful methods of disease control. It is crucial to acknowledge the presence of induced resistance to a natural phenomenon that functions to safeguard plants against diseases. Elicitors are substances that stimulate the activation of chemical defenses within plants. In treated plants, diverse biosynthetic pathways are activated contingent upon the specific compound utilized. Chemical elicitors that are often studied encompass benzothiadiazole, chito-

E. H. Khan (✉)
National Key Laboratory for Germplasm Innovation and Utilization of Horticultural Crops, The Center of Crop Nanobiotechnology, College of Plant Science and Technology, Huazhong Agricultural University, Wuhan, China
e-mail: ehtishamhassan@aup.edu.pk

W. Liaqat (✉)
Department of Field Crops, Faculty of Agriculture, Institute of Natural and Applied Sciences, Cukurova University, Adana, Turkey

M. F. Jan
College of Agriculture, Northeast Agricultural University, Harbin, China

M. T. Altaf
Department of Field Crops, Faculty of Agriculture and Natural Sciences, Recep Tayyip Erdoğan University, Rize, Turkey

A. Ali
Department of Plant Protection, Faculty of Agricultural Sciences and Technologies, Sivas University of Science and Technology, Sivas, Turkey

H. Ahmad
Guangxi Key Laboratory of Agro-environment and Agro-products Safety, Key Laboratory of Crop Cultivation and Physiology, College of Agriculture, Guangxi University, Nanning, China

K. A. Abd-Elsalam, H. I. Mohamed (eds.), *Elicitors for Sustainable Crop Production*, Sustainability Sciences in Asia and Africa,
https://doi.org/10.1007/978-981-96-6569-3_2

san, benzoic acid, methyl salicylate, salicylic acid, and various factors that affect the synthesis of phenolic compounds, as well as the activation of various enzymes associated with plant defense mechanisms. The incorporation of these elicitors into farming practices has the potential to reduce the dependency on chemical control methods, thereby facilitating the advancement of sustainable agriculture.

Keywords Disease resistance · Immunity · Stress signaling · Defense mechanism · Stress responses

2.1 Introduction

Plants are perpetually subjected to attacks and utilized as hosts for numerous potential pathogenic organisms, the living beings that inhabit their vicinity (Dongre et al. 2023). The endeavor for survival constitutes an ongoing and inherent phenomenon. The survival of microbial organisms is contingent upon their reliance on plants for nutrition; conversely, plants utilize various mechanisms to withstand and endure microbial attacks (Keswani et al. 2019a, b; Enebe and Babalola 2019). Among the various strategies accessible to farmers for the cultivation of healthy crops, crop rotation implementation, the employment of tolerant varieties, application of chemical pesticides, and biological control methods utilization are the most prevalent approaches. The utilization of synthetic pesticides in combating pests and pathogens is both costly and complex, with associated carcinogenic risks (Parra-Arroyo et al. 2022). Plants activate a variety of defense mechanisms when confronted with biotic stress resulting from various herbivores and pathogens (Zebelo 2020). When biotic stress is encountered, intricate signaling processes are activated in crops, coordinating various response mechanisms (Chen et al. 2020). Biotrophic creatures, which exhibit a high degree of dependence on host plant cells as well as hemibiotrophic pathogens that shift from a predominantly biotrophic phase to a following necrotrophic phase, include various phloem-feeding insects (Ratnadass and Deguine 2020), fungi (Fei and Liu 2023; Wu et al. 2013), viruses (Zhu et al. 2014), and bacteria (Keswani et al. 2019b). These entities are known for their role in initiating systemic acquired resistance (SAR) defense responses within plant systems. The intricate transcriptional regulators, signaling pathways, and elicitor substances are modulated by defense responses in the plants (Meraj et al. 2020).

The regulation of induced defense responses is primarily attributed to the plant hormones such as salicylic acid (SA), jasmonic acid (JA), and ethylene (ET) (Li and Ahammed 2023). In response to the invasion of the pathogen, the initial defense mechanism is characterized by a hypersensitive response, serving as a confined barrier of defense and at the site of infection modulating cellular death (Nadarajah 2024). The accumulation of SA subsequently activates defense mechanisms in plants, which activates the biochemical defense responses and contributes to the mitigation of additional infections in far-away tissues (Venegas-Molina et al. 2020). When inducers or elicitors initiate the activation of defensive genes in plants, the

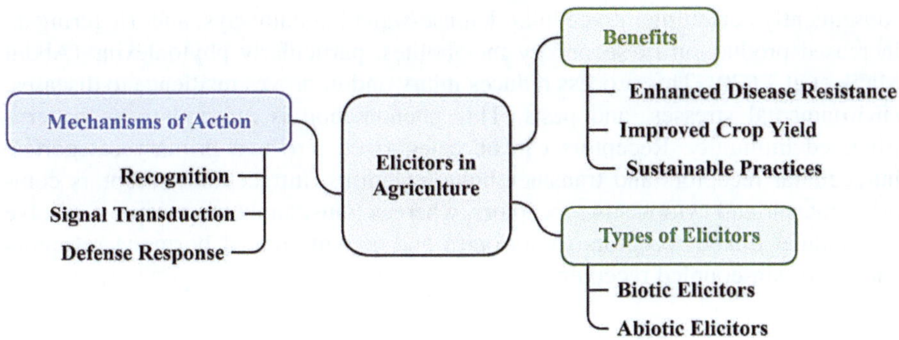

Fig. 2.1 Elicitor's role in agriculture

establishment of induced systemic resistance occurs (Zehra et al. 2021). The receptor and elicitor molecules represent two essential components required for the operation of the immune system and defense signaling pathways (Abdul Malik et al. 2020). Elicitors are substances that activate systemic acquired resistance or generate systemic resistance by enhancing the expression of pathogenesis-related genes in the host (Fig. 2.1). The receptors located on or within plants recognize these molecules that trigger the defense signals subsequently (Abdul Malik et al. 2020). The elicitor's molecules differ according to their origin, characteristics, and composition, consisting of molecules such as proteins, lipids, oligosaccharides, and peptides (Pršić and Ongena 2020; Zheng et al. 2020). This chapter intends to furnish a thorough examination of elicitors, their types, and roles in plant defense mechanisms in improving crop resilience and disease management.

2.2 What Are Elicitors?

Elicitors are chemical compounds from various sources that stimulate morphological and physiological responses, as well as the accumulation of phytoalexins in the targeted organisms. Initially, the term "elicitor" was employed to denote specific molecules that could trigger the synthesis of phytoalexins; however, it has since evolved to encompass compounds that stimulate various forms of defensive responses in plants (Urban et al. 2022; Alvarado et al. 2019). Ultimately, the activation of defense mechanisms may result in improved resistance. Elicitors include both substances derived from pathogenic organisms (exogenous elicitors) and chemicals emitted by plants in reaction to pathogen activity (endogenous elicitors) (Meena et al. 2022; Zehra et al. 2021; Patel et al. 2020). Elicitor substances interact with specific receptor proteins located on the membranes of plant cells (Chamkhi et al. 2021). A receptor is a molecule of protein that functions to receive chemical signals that originate outside of the cell. A ligand is a chemical substance that engages with a receptor. The receptors can identify the elicitor's molecular pattern, subsequently activating intracellular defense-signaling pathways, and triggering an increased production of secondary metabolites, particularly phytoalexins (Abdul Malik et al. 2020). This process reduces injury and improves resilience to diseases, environmental stresses, and pests. This phenomenon is referred to as pattern-triggered immunity. Receptors can be categorized into two primary categories: intracellular receptors and transmembrane receptors. Intracellular receptors comprise nuclear and cytoplasmic receptors, whereas transmembrane receptors involve ion-channel linked receptors (ionotropic) and enzyme-linked hormone receptors and G-protein-coupled receptors.

2.2.1 Application of Elicitors in Defense Mechanisms

The global population is experiencing a significant increase and is projected to rise to approximately 9.7 billion by the year 2050 (Gu et al. 2021). However, the efficiency of agricultural output is declining because of various biotic and abiotic stresses. The mitigation of such losses constitutes a critical issue for all countries as they seek to meet the increasing demands for food. Farmers have access to a variety of options for the protection of their crops. Potential strategies encompass the implementation of biological control methods, crop rotation utilization, resistant cultivars development, application of agrochemicals, and tillage practices adaptation. Almost all chemical fungicides and pesticides possess a direct antibiotic mechanism. Nonetheless, their application in a commercial context is regarded as economically unviable, their implementation is challenging, and certain substances have been demonstrated to possess carcinogenic properties (Lalaleo et al. 2016). As a result, considerable initiatives have been undertaken to formulate sustainable approaches for the management of phytopathologies, thereby protecting human health from potential hazards. The improved comprehension of plant signaling processes has led to the identification of both synthetic and natural substances, referred to as elicitors, that provoke defense responses in crops (Abdul Malik et al. 2020; Zehra et al. 2021). Elicitors are capable of enhancing a plant's resistance to various pathogens by triggering its innate defense mechanisms (Table 2.1). These substances work by triggering pathways such as the production of reactive oxygen species (ROS), hypersensitive response, and synthesis of secondary metabolites, all of which bolster the plant's ability to resist infections (Singh et al. 2024; Song et al. 2023). As an eco-friendly alternative to chemical treatments, elicitors present a promising approach for sustainable agriculture, reducing reliance on harmful pesticides while improving crop resilience (Gaddam et al. 2024). Future research on the optimization of elicitor formulations and their large-scale application could significantly contribute to global food security and sustainable farming practices.

The defense system of plants is fundamentally based on prevention, resistance, or tolerance (Cipollini et al. 2014). These processes are activated in various manners, contingent upon the category of pathogens assailing the plant (or biotrophs manifesting necrotrophic behavior) (Jones et al. 2019). These mechanisms may be

Table 2.1 Various elicitor molecules and their functions in plants

Elicitor	Source of elicitor	Effects	References
Reticine A	Fruit peel of *Citrus reticulata*	Enhance systemic acquired resistance against TMV in tobacco plants by upregulating genes involved in salicylic acid biosynthesis	Wang et al. (2021)
C6 zinc finger protein-like elicitor (Thc6)	*Trichoderma harzianum*	Stimulates ROS production, increases calcium levels in the leaf, and upregulates the expression of JA and ET genes	Saravanakumar et al. (2016)
Acetoin	*Bacillus subtilis*	Triggers systemic resistance in *Arabidopsis thaliana* against *Pseudomonas syringae* pv. *tomato* DC3000	Shen et al. (2019)
Proteinaceous elicitor PeFOC1	*Fusarium oxysporum*	Triggers hypersensitive responses (HRs) in tobacco plants against TMV and *Pseudomonas syringae* pv. *tabaci*	Li et al. (2019)
Peptide PIP1	Insects	Enhances immunity in *Arabidopsis* via receptor-like kinase 7 (RLK7)	Hou et al. (2014)
StPep1	*Bacillus subtilis*	Reduces root galling in potatoes caused by *Meloidogyne chitwoodi* infection	Zhang and Gleason (2020)
PemG1	*Magnaporthe grisea*	SAR in rice through the transient induction of pathogenesis-related (PR) genes and increased hydrogen peroxide accumulation	Peng et al. (2011)

either triggered or constitutive. The first line of defense against pathogens involves physical barriers or pre-existing biochemical compounds, while induced resistance is triggered only upon pathogen invasion (Kamle et al. 2020). A plant's defense efficiency depends on the speed at which it detects a pathogen (Zhang et al. 2022). This detection involves activation signals generated by general elicitors and specific molecules, which are initially identified by the plant to trigger a signaling cascade, ultimately leading to a defense response (Patel et al. 2020).

The timing of defense responses plays a critical role in determining whether a plant successfully resists or succumbs to the threat of a parasite or pathogen. Two forms of induced resistance, systemic acquired resistance (SAR) and induced systemic resistance (ISR), are activated following prior infection or treatment, providing resistance (or tolerance) to future pathogen or parasite attacks (Kamle et al. 2020; Meena et al. 2020). These mechanisms have been extensively explored for their signaling pathways and potential applications in the protection of plants. In response to a localized infection, plants initiate a salicylic acid-dependent signaling pathway, leading to systemic, long-lasting resistance effective against a broad spectrum of pathogens, including viruses, fungi, and bacteria (Saini et al. 2023).

Table 2.2 Example of elicitors employed to induce resistance in plants for diseases control

Disease	Causal agent	Crop	Inducer	References
Bacillus subtilis	*Pseudomonas syringae* pv. *tomato DC3000*	*Arabidopsis thaliana*	Acetoin	Shen et al. (2019)
Bacillus subtilis	*Meloidogyne chitwoodi*	Potato	StPep1	Zhang and Gleason (2020)
Fusarium oxysporum	*TMV* and *Pseudomonas syringae* pv. *tabaci*	Tobacco plants	Proteinaceous elicitor PeFOC1	Li et al. (2019)
Fruit peel of *Citrus reticulata*	TMV virus	Tobacco	Reticine A	Wang et al. (2021)
Trichoderma harzianum	*Curvularia* leaf spot	Maize	C6 zinc finger protein-like elicitor (Thc6)	Saravanakumar et al. (2016)

Resistance is linked to alterations in the production of pathogenesis-related proteins like glucanases and chitinases, cell wall composition, and phytoalexins synthesis, although additional defensive molecules likely exist but have yet to be recognized (Le Mire 2018; Kaur et al. 2022; Hückelhoven and Schouten 2024). The benefits of induced resistance (IR) have been investigated in both monocotyledonous and dicotyledonous plants. Table 2.2 provides examples of IR applications for managing plant diseases.

2.2.2 *Classification of Elicitors*

Elicitors are categorized as exogenous and endogenous elicitors. Molecules produced by pathogens are referred to as exogenous elicitors, while endogenous elicitors are secreted by plants in reaction to the attack of pathogens (Ramirez-Estrada et al. 2016; Patel et al. 2020). Elicitors are either naturally occurring or synthesized substances secreted by microorganisms or formed through elemental combinations, and they take part in a crucial part in inducing systematic resistance within plants (Alvarado et al. 2019). The presence of pathogenic organisms elevates the intracellular levels of ROS and raises calcium concentrations in plant cells, leading to programmed cell death or apoptosis (Ram et al. 2019; Li et al. 2018). Based on their origin, elicitors are categorized into biotic and abiotic types.

2.2.2.1 Biotic Elicitors

The global risk of hunger has increased as a result of the decline in agricultural products attributed to biotic stress (Keswani et al. 2022; Moustafa-Farag et al. 2019). Throughout their lifespans, plants face a range of biological stresses and adverse conditions, including competition, nutrient deficiencies, fluctuating temperatures, interactions with phytopathogens, and extreme weather (Mertens et al. 2021). Biological stress is caused by harmful interactions with insects, bacteria, viruses, parasitic nematodes, fungi, and weeds (Moustafa-Farag et al. 2019). To counter these biotic stresses, plants deploy various molecular, biochemical, and morphological strategies. For example, bacteria and fungi can invade plant tissues, causing diseases such as leaf spots, cankers, and vascular wilts (Meena et al. 2016). Biotic elicitors, derived from living organisms or their by-products, include polysaccharides such as pectin and cellulose, as well as chitin, chitosan, alginate, and glucans. Other biotic elicitors are oligosaccharides like mannuronate, mannan, galacturonides, and guluronate, and proteins like oligandrin, cellulose, glycoproteins, cryptogein, lactoferrin, and pectolyase. Pathogen-derived toxins, such as coronatine, also function as biotic elicitors.

2.2.2.2 Abiotic Elicitors

Abiotic elicitors, on the other hand, originate from nonliving sources and can be categorized into chemical and physical types (Baenas et al. 2014). Chemical elicitors include silicon, benzothiadiazole, ethanol, acetic acid, ethene, prohexadione, and inorganic salts like VSO4, $HgCl_2$, $CuSO_4$, and $CaCl_2$, as well as metal ions such as Ag^{2+}, Co^{2+}, Cu^{2+}, Mn^{2+}, Zn^{2+}, Fe^{2+}, Pb^{2+}, Al^{3+}, and Cd^{2+}. These chemical elicitors stimulate the production of phenolic substances essential for the plant's defense mechanisms (Heimpel and Mills 2017).

Physical elicitors include environmental factors such as altered chilling, extreme temperature shocks, changes in gas composition, drought, osmotic stress, saline stress, and ultraviolet irradiation. Other physical elicitors include mechanical damage (wounding) and plant hormones like methyl jasmonate, methyl salicylate, polyamines, ethylene, triacontanol, salicylic acid, abscisic acid, brassinosteroids, and jasmonic acid, which are all of significant interest in plant physiological studies.

2.3 Classification of Elicitors According to Their Interactions with the Host Plant

Elicitors can be categorized into two distinct groups based on their interaction with the host plant: race-specific elicitors and general elicitors (Montesano et al. 2003; Meena et al. 2022). General elicitors are capable of activating defense mechanisms to respond to the perceived existence of pathogenic organisms within nonhost and host plants. In contrast, race-specific elicitors play a crucial part in facilitating the signaling of resistance genes (Meena and Samal 2019; Dalio et al. 2017). The specific resistance of a host cultivar to a particular pathogen race is determined by a pair of complementary genes, exemplifying a gene-for-gene interaction. Thus, an elicitor specific to a particular race, either produced or encoded by the activity of an avirulence gene in a specific pathogen, will activate a resistance response exclusively in the host plant that possesses the corresponding resistance R gene. The lack of a particular gene product frequently leads to the manifestation of diseases (Nürnberger and Scheel 2001; Nimchuk et al. 2001). Conversely, general elicitors indicate the existence of possible pathogens in both nonhost as well as host plants. General elicitors with a nonspecific nature are, however, relative, as certain elicitors are recognized exclusively by a limited number of plant species (Shibuya and Minami 2001). Additionally, it is important to note that although general elicitors are broadly effective, their defense responses tend to be weaker and less rapid compared to race-specific elicitors. This distinction between elicitors provides a foundation for developing targeted plant protection strategies, where the identification of specific R genes and their corresponding elicitors can enhance resistance breeding programs. Furthermore, the ability of general elicitors to activate broad-spectrum defense responses offers the potential to manage emerging pathogen threats in diverse crop species (Appu et al. 2021; Abbasi et al. 2021).

2.3.1 Specific Elicitors (Effectors)

These are pathogen-derived molecules that operate exclusively in plant varieties possessing the corresponding disease-resistance genes. Effectors typically function in the secondary immune response, which is activated upon detection by intracellular immune receptors, such as nucleotide-binding leucine-rich repeat (NLR) proteins, thereby triggering more specialized defense mechanisms (Lolle et al. 2020; Jubic et al. 2019).

Elicitors can be categorized into two distinct groups based on their interaction with host plants: race-specific elicitors and general elicitors (Montesano et al. 2003; Jain et al. 2024). General elicitors activate defense mechanisms in response to the perceived presence of pathogenic organisms in both nonhost and host plants. In contrast, race-specific elicitors are crucial in signaling resistance genes in host plants (Meena and Samal 2019; Dalio et al. 2017). Race-specific resistance, often governed by a gene-for-gene interaction, depends on a complementary pair of genes—one in the host and one in the pathogen. For instance, an elicitor specific to a particular pathogen race, either encoded by or resulting from the activity of an avirulence (Avr) gene, will trigger a resistance response only in plants possessing the corresponding resistance (R) gene. The absence of this R gene product frequently results in susceptibility to disease (Nürnberger and Scheel 2001; Nimchuk et al. 2001).

Conversely, general elicitors signal the presence of potential pathogens in both nonhost and host plants, initiating broad-spectrum defense responses. However, the term "nonspecific" is relative since certain general elicitors are recognized only by a limited range of plant species (Shibuya and Minami 2001). Recent research has revealed significant parallels between the defense mechanisms activated by general elicitors and innate immune responses in animals. This suggests that the identification of general elicitors may play a role in triggering plant innate immunity, which offers exciting possibilities for enhancing plant defense strategies (Nürnberger and Brunner 2002). It is also worth noting that general elicitors often activate slower and less intense defense responses compared to race-specific elicitors. This distinction provides a framework for targeted crop protection strategies, where specific R genes and their associated elicitors can be leveraged in resistance breeding programs. Furthermore, the capacity of general elicitors to induce broad-spectrum responses could be instrumental in addressing emerging pathogen threats across various crops (Riseh and Vazvani 2024; Hönig et al. 2023).

2.3.2 General Elicitors

These elicitors induce broad defense responses that do not vary significantly across different cultivars within a plant species and are linked to primary innate immunity. Their components include chemicals, damage-associated molecular patterns (DAMPs), derived from the plant itself, as well as pathogen-associated molecular patterns (PAMPs) and microbe-associated molecular patterns (MAMPs) from disease-causing and nondisease-causing microorganisms, respectively. While elicitor activity is often described as being mediated by specific receptors, only a limited number of binding sites for these receptors have been characterized to date (Gowthami 2018).

Benefits of Using Elicitors

- It mitigates the harm caused by pests, fungi, insects, and herbivores (Zehra et al. 2021).
- Elicitors have a direct impact on crop plants, and their toxic effect on organisms that are not targeted is considerably reduced than that of conventional pesticides, thereby mitigating environmental risks (Alagna et al. 2020).
- Elicitors may be used alongside current spraying technologies, similar to other agrochemicals Caicedo-López et al. 2021).
- Elicitors can function as a viable alternative to genetically modified (GM) crops in attracting natural enemies, thus aiding in the reduction of pest populations (Krishnamoorthi et al. 2024; Francis et al. 2020).
- Plants that have been treated with elicitors present lower ecological risks in comparison to GM plants (Caicedo-López et al. 2021). Various advantages of elicitors are presented in Fig. 2.2.

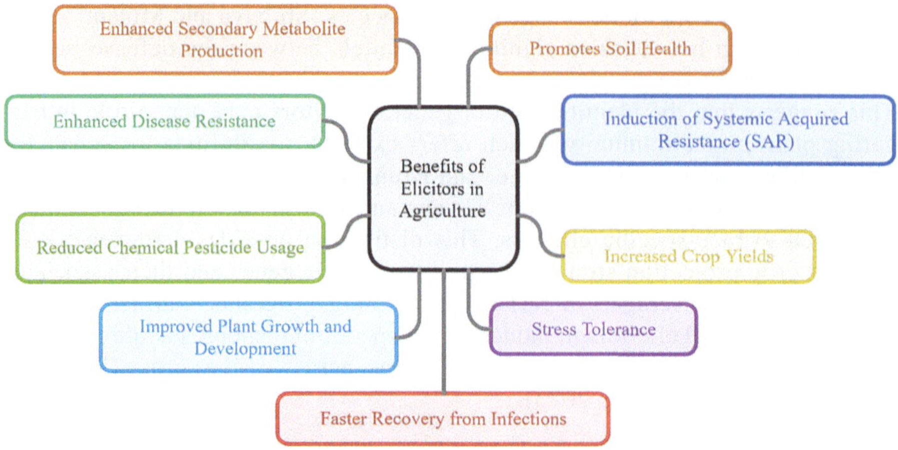

Fig. 2.2 Some advantages of elicitors in agriculture

2.4 Effects Induced by Elicitors in Plant Cells

The application of elicitors induces various effects in plant cells, as evidenced by alterations in cellular metabolism. These effects include augmentation of secondary metabolite synthesis, biosynthesis of phytoalexins, and augmentation of the activity of defensive enzymes. The process induces systemic acquired resistance and local acquired resistance through the aggregation of pathogenesis-related PR proteins and differential gene expression. This mechanism subsequently results in the formation of enzymes that are involved in the formation of polysaccharides, including hydroxyproline-rich glycoproteins and callose in cell walls, enabled by the activation of proline hydroxylase. Furthermore, polyphenolics and lignin are incorporated into the cell walls as a component of this response. The binding of elicitors to the receptor located on the cell wall initiates the activation of signal transduction pathways, leading to the expression of defense genes. The defense genes produce various gene products, including secondary metabolites associated with abiotic stress, chitinases, phytoalexins, and pathogenesis-related (PR) proteins, linked to biotic stress; collectively, the elements in question are integral to the primary immune response observed in plants. Elicitors such as JA and SA activate their corresponding dependent pathways, namely the JA-dependent pathway and SA-dependent pathway (Jeyasri et al. 2023). Salicylic acid induces SAR, a phenomenon in which an attack by a pathogen in one area of the plant triggers a defensive response in other regions, enhancing their resistance to future pathogen attacks (Mishra et al. 2024; Abdul Malik et al. 2020). Salicylic acid initiates the activation of NPR1 (non-expresser of pathogenesis-related (PR) genes), which then in turn facilitates the release of pathogenesis-related (PR) proteins (PR (1), PR (2), and PR (5), ultimately resulting in the development of systemic acquired resistance. JA induces LAR, whereby, upon pathogen invasion, JA facilitates the programmed cell death of the affected tissues, thereby impeding further pathogen proliferation. JA initiates

MYC2, a transcription factor that subsequently releases PDF 1.2 (protodermal factor 2), PR-3, and PR-4, leading to the occurrence of LAR (Enoki and Suzuki 2016).

2.4.1 Systemic Acquired Resistance (SAR)

The phenomenon of induced resistance in plants signifies a crucial mechanism through which the defensive capabilities of the plant are augmented in reaction to various insect pests and pathogens following appropriate stimulus. This augmented resistance, which arises from the activation of pathogenic agents, is referred to as SAR (Swapnil et al. 2021; Vallad and Goodman 2004). The defensive responses to pathogenic infections manifest not just at the initial point of infection as well as in distinct, anatomically separate tissues. SAR represents a mechanism of induced defense that offers potentially enduring defense against a diverse range of microorganisms and has been recorded in various plant species. Upon the identification of pathogens by a host, an array of initial reactions is triggered, involving phosphorylation and ion fluxes, which take place before the accumulation of nitric oxide, ROS, SA, and the transcriptional activation of genes associated with defense mechanisms.

2.4.1.1 PR Proteins

Pathogenesis-related (PR) proteins constitute a group of proteins encoded by plant genomes. These substances possess low molecular weights and can be extracted at low pH levels. The majority of them are located within the apoplast. Their output was produced in response to stressful conditions. These compounds are widely found in plants in minimal amounts; however, their synthesis increases significantly following pathogen invasion (Dos Santos and Franco 2023).

Importance of PR Proteins

Their antifungal and various antimicrobial activities exhibit notable strength. The observed phenomena are linked to the reinforcement of the host cell wall, encompassing its extensions and papillae. Certain pathogenesis-related proteins, including ß- 1, 3-glucanase, and chitinase, possess the capability to break down components of fungal cell walls, thereby releasing elicitors in the form of oligosaccharides, from the pathogen cell wall (Ochoa-Meza et al. 2021).

2.4.2 Local Acquired Resistance (LAR)

Induced resistance initially manifests as a localized reaction in the vicinity of the necrotic area caused by chemical exposure or pathogen infection, a phenomenon known as locally acquired resistance. Local acquired resistance is defined by increased concentrations of antimicrobial substances and defensive enzymes. LAR demonstrates efficacy against a variety of pathogenic organisms (Kamle et al. 2020).

2.4.2.1 Synthetic Elicitors

Synthetic elicitors are diminutive, druglike substances that stimulate defense mechanisms of plants; however, they are separate from recognized natural elicitors that induce immune system of plants. Examples of these elicitors encompass jasmonic acid, polyamines, brassinosteroids, chitosan, salicylic acid, benzothiadiazole, 2,6-dichloro-isonicotinicacid, probenazole, tricontinol, clopamide, butamide, isotianil, and sulfanilamide (Shaukat et al. 2024; Abd-Elsalam and Mohamed 2024; Mrid et al. 2021; Bektas and Eulgem 2015).

(A) *Polyamines*

Polyamines are organic polycations characterized by their diverse chains of hydrocarbons and the existence of more than one primary amino group. These substances are frequently present in biological systems and are found in higher concentrations within cells that are actively undergoing division. The principal forms of polyamines include spermine, spermidine, putrescine, and cadaverine.

Function of Polyamines
In plants, polyamines are essential for numerous functions, including development and growth, due to their impact on cell differentiation and division, as well as their role in flowering and fruit ripening. The regulatory effects of these polyamines on enhancing plant productivity have been documented (Blázquez 2024). The identified substances function as agents that counteract senescence and demonstrate efficacy in postponing the softening process in a range of fruits, through the improvement of the cell membrane's integrity (Serrano and Valero 2018). They are recognized as important reservoirs of nitrogen and carbon, constituting a unique category of growth regulators that operate as secondary hormonal messengers. Polyamines are crucial in addressing biotic and abiotic stress. They aid in alleviating chilling injury, drought, salinity, and heat stress and disease management (Shao et al. 2022). Furthermore, polyamines help to reduce the adverse impacts of ozone on plant health (Khajuria et al. 2018).

(B) *Salicylic Acid*

Salicylic acid, a phenolic derivative, is present in various plant species across a broad spectrum. The synthesis occurs by utilizing the amino acid phenylalanine. Synthetic analogs of SA comprise 2,6-dichloroisonicotinic acid and BTH. The process of decarboxylating trans-cinnamic acid to yield benzoic acid, subsequently followed by 2-hydroxylation, the end product is the synthesis of salicylic acid.

Function of Salicylic Acid
Phenolic compounds impact numerous biochemical and physiological processes, that is, membrane permeability, photosynthesis, enzyme activity, ion absorption,

flowering, and plant growth and development. Bion, or Actigard, features benzothiadiazole as its active component and serves as a commercially available plant health promoter that activates the defense pathway reliant on salicylic acid. Salicylic acid serves an essential function in internal signaling networks that regulate plant defense mechanisms against pathogens (Mishra et al. 2024; Chen et al. 2020). It improves resistance to infections through the stimulation of pathogenesis-related protein production. This process is integral to SAR, wherein the attack of the pathogen on a specific region elicits a resistance response in other regions of the plant. This signal can disseminate to adjacent plants through the transformation of SA into the volatile ester, methyl salicylate.

(C) *Jasmonic Acid*

Jasmonic acid is produced from the fatty acid known as linolenic acid. Methyl jasmonate is a volatile chemical molecule that originates from jasmonic acid. JA and its metabolic by-products are referred to as jasmonates.

Function of Jasmonic Acid

The mechanisms that regulate plant development and growth encompass senescence, leaf abscission, growth inhibition, and floral development. JA is essential for the production of tuber in species, including yams, onions, and potatoes. It is crucial in the response to plant injury and SAR. Jasmonic acid concentrations elevate in reaction to tissue injury (Ruan et al. 2019). JA activates the transcription of numerous genes linked to the defense mechanism of plants (Wang et al. 2020).

(D) *Chitosan*

Chitosan, obtained from deacetylated chitin, is now extracted from the exoskeletons of various crustaceans, including shrimps, crabs, and krills. Chitin and chitosan are polysaccharides that exhibit chemical similarities to cellulose, with the primary distinction being the absence or presence of nitrogen.

Function of Chitosan

The application of chitosan as a carbon source has the potential to promote the growth of beneficial microorganisms within the soil ecosystem. It facilitates the transformation of inorganic substances into organic compounds, enhances the root system of the plant, and increases nutrient absorption from the soil, thereby facilitating the growth of plants (Rassaei 2024). Chitosan is employed in agricultural practices as a biopesticide and treatment for seeds, facilitating the enhancement of plants' resistance to fungal diseases (Karamchandani et al. 2022).

(E) *Brassinosteroids*

Brassinosteroids represent a category of plant polyhydroxysteroids recognized as a novel class of phytohormones. BR has been identified in nearly all regions of plant life. Immature seeds and pollen are the predominant sources of BR, with concentrations varying from 1 to 100 μg/kg in fresh tissue. Conversely, leaves and shoots generally possess reduced amounts in fresh tissues, roughly 0.01–0.1 μg/kg. Approximately 70 BR were successfully extracted from various plant species.

Function of Brassinosteroids

Brassinosteroids can regulate cellular elongation, differentiation and division, at the cellular level. Brassinosteroids can affect leaf development, male fertility, root and shoot development, hypocotyl elongation, senescence, and responses to abiotic and biotic stimuli at the whole plant level (Ahammed et al. 2021; Hossain et al. 2022).

(F) *Triacontanol*

Triacontanol, also a fatty alcohol, is characterized by the molecular formula C30H62O, and is also referred to as myricyl alcohol or melissyl alcohol. It is present in beeswax as well as in the cuticular waxes of various plant species.

Role of Triacontanol

TRIA demonstrates enhancements in protein synthesis, photosynthesis, yield, growth, nutrient uptake and water, enzyme activities, and nitrogen fixation (specifically malate dehydrogenase, a respiratory enzyme), as well as the concentrations of reducing soluble protein, sugars, free amino acids, and active components of essential oils across a variety of crops. It functions as a growth enhancer for various plant species (El-Kinany 2020; Zaid et al. 2020) (Table 2.3).

2.5 Commercialized Products

The application of elicitors is still in its nascent phase; however, certain drugs have been employed commercially in specific countries. They may also be utilized alongside chemical fungicides, depending on their effectiveness and compatibility (Zehra et al. 2021; Ons et al. 2020). The elicitor, sold as Bion or Actigard, functions as a booster of plant health in annual crops. Chemical inducers, such as benzo (1,2,3)-thiadiazole-7-carbothioic acid-S-methyl ester (known as ASM or acibenzolar-S-methyl or BTH, Bion), have been formulated as effective enhancers of SAR in different species of plants. BTH serves as a chemical analog of SA and has been effectively utilized to enhance resistance against various diseases in crops (Faize and Faize 2018; Tripathi et al. 2019). The non-protein amino acid β-aminobutyric acid (BABA) confers protective benefits to several plant species against numerous diseases (Li et al. 2019; Decsi et al. 2024). A variety of substances have been employed as resistance inducers in plants against infections, including salicylic acid analogs and chitosan, both living and processed seaweed extracts and fungal products. Various synthetic chemicals, including 2,6-dichloroisonicotinic acid, and potassium salts have been documented in plants to elicit systemic acquired resistance (Urban et al. 2022; Kashyap et al. 2022).

Table 2.3 Categories of elicitors utilized in various crops

Type of plants	Elicitor utilized	Impacts	References
Blackberry (*Rubus* spp. Var. loch ness)	Metabolic elicitors	Enhanced plant immunity system, and improved their fitness by decreasing oxidative stress, increasing photosynthesis, and activating photogenesis-related proteins	Martin-Rivilla et al. (2020)
Gladiolus (sword lilly)	Salicylic acid	Enhances the quality of cut spikes and prolongs petal senescence	Hatamzadeh et al. (2012)
Butterhead lettuce (*Lactuca sativa* var. *capitata*)	Salicylic acid (SA), arachidonic acid (AA), Harpin protein (HP), and methyl jasmonate (MJ)	Increase in phytochemical content, rise in antioxidant content, increase in the content of phenolic acids	Moreno-Escamilla et al. (2017)
Tobacco and *Arabidopsis*	Cryptogein, oligogalacturonides, and flagellin	Increase of plasma membrane order	Sandor et al. (2016)
Grapewine (*Vitis vinifera*)	Benzothiadiazole (BTH), methyl jasmonate (MeJA), and phosphonates (PHOS)	The accumulation of transcripts aligned with the signaling pathways specific to the elicitors, with salicylic acid being associated with BTH and jasmonic acid with MeJA. PHOS appeared to regulate defense responses similarly to BTH, as leaves pre-treated with PHOS and BTH showed elevated production of pterostilbene. Following MeJA treatment, increased levels of piceids and ε-viniferin were observed compared to leaves treated with elicitors alone in the absence of inoculation	Burdziej et al. (2021)
Monastrell grapes	Methyl jasmonate (MeJ), benzothiadiazole (BTH), chitosan from fungi (CHSf), chitosan from seafood (CHSs)	Altered the complex carbohydrate profile of the resulting wine	Apolinar-Valiente et al. (2018)

2.6 Conclusion

The administration of plant diseases primarily involves the application of fungicides, which provide rapid and effective outcomes. However, this approach is associated with several drawbacks, including the potential for phytotoxicity resulting from excessive dosages, the harmful residual effects of fungicides on agricultural produce that pose risks to human health, the emergence of pathogen resistance to specific fungicides, and the escalating chemical burden on the environment, which contributes to environmental pollution. To address the various challenges associated with the utilization of chemical pesticides, it is imperative to explore alternative strategies, with the application of elicitors emerging as a particularly potential alternative for the management of plant diseases.

References

Abbasi S, Sadeghi A, Omidvari M, Tahan V (2021) The stimulators and responsive genes to induce systemic resistance against pathogens: an exclusive focus on tomato as a model plant. Biocatal Agric Biotechnol 33:101993

Abd-Elsalam K, Mohamed HI (2024) Plant growth regulators to manage biotic and abiotic stress in agroecosystems. CRC Press, Boca Raton

Abdul Malik NA, Kumar IS, Nadarajah K (2020) Elicitor and receptor molecules: orchestrators of plant defense and immunity. Int J Mol Sci 21:963

Ahammed GJ, Sharma A, Yu J (2021) Brassinosteroids in plant developmental biology and stress tolerance. Academic Press, London

Alagna F, Balestrini R, Chitarra W, Marsico AD, Nerva L (2020) Getting ready with the priming: innovative weapons against biotic and abiotic crop enemies in a global changing scenario. In: Priming-mediated stress and cross-stress tolerance in crop plants. Academic Press, London, pp 35–56

Alvarado AM, Aguirre-Becerra H, Vázquez-Hernández MC, Magaña-Lopez E, Parola-Contreras I, Caicedo-Lopez LH, Feregrino-Perez AA (2019) Influence of elicitors and eustressors on the production of plant secondary metabolites. Nat Bioact Compd 1:333–388

Apolinar-Valiente R, Ruiz-García Y, Williams P, Gil-Muñoz R, Gómez-Plaza E, Doco T (2018) Preharvest application of elicitors to Monastrell grapes: impact on wine polysaccharide and oligosaccharide composition. J Agric Food Chem 66:11151–11157

Appu M, Ramalingam P, Sathiyanarayanan A, Huang J (2021) An overview of plant defense-related enzymes responses to biotic stresses. Plant Gene 27:100302

Baenas N, García-Viguera C, Moreno DA (2014) Elicitation: a tool for enriching the bioactive composition of foods. Molecules 19:13541–13563

Bektas Y, Eulgem T (2015) Synthetic plant defense elicitors. Front Plant Sci 5:804

Blázquez MA (2024) Polyamines: their role in plant development and stress. Annu Rev Plant Biol 75:95–117

Burdziej A, Bellée A, Bodin E, Valls Fonayet J, Magnin N, Szakiel A, Corio-Costet MF (2021) Three types of elicitors induce grapevine resistance against downy mildew via common and specific immune responses. J Agric Food Chem 69:1781–1795

Caicedo-López LH, Aranda ALV, Sáenz de la OD, Gómez CEZ, Márquez EE, Zepeda HR (2021) Elicitors: bioethical implications for agriculture and human health. Rev Bioética 29:76–86

Chamkhi I, Benali T, Aanniz T, El Menyiy N, Guaouguaou FE, El Omari N, Bouyahya A (2021) Plant-microbial interaction: the mechanism and the application of microbial elicitor induced secondary metabolites biosynthesis in medicinal plants. Plant Physiol Biochem 167:269–295

Chen YL, Fan KT, Hung SC, Chen YR (2020) The role of peptides cleaved from protein precursors in eliciting plant stress reactions. New Phytol 225:2267–2282

Cipollini D, Walters D, Voelckel C (2014) Costs of resistance in plants: from theory to evidence. Annu Plant Rev Online 47:263–307

Dalio RJ, Magalhaes DM, Rodrigues CM, Arena GD, Oliveira TS, Souza-Neto RR, Machado MA (2017) PAMPs, PRRs, effectors and R-genes associated with citrus–pathogen interactions. Ann Bot 119:749–774

Decsi K, Ahmed M, Rizk R, Abdul-Hamid D, Kovács GP, Tóth Z (2024) Emerging trends in non-protein amino acids as potential priming agents: implications for stress management strategies and unveiling their regulatory functions. Int J Mol Sci 25:6203

Dongre MA, Ashraf MS, Koshariya AK, Reddypriya P (2023) Plant pathology and plant diseases. AG publishing house (AGPH Books), Bhopal
Dos Santos C, Franco OL (2023) Pathogenesis-related proteins (PRs) with enzyme activity activating plant defense responses. Plan Theory 12:2226
El-Kinany RG (2020) The beneficial role of salicylic acid, triacontanol and δ-aminolevulinic acid on the growth, flowering and chemical composition of pansy (Viola wittrockiana gams) under salt stress conditions. Hortsci J Suez Canal Univ 9:13–30
Enebe MC, Babalola OO (2019) The impact of microbes in the orchestration of plants' resistance to biotic stress: a disease management approach. Appl Microbiol Biotechnol 103:9–25
Enoki S, Suzuki S (2016) Pathogenesis-related proteins in grape. Grape Wine Biotechnol 43:46–49
Faize L, Faize M (2018) Functional analogues of salicylic acid and their use in crop protection. Agronomy 8:5
Fei W, Liu Y (2023) Biotrophic fungal pathogens: a critical overview. Appl Biochem Biotechnol 195:1–16
Francis F, Jacquemyn H, Delvigne F, Lievens B (2020) From diverse origins to specific targets: role of microorganisms in indirect pest biological control. Insects 11:533
Gaddam NR, Devi TM, Rupali JS, Darapureddy NSSS, Reddy GR (2024) Exploiting induced plant resistance for sustainable Pest management: mechanisms, elicitors, and applications: a review. J Exp Agric Int 46:586–599
Gowthami L (2018) Role of elicitors in plant defense mechanism. J Pharm Phytochem 7:2806–2812
Gu D, Andreev K, Dupre ME (2021) Major trends in population growth around the world. China CDC Wkly 3:604–613
Hatamzadeh A, Hatami M, Ghasemnezhad M (2012) Efficiency of salicylic acid delay petal senescence and extended quality of cut spikes of Gladiolus grandiflora cv 'wing's sensation. Afr J Agric Res 7:540–545
Heimpel GE, Mills NJ (2017) Biological control. Cambridge University Press, Cambridge
Hönig M, Roeber VM, Schmülling T, Cortleven A (2023) Chemical priming of plant defense responses to pathogen attacks. Front Plant Sci 14:1146577
Hossain A, Venugopalan VK, Rahman MA, Alam MJ, Al-Mahmud A, Islam MA, Aftab T (2022) Physiological, biochemical, and molecular mechanisms of plant steroid hormones brassinosteroids under drought-induced oxidative stress in plants. In: Emerging plant growth regulators in agriculture. Academic Press, London, pp 99–130
Hou S, Wang X, Chen D, Yang X, Wang M, Turrà D, Zhang W (2014) The secreted peptide PIP1 amplifies immunity through receptor-like kinase 7. PLoS Pathog 10(9):e1004331
Hückelhoven R, Schouten A (2024) Plant immunity and plant defense. In: Agrios' plant pathology. Academic Press, London, pp 161–210
Jain D, Bisht S, Parvez A, Singh K, Bhaskar P, Koubouris G (2024) Effective biotic elicitors for augmentation of secondary metabolite production in medicinal plants. Agriculture 14:796
Jeyasri R, Muthuramalingam P, Karthick K, Shin H, Choi SH, Ramesh M (2023) Methyl jasmonate and salicylic acid as powerful elicitors for enhancing the production of secondary metabolites in medicinal plants: an updated review. Plant Cell Tissue Organ Cult 153:447–458
Jones P, Garcia BJ, Furches A, Tuskan GA, Jacobson D (2019) Plant host-associated mechanisms for microbial selection. Front Plant Sci 10:862
Jubic LM, Saile S, Furzer OJ, El Kasmi F, Dangl JL (2019) Help wanted: helper NLRs and plant immune responses. Curr Opin Plant Biol 50:82–94
Kamle M, Borah R, Bora H, Jaiswal AK, Singh RK, Kumar P (2020) Systemic acquired resistance (SAR) and induced systemic resistance (ISR): role and mechanism of action against phytopathogens. In: Fungal biotechnology and bioengineering. Springer, Cham, pp 457–470
Karamchandani BM, Chakraborty S, Dalvi SG, Satpute SK (2022) Chitosan and its derivatives: promising biomaterial in averting fungal diseases of sugarcane and other crops. J Basic Microbiol 62:533–554
Kashyap PL, Kumar S, Jasrotia P, Kumar RS, Sharma A, Tripathi R, Singh GP (2022) Induced resistance for sustainable management of wheat diseases. In: Handbook of research on green

technologies for sustainable management of agricultural resources. IGI Global, Hershey, pp 385–408

Kaur S, Samota MK, Choudhary M, Choudhary M, Pandey AK, Sharma A, Thakur J (2022) How do plants defend themselves against pathogens-biochemical mechanisms and genetic interventions. Physiol Mol Biol Plants 28:485–504

Keswani C, Dilnashin H, Birla H, Singh SP (2019a) Unravelling efficient applications of agriculturally important microorganisms for alleviation of induced inter-cellular oxidative stress in crops. Acta Agric Slov 114:121–130

Keswani C, Prakash O, Bharti N, Vílchez JI, Sansinenea E, Lally RD, Singh HB (2019b) Re-addressing the biosafety issues of plant growth promoting rhizobacteria. Sci Total Environ 690:841–852

Keswani C, Singh HB, Hermosa R, García-Estrada C, Caradus J, He YW, Sansinenea E (2019c) Antimicrobial secondary metabolites from agriculturally important fungi as next biocontrol agents. Appl Microbiol Biotechnol 103:9287–9303

Keswani C, Singh SP, García-Estrada C, Mezaache-Aichour S, Glare TR, Borriss R, Sansinenea E (2022) Biosynthesis and beneficial effects of microbial gibberellins on crops for sustainable agriculture. J Appl Microbiol 132:1597–1615

Khajuria A, Sharma N, Bhardwaj R, Ohri P (2018) Emerging role of polyamines in plant stress tolerance. Curr Protein Pept Sci 19:1114–1123

Krishnamoorthi A, Patel PK, Kumar S, Khambhu CV, Kumar V, Selvam H (2024) The future of farming with advances in biological control techniques for crop health. Microbiol Res J Int 34:93–112

Lalaleo L, Khojasteh A, Fattahi M, Bonfill M, Cusido RM, Palazon J (2016) Plant anti-cancer agents and their biotechnological production in plant cell biofactories. Curr Med Chem 23:4418–4441

Le Mire G (2018) Identification of elicitors inducing resistance in wheat against Zymoseptoria tritici and characterization of the subsequent triggered defense-signaling pathways. Universite de Liege, Liege

Li Z, Ahammed GJ (2023) Salicylic acid and jasmonic acid in elevated CO2-induced plant defense response to pathogens. J Plant Physiol 286:154019

Li Y, Li Q, Hong Q, Lin Y, Mao W, Zhou S (2018) Reactive oxygen species triggering systemic programmed cell death process via elevation of nuclear calcium ion level in tomatoes resisting tobacco mosaic virus. Plant Sci 270:166–175

Li S, Nie H, Qiu D, Shi M, Yuan Q (2019) A novel protein elicitor PeFOC1 from Fusarium oxysporum triggers defense response and systemic resistance in tobacco. Biochem Biophys Res Commun 514:1074–1080

Lolle S, Stevens D, Coaker G (2020) Plant NLR-triggered immunity: from receptor activation to downstream signaling. Curr Opin Immunol 62:99–105

Martin-Rivilla H, Garcia-Villaraco A, Ramos-Solano B, Gutierrez-Manero FJ, Lucas JA (2020) Improving flavonoid metabolism in blackberry leaves and plant fitness by using the bioeffector Pseudomonas fluorescens N 21.4 and its metabolic elicitors: a biotechnological approach for a more sustainable crop. J Agric Food Chem 68:6170–6180

Meena M, Samal S (2019) Alternaria host-specific (HSTs) toxins: an overview of chemical characterization, target sites, regulation and their toxic effects. Toxicol Rep 6:745–758

Meena M, Zehra A, Dubey MK, Aamir M, Gupta VK, Upadhyay RS (2016) Comparative evaluation of biochemical changes in tomato (Lycopersicon esculentum Mill.) infected by Alternaria alternata and its toxic metabolites (TeA, AOH, and AME). Front Plant Sci 7:1408

Meena M, Swapnil P, Divyanshu K, Kumar S, Harish Tripathi YN, Upadhyay RS (2020) PGPR-mediated induction of systemic resistance and physiochemical alterations in plants against the pathogens: current perspectives. J Basic Microbiol 60:828–861

Meena M, Yadav G, Sonigra P, Nagda A, Mehta T, Swapnil P, Marwal A (2022) Role of elicitors to initiate the induction of systemic resistance in plants to biotic stress. Plant Stress 5:100103

Meraj TA, Fu J, Raza MA, Zhu C, Shen Q, Xu D, Wang Q (2020) Transcriptional factors regulate plant stress responses through mediating secondary metabolism. Genes 11:346

Mertens D, Boege K, Kessler A, Koricheva J, Thaler JS, Whiteman NK, Poelman EH (2021) Predictability of biotic stress structures plant defence evolution. Trends Ecol Evol 36:444–456

Mishra S, Roychowdhury R, Ray S, Hada A, Kumar A, Sarker U, Das R (2024) Salicylic acid (SA)-mediated plant immunity against biotic stresses: an insight on molecular components and signaling mechanism. Plant Stress 11:100427

Montesano M, Brader G, Palva ET (2003) Pathogen derived elicitors: searching for receptors in plants. Mol Plant Pathol 4:73–79

Moreno-Escamilla JO, Alvarez-Parrilla E, de la Rosa LA, Núñez-Gastélum JA, González-Aguilar GA, Rodrigo-García J (2017) Effect of different elicitors and preharvest day application on the content of phytochemicals and antioxidant activity of butterhead lettuce (Lactuca sativa var. capitata) produced under hydroponic conditions. J Agric Food Chem 65:5244–5254

Moustafa-Farag M, Almoneafy A, Mahmoud A, Elkelish A, Arnao MB, Li L, Ai S (2019) Melatonin and its protective role against biotic stress impacts on plants. Biomol Ther 10:54

Mrid RB, Benmrid B, Hafsa J, Boukcim H, Sobeh M, Yasri A (2021) Secondary metabolites as biostimulant and bioprotectant agents. Sci Total Environ 777:146204

Nadarajah KK (2024) Defensive strategies of ROS in plant–pathogen interactions. In: Plant pathogen interaction. Springer, Singapore, pp 163–183

Nimchuk Z, Rohmer L, Chang JH, Dangl JL (2001) Knowing the dancer from the dance: R-gene products and their interactions with other proteins from host and pathogen. Curr Opin Plant Biol 4:288–294

Nürnberger T, Brunner F (2002) Innate immunity in plants and animals: emerging parallels between the recognition of general elicitors and pathogen-associated molecular patterns. Curr Opin Plant Biol 5:318–324

Nürnberger T, Scheel D (2001) Signal transmission in the plant immune response. Trends Plant Sci 6:372–379

Ochoa-Meza LC, Quintana-Obregón EA, Vargas-Arispuro I, Falcón-Rodríguez AB, Aispuro-Hernández E, Virgen-Ortiz JJ, Martínez-Téllez MÁ (2021) Oligosaccharins as elicitors of defense responses in wheat. Polymers 13:3105

Ons L, Bylemans D, Thevissen K, Cammue BP (2020) Combining biocontrol agents with chemical fungicides for integrated plant fungal disease control. Microorganisms 8:1930

Parra-Arroyo L, González-González RB, Castillo-Zacarías C, Martínez EMM, Sosa-Hernández JE, Bilal M, Parra-Saldívar R (2022) Highly hazardous pesticides and related pollutants: toxicological, regulatory, and analytical aspects. Sci Total Environ 807:151879

Patel ZM, Mahapatra R, Jampala SSM (2020) Role of fungal elicitors in plant defense mechanism. In: Molecular aspects of plant beneficial microbes in agriculture. Academic Press, Amsterdam, pp 143–158

Peng DH, Qiu DW, Ruan LF, Zhou CF, Sun M (2011) Protein elicitor PemG1 from Magnaporthe grisea induces systemic acquired resistance (SAR) in plants. Mol Plant-Microbe Interact 24:1239–1246

Pršić J, Ongena M (2020) Elicitors of plant immunity triggered by beneficial bacteria. Front Plant Sci 11:594530

Ram RM, Tripathi R, Birla H, Dilnashin H, Singh SP, Keswani C (2019) Mixed PGPR consortium: an effective modulator of antioxidant network for management of collar rot in cauliflower. Arch Phytopathol Plant Protect 52:844–862

Ramirez-Estrada K, Vidal-Limon H, Hidalgo D, Moyano E, Golenioswki M, Cusidó RM, Palazon J (2016) Elicitation, an effective strategy for the biotechnological production of bioactive high-added value compounds in plant cell factories. Molecules 21:182

Rassaei F (2024) Chitosan as an organic amendment to improve soil properties and plant growth in the presence of polystyrene microplastics. Environ Prog Sustain Energy 43:14301

Ratnadass A, Deguine JP (2020) Three-way interactions between crop plants, phytopathogenic fungi, and mirid bugs. A review. Agron Sustain Dev 40:46

Riseh RS, Vazvani MG (2024) Unveiling methods to stimulate plant resistance against pathogens. Front Biosci Landmark 29:188

Ruan J, Zhou Y, Zhou M, Yan J, Khurshid M, Weng W, Zhang K (2019) Jasmonic acid signaling pathway in plants. Int J Mol Sci 20:2479

Saini S, Lohani S, Khati P, Rani V (2023) PGPR-mediated mitigation of biotic and abiotic stress in plants. In: Advanced microbial technology for sustainable agriculture and environment. Academic Press, London, pp 199–227

Sandor R, Der C, Grosjean K, Anca I, Noirot E, Leborgne-Castel N, Gerbeau-Pissot P (2016) Plasma membrane order and fluidity are diversely triggered by elicitors of plant defence. J Exp Bot 67:5173–5185

Saravanakumar K, Fan L, Fu K, Yu C, Wang M, Xia H, Chen J (2016) Cellulase from Trichoderma harzianum interacts with roots and triggers induced systemic resistance to foliar disease in maize. Sci Rep 6:35543

Serrano M, Valero D (2018) Application of polyamines to maintain functional properties in stored fruits. In: Polyamines: methods and protocols. Springer, New York, pp 449–458

Shao J, Huang K, Batool M, Idrees F, Afzal R, Haroon M, El Sabagh A (2022) Versatile roles of polyamines in improving abiotic stress tolerance of plants. Front Plant Sci 13:1003155

Shaukat S, Mustafa G, Cheng K (2024) Phytohormones for combating global challenges: an eco-friendly approach. In: Plant growth regulators: resilience for sustainable agriculture. Springer, Singapore, pp 267–284

Shen Y, Li J, Xiang J, Wang J, Yin K, Liu Q (2019) Isolation and identification of a novel protein elicitor from a Bacillus subtilis strain BU412. AMB Express 9:1–9

Shibuya N, Minami E (2001) Oligosaccharide signalling for defence responses in plant. Physiol Mol Plant Pathol 59:223–233

Singh R, Choudhary P, Kumar S, Daima HK (2024) Mechanistic approaches for crosstalk between nanomaterials and plants: plant immunomodulation, defense mechanisms, stress resilience, toxicity, and perspectives. Environ Sci Nano 11:2324–2351

Song W, Shao H, Zheng A, Zhao L, Xu Y (2023) Advances in roles of salicylic acid in plant tolerance responses to biotic and abiotic stresses. Plan Theory 12:3475

Swapnil P, Meena M, Singh SK, Dhuldhaj UP, Marwal A (2021) Vital roles of carotenoids in plants and humans to deteriorate stress with its structure, biosynthesis, metabolic engineering and functional aspects. Curr Plant Biol 26:100203

Tripathi D, Raikhy G, Kumar D (2019) Chemical elicitors of systemic acquired resistance—salicylic acid and its functional analogs. Curr Plant Biol 17:48–59

Urban L, Lauri F, Ben Hdech D, Aarrouf J (2022) Prospects for increasing the efficacy of plant resistance inducers stimulating salicylic acid. Agronomy 12:3151

Vallad GE, Goodman RM (2004) Systemic acquired resistance and induced systemic resistance in conventional agriculture. Crop Sci 44:1920–1934

Venegas-Molina J, Proietti S, Pollier J, Orozco-Freire W, Ramirez-Villacis D, Leon-Reyes A (2020) Induced tolerance to abiotic and biotic stresses of broccoli and Arabidopsis after treatment with elicitor molecules. Sci Rep 10:10319

Wang J, Song L, Gong X, Xu J, Li M (2020) Functions of jasmonic acid in plant regulation and response to abiotic stress. Int J Mol Sci 21:1446

Wang D, Liu B, Ma Z, Feng J, Yan H (2021) Reticine A, a new potent natural elicitor: isolation from the fruit peel of citrus reticulate and induction of systemic resistance against tobacco mosaic virus and other plant fungal diseases. Pest Manag Sci 77:354–364

Wu Y, Yi G, Peng X, Huang B, Liu EE, Zhang J (2013) Systemic acquired resistance in Cavendish banana induced by infection with an incompatible strain of Fusarium oxysporum f. sp. cubense. J Plant Physiol 170:1039–1046

Zaid A, Asgher M, Wani IA, Wani SH (2020) Role of triacontanol in overcoming environmental stresses. In: Protective chemical agents in the amelioration of plant abiotic stress: biochemical and molecular perspectives. Wiley, Newark, pp 491–509

Zebelo SA (2020) Decrypting early perception of biotic stress on plants. In: Co-evolution of secondary metabolites. Springer, Cham, pp 577–592

Zehra A, Raytekar NA, Meena M, Swapnil P (2021) Efficiency of microbial bio-agents as elicitors in plant defense mechanism under biotic stress. Curr Res Microbiol Sci 2:100054

Zhang L, Gleason C (2020) Enhancing potato resistance against root-knot nematodes using a plant-defence elicitor delivered by bacteria. Nat plants 6:625–629

Zhang S, Li C, Si J, Han Z, Chen D (2022) Action mechanisms of effectors in plant-pathogen interaction. Int J Mol Sci 23:6758

Zheng F, Chen L, Zhang P, Zhou J, Lu X, Tian W (2020) Carbohydrate polymers exhibit great potential as effective elicitors in organic agriculture. Carbohydr Polym 230:115637

Zhu F, Xi DH, Yuan S, Xu F, Zhang DW, Lin HH (2014) Salicylic acid and jasmonic acid are essential for systemic resistance against tobacco mosaic virus in Nicotiana benthamiana. Mol Plant-Microbe Interact 27:567–577

Chapter 3
Actinobacteria's Role in Plant Defense and Their Effectiveness as Biocontrol Agents Against Plant Pathogens

Hiran Kanti Santra and Debdulal Banerjee

Abstract Agriculture is one of the most critical domains of healthy living and a basic need of many agro-product-based countries' economic upliftment. With the increasing problems of global warming, anthropogenic interference, and the introduction of high-yielding varieties, and application of chemical fertilizers–pesticides/weedicides, agricultural productivity is backfiring, and the biotic stress caused by pathogenic fungi and bacteria is going logarithmic. Bio-based novel solutions are highly solicited to reduce the events of agricultural catastrophes, and biocontrol agent/microbe-mediated options are the safest ones to go for. Soil fungi or actinobacteria and their endophytic counterparts can sometimes favor plant growth and provide a strong defense against plant pathogens. Actinobacteria or ray fungi are one step ahead in this respect due to their unique production abilities of diverse antibiotics and high guanine and cytosine contents in their core genome. They are an integral part of the plant–rhizospheric microbiome and are critical in plant growth and protection from biotic and abiotic stress. Thus, they contribute to developing sustainable agricultural practices and act as a biocontrol agent to tackle devastating phytopathogens. Their synthesized exo-metabolomes, exopolysaccharides, or low-molecular-weight volatile and nonvolatile secondary metabolites are effective and perform different plant growth-promoting attributes. Also, the direct introduction of actinobacterial consortium or synthetic communities are practical plant nurturers. The chapter focuses on novel actinobacterial isolates from rhizospheric and endophytic origin, along with their utility as plant-protecting agents against detrimental phytopathogens.

H. K. Santra
Department of Botany and Forestry, Microbiology and Microbial Biotechnology Laboratory ,
Vidyasagar University, Midnapore, West Bengal, India

D. Banerjee (✉)
Department of Botany and Forestry, Microbiology and Microbial Biotechnology Laboratory,
Vidyasagar University, Midnapore, West Bengal, India

Centre for Life Sciences, Vidyasagar University, Midnapore, West Bengal, India
e-mail: db@mail.vidyasagar.ac.in

K. A. Abd-Elsalam, H. I. Mohamed (eds.), *Elicitors for Sustainable Crop Production*,
Sustainability Sciences in Asia and Africa,
https://doi.org/10.1007/978-981-96-6569-3_3

Keywords Actinobacteria · Novel biocontrol agents · Sustainability · Nature-based solutions

3.1 Introduction

One of the major problems of the twenty-first century is logarithmic population growth uneven urbanization and industrialization, which ultimately leads to the abrupt decrease in the quantity of agricultural land. So, the burden of overproduction of essential agricultural commodities from a small land over the shortest span has become the trend. These things have triggered the use of chemical fertilizers, pesticides, xenobiotic fungicides, and other harmful chemicals, which ultimately leads to the irreversible deterioration of soil health, that is, both the chemical (pH, change in carbon and nitrogen profile) and physical (soil porosity, water-holding capacity) characteristics of a standard farmable soil get disrupted. So, instead of chemical fertilizers, nowadays farmers are implementing live organisms (in terms of biofertilizer), that is, microbial inoculants such as blue-green algae (cyanobacteria), water fern *Azolla* sp. on the agricultural fields to enhance soil productivity. This may ease up things and can to some extent meet the global need. But not only chemical fertilizers, the issue is more severe when it comes to pesticides or fungicides application. They impede irreparable impacts on the ecosystem and get biomagnified when consumed through the food chain or food web. They cannot be put aside as fungal or bacterial infections, and pest attacks in crops/vegetables/fruits cause a severe drop in production and create a pandemonium for agri-management. So, sustainable solutions are in high demand and any bio-based or nature-based solutions (NBS) are the first to be welcomed (Boubekri et al. 2022). The role of microbial inoculants or their derived metabolites as an alternative to chemical ones is given next. On the one hand, they can act as plant growth promoters, and on the other hand, they can serve the purpose of deviating from the biotic stresses, namely the infectious agents and insect pests. Also, this ecofriendly consortium directly or their extracts indirectly can eliminate abiotic stressors, making the agri-environment a suitable place to cultivate through their multifunctional traits. The broad-spectrum use of microbial metabolites or live microbial bioformulation can nurture the needs of agriculture and maintain the ecosystem components in a suitable harmony. Microbes are always beneficial for these purposes as they can perform multiple functions, that is, from nitrogen fixation to phosphorus, potassium solubilization, synthesizing auxin-like compounds (indole 3 acetic acid, indole 3 butyric acid, indole 3 pyruvic acid, gibberellins, cytokinins) to hydrolyzing enzymes (lipases, proteases, xylanases, glucanases, chitinase, cellulase, etc.) and be engaged in a variety of ecological interactions from symbiotic to mutualism (Mitra et al. 2022). In this chapter, we have focused on discussing the various intriguing roles of a particular type of microbes–actinobacteria in this aspect.

Actinobacteria, also known as "ray fungi," are Gram-positive, free-living branched filamentous microbes, with high GC contents (50–70%), that are ubiquitous in distribution and potent contributors of pharmaceutically valuable compounds, such as antibiotics nocobactin, formobactin, amamistatin, and brasilibactin (Bahrami et al. 2022). They can be considered the never-ending source of biotechnologically relevant compounds (Al-Quwaie 2024). They are popular for their earthy smell, which is called geosmin. Their prevalence has been recorded in the soil, water, sands, extreme environments of high salinity, hyper acidic, volcanic zones, uranium-contaminated environments and guts, and stools of mammals, and even in ancient paintings (Hozzein and Goodfellow 2008; Jose and Jebakumar 2014; Ouchari et al. 2019; Cimermanova et al. 2021). They have been widely reported as endophytes of various plant taxa and are known to produce lots of pharmaceutically effective metabolites (Anavadiya et al. 2024). They are all square in action and can produce secondary metabolites (i.e., both volatiles and nonvolatiles) of diverse origin and mostly are reported to be in the one strain many active compounds (OSMAC) approach. In the recent past, they have caught attention as promising biological control agents (BCA) in combating agricultural pathogens (Giacomelli Ribeiro and Teresinha Van Der Sand 2024). In comparison to bacteria, actinomycetes represent a potentially large genome that makes them functionally stable and is the main reason for the production of such a huge array of bioactive secondary metabolites, that is, industrially valuable enzymes, antimicrobials, antioxidative, cytotoxic, anti-neoplastic, plant growth promoters, agriculturally valuable compounds, etc.

The presence of nonribosomal peptide synthases (NRPS), polyketide synthase (PKS), hybrid NRPS-PKS, cluster of genes, and post-translationally modified peptides (RiPPs) influences the production of potent secondary metabolites (Komaki et al. 2014). In particular, from an agricultural perspective, actinomycetes can solubilize phosphate, assimilate nitrogen, synthesize hydrogen cyanide (HCN), and siderophore (catecholate, hydroxamate, carboxylate, nocardichelins, asterobactin, and nocardimicins), synthesize 1-aminocyclopropane-1-carboxylate (ACC) deaminase and auxins (indole acetic acid), gibberellin, and cytokinin (Wang et al. 2014). Thus, they act as biostimulants and can bioavailable the essential minerals and improve the plants' growth in terms of respiration and photosynthesis (Silva et al. 2022; Oyedoh et al. 2023). The NRPS genes are responsible for the production of siderophores, which are natural ligands of iron that promote the capture and translocation of iron from the environmental components and finally make it available to the soil. Siderophores help in blocking phytopathogens and improve nutrient acquisition. The major reports on plant growth-promoting/anti-phytopathogenic actinobacteria belong to *Streptomyces* species but different species of *Actinoplanes* and Micromonospora (family Micromonosporaceae), *Arthrobacter* (family Micrococcaceae), *Microbacterium* (family Microbacteriaceae), *Nocardia,* and *Rhodococcus* (Nocardiaceae) are also potent contributors. Major members of actinobacteria, that is, biocontrol agents, *Streptomyces* spp., *Actinoplanes campanulatus*, *Amycolatopsis* sp.; plant growth promotors and elicitors of plant defense, *Streptomyces* spp., *Micromonospora* sp.; symbionts, *Frankia* sp.; and endophytes,

Microbispora sp., are particularly involved in plant growth promotion and development of disease resistance. Different actinobacterial metabolites have been commercialized by different multinational companies in Thailand, China, Tokyo, Mexico, Croatia, Germany, Japan, Saudi Arabia, South Korea, Finland, and the United States for their agricultural use (Copping and Duke 2007; Saxena and Pandey 2001; Kabaluk et al. 2010; Aggarwal et al. 2016), and in different countries (Canada, the European Union, and the United States), they are being applied on a large scale (Table 3.1).

In the present context, agricultural productivity is largely affected by the attack of pathogenic microorganisms, nematodes, and several other devastating pests/weeds, etc. The most common control measures include the utilization of chemical pesticides/weedicides/microbicidal agents that bring a variety of detrimental side effects from cellular to organismal levels and disrupt human and ecosystem health. Thus, sustainable agricultural solutions are highly solicited to solve this serious issue. The utility of nature-based solutions obtained from novel microorganisms and actinobacteria are the most prominent candidates in this respect.

- Thus, in this chapter, an effort has been made to summarize the recent research on the potential of actinobacterial metabolites in controlling agriculturally impactful bacterial and fungal pathogens.
- Also, the biocontrol potentials of actinobacteria have been focused. A glimpse of commercially available actinobacterial products is added.
- The utility of both the soil/rhizospheric and endophytic actinobacteria in tackling biotic and abiotic stressors in agricultural settings has been focused on.
- The in vitro and in vivo/in planta application of actinobacteria and its extracellular enzymes, and metabolite (both volatile and nonvolatile) against parasitic nematodes, fungi, and bacteria have been emphasized.
- The ability of the actinobacteria to induce disease resistance in agriculturally valuable crops has been summarized.

3.2 Role of Actinobacteria in Plant Defense

Plant growth and vigor are the result of multifactorial cross-talks of different interacting factors, that is, both biotic and abiotic. The susceptibility of a plant toward disease is influenced by the availability of minerals phosphate and calcium, nitrogen, indole-like compounds, other phytohormones cytokinin, gibberellin, iron chelation, concentrations of exopolysaccharides, osmolytes, the concentration of defense-related compounds (salicylate and jasmonate) in plant tissues, etc. (Ebrahimi-Zarandi et al. 2022). Also, the presence of abiotic stress components like drought, salinity, heavy metals, and water logging can deteriorate plant health and primary defense factors. To tackle all these components; microbial formulations can be a valid option and actinobacteria can perform all the above-mentioned aspects, leading to a safe condition for agriculture. The metabolites of actinomycetes are

Table 3.1 A brief representation of different antifungal agents, obtained from actinobacterial sources, used worldwide to control agricultural infection

Information about the antifungal agents	Target pathogens	References
Active substance, natamycin; trade name Delvolan Obtained from *Streptomyces natalensis* and *Streptomyces chattanoogensis*	Used in treating *fusarium oxysporum* infection (basal rots) in *narcissus* spp. (daffodils) and other floriculture-relevant plants	Elsayed et al. (2013)
Active substance, polyoxorim; trade name endorse, Stopit, Polyoxin AL and Z, Polybelin obtained from *Streptomyces cacaoi* var. *asoensis*	Used in treating *Alternaria alternata, Botrytis cinerea, Corynespora melonis, Cochliobolus miyabeanus, Helminthosporium* sp., *Sphaerotheca* spp., *Rhizoctonia solani, Sclerotinia sclerotiorum* infections in fruits like *Pyrus malus* (apple), *Pyrus communis* (pear), vines, vegetables, grains like *Oryza sativa* (rice)	Chen et al. (2013)
Trade name Mycostop, Verdera obtained from *Streptomyces griseoviridis* K61	Effective in curing root and wilt infections caused by *Alternaria alternata, Rhizoctonia solani, Fusarium* sp., *Phytophthora* sp., *Pythium* sp.	Minuto et al. (2006)
Trade name Mykocide KIBC obtained from *Streptomyces colombiensis*	Effectively reduces infections like powdery mildews in the leaf, gray mold in fruits like strawberry (*Fragaria* sp.), and brown patches in the stem and leaf	Meidani et al. (2020)
Active substance, Phytomycin; trade name Mycoshield, Cuprimic 100 and 500, Mycoject obtained from *Streptomyces rimosus*	Treats bacterial diseases (majorly blight) in crops and vegetables, i.e., infections of *Erwinia amylovora*, Mycoplasma-like organisms (MLOs), *Pseudomonas* sp., and *Xanthomonas* sp.	Copping and Duke (2007)
Trade name Actinovate from Novozymes obtained from *Streptomyces lydicus* WYEC 108	Reduces common but devastating soil-borne fungal infections, and phyllospheric pathogens in agriculturally valuable cash crops, i.e., different species of *Pythium, Fusarium, Phytophthora, Rhizoctonia*, and *Verticillium, Botrytis* (powdery and downy mildew pathogens), *Alternaria, Postia, Geotrichum*, and *Sclerotinia*, respectively	Quintana-Jones (2011)
Trade name Bactophil obtained from *Streptomyces albus*	Checks pathogenic infections during the early stages of sapling growth, i.e., during the events of seed germination.	Ngamcharungchit et al. (2023)
Trade name Mykocide obtained from *Streptomyces colombiensis*	Reduces the infection of common pathogenic fungi causing powdery mildews, and also blocks the growth of gray mold, and brown patch pathogen, i.e., *Rhizoctonia solani*	Meidani et al. (2020)
Trade name Safegrow *Streptomyces kasugaensis*	Fights against the sheath blight and large patch pathogens (e.g., *Rhizoctonia solani*) and protects the agricultural products	Hamada et al. (1995)

(continued)

Table 3.1 (continued)

Information about the antifungal agents	Target pathogens	References
Trade name Blasticidin-S *Streptomyces griseochromogenes*	Reduces the infection of *Pyricularia oryzae*, i.e., rice blast disease	Cone et al. (2003)
Active substance, validamycin; trade name Validacin, Valimun, Dantotsupadanvalida, Mycin, hustler, Valida, Sheathmar obtained from *Streptomyces hygroscopicus*	Controls infection of *Rhizoctonia* in rice (*Oryza sativa*), potatoes (*Solanum tuberosum*), vegetables, strawberries (*Fragaria* sp.), tobacco (*Nicotiana tabacum*), ginger (*Zingiber officinale*), cotton (*Gossypium herbaceum*), sugar beet (*Beta vulgaris*), etc.	Bai et al. (2006)
Active substance, streptomycin; trade name Agrimycin, Paushak, Cuprimicin 17, AAstrepto 17, AS-50, Dustret, Cuprimic 100 and 500 obtained from- *Streptomuces griseus*	Checks the invasion of bacterial pathogens- *Xanthomonas oryzae*, *Xanthomonas citri*, and *Pseudomonas tabaci* of pome fruit (*Punica granatum*), stone fruit (*Aegle marmelos*), citrus (*Citrus citri*), olives (*Olea europaea*), vegetables, potatoes (*Solanum tuberosum*), tobacco (*Nicotiana tabacum*), cotton (*Gossypium herbaceum*)	Saxena and Pandey (2001)
Trade name Nigericin obtained from *Streptomyces* sp. YCED9	Anti-*Rhizoctonia solani* and *Sclerotinia homeocarpa* activity and protects pathogenic infections in turfgrass	Trejo-Estrada et al. (1998)
Trade name Buramycin obtained from *Streptomyces* sp. JCK-6131	Inhibits bacterial wilt pathogen *Pseudomonas solanacearum*	Kabaluk et al. (2010)
Trade name manumycin obtained from *Streptomyces flaveus* A11	Effective against a broad-spectrum fungal pathogen *Alternaria* sp., *Magnaporthe grisea*, *Cladosporium cucumerinum*, *Phytophthora capsici*	Hwang et al. (1996)
Trade name rapamycin obtained from *Streptomyces hygroscopicus*	Antifungal against *Verticillium dahliae*	Kim et al. (2014)
Trade name anisomycin obtained from *Streptomyces* sp.	Antifungal against *Erysiphe polygoni* and reduces the infections in crabgrass, bean mildew	Sholkamy et al. (2020)

both strong antimicrobials and can fight fungal, bacterial, and nematode infections (Aallam et al. 2022). Also, it removes the weeds/pests, and pathogenic larvae both in vitro and in vivo. Being hyphal in appearance, they can parasitize on plant pathogens and secrete hydrolytic enzymes to eliminate aggressive invaders; the emitted volatiles can affect crop health by eliminating pathogens (Abbasi et al. 2019). They can synthesize plant growth-promoting substances and can impact the soil microbial flora, which provides a natural defense against parasitic microbes. Also, they have potent phytoremediation action and can assure plants' survival in drought and salt stress by inducing the plant to synthesize prolines, high chlorophyll contents, antioxidant defense enzymes, increased dry and fresh weight, etc. (Behera and Das 2023). They either directly protect crops upon application of soluble metabolites or can indirectly switch on the expression of plant defense-related genes.

3.2.1 Actinobacteria Offer Protection from Abiotic Stress

Different isolates of *Streptomyces* can break down toxic components and assure plant health. Hydrocarbon components (petroleum, diesel), herbicides (atrazine, cypermethrin, 3-phenoxybenzaldehyde), lindane, heavy metals (chromium, cadmium, nickel), and common metal contaminants (zinc, copper, aluminum) can be broken down efficiently by the actinobacterial action and thus make the agricultural land suitable for cultivation (Ali et al. 2021; Silambarasan et al. 2022). They remediate these components through accumulation, chelation, and biotransformation-based degradation. In water stress conditions, in the case of rice, and tomato cultivation, the inoculation of *Streptomyces* species ensures superior survival chances and less hamper to standard harvests. The treated plants are known to have high contents of proline, polysaccharides, ectoine, soluble sugars (fructose), improved removal of reactive oxygen species, less occurrence of ethylene, enhanced absorption of available nutrients, and supportive levels of phytohormones (Abbasi et al. 2020; Niu et al. 2022). The issues of high salinity in the cultivable soil can be tackled through the application of *Streptomyces albidoflavus* OsiLf-2, *Streptomyces paradoxus* D2-8, *Streptomyces venesuelae* ATCC 10712, and other species in case of corn, rice, vine, soybean, tomato, etc., cultivations. These isolates produce indole-like compounds, phenazines, choline, 1-aminocyclopropane-1-carboxylate deaminase, organic acids, glycine betaine, siderophores, solubilizes phosphates and maintain the plants' healthy physiology. The antioxidant-related enzymes like catalase, peroxidase, and malondialdehyde contents were in an improved state in the treated plants than in the nontreated control sets. The root and shoot length remained normal, the rate of photosynthesis was standard, and the fresh and dry weight was maintained at healthy values (Nozari et al. 2022; Karimi and Noori 2022; Niu et al. 2022; Nazari et al. 2023).

3.2.2 Role of Actinobacteria in Managing Biotic Stress in Crops

Actinobacterial metabolites are effective biocontrol agents and can provide broad-spectrum protection to cash crops. This topic is discussed in the following section.

3.2.2.1 Biocontrol Abilities of Actinobacteria Toward Phytopathogens

3.2.2.1.1 Antibacterial Activity

Actinomycetes, especially different species of *Streptomyces,* are potent contributors of antibiotics (>75%), bactericides, insecticides, fungicides, acaricides, and herbicides used randomly in both agriculture and pharmaceuticals (Berdy 2005; Chouyia et al. 2022). They are one of the most critical contributors to the discovery of novel natural products (Khan et al. 2023). As a biocontrol agent to combat bacterial infection actinobacterial spores in the form of cell-free culture extracts, suspensions, granulated/coated forms, in the form of emulsions or dry powders can be applied on soil/agricultural fields or in leaf tissues/stems, and even in growing flowers/fruits, post-harvest storage/transportation conditions can be applied.

Actinomycetes are thread-like in appearance in their vegetative structure, and this structural uniqueness has made them functionally stable to successfully colonize rhizospheric soils; thus, they can invade plant cells proliferating hosts' growth, nutrient acquisition, and defense against harmful pathogens (Al Hamad et al. 2021). Aloesaponarin II obtained from *Streptomyces termitum* ATC-2 effectively stops the growth of *Xanthomonas oryzae* pv. *oryzae*; responsible for bacterial blight of rice. Another isolate of *Streptomyces* RM 365 exhibited antibacterial activity against *Xanthomonas campestris* pv. glycines, a causal agent of pustules in *Glycine max.* The causal agent of soft rots of potato (*Solanum tuberosum*) *Pectobacterium carotovorum* and *Pectobacterium atrosepticum* was effectively blocked by the action of *Streptomyces* sp. OE7. Actinobacteria synthesizes antibacterial peptides valinomycin (cyclipeptides) and kutznerides with strong effectivity (Zhang et al. 2020).

Kaari et al. (2022) prepared a farmyard manure supplemented with *Streptomyces* isolates and reported that the formulation effectively blocks the infection of bacterial wilt pathogen *Ralstonia solanacearum* in tomato seedlings. Plants provided with *Streptomyces* inoculated manure exhibited a higher degree of growth and elevated levels of defense-related enzymes peroxides, polyphenol oxidase, and phenylalanine ammonia-lyase (PAL). The root and shoot length improved, and the chlorophyll contents and phenol contents also enhanced. Thus, the isolate UP1-A when fortified to farm yard manure is an effective ecofriendly bio-inoculant to alleviate bacterial wilt disease in cash crops like tomato.

Streptomyces sp. JCK-6131 produced some potent antimicrobial compounds streptothricin E, streptothricin D, and 12-carbamoyl streptothricin (as detected from reverse-phase high-performance liquid chromatography and electron spray ionization-mass spectrometry mass spectrometry analysis) that exhibited strong antibacterial activity against a panel of microorganisms *Acidovorax avenae* subsp. *Cattleyae* (causing-seedling blight and bacterial fruit blotch in plants of Cucurbitaceae family), *Acidovorax konjaci* (causal agent of the blight of leaf in *Amorphophallus rivieri*), *Agrobacterium tumefaciens* (causes crown gall disease), *Burkholderia glumae* (panicle blight in rice), *Clavibacter michiganensis* subsp. *Michiganensis* (bacterial wilt and canker of tomato), *Erwinia pyrifoliae* (cause Asian pear blight), *Erwinia amylovora* (fire blight in members of Rosaceae), *Pectobacterium carotovorum* subsp. *carotovorum* (cause bacterial soft rot in vegetables), *Pectobacterium chrysanthemi* (vascular wilt pathogen), *Pseudomonas syringae* pv. *lachrymans* (develop angular leaf spot in cucumber), *Pseudomonas syringae* pv. *actinidiae* (cause bacterial canker), *Ralstonia solanacearum* (bacterial wilt pathogen), *Xanthomonas euvesicatoria* (bacterial leaf spot of tomato and pepper), *Xanthomonas axonopodis* pv. *citri* (citrus canker), *Xanthomonas arboricola* pv. *pruni* (bacterial spot of stone fruits), and *Xanthomonas oryzae* pv. *oryzae* (bacterial blight of rice) with a minimum inhibitory concentration of 0.31–10%. *In planta* application (foliar spray) of the strain caused a sharp decrease in the incidents of fire blight of apple and bacterial wilt of tomato by uplifting the disease resistance in the treated seedlings. There was a surge in the pathogenesis-related proteins (PR) PR1, PR3, PR5, PR12 and switching on of the salicylate and jasmonate signalings. Thus, JCK-6131 acts both in the form of antibiosis and induction of plants' internal defense (Le et al. 2021). *Streptomyces* sp. J145, *Streptomyces termitum* ATC-2, *Streptomyces humidus* through its synthesized metabolites α-1-Sorbofuranose and β-D-altrofuranose, aloesaponarin II, and phenylacetic acid and sodium phenylacetate blocked *Xanthomonas campestris* pv. *campestris*, *Xanthomonas oryzae* pv. *oryzae*, and *Pseudomonas syringae* pv. *syringae*; pathogens of cabbage black rot, bacterial blight of rice, and bacterial canker of almonds, respectively (Solanki et al. 2016). *Streptomyces diastatochromogenes* inhibited the causal agent of bacterial soft rot in potato, carrot, and cabbage *Erwinia carotovora* through its synthesized component oligomycin (Doolotkeldieva et al. 2016). Components like actinomycin D, cycle (I-Pro-Tyr) and cycle (α-Pro-I-Tyr), 2-methyl-N-(2- phenylethyl)-butanamide, and 3-methyl-N-(2- phenylethyl)- butamide produced from *Streptomyces hydrogenans* IB310, *Streptomyces* strain 22–4, and *Streptomyces* sp. PNM-9 are strong antibacterial against cabbage (black rot, soft rot), almond (crown gall), rice (panicle blight), and tomato (bacterial wilt) pathogens (*Agrobacterium tumefaciens*, *Xanthomonas axonopodis*, *Ralstonia solanacearum*, *Calvibacter michiganensis*, *Pseudomonas syringae*, *Xanthomonas campestris*, *Burkholderia glumae* and *Burkholderia gladioli*), respectively (Kulkarni et al. 2017; Wattana-Amorn et al. 2016; Betancur et al. 2019). Antibacterial components α-L-sorbofuranose (3–2)- β-D-altrofuranose, n-hexadecanoic acid obtained from *Streptomyces* strain FJAT-31547, and *Streptomyces* strain JJ45 inhibited the growth of *Ralstonia solanacearum* (causal agent of bacterial wilt of tomato) and *Xanthomonas campestris* pv. *campestris* (causal agent of black rot of cabbage), respectively (Muangham et al. 2015; Zheng et al. 2019). Le et al. (2021, 2022) reported anti-*Ralstonia solanacearum*, *Xanthomonas euvesicatoria*, *Sclerotinia homeocarpa* activity of dimethyl sulfide, trimethyl sulfide, streptothricin E acid, streptothricin D and 12-carbamoyl streptothricin D produced from *Streptomyces* sp. AN090126, and *Streptomyces* sp. JCK-6131, respectively. These metabolites eliminate the infections related to leaf spot of pepper, bacterial wilt of tomato, dollar spot of bentgrass, and blight of apple. Cyclo-tetrapeptides, diketopiperazines (produced from *Streptomyces* sp. 161a) were effective against bacterial blight of rice pathogen *Burkholderia glumae* and *Burkholderia gladioli* (Betancur et al. 2019).

3.2.2.1.2 Antifungal Activity

Actinobacteria are potent biocontrol agents, and their synthesized metabolites are effective antimicrobial components; thus, they can be commercialized for sustainable agricultural production (Nazari et al. 2023). They are prolific producers of bioactive metabolites. Majorly, the antibiotics produced by the actinobacteria are the products of polyketide synthatase genes. They counteract the pathogens in different ways by synthesizing antibiotics, secreting enzymes, parasitizing on other microbes, competing for necessary nutrients, and internally modifying plants' defense to combat stress (Kaur et al. 2023). Because of the many modes of action to combat parasitic infection, actinobacteria are known to be promising biocontrol agents for sustainable agricultural practice (Giacomelli Ribeiro and Teresinha Van Der Sand 2024). Azalomycin, a natural polyhydroxy macrolide synthesized by different *Streptomyces* strains, exhibits broad-spectrum antimicrobial action. Cheng et al. (2010) isolated azalomycin from *Streptomyces malaysiensis* MJM1968 that expressed strong antifungal activity against agricultural pathogens *Fusarium oxysporum*, *Fusarium chlamydosporum* (cause infection in tropical fruits such as mango, banana, papaya, pineapple, and avocado), *Rhizoctonia solani* (causal agents of root rot, damping off, and wire stem in Solanaceae members), *Cladosporium cladosporioides* (majorly *Cladosporium* rot in red wine grape vines), *Colletotrichum gloeosporioides* (anthracnose in trees and grasses), *Pestalotia* spp. (leaf spot and necrosis), and *Alternaria mali* (causing *Alternaria* blotch of apple orchards) (Zakaria 2023; Adeniyi and Asogwa 2023; Li et al. 2019a, b). Different products have been commercialized from actinobacteria with antifungal activity. For instance, Mycostop, a product developed in Finland from *Streptomyces griseoviridis* K61, exhibits anti-*Alternaria* sp., *Rhizoctonia solani*, *Fusarium* sp., *Phytophthora* sp., *Pythium* sp. activity protecting root diseases, and damping off. *Streptomyces lydicus* WYEC 108 synthesized metabolites are effective against major soil-borne pathogens, powdery-downy mildews *Pythium* sp., *Fusarium* sp., *Phytopthora* sp., *Rhizoctonia* sp., *Verticillium* sp., *Geotrichium* sp., *Sclerotinia* sp. and commercialized as Actinovate, Novozymes (US-based product). Another product, Mykocide KIBC (formulated in Korea), is prepared from *Streptomyces colombiensis* and is heavily effective in controlling powdery mildews, gray molds, and leaf-brown patches. Another South Korean product, Safegrow KIBC, obtained from *Streptomyces kasugaensis* is effective against sheath blight, and large patch diseases in agricultural production. Another product Kasumin, Kasurab (active principle Kasugamycin), obtained from *Streptomyces albus*, controls leaf spots, scabs, and root rots in sugar beet, celery, pears and apples, and soybean, respectively. Other two species *Streptomyces albus* and *Streptomyces griseochromogenes* were exploited for the production of Bactophil, Blasticidin-S, and these can effectively control seed germination-related diseases, and also *Pyricularia oryzae* infection. The most common antibiotic streptomycin, marketed in the name of Agrimycin, Paushak, Cuprimycin 17, AAstrepto 17, AS-50, Dustret, Cuprimic 100, and 500, was obtained from *Streptomyces griseus* and protects the crops from bacterial rots, canker (i.e., pathogenic infections of *Xanthomonas oryzae*, *Xanthomonas citri*, *Pseudomonas tabaci*) in fruits like pomegranate, stone fruit, citrus, and olives, vegetables like potatoes, tobacco, cotton, and ornamentals. The formulation of *Streptomyces* sp. Di-944 reduces damping off in *Lycopersicum esculentum* caused due to *Rhizoctonia solani* infections (Sabaratnam and Traquair 2002). Phytomycin is known to act as a potent antimicrobial against fire blight pathogen *Erwinia amylovora*, *Pseudomonas* sp., *Xanthomonas* sp., and also mycoplasma-like organisms. Other popular actinobacterial products validamycin, polyoxorim, and natamycin with the trade names of Validacin, Valium, Endorse, Polybelin, and Delvolan produced from *Streptomyces cacaoi* var. *asoensis*, *Streptomyces natalensis*, *Streptomyces chattanoogensis,* and block the infection of phyto-pathogens daffodil basal rot agent *Fusarium oxysporum*, *Sphaerotheca* spp., powdery mildew pathogens *Botrytis cinerea*, ergot disease of rye- *Sclerotinia sclerotiorum*, *Corynespora melonis*, Tikka disease of ground nut *Cochliobolus miyabeanus,* respectively. Mildiomycin obtained from *Streptomyces rimofacienes* exhibits antifungal activity against powdery mildew pathogens *Erysiphe* sp., *Microsphaera* sp., *Oidium* sp., *Leveillula* sp., and *Sphaerotheca* sp. (Kishimoto et al. 1996). Faerifungin isolated from *Streptomyces griseus* blocks *Fusarium oxysporum* and protects *Asparagus* root disease (Kobinata et al. 1993). Malayamycin and Gopalamycin were obtained from *Streptomyces malaysiensis,* and *Streptomyces hygroscopicus* is antifungal against *Mycosphaerella graminicola*, *Puccinia triticina,* and *Phytophthora capsica* and protects the wheat and pea plants from brown stem rust and root rot disease, respectively (Li et al. 2008; Sheng et al. 2015). The antifungal activity of some recently selected actinobacterial isolates is presented in Table 3.2.

Khilyas et al. (2024) reported the antifungal abilities of cell-free metabolites of *Nocardia mangyaensis* NH1 against phytopathogenic *Colletotrichum coccodes* MF 16-014, *Rhizoctonia solani* MFP 936011, *Fusarium oxysporum* PR 57, and *Alternaria* sp. Also, the metabolites (i.e., siderophores hydroxamate and catechol, indole-3-acetic-acid, lipo, and cyclic peptides) are antioxidative in action and can chelate irons. Thus, it can be used as an external applicant to improve plant growth and development, modulate productivity of soil, and also make available the macro- and micronutrients. Abbasi et al. (2019, 2020) reported the plant growth-promoting and biocontrol abilities of a *Streptomyces* species *Streptomyces rochei* Y28 against a detrimental fungal pathogen of tomato wilt *Fusarium oxysporum* f.sp. lycopersici race 3 and documented the drought stress alleviation property through uplifting the transcription factors ERF1 and WRKY70. *Streptomyces vinaceusdrappus* is antagonistic to two major pathogens of tomato and rice, that is, *Rhizoctonia solani* and *Pyricularia oryzae,* respectively (Yandigeri et al. 2015). It produces chitinolytic substances and upon supplementation of an external chitin source, the isolate protects the host plant. 2,4-ditert-butylphenol, 1-ethylthio-3-methyl-1,3-butadiene, 2-(chloromethyl)-2-cyclopropyl oxirane, (Z)-5-(2-methylphenyl)-4-pentenoic acid, (Z)-7-(2-methylphenyl)-6-heptenoic acid, anguinomycin A, leptomycin A, actinopyrone A, (4E,8E)-4,10,12-trimethyl-4,8-diene-13-oxo tetradecanoic acid, salvianolic acid B, 5-hydroxymethyl-2- furancarboxaldehyde, 1-Nonadecene, methyl anthranilate,,2′-methylenebis [6-(1,1-dimethylethyl)-4-methyl-, N-hexadecanoic acid, Azalomycins F4a and Azalomycins F5a, bafilomycin B1, actinolactomycin,

Table 3.2 Biocontrol potential of different actinobacterial isolates against plant pathogenic fungi

Name of the actinobacteria	Controlling pathogen	References
Streptomyces flavotricini 25, *Streptomyces globisporus* subsp. *globisporus* C28, *Streptomyces globisporus* 7, *Streptomyces senoensis* 420, *Streptomyces pactum* Act12	*Sclerotium rolfsii*	Li et al. (2017)
Streptoverticillium morookaense	*Ustilaginoidea virens*, *Rhizoctonia solani*, *Bipolaris maydis*	Andargie and Li (2019)
Curtobacterium sp. XA15–35	*Botrytis cinerea*, *Sclerotinia sclerotiorum*	Xu et al. (2019)
Streptomyces violaceusniger XL-2	Leaf blight pathogen- *Coriolus versicolor*, *Phanerochaete chrysosporium*	Chaiharn et al. (2020)
Streptomyces sp. A17 (produced metabolite antimycin)	Soybean blight pathogen- *Xanthomonas campestris* pv. *glycines*	Chaiharn et al. (2020)
Streptomyces sp. PC 12, *Streptomyces* sp. D 4.1, *Streptomyces* sp. D 4.3, *Streptomyces* sp. W1 (produced metabolite-pyrroles (Pyr-roles [1,2- a] pyrazine-1,4-dione	Brown spot pathogen- *Pyricularia* sp.	Chaiharn et al. (2020)
Streptomyces sp. YYS-7	*Fusarium oxysporum* f. sp. cubense race 4	Wei et al. (2020)
Streptomyces albidoflavus OsiLf-2	*Magnaporthe oryzae*	Gao et al. (2020)
Streptomyces nigrogriseolus GanoSA1	*Ganoderma boninense*	Shariffah-Muzaimah et al. (2020)
Streptomyces violaceusniger JBS5-6	*Fusarium oxysporum* race 4, *Colletotrichum acutatum*, *Curvularia fallax*, *Fusarium oxysporum* sp. *cucumebrium*, *Pyricularia oryzae*, *Colletotrichum gloeosporioides*, *Fusarium graminearum*, *Botryosphaeria dothidea*, *Curvularia lunata*, *Colletotrichum fragariae*, *Botrytis cinerea*	Jing et al. (2020)
Streptomyces sp. M4	*Alternaria brassicicola*, *Alternaria solani*, *Fusarium oxysporum*, *Colletotrichum acutatum*, *Colletotrichum gloeosporioides*	Sharma and Manhas (2020)
Streptomyces violaceusniger G10 (produced metabolite propanoic acid)	Banana wilt pathogen wilt *Fusarium oxysporum*	Chaiharn et al. (2020)
Streptomyces sp. CACIS-1.5CA	*Rhizoctonia* sp., *Rhizopus* sp., *Colletotrichum* sp., *Alternaria* sp., *Aspergillus* sp., *Botrytis* sp.	Evangelista-Martínez et al. (2020)
Streptomyces yanglinensis	*Aspergillus flavus*, *Aspergillus parasiticus*	Lyu et al. (2020)

(continued)

Table 3.2 (continued)

Name of the actinobacteria	Controlling pathogen	References
Nocardiopsis aegyptica H14	*Fusarium oxysporum* f. sp. *radicis* lycopersici, *Rhizoctonia solani*	Djebaili et al. (2021)
Streptomyces salmonis	*Colletotrichum gloeosporioides*	Boukaew et al. (2021)
Streptomyces sp. TP199	*Pectobacterium carotovorum* subsp. *Carotovorum*, *Pectobacterium atrosepticum*	Padilla-Gálvez et al. (2021)
Streptomyces sp.	*Colletotrichum fragariae*	Li et al. (2021a, b)
Amycolatopsis sp. 1119	*Pythium ultimum*, *Rhizoctonia solani*, *Fusarium oxysporum*	Alipour Kafi et al. (2021)
Streptomyces viridodiasticus (produced metabolite natamycin; 4-phenyl-3-butenoic acid)	Basal drop of lettuce *Sclerotinia minor*	Gao et al. (2021)
Streptomyces sp. A1022 (produced metabolite lucensomycin)	Anthracnose of pepper and cherry tomato *Colletotrichum gloeosporioides*	Gao et al. (2021)
Streptomyces spp. 47 W08, 47 W10 (produced metabolite avermectins)	Pepper blight pathogen *Phytophthora capsici*	Yandigeri (2021)
Streptomyces sp. S30 (produced metabolite rapamycin)	Damping off of tomato *Rhizoctonia solani*	Gao et al. (2021)
Streptomyces violaceusniger YCED9 (produced metabolite Natamycin; 4-Phenyl-3-butenoic acid)	Crown disease of turfgrass *Rhizoctonia solani*	Gao et al. (2021)
Streptomyces sp. H4	*Colletotrichum fragariae*	Li et al. (2021a, b)
Streptomyces sp. (PNM-149)	*Colletotrichum gloeosporioides*	Gómez et al. (2021)
Streptomyces albidoflavus H12	*Fusarium oxysporum* f. sp. *radicis lycopersici* and *Rhizoctonia solani*	Djebaili et al. (2021)
Streptomyces spp. M2A2	*Rhizoctonia solani*	Zambrano et al. (2021)
Streptomyces exfoliatus	*Aspergillus flavus*	El-Shanshoury et al. (2022)
Streptomyces cangkringensis CIAD-CA07, *Streptomyces misionensis* CIAD-CA27, *Streptomyces kanamyceticus* CIAD-CA45	*Fusarium equiseti*, *Fusarium oxysporum*, *Alternaria alternata*, *Sclerotium rolfsii*	Pérez-Corral et al. (2022)
Amycolatopsis lexingtonensis CIAD-CA13, *Amycolatopsis lurida* CIAD-CA30	*Alternaria alternata*, *Sclerotium rolfsii*	Pérez-Corral et al. (2022)
Saccharopolyspora shandongensis CIAD-CA15	*Sclerotium rolfsii*	Pérez-Corral et al. (2022)
Nocardioides albertanoniae CIAD-CA20	*Fusarium oxysporum*, *Alternaria alternata*, *Sclerotinia sclerotiorum*	Pérez-Corral et al. (2022)

(continued)

Table 3.2 (continued)

Name of the actinobacteria	Controlling pathogen	References
Kocuria rhizophila PT10	*Botrytis cinerea, Fusarium graminearum*	Guesmi et al. (2022)
Streptomyces setonii WY228	*Ceratocystis fimbriata*	Gong et al. (2022)
Streptomyces sp. AN090126	*Ralstonia solanacearum, Xanthomonas euvesicatoria, Sclerotinia homoeocarpa*	Le et al. (2022)
Streptomyces fulvissimus Uts22	*Pythium aphanidermatum, Gaeumannomyces graminis* var. *tritici*	Saberi-Risesh and Moradi-Pour (2021), Saberi-Risesh et al. (2022)
Streptomyces malaysiensis 8ZJF-21	*Fusarium oxysporum* f.sp. *cubense*	Zhang et al. (2022)
Streptomyces spp.	*Fusarium* sp.	Aallam et al. (2022)
Streptomyces albidoflavus	*Fusarium oxysporum*	Du et al. (2022)
Streptomyces bikiniensis	*Magnaporthe oryzae*	Liu et al. (2022)
Streptomyces malaysiensis	*Colletotrichum gloeosporioides*	Zhou et al. (2022)
Streptomyces sp.	*Fusarium oxysporum, Ganoderma boninense*	Islamiati et al. (2022)

2,4-bis (1,1-dimethylethyl)-phenol, (Z)-13-docosenamide, oleic acid, tetradecanoic acid, dimeric dynactin, tetranactin, dynactin, maremycin G, and its analog are some of the common antifungal metabolites isolated from efficient actinobacterial isolates (majorly different species of *Streptomyces S. palmae* CMU-AB204T, *S. violaceusniger* JBS5–6, *Streptomyces* sp. SCA3–4, SN0280, M4, FJAT-31547, PNM-149, H4, YYS-7 2, *S. sporoclivatus* DAAG3-11, *S. cavourensis* DGS1-1, *S. pratensis* DGS3-15, *S. capitiformicae* DDPA2-14) (Giacomelli Ribeiro and Teresinha Van Der Sand 2024).

3.2.2.1.3 Role of Actinobacteria in Tackling Plant Pathogenic Virus

Actinobacteria are all-square in action and are effective producers of antiviral compounds that protect crops from serious diseases (Trojak-Goluch 2024). Actinobacteria function in multiple strategies to combat viral infection by producing antiviral secondary metabolites, inducing systemic resistance, producing hydrolytic enzymes, inhibiting viral enzymes, outcompeting the viral population, forming biofilm, and colonizing rapidly, etc. Cell-free metabolites of *Streptomyces* sp. can inhibit the infection of potato virus Y and tobacco mosaic virus in *Chenopodium quinoa* and *Nicotiana glutinosa,* respectively. Han et al. (2014) reported the anti-TMV action of a cytosine nucleotide peptide ningnanmycin, isolated from *Streptomyces noursei* var. xichangensis. This antibiotic is strongly antifungal, and antibacterial and also blocks the proliferation of plant pathogenic virus. This potent antibiotic induces systemic acquired resistance (by accumulating the pathogenesis-related proteins) in tobacco plants and can positively modulate the treated seedlings' pigment profile (chlorophyll and carotenoid). An elevated carotenoid content assures the protection of photosynthetic apparatus and ningnanmycin blocks the polymerization of viral CP protein. The different defense-related signaling cascades (upregulation of NPR1, JAZ3, and jasmonic acid signaling) were activated upon application of ningnanmycin. The isolate is non-phytotoxic, and further field studies are required for application as an anti-TMV tool. The cell-free metabolites from the broth culture of *Streptomyces* sp. ZX01, *Streptomyces erythraeus*, *Streptomyces naganishii*, and *Streptomyces michiganensis* have been reported to block TMV (Mohamed and Galal 2005; Zhang et al. 2015). Shafie et al. (2016) reported the potential of culture filtrates of *Streptomyces griseorebens* and *Streptomyces cavourensis* (through foliar application) to protect the cucumber plants from the cucumber mosaic virus. The major inhibition of the virus was detected upon soaking of seeds with the cell-free metabolites. There was a surge in phenol contents, peroxidase, and polyphenol oxidase action.

Streptomyces cellulosae isolate Actino 48 is known to delay the development of tobacco mosaic virus infection in tomato plants (Abo-Zaid et al. 2020). The application of cellular suspension of Actino 48 on the infected tomato plants is reported to increase the antioxidant enzyme activities (peroxidase) and defense compounds (like phenol). It has elevated defense-related gene expressions (i.e., systemic acquired resistance development) PAL, CHS, and PR1. *Streptomyces sampsonii*, *Streptomyces rochei*, and *Streptomyces griseus* exhibit anti-Zucchini yellow mosaic virus activity and protect squash (*Cucurbita pepo* L.) from viral infections (Ghanem et al. 2023). The isolates induced systemic acquired resistance by elevating the levels of proline, phenols, and stress marker enzymes like catalase, superoxide dismutase, and glutathione peroxidase. The cell-free extracts of these actinobacteria are rich in docosanol and downregulate P1 gene response, which is directly linked with the events of the central dogma of the virus.

Streptomyces chromofuscus Strain RFS-23 is known to induce systemic acquired resistance in *Nicotiana benthamiana* against tomato yellow leaf curl virus (Chen et al. 2022). The defense-related genes (NbCIS, NbNCED) were upregulated, and genes associated with the genesis of chlorophyll (CLH1, HEMA1, PORA), abscisic acid, and salicylic acid were also switched on. Catalase, peroxidase, superoxide dismutase, and malondialdehyde contents were modulated in favor of the plant's survival. *Streptomyces pactum* Act12 inhibits the infection of tomato yellow leaf curl virus in *Lycopersicum esculentum* plants (Li et al. 2019a, b). Rhizospheric pre-inoculation of this strain upregulates the specific genes (PR-I, SIPI-II) related to systemic resistance (SIPer1, SIVRSLip) from the viral pathogen. In vivo experimentation confirms the enhancement in peroxidase, polyphenol oxidase, and phenylalanine ammonia-lyase action of enzymes as detected in tomato leaves. The levels of H_2O_2 and O^{-}_2 were reduced in treated plants. The application of Act12 increased the rhizospheric bacterial diversity and a higher occurrence of effective microorganisms *Trichoderma* sp. and *Bacillus* sp. were detected. This emphasizes the fact that actinobacteria act as the plant probiotic that can both directly and indirectly influence plants' survival against viral infection. Taha et al. (2021) reported the anti-tomato mosaic virus effect of cell-free metabolites of *Streptomyces ovatisporus* LC597360 in *Solanum lycopersicum* cv. Supra. There was an upliftment in plant growth, and vigor and an increase in defense-related enzymes, that is, ascorbate oxidase, catalase, peroxidase, and polyphenol oxidase. *Streptomyces ahygroscopicus*, *Streptomyces* sp. ZX01, *Streptomyces noursei* var. *xichangensis* exhibit anti-*TMV* action in tobacco plants (Manjunatha et al. 2022). ε-poly-l-lysine (ε-PL) is synthesized by *Streptomyces ahygroscopicus* (Xiang et al. 2010). Latake and Borkar (2017) reported the anti-cucumber mosaic virus activity of 28 actinomycetes. *Streptomyces olivaceus* effectively blocks the infection of cauliflower mosaic virus both in vitro and in vivo. *Streptomyces rochei*-IS-02, *Streptomyces caeaoi*-IS-03, and *Streptomyces aureocirculatus*-IS-01 are strong antiviral producers against tobacco mosaic viruses (Akl and El-Masry 2024). *Streptomyces albulus* SN40 produces anisomycin with strong antiviral activity against potato virus Y (Shu et al. 2024). Thus, it could be concluded that both the metabolites from actinomycetes and their direct application can protect agricultural crops from viral pathogens.

3.2.2.2 Actinobacteria Against Plant Parasitic Nematodes

Nematodes are one of the most abundant invertebrates and account for four-fifths of all animals in terrestrial ecosystems. Though they are major contributors to nutrient cycling in the rhizosphere but sometimes also appear to be a potential threat to agricultural production, causing a loss of USD173 billion (Pires et al. 2022). Chemical-based treatments are not supported nowadays as they come with irreversible impacts on human and ecosystem health. Bacterial and fungal members have been randomly studied to control nematicidal infections (against major phytonematodes, i.e., *Radopholus similis*, *Meloidogyne* spp., *Rotylenchulus reniformis*, and *Ditylenchus dipsaci,* etc.) in crops as a novel bio-based technique, but a few reports are available regarding actinomycetes. Atif et al. (2023) reported five efficient chitinase-producing actinobacterial isolates, that is, *Streptomyces levis* ST01, *Streptomyces enissocaesilis* ST07, *Streptomyces plicatus* ST10, *Streptomyces rochei* ST15, and *Streptomyces sparsus* ST16 that can reduce the infection of root-knot nematode *Meloidogyne incognita* (cause drop of production in brinjal *Solanum melongena*, tomato *Solanum lycopersicum*, potato *Solanum tuberosum*, carrot *Daucus carota*, watermelon *Citrullus vulgaris*) in eggplant. Actinobacterial metabolites can tackle the nematode infection both by direct interaction or by indirectly developing host plants' resistance and decreasing their invasion. Khanna et al. (2019a, b) reported anti-*Meloidogyne incognita* action of plant growth-promoting rhizobacteria through modulating host plants' (tomato *Lycopersicum esculentum*) defense. Hu et al. (2022) reported the nematicidal (against root-knot nematode) ability of the cell-free metabolites (polar macromolecules sugar, and phenolic hydroxyl compounds) of *Streptomyces albogriseolus* ND41 and *Streptomycces fimicarius* D153 with a mortality rate of 82% (post 12 h incubation). In the pot experiment, there was a 68.89% reduction in the nematode population after 1 month of treatment (100 mL inoculation per pot) with ND41 and a 74.44% reduction upon treatment with D153. Metabolites of *Streptomyces avermitilis* Manp blocked the growth of *M. incognita* up to 89.54% upon 72 h of exposure (Nimnoi and Ruanpanun 2020). Meidani et al. (2020) evaluated the nematicidal effect (against second-stage juveniles of *Meloidogyne javanica* and *Meloidogyne incognita*) of culture extracts of three actinomycetes and *Streptomyces monomycini* ATHUBA 220, *Streptomyces colombiensis* ATHUBA 438 had higher rates of inhibition in comparison to *Streptomyces youssoufensis* ATHUBA 546. Initially, they screened 461 *Streptomyces* species obtained from Greek rhizosphere soils. Cell-free extracts not only paralyzed the larvae but also their life cycle was found to be arrested when checked in vivo in model plants (*Arabidopsis thaliana*) wild type Col-0 and the katanin mutant fra2 (susceptible to *M. incognita*). Metabolites were slightly effective against *M. javanica* infection in comparison to *M. incognita*. Actinomycin was produced from *Streptomyces antibioticus* and was known to have anti-nematode action (Sharma et al. 2019). Also, different isolates *Streptomyces cacaoi* and *Streptomyces albogriseolus* produce 3-benzyl-1,4-diaza-2,5- dioxobicyclo[4.3.0]nonane, fungichromin with strong nematicidal action (Yoon et al. 2012; Zeng et al. 2013). *Streptomyces enissocaesilis* OM182843, with its chitinolytic, proteolytic, and gelatinolytic activity, is cidal to root-knot nematode (El-Akshar et al. 2022).

Not only different larval stages but also the pre-hatching condition (i.e., eggshells of nematode) and surface coats are the primary targets for major biocontrol agents. As their eggshells are made up of cuticles, lipids, chitin, and proteins (Raja and Dhanasekaran 2022), actinobacteria with chitinolytic action and hydrolytic enzyme-producing abilities can be a good option to be used as biocontrol agents for nematode control. *Streptomyces nigrescens* AMV 1033 is a chitinase-synthesizing one and can reduce the nematode population by breaking down nematode eggshells (Eberlein and Westphal 2023). Actinobacteria can colonize plant roots and can parasite on the eggs of nematodes, and they can parasitize nematode eggs. Jin et al. (2017) documented *Streptomyces rubrogriseus* from *M. incognita* eggs, which describes the parasitic ability of the species on nematode eggs. Also, the culture filtrates of *Streptomyces rubrogriseus* and *Streptomyces cacaoi* encountered egg hatching through chitinase action. Teleocidin B4, spectinabilin, aureothin, alloaureothin from *Streptomyces* sp. 680,560, *Streptomyces* sp. AN091965, and endophytic *Streptomyces* sp. AE170020 had anti-pine wood nematode activity *Bursaphelenchus xylophilus* (Liu et al. 2019; Kang et al. 2021, 2022). Other metabolites, that is, fervenulin from *Streptomyces* sp. CNU-MHO21, are also effective nematicidals (Ruanpanun et al. 2011a, b). Avermectin obtained from *Streptomyces avermitilis* G8-17 is a strong nematicidal component against a variety of pathogens *Meloidogyne incognita* (root-knot nematode), *Pratylenchus zeae* (pathogenic to crops like *Solanum tuberosum, Triticum aestivum, Nicotiana tabacum, Coffea arabica, Saccharum officinarum*), *Heterodera schachtii* (beet cyst eel worm-pathogenic to *Beta vulgaris*), *Tylenchulus semipenetrans* (pathogenic to citrus plants–root nematode), *Radopholus similis* (burrowing nematode infecting *Persea americana*, *Cocos nucifera*, *Musa acuminata*), and *Bursaphelenchus xylophilus* (pine wilt nematode infects *Pinus* sp.). Also, it is effective against pests, insects, and mites of different taxon *Coleoptera*, *Hymenoptera*, *Diptera*, *Orthoptera*, *Isoptera*, and *Lepidoptera*, (Cerna-Chávez et al. 2024). Nemadectin, actinomycins, milbemycin, 3-methoxy-2-methyl-carbazole-1,4-quinone, carbazomycins D and F, angumycinones, fungichromin B extracted from *Streptomyces microflavus* NEAU3 *Streptomyces antibiotics* M7, *Streptomyces bingchenggensis* BCJ60, *Streptomyces* sp. CMU-JT005, *Streptomyces* sp. P294, *Streptomyces albireticuli* HA10002, *Streptomyces* sp. SH2 have acaricidal and nematicidal (*Tetranychus cinnabarinus*, *Bursaphelenchus xylophilus*) activity, respectively (Xiang et al. 2010; Ruanpanun et al. 2011a, b; Zeng et al. 2013; Kaur et al. 2016; Sharma et al. 2019).

Other than larvicidal or antimicrobial action the actinomycetous metabolites are potent phytotoxic and insecticidal agents also. Some examples have been presented here. *Streptomyces* sp. KR0006 and *Streptomyces* sp. KRA18–249 are insecticidal against *Plutella xylostella* and herbicidal against *Digitaria ciliaris,* respectively (Kim et al. 2022; Umurzokov et al. 2022). The metabolites are antimycins A1a, A8a, and A3a, and rubiginone D2. Cho et al. (2022) also reported the herbicidal (toward *Digitaria ciliaris*) action of three isolates *Streptomyces* sp. W12, W177, W200. Insecticidal (*Aedes albopictus* and *Plutella xylostella*) activity of *Streptomyces* sp. AN120537 was reported due to the presence of antimicyns A1, A2, A3, A4, A5 in the synthesized culture filtrate (Kim et al. 2020). Won et al. (2016) demonstrated the herbicidal action of *Streptomyces scopulidiris* toward *Echinochloa oryzoides*, *Digitaria ciliaris*, *Abutilon theophrasti*, and *Amaranthus retroflexus*. Priyadharsini et al. (2013) reported that N-phenylpropanamide obtained from *Streptomyces* sp. KA1-3 is phytotoxic toward *Celtis occidentalis* and *Cyperus rotundus*. *Echinochloa crusigalli*, *Amaranthus spinosus*, and *Cyperus rotundus* were found to be inhibited by the metabolites (diethyl 7-hydroxytrideca-2,5,8,11-tetraenedioate) of *Streptomyces* sp. DDBH019 (Priyadharsini et al. 2017). *Streptomyces diastatochromogenes* is insecticidal (through the application of synthesized metabolites Macrolide 5′-epi-SPA-6952 A) toward *Mythimna separata* with an LC_{50} of 10.26 mg L^{-1} (Cao et al. 2019). Plant parasitic nematodes (PPNs) are suppressed by bacteria through a variety of processes, which can help create efficient biocontrol methods to prevent nematode infestations. As seen in Fig. 3.1a, b, these pathways entail both direct and indirect antagonistic interactions between PPNs and bacteria (Ayaz et al. 2024). Through colonization, parasitism, and antibiosis—which includes the generation of lytic enzymes, antibiotics, poisons, and volatile organic compounds—studies have demonstrated that bacteria directly affect PPNs (Migunova and Sasanelli 2021).

Antimicrobial Activity Toward Phytopathogens Through Hyperparasitism or Secreting Hydrolytic Enzymes

Actinobacteria adopt various types of mechanisms to combat fungal or bacterial infections to protect crop plants. The production of antibiotics to eliminate

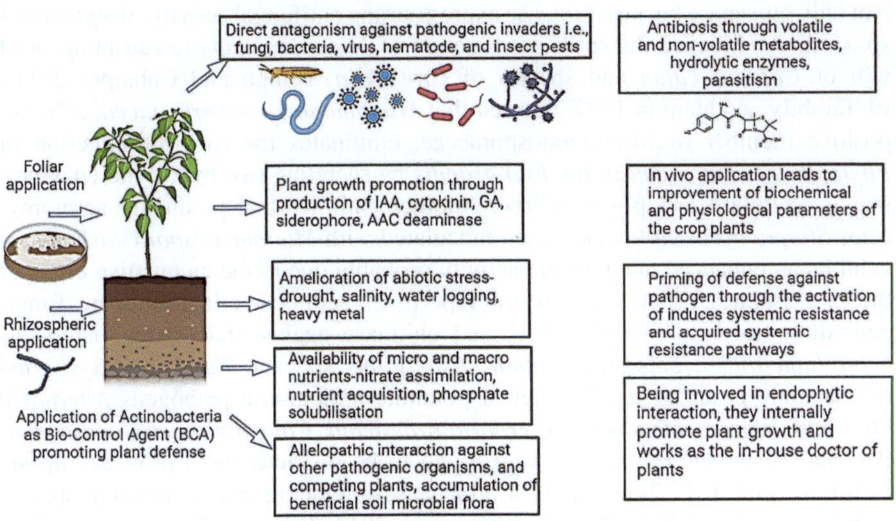

Fig. 3.1 Examples of bacterial and fungal biocontrol of PPNs. (**a**) Rhizospheric complex interaction and PPN biocontrol in infected plants. (**b**) Bacterial mechanisms of resistance to nematodes (**a–c**), (**a**) endoparasitic bacteria, (**b**) opportunistic bacteria that form crystals, and (**c**) rhizobacteria. The following are the ways that fungi work against nematodes: (**d**) nematode capturing or predatory fungi; (**e**) fungi that produce toxins; (**f**) endopafigrasitic fungi; (**g**) fungi that produce lytic enzymes for the inhibition of PPNs; and (**g**) opportunistic fungi that parasitize eggs and cysts. (Cited from Ayaz et al. 2024)

competitors is a common way of tackling infections. Apart from this, hydrolytic enzymes are also a technically effective way to counteract invading pathogens by degrading the cell wall, cell membrane, associated proteins, and extranuclear or extracellular virulence factors (Pal and Gardener 2006). The major cell wall degrading enzymes reported from actinobacterial sources are β-1, 3-glucanase, chitinase, amylase, pectinase, cellulase, xylanase, lipase, phospholipase, and protease, and these features make them suitable candidates to be employed as biocontrol agents (Rajendran et al. 2024). Anthracnose-causing fungi *Colletotrichum* sp. was reported to be inhibited in bell-pepper infection upon the implication of glucanase-producing actinomycetes *Streptomyces cavourensis* SY224 (Lee et al. 2012). Another report made by Shekhar and coworkers details the inhibition of the growth of a wood-rotting pathogenic fungi upon application of a chitinase-producing *Streptomyces* isolate *S. violaceusniger* XL-2 (Shekhar et al. 2006). The hunt for glucanase-producing actinomycetes was reported a long ago in 1997 and was known to inhibit the phytopathogen *Phytophthora fragariae* var. rubi. El-Tarabily and colleagues, while working on three different species of actinomycetes *Actinoplanes campanulatus*, *Micromonospora chalcea*, and *Streptomyces spiralis*, elucidate that the consortium ameliorates infections of *Pythium aphanidermatum* in vivo in cucumber seedlings. Also, the symptoms of damping off, crown, and root rot were removed to some extent even better than the utilization of metalaxyl, a popular chemical fungicide.

Not only glucanase but also protease express strong antifungal activity. *Streptomyces* sp. strain A6 blocked the spore germination of *Fusarium udum* (a causal agent of wilt of *Cajanus cajan* and species of *Crotalaria*) (Singh and Chhatpar 2011). El-Tarabily and team in 1997 reported that *Micromonospora carbonacea*, a Gram-positive member of Micromonosporaceae, eliminates the root rot infection of *Phytopthora cinnamomi* in *Banksia grandis* by secreting two hydrolytic enzymes, that is, cellulase and β-glucosidase. Another antimicrobial-producing actinomycetes *Streptomyces violascens* when inoculated with *Micromonospora carbonacea* exhibits an enhanced antagonistic activity probably due to the cumulative effort of the two isolates. There are several reports of actinobacteria producing fungi-degrading enzymes like glucanase and chitinase against devastating pathogens *Sclerotium rolfsii*, *Sclerotinia minor*, *Aspergillus* sp., and *Fusarium oxysporum* (El-Saadony et al. 2022). A panel of potential agricultural pathogens *Alternaria alternata*, *Rhizoctonia solani*, *Alternaria brassicola*, *Colletotrichum gloeosporioides*, and *Phytophthora capsici* were found to be restricted with protease, lipase, chitinase, and β-1, 3 and β-1, 4 glucanase-producing actinobacterial species (Srividya et al. 2012). In Ashokvardhan et al. 2014, Ashokvardhan and coworkers elucidated the antifungal potential of chitinase-synthesizing isolates obtained from the rhizospheric section of the chili plant. Not only pulses or grains but actinobacterial isolates can offer protection to highly demanding vegetables also, that is, fungal wilt of *Cucumis sativa* (cucumber) can be restricted by *Streptomyces bikiniensis* strain HD-087 by inducing the switching on of the defense-related proteins β-1,3, glucanase, phenylalanine ammonia lyase (PAL) and peroxidase in the host (Zhao et al. 2012).

There are also some reports on hyper-parasitism of *Streptomyces* species on potent crop pathogens. Actually, mycoparasitism from actinobacteria is a combined effect of both the hydrolytic enzymes and secreted antibiotics. Also, a single actinobacterial isolate exhibits different types of modes of action in vitro and in soil/in planta conditions. *Streptomyces griseus* confronts *Colletotrichum lindemuthianum* and grows on its mycelia, which leads to the development of multiple blebs, degeneration of hyphae, and perforations (Tu 1988). Also, there are reports regarding the production of multiple types of enzymes from a single strain. *Streptomyces phaeopurpureus* ExPro138 secretes a combination of proteases that blocks the adhesion, germination of spores of *Colletotrichum coccodes*, and also reduces the formation of the appressorium and eliminates the anthracnose in tomato (Palaniyandi et al. 2013).

Streptomyces griseoviridis K61 parasitizes on *Pythium* sp., *Rhizoctonia solani*, *Fusarium oxysporum*, and *Alternaria* sp. by perforating fungal hyphae and inhibiting conidial germination (Tapio and Pohto-Lahdenperä 1991). Different species of *Streptomyces,* that is, *S. cyaneofuscatus* ZY-153, *S. kanamyceticus* B-49, *S. rochei* X-4, *S. flavotricini* Z-13 effectively block parasitic fungi through coiling over the fungal pathogens and secreting lytic enzymes (Xu et al. 2019). Non-streptomycete strains like *Nocardiopsis dassonvillei* and *Micromonospora globosa* secrete antimicrobials, hydrolysable enzymes, and also parasitize *Fusarium oxysporum* f.sp.

albedinis, and *Fusarium oxysporum* through coiling and penetrating inside the hyphal cell wall. Also, other different strains of *Actinoplanes philippinensis*, *Actinoplanes missouriensis*, *Actinoplanes utahensis*, *Amorphosporangium auranticolo*, *Ampullariella regularis*, *Spirillospora albida*, and *Micromonospora* sp. inhibit the growth of common agricultural pathogens *Pythium coloratum*, *Phytophthora megasperma* var. *glycinea*, *Aphanomyces cochlioides*, *Aphanomyces euteiches*, *Phytophthora citrophthora*, *Phytophthora aphanidermatum*, *Pythium ultimum*, *Pythium megasperma* var. *glycinea* through antibiotics production, hydrolytic enzyme secretion, and ultimately performing parasitisim (Sutherland et al. 1984; El-Tarabily et al. 1997).

3.2.2.3 Volatile Organic Compounds (VOCs) from Actinomycetes Serve as Potent Antimicrobials Against Agricultural Pathogens

Actinobacteria are all-square in terms of antimicrobial production from cell-free culture extracts, that is, water-soluble metabolites to volatile organic compounds (Anavadiya et al. 2024). Though the mechanisms are similar, the degree of effectivity between the cell-free secondary metabolites and emitted volatiles differs slightly. VOCs are smoothly diffusible (at a room temperature of 28 ± 2 °C) low-molecular-weight lipophilic and slightly hydrophilic secondary metabolites that act as a chemical signaling between distant organisms (del Rosario et al. 2019). VOCs, for example, primary, secondary, and tertiary alcohols, aldehydes, ketones, alkanes-alkenes, aromatic hydrocarbons, esters, ethers, and furans, are abundant and are involved in various types of interaction in both aqueous, solid, and airy mediums, that is, attractant, repellent, symbiotic, competitive, growth inducing, allelopathic, antibiosis, and growth retarding, (Carmona-Hernandez et al. 2019). VOCs emitted from actinomycetes exhibit different attributes, that is, antifungal, plant growth promotor, nematicidal, and act as an elicitor of systemic resistance in treated green plants (Gebily et al. 2021; Gong et al. 2022). These components can be considered environmentally safe bio-fumigants that can tackle postharvest infections or storage-related contaminations. *Streptomyces albolongus* strain S9 synthesized propyl ester of octadecanoic acid that blocked the growth of tomato leaf spot pathogen *Corynespora cassiicola* (Devi and Rao 2017). Another isolate *Streptomyces platensis* F-1 showed static action against *Rhizoctonia solani*, *Sclerotinia sclerotiorum*, and *Botrytis cinerea* in vivo in rice, oilseed rape, and strawberries, respectively (Wan et al. 2008). VOCs (2-ethyl-5-methylpyrazine and dimethyl disulfide as detected from gas chromatographic analysis) synthesized from *Streptomyces setonii* WY228 exhibit strong antibiotic activity against *Ceratocystis fimbriata* (black spot pathogen of sweet potato *Ipomea batatas*) by inhibiting both the elongation of mycelia and germination of spores. VOCs actually blocked the expression of genes related to ribosome synthesis, spore formation, development of cell membrane and mitochondrial operation, and also synthesis of hydrolyzing enzymes and relevant toxin proteins of the devastating pathogen *C. fimbriata* (as confirmed from transcriptomic study). Also, it upregulated the 26 s proteasome machinery. In planta application of VOCs of WY228 for a period of 20 consecutive days provides a sharp change in the vigor of the treated plants in comparison to the non-VOC-treated ones. The contents of antioxidants increased sharply, defense-related enzymes were found to be elevated, and the flavonoids and starch contents boosted. The application of 2-ethyl-5-methylpyrazine at a dosage of 100 $\mu L\ L^{-1}$ for a period of 10 days totally eliminated the infection of *C. fimbriata* in sweet potato. Thus, actinobacteria as a bioinoculant or its derived volatiles as formulation can be a potent alternative for chemical fungicides even in storage or post-harvest pathogen (food–vegetables–fruits–dry fruits, etc.) tackling (Gong et al. 2022). Another report from Widada et al. (2021) elucidates the antifungal activity (against *Ganoderma boninense*; a devastating pathogen of oil palm *Elaeis guineensis*) of the volatiles S-methyl ethanethioate, 1,2-dimethyldisulfane, 2-methyl propanoic acid, acetic acid, 3-methyl-butanoic acid, undecan-2-one, nonan-2-one, and 2-isopropyl-5-methylcyclohexan-1-ol emitted from *Nocardipsis alba*. Gebily et al. (2021) documented the anti-*Sclerotinia sclerotiorum* activity of both the volatile metabolites and culture filtrate of *Streptomyces griseus* DG5 (nearly 50%), *Streptomyces rochei* DG4 (an inhibition of 50–70%), *Streptomyces sampsonii* DG1 (>80% inhibition of mycelial growth). The hyphae of the fungal pathogen had prominent deformities, zones of contractions, and areas of partial to complete collapse. There was a remarkable degree of inhibition of germination from the sclerotia upon treatment with cell-free metabolites of the isolates upon 10–30 days of treatment. *Streptomyces griseus* produced volatile carvacrol (IUPAC name phenol, 2-methyl 5-(1-methylethyl), cyclohexanol, benzaldehyde, albocarbon-napthalene was found to be antifungal against harmful phytopathogens *Penicillium glabrum* (forms post-harvest fruit rot in pomegranate), *Phytophthora capsici* (causal agent of light and fruit rot of peppers), *Rhizoctonia solani* (majorly cause root rot, damping off in Solanaceous plants), *Fusarium moniiliforme* (infections in maize crop), *Sclerotinia sclerotiorum* (infects a variety of hosts through the white mold, necrotic lesions, cottony and watery soft rot, stem and crown rot, drop, and also blossom blight), and *Cladosporium herbarum* (infectious to *Vicia faba* faba bean) (Danaei et al. 2014). Another prime pathogen of *Aspergillus ochraceus* (producer of ochratoxin A) was found to be blocked by volatiles of *Streptomyces alboflavus* TD-1 (Yang et al. 2018). Alwahshi et al. (2022) and Al Hamad et al. (2021) reported a series of volatiles cyclohexanol, decanol, 2-ethyl-1-hexanol, nonanol, benzothiazole from *Streptomyces violaceoruber* UAE1, *Streptomyces griseorubens* UAE2 with anti-*Sclerotinia sclerotiorum* activity.

Terpenoid-rich VOCs, 3-carene 2,5-dione, geosmin, beta-cubebene, and one phenolic compound phenol, 2-(1,1-dimethylethyl)-6-methyl-, produced from *Streptomyces lydicus* S97, were reported to have strong antibiotic efficacy of 87%, 95%, 69%, and 20% against *Botrytis cinerea*, *Rhizoctonia solani*, *Phoma medicaginis*, *Fusarium solani*, *Fusarium oxysporum*, and *Sclerotium rolfsii*, respectively (Ayed et al. 2021). The fumigated volatile altered the standard morphology of the mycelia, leading to swelling of the hyphae, rupturing of conidia, and disintegration of the cell wall (as confirmed from light and scanning microscopic observations).

Ripe strawberries infected with *B. cinerea* with prominent gray mold symptoms were found to be minimized upon treatment with S97 volatiles. This type of investigational outcome emphasizes the utility of actinobacterial volatile metabolites as a tool for sustainable disease management.

3.2.3 Role of Actinobacteria in Promoting Disease Resistance

Actinobacteria can tackle pathogenic infections through synthesized metabolites, plant growth promotion (phosphate solubilization, IAA, siderophore, and ACC deaminase production) and also induction of systemic resistance in the host. This makes them suitable agents for sustainable agriculture and alternatives to synthetic chemicals. They can synthesize different phytohormones and can have growth-promoting activities both in vitro and in vivo. The multipronged strategy of actinobacteria in developing plant defense through, growth promotion, and protection from biotic and abiotic stress is discussed in Fig. 3.2.

The application of actinomycetes promotes the surge of jasmonic acid and ethylene in the plant tissues and ultimately switches on the pathways of systemic acquired resistance (SAR) responses. Application of non-pathogenic strains of *Micromonospora* sp., and *Streptomyces* sp. on wheat seedlings exhibits these responses (Conn et al. 2008). Other than direct application of live strains, spraying of cell-free metabolites exhibits induced systemic resistance in plants. Mild elicitation of salicylate, ethylene, and jasmonate pathways involving both secondary metabolic pathways and pathogenesis-related protein synthesis is the basis of priming. Priming with secondary metabolites of *Streptomyces bikiniensis* HD-087 on cucumber exhibits resistance against *Fusarium* pathogen and the peroxidase, PAL, glucanase activities were also reported (Zhao et al. 2012). *Streptomyces* species are known to act (elicitation of systemic resistance) in the same way (surge in cytosolic calcium ions and oxidative stress against pathogens) as *Pseudomonas* sp. and *Bacillus* sp. (Baz et al. 2012). Patil et al. (2011) reported the improvement in phenolic contents of a tomato plant upon inoculation with *Streptomyces avidinii*, *Streptomyces toxytricini* in *Rhizoctonia* infections. Also, common enzymatic antioxidantsperoxidase, polyphenol oxidase, PAL, and flavonoids were found to be uplifted in inoculated plants and the genes (i.e., pathogenesis-related *OsPR1a*, salicylic acid-responsive *Oswrky45*) responsible for encoding SOD, PAL, PO, APX, CAT, CHI, GLU were found to be upregulated (Singh et al. 2017; Vilasinee et al. 2019). Transcriptomic analysis of treated plants (*Microsphaera alphitoides*-treated oak plant-primed with *Streptomyces* sp.) suggests the involvement of JA, ET, SA, ABA, phenylpropanoid synthesis, and plants' secondary metabolic pathways (Kurth et al. 2014). The activation of these pathways is not momentary but long-lasting as reported from *Botrytis*-infected tomato plants treated with *Micromonospora* sp. (Singh and Gaur 2017). The involvement of other components, that is, oxidative burst (higher accumulation of H_2O_2), reduced events of lipid peroxidation,

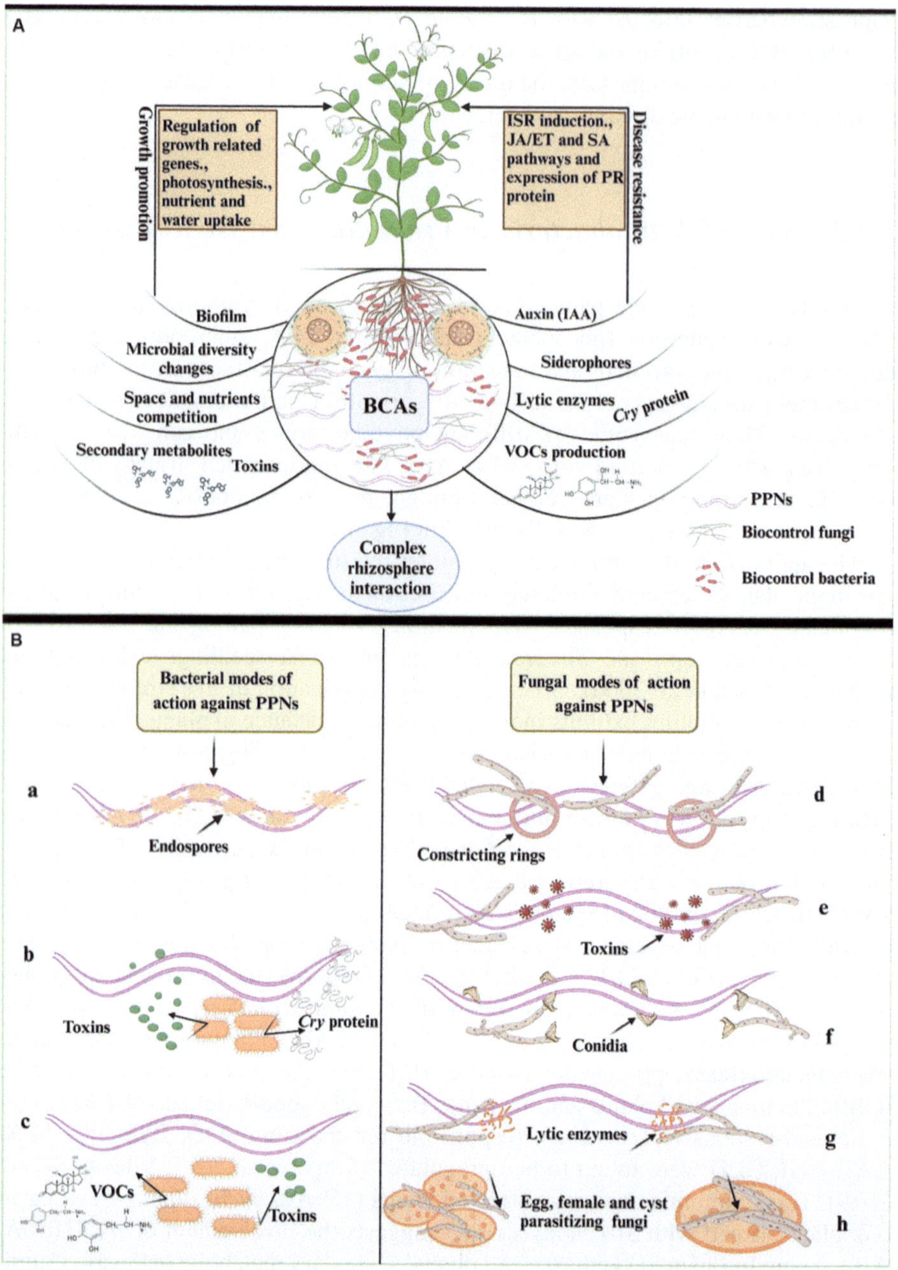

Fig. 3.2 Role of actinobacteria as a biocontrol agent in maintaining plant health and productivity

alkalization of extracellular components, activation of protein kinase and chalcone synthase, and increased concentration of phytoalexins have also been noticed (Vatsa-Portugal et al. 2017). Both the rhizospheric application of actinobacteria and inoculation in seed are effective but provide improved priming (elicitation of PR-1a, Chi3, Chi9, CEVI-1, PAL1, LOXB) outcome upon cumulative application (Vilasinee et al. 2019). Priming through different species of *Streptomyces* on rice, grapevine, tomato, pepper, apple fruit, cucumber, and tobacco plants has been shown to protect against pathogens, species of *Alternaria*, *Fusarium*, *Botrytis*, *Sclerotinia*, bacterial blight, and tobacco mosaic virus (Ebrahimi-Zarandi et al. 2022). Endophytic actinobacteria also serve the purpose of biocontrol agents as they produce both plant protective and antimicrobial compounds (Table 3.3).

Table 3.3 Endophytic actinobacteria and their role in plant defense

Endophytic actinobacteria	Bioactivity	Metabolites produced	References
Streptomyces albidoflavus (an isolate of Chinese medicinal herb *Thymus roseus* Schipcz)	Antifungal against *Fusarium oxysporum*, *Fusarium moniliforme*, *Verticillium dahliae Kleb*, *Fusarium graminearum* Also exhibits plant growth-promoting potentials	Tetrangulol, 4-hydroxybenzaldehyde, and cyclohexane Additionally, we identified plant growth-regulating metabolites such as N-succinyl-L-glutamate, nicotinic acid, L-aspartate, and indole-3-acetamide	Abdelshafy Mohamad et al. (2024)
Streptomyces sp. NEAU-DD186 (an isolate from moss)	Antibacterial against soilborne phytopathogenic bacteria *Ralstonia solanacearum*, *Bipolaris sorokiniana*	25-O-malonylguanidylfungin A and 23-O-malonylguanidylfungin A	Gao et al. (2024)
Streptomyces sp., *Nocardiopsis* sp. (an isolate from wheat plant)	Antifungal against *Fusarium graminearum*, *Bipolaris sorokiniana*, *Alternaria* sp., *Tilletia indica*	Produces hydrolytic enzymes	Sharma et al. (2023)
Streptomyces sp. GLL-9 (an isolate from a navel orange tree)	Antimicrobial against *Xanthomonas citri* susp. *citri*, *Penicillium italicum*	Produced three actinomycins: Actinomycin X_2 (Act-X_2), actinomycin D (Act-D), actinomycin $X_{O\beta}$ (Act-$X_{O\beta}$)	Gao et al. (2023)

(continued)

Table 3.3 (continued)

Endophytic actinobacteria	Bioactivity	Metabolites produced	References
Streptomyces endophytica sp. nov. (an isolate from *Dioscorea opposita*)	Antifungal against yam anthracnose pathogen *Colletotrichum gloeosporioides* and other fungal pathogens *Fusarium oxysporum* f. sp. cubense race 4, *Magnaporthe oryzae, Lasiodiplodia theobromae, Neoscytalidium dimidiatum, Colletotrichum capsici, Stagonosporopsis cucurbitacearum* Exhibits plant growth-promoting potentials, solubilizes phosphate, produces IAA, and exhibits ACC deaminase activity	Produces chitinase, protease, and glucanase	Zhou et al. (2023)
Streptomyces albus	Antimicrobial against *Scopulariopsis gossypii* Co1, *Fusarium fujikuroi* L3, *Ralstonia solanacearum, Xanthomonas oryzae*	Not reported	Quach et al. (2023)
Streptomyces tuirus isolated from chili	Antagonistic against fruit rot pathogens *Colletotrichum scovillei, Colletotrichum truncatum, Fusarium oxysporum*	Volatile and nonvolatile metabolites	Renuka et al. (2023)

3.2.4 Actinobacteria Improvise Soil Microbial Community: A Modified Way of Disease Resistance

The susceptibility of a host toward a pathogen is not only influenced by the hosts' physiology or internal biochemistry but also is epigenetically modified by the environmental parameters, that is, temperature, humidity, sun rays, wind velocity, etc. Here, the rhizospheric microbial flora and phylospheric microbial load are crucial and strongly influence disease resistance/development. The application of biocontrol agents in the form of microbial inoculants can not only impact the host plants' resistance but also improve the rhizospheric microbial composition, which can differ in the disease's progression. The efficiency of biocontrol agents is associated with the cumulative interaction between plant–pathogen-associated microbes and the environment (Gouda et al. 2018). Reports include that an uplifted soil bacterial diversity is negatively proportionate to the relative abundance of disease-causing pathogens (Van Elsas et al. 2012). This is a next-generation technique for tackling microbial infections in agricultural fields. A rich composition of the native microbial flora of the rhizosphere reduces the aggressiveness of the pathogen and minimizes the process of disease development and improves the effectivity of the applied biocontrol agents (Chen et al. 2013; Wang et al. 2017; Yang et al. 2017).

Reports include that an extended duration of application of organic fertilizers or biofertilizers can modulate the soil microbial community. A case study reveals that *Bacillus* as antagonists remarkably suppressed the population of the pathogen *Phytophthora capsici*; a causal agent of blight of pepper. This type of investigation incorporating actinobacterial species is quite rare. Abbasi et al. (2021, 2022) came up with an innovative approach and initially screened different *Streptomyces* isolates for their anti-*Phytophthota capsica* action and checked the degree of disease suppression (pepper blight caused by the same pathogen) both in vitro and in vivo. The third but most unique part was documenting the change in the rhizospheric microbial community through culture-dependent and metagenomic approaches. They investigated the impact of phosphate solubilizing *Streptomyces* sp. on the abundance of native microflora of soil in relation to the disease severity of *Phytophthora capsici*. Two strains of *Streptomyces*—SS14 (melanin producer) and IT20 (phosphate solubilizer)—were implicated in the bell pepper plants, and it was found that IT20 inoculated one had a higher diversity of cyanobacteria (blue-green algae) than SS14 treated and untreated (control) one (obtained from the next-generation sequencing results). The higher occurrence of cyanobacteria assures the rich percentage of soluble nitrogen (in the form of nitrate and nitrite ions) to the soil that may in turn impact the health of the surrounding plants/crops positively. It was found from the Illumina Mi Seq data (results of high-throughput sequencing) that IT20 has more microbial responders than the SS14. It is probably because IT20 being a phosphate solubilizer increases the amount of available phosphate in the surrounding soil and other microbes reach out (communicated through quorum sensing) to utilize that phosphorus in the form of rewards. Also, it was documented that different phylotypes, that is, Devosia, Promicromonospora, Kribbella, Microbacterium, Amylocolatopsis, and Pseudomonas responded in favor of the host plant and act as antagonists to the pathogen. The alpha and beta diversity also changed upon the IT20 application. Finally, it was seen that the IT20-treated set exhibits a higher survival rate against the pathogen. So, from these experimental outcomes, it could be concluded that proper preparation/formulation of a synthetic community of plant growth-promoting agonist bacteria can be applied to agricultural fields with an aim to modulate plant growth and defense in terms of modulation of the soil microbial community.

3.3 Conclusion

In the present scenario of world agriculture concerning the growth of population and lofty demand for agro-products, assuring healthy and safe, environmentally friendly, sustainable wayouts are the priorities. In this respect, nature-based solutions, especially from actinobacterial sources, are a possible alternative. Actinomycetes and their synthesized metabolites hold the potential to induce plant defense and effectively act as a biocontrol agent against devastating phytopathogens. These are nonhazardous to the interacting ecosystem, act target-oriented, do not leave any harmful residues, and are superior in action to the chemically derived ones. They modulate plants' internal biochemical machinery and trigger the basic defense against invading pathogens, also nurture the rhizospheric nutrient cycling through mineral solubilization, and exhibit repellent action toward disease-causing pathogens. Upon successful tackling of some critical bottlenecks related to the compatibility of the consortia members, longer shelf life, improved stability, dormant carrier, effective transport and storage strategies, and higher microbial load, actinobacterial colonies can be commercialized for bio-inoculants. Despite their comparatively slow growth rate than bacteria and a bit high investment of the economy, their possible use as next-generation bio-inoculants upon synchronized activity in both in vitro and in vivo/in planta conditions can be a deep ecology-driven step for sustainable agricultural practices to tackle agricultural diseases.

Acknowledgments The authors are thankful to biorender.com, and the figures used in this chapter were prepared using the same.

References

Aallam Y, Dhiba D, El Rasafi T, Lemriss S, Haddioui A, Tarkka M, Hamdali H (2022) Growth promotion and protection against root rot of sugar beet (*Beta vulgaris* L.) by two rock phosphate and potassium solubilizing *Streptomyces* spp. under greenhouse conditions. Plant Soil 1:1–4

Abbasi S, Safaie N, Sadeghi A, Shamsbakhsh M (2019) *Streptomyces* strains induce resistance to *Fusarium oxysporum* f. sp. lycopersici race 3 in tomato through different molecular mechanisms. Front Microbiol 3(10):1505

Abbasi S, Sadeghi A, Safaie N (2020) *Streptomyces* alleviate drought stress in tomato plants and modulate the expression of transcription factors ERF1 and WRKY70 genes. Sci Hortic 30(265):109206

Abbasi S, Spor A, Sadeghi A, Safaie N (2021) Streptomyces strains modulate dynamics of soil bacterial communities and their efficacy in disease suppression caused by *Phytophthora capsici*. Sci Rep 11(1):9317

Abbasi S, Kafi SA, Karimi E, Sadeghi A (2022) *Streptomyces* consortium improved quality attributes of bell pepper fruits, induced plant defense priming, and changed microbial communities of rhizosphere under commercial greenhouse conditions. Rhizosphere 1(23):100570

Abdelshafy Mohamad OA, Liu YH, Huang Y, Kuchkarova N, Dong L, Jiao JY, Fang BZ, Ma JB, Hatab S, Li WJ (2024) Metabonomic analysis to identify exometabolome changes underlying antifungal and growth promotion mechanisms of endophytic Actinobacterium *Streptomyces albidoflavus* for sustainable agriculture practice. Front Microbiol 30(15):1439798

Abo-Zaid GA, Matar SM, Abdelkhalek A (2020) Induction of plant resistance against tobacco mosaic virus using the biocontrol agent *Streptomyces cellulosae* isolate Actino 48. Agronomy 10(11):1620

Adeniyi DO, Asogwa EU (2023) Dynamics of diseases and insect pests of cashew tree. Forest Microbiol 1:265–284

Aggarwal N, Thind SK, Sharma S (2016) Role of secondary metabolites of actinomycetes in crop protection. In: Plant growth promoting actinobacteria: a new avenue for enhancing the productivity and soil fertility of grain legumes, pp 99–121

Akl B, El-Masry SS (2024) Genetic characterization of three soil Streptomyces isolates exhibiting antiviral activity against TMV; insights from 16S rRNA gene sequencing and RAPD analysis. Egypt Acad J Biol Sci G Microbiol 16(1):33–53

Al Hamad BM, Al Raish SM, Ramadan GA, Saeed EE, Alameri SS, Al Senaani SS, AbuQamar SF, El-Tarabily KA (2021) Effectiveness of augmentative biological control of *Streptomyces griseorubens* UAE2 depends on 1-aminocyclopropane-1-carboxylic acid deaminase activity against *Neoscytalidium dimidiatum*. J Fungi 7(11):885

Ali A, Li Y, Jeyasundar PG, Azeem M, Su J, Wahid F, Mahar A, Shah MZ, Li R, Zhang Z (2021) *Streptomyces pactum* and *Bacillus* consortium influenced the bioavailability of toxic metals, soil health, and growth attributes of *Symphytum officinale* in smelter/mining polluted soil. Environ Pollut 15(291):118237

Alipour Kafi S, Karimi E, Akhlaghi Motlagh M, Amini Z, Mohammadi A, Sadeghi A (2021) Isolation and identification of *Amycolatopsis* sp. strain 1119 with potential to improve cucumber fruit yield and induce plant defense responses in commercial greenhouse. Plant Soil 468:125–145

Al-Quwaie DA (2024) The role of *Streptomyces* species in controlling plant diseases: a comprehensive review. Australas Plant Pathol 53(1):1–4

Alwahshi KJ, Purayil GP, Saeed EE, Abufarajallah HA, Aldhaheri SJ, AbuQamar SF, El-Tarabily KA (2022) The 1-aminocyclopropane-1-carboxylic acid deaminase-producing *Streptomyces violaceoruber* UAE1 can provide protection from sudden decline syndrome on date palm. Front Plant Sci 27(13):904166

Anavadiya B, Chouhan S, Saraf M, Goswami D (2024) Exploring endophytic actinomycetes: a rich reservoir of diverse antimicrobial compounds for combatting global antimicrobial resistance. Microbe 29:100110

Andargie M, Li J (2019) Antifungal activity against plant pathogens by compounds from *Streptoverticillium morookaense*. J Plant Pathol 15(101):547–558

Ashokvardhan T, Rajithasri AB, Prathyusha P, Satyaprasad K (2014) Actinomycetes from *Capsicum annuum* L. rhizosphere soil have the biocontrol potential against pathogenic fungi. Int J Curr Microbiol App Sci 3(4):894–903

Atif AM, Elzamik FI, Mohamed GM, Al-Quwaie DA, Ashkan MF, Alqahtani FS, Motwali EA, Alomran MM, Alharbi NK, El-Tarabily KA, Abdelbasit HM (2023) Biological control of the root-knot nematode (*Meloidogyne incognita*) on eggplants with various chitinase-producing *Streptomyces* strains. Eur J Plant Pathol 167(3):371–394

Ayaz M, Zhao JT, Zhao W, Chi YK, Ali Q, Ali F et al (2024) Biocontrol of plant parasitic nematodes by bacteria and fungi: a multi-omics approach for the exploration of novel nematicides in sustainable agriculture. Front Microbiol 15:1433716

Ayed A, Kalai-Grami L, Ben Slimene I, Chaouachi M, Mankai H, Karkouch I, Djebali N, Elkahoui S, Tabbene O, Limam F (2021) Antifungal activity of volatile organic compounds from *Streptomyces* sp. strain S97 against *Botrytis cinerea*. Biocontrol. Sci Technol 31(12):1330–1348

Bahrami Y, Delbari Y, Buzhani KR, Kakaei E, Mohassel Y, Bouk S, Franco CM (2022) Endophytic actinobacteria in biosynthesis of bioactive metabolites and their application in improving crop yield and sustainable agriculture. In: Natural products from actinomycetes: diversity, ecology and drug discovery. Singapore, Springer Singapore, pp 119–150

Bai L, Li L, Xu H, Minagawa K, Yu Y, Zhang Y, Zhou X, Floss HG, Mahmud T, Deng Z (2006) Functional analysis of the validamycin biosynthetic gene cluster and engineered production of validoxylamine A. Chem Biol 13(4):387–397

Baz M, Tran D, Kettani-Halabi M, Samri SE, Jamjari A, Bernadette B, Meimoun P et al (2012) Calcium-and ROS-mediated defence responses in BY2 tobacco cells by nonpathogenic Streptomyces sp. J App Microbiol 112:782–792

Behera S, Das S (2023) Potential and prospects of Actinobacteria in the bioremediation of environmental pollutants: cellular mechanisms and genetic regulations. Microbiol Res 1(273):127399

Berdy J (2005) Bioactive microbial metabolites. J Antibiot 58(1):1–26

Betancur LA, Forero AM, Romero-Otero A, Sepúlveda LY, Moreno-Sarmiento NC, Castellanos L, Ramos FA (2019) Cyclic tetrapeptides from the marine strain *Streptomyces* sp. PNM-161a with activity against rice and yam phytopathogens. J Antibiot 72(10):744–751

Boubekri K, Soumare A, Mardad I, Lyamlouli K, Ouhdouch Y, Hafidi M, Kouisni L (2022) Multifunctional role of Actinobacteria in agricultural production sustainability: a review. Microbiol Res 1(261):127059

Boukaew S, Cheirsilp B, Prasertsan P, Yossan S (2021) Antifungal effect of volatile organic compounds produced by *Streptomyces salmonis* PSRDC-09 against anthracnose pathogen *Colletotrichum gloeosporioides* PSU-03 in postharvest chili fruit. J App Microbiol 131(3):1452–1463

Cao H, Liu W, Chang Y, Chen H, Zhang Y, Xu W, Tao L (2019) 5′-Epi-SPA-6952A, a new insecticidal 24-membered macrolide produced by *Streptomyces diastatochromogenes* SSPRC-11339. Nat Prod Res 33(5):659–664

Carmona-Hernandez S, Reyes-Pérez JJ, Chiquito-Contreras RG, Rincon-Enriquez G, Cerdan-Cabrera CR, Hernandez-Montiel LG (2019) Biocontrol of postharvest fruit fungal diseases by bacterial antagonists: a review. Agronomy 9(3):121

Cerna-Chávez E, Rodríguez-Rodríguez JF, García-Conde KB, Ochoa-Fuentes YM (2024) Potential of *Streptomyces avermitilis*: a review on avermectin production and its biocidal effect. Metabolites 14(7):374

Chaiharn M, Theantana T, Pathom-Aree W (2020) Evaluation of biocontrol activities of *Streptomyces* spp. against rice blast disease fungi. Pathogens 9(2):126

Chen F, Wang M, Zheng Y, Li S, Wang H, Han D, Guo S (2013) The effect of biocontrol bacteria on rhizosphere bacterial communities analyzed by plating and PCR-DGGE. Curr Mcrobiol 67:177–182

Chen D, Ali MN, Kamran M, Magsi MA, Mora-Poblete F, Maldonado C, Waris M, Aljowaie RM, Zehri MY, Elshikh MS (2022) The *Streptomyces chromofuscus* strain RFS-23 induces systemic resistance and activates plant defense responses against tomato yellow leaf curl virus infection. Agronomy 12(10):2419

Cheng J, Yang SH, Palaniyandi SA, Han JS, Yoon TM, Kim TJ, Suh JW (2010) Azalomycin F complex is an antifungal substance produced by *Streptomyces malaysiensis* MJM1968 isolated from agricultural soil. J Korean Soc Appl Biol Chem 53:545–552

Cho KM, Shin SC, Bo AB, Umurzokov M, Jia W, Park KW, Choi JS, Kim YS (2022) Isolation, screening, and molecular identification of *Streptomyces* sp. W-200 and its bioherbicidal activity in weed control. Weed &Turfgrass Sci 11(3):277–287

Chouyia FE, Ventorino V, Pepe O (2022) Diversity, mechanisms and beneficial features of phosphate-solubilizing *Streptomyces* in sustainable agriculture: a review. Front Plant Sci 6(13):1035358

Cimermanova M, Pristas P, Piknova M (2021) Biodiversity of actinomycetes from heavy metal contaminated technosols. Microorganisms 9(8):1635

Cone MC, Yin X, Grochowski LL, Parker MR, Zabriskie TM (2003) The blasticidin S biosynthesis gene cluster from *Streptomyces griseochromogenes*: sequence analysis, organization, and initial characterization. Chembiochem 4(9):821–828

Conn VM, Walker AR, Franco CM (2008) Endophytic actinobacteria induce defense pathways in *Arabidopsis thaliana*. Mol Plant-Microbe Interact 21(2):208–218

Copping LG, Duke SO (2007) Natural products that have been used commercially as crop protection agents. Pest Manag Sci 63(6):524–554

Danaei M, Baghizadeh A, Pourseyedi S, Amini J, Yaghoobi MM (2014) Biological control of plant fungal diseases using volatile substances of *Streptomyces griseus*. Eur J Exp Biol 4(1):334–339

del Rosario CL, Chiappero J, Banchio E (2019) Invisible signals from the underground: a practical method to investigate the effect of microbial volatile organic compounds emitted by rhizobacteria on plant growth. Biochem Mol Biol Educ 47(4):388–393

Devi SS, Rao KB (2017) Exploration of antimicrobial compounds from Streptomyces S9 against a phytopathogen, *Corynespora cassiicola* (Berk & Curtis). J Biopest 10(1):1–9

Djebaili R, Pellegrini M, Ercole C, Farda B, Kitouni M, Del Gallo M (2021) Biocontrol of soil-borne pathogens of *Solanum lycopersicum* L. and *Daucus carota* L. by plant growth-promoting actinomycetes: in vitro and in planta antagonistic activity. Pathogens 10(10):1305

Doolotkeldieva T, Saykal B, Ayzat S (2016) Biological control of Erwinia carotovora ssp. carotovora by Streptomyces species. Advances in Microbiology 6:104

Du Y, Wang T, Jiang J, Wang Y, Lv C, Sun K, Sun J, Yan B, Kang C, Guo L, Huang L (2022) Biological control and plant growth promotion properties of *Streptomyces albidoflavus* St-220 isolated from *Salvia miltiorrhiza* rhizosphere. Front Plant Sci 30(13):976813

Eberlein C, Westphal A (2023) Suppression of Meloidogyne incognita by co-application of chitin and *Streptomyces nigrescens* (AMV1033). Biocontrol Sci Technol 33(5):484–498

Ebrahimi-Zarandi M, Saberi Riseh R, Tarkka MT (2022) Actinobacteria as effective biocontrol agents against plant pathogens, an overview on their role in eliciting plant defense. Microorganisms 10(9):1739

El-Akshar EA, Elmeihy RM, Tewfike TA, Abou-Aly HE (2022) Endophytic *Streptomyces enissocaesilis* as a nematicidal and biostimulant agent. Egypt Acad J Biol Sci 14(2):123–133

El-Saadony MT, Saad AM, Soliman SM, Salem HM, Ahmed AI, Mahmood M, El-Tahan AM, Ebrahim AA, Abd El-Mageed TA, Negm SH, Selim S (2022) Plant growth-promoting microorganisms as biocontrol agents of plant diseases: mechanisms, challenges and future perspectives. Front Plant Sci 6(13):923880

Elsayed EA, Farid MA, El Enshasy HA (2013) Improvement in natamycin production by *Streptomyces natalensis* with the addition of short-chain carboxylic acids. Process Biochem 48(12):1831–1838

El-Shanshoury AE, Metwally MA, El-Sabbagh SM (2022) Biocontrol of aspergillus flavus producing aflatoxin B1 by *Streptomyces exfoliatus*. Egypt J Bot 62(2):457–473

El-Tarabily KA, Hardy GE, Sivasithamparam K, Hussein AM, Kurtböke DI (1997) The potential for the biological control of cavity-spot disease of carrots, caused by *Pythium coloratum*, by streptomycete and non-streptomycete actinomycetes. New Phytol 137(3):495–507

Evangelista-Martínez Z, Contreras-Leal EA, Corona-Pedraza LF, Gastélum-Martínez É (2020) Biocontrol potential of *Streptomyces* sp. CACIS-1.5 CA against phytopathogenic fungi causing postharvest fruit diseases. Egypt J Biol Pest Control 30(1):117

Gao Y, Ning Q, Yang Y, Liu Y, Niu S, Hu X, Pan H, Bu Z, Chen N, Guo J, Yu J (2021) Endophytic *Streptomyces hygroscopicus* OsiSh-2-mediated balancing between growth and disease resistance in host rice. MBio 12(4):10–128

Gao L, Kumaravel K, Xiong Q, Liang Y, Ju Z, Jiang Y, Zhang J (2023) Actinomycins produced by endophyte *Streptomyces* sp. GLL-9 from navel orange plant exhibit high antimicrobial effect against *Xanthomonas citri* susp. citri and *Penicillium italicum*. Pest Manag Sci 79(11):4679–4693

Gao C, Wang Z, Wang C, Yang J, Du R, Bing H, Xiang W, Wang X, Liu C (2024) Endophytic *Streptomyces* sp. NEAU-DD186 from moss with broad-spectrum antimicrobial activity: biocontrol potential against Soilborne diseases and bioactive components. Phytopathology 114(2):340–347

Gao Y, Zeng XD, Ren B, Zeng JR, Xu T, Yang YZ, Hu XC et al (2020) Antagonistic activity against rice blast disease and elicitation of host-defence response capability of an endophytic Streptomyces albidoflavus OsiLf-2. Plant pathology 69:259–271

Gebily DA, Ghanem GA, Ragab MM, Ali AM, Soliman NE, Abd El-Moity TH (2021) Characterization and potential antifungal activities of three Streptomyces spp. as biocontrol agents against *Sclerotinia sclerotiorum* (Lib.) de Bary infecting green bean. Egypt J Biol Pest Control 31:1–5

Ghanem GA, Mahmoud AM, Kheder AA, Gebily DA, Osamy A (2023) Antiviral activities of three Streptomyces spp. against Zucchini yellow mosaic virus (ZYMV) infecting squash (*Cucurbita pepo* L.) plants. Egypt J Biol Pest Control 33(1):113

Giacomelli Ribeiro H, Teresinha Van Der Sand S (2024) Exploring the trends in actinobacteria as biological control agents of phytopathogenic fungi: a (mini)-review. Indian J Microbiol 64(1):70–81

Gómez ÁG, Ramos FA, Sinuco DC (2021) Screening of volatile organic compounds from actinobacteria for the control of phytopathogen *Colletotrichum gloeosporioides*. Biocontrol Sci Technol 31(10):1067–1079

Gong Y, Liu JQ, Xu MJ, Zhang CM, Gao J, Li CG, Xing K, Qin S (2022) Antifungal volatile organic compounds from *Streptomyces setonii* WY228 control black spot disease of sweet potato. App Environ Microbiol 88(6):e02317–e02321

Gouda S, Kerry RG, Das G, Paramithiotis S, Shin HS, Patra JK (2018) Revitalization of plant growth promoting rhizobacteria for sustainable development in agriculture. Microbiol Res 1(206):131–140

Guesmi S, Mahjoubi M, Pujic P, Cherif A, Normand P, Sghaier H, Boubakri H (2022) Biotechnological potential of *Kocuria rhizophila* PT10 isolated from roots of *Panicum turgidum*. Int J Environ Sci Technol 1:1–4

Hamada M, Kinoshita N, Hattori S, Yoshida A, Okami Y, Higashide K, Sakata N, Hori M (1995) *Streptomyces kasugaensis* sp. nov.: a new species of genus *Streptomyces*. Actinomycetologica 9(1):27–36

Han Y, Luo Y, Qin S, Xi L, Wan B, Du L (2014) Induction of systemic resistance against tobacco mosaic virus by Ningnanmycin in tobacco. Pestic Biochem Physiol 1(111):14–18

Hozzein WN, Goodfellow M (2008) *Nocardiopsis arabia* sp. nov., a halotolerant actinomycete isolated from a sand-dune soil. Int J Syst Evol Microbiol 58(11):2520–2524

Hu Q, Yang M, Bo T, Li Y, Wu C, Mo M, Liu Y (2022) Soluble macromolecules from two Streptomyces strains with potent nematicidal activity against *Meloidogyne incognita*. Rhizosphere 22:100529

Hwang BK, Lee JY, Kim BS, Moon SS (1996) Isolation, structure elucidation, and antifungal activity of a manumycin-type antibiotic from *Streptomyces flaveus*. J Agric Food Chem 44(11):3653–3657

Islamiati ED, Widada J, Wahyuningsih TD, Widianto D (2022) Volatile organic compounds of *Streptomyces* sp. GMR22 inhibit growth of two plant pathogenic fungi. Agric Nat Resour 56(5):899–908

Jin N, Xue H, Li WJ, Wang XY, Liu Q, Liu SS, Liu P, Zhao JL, Jian H (2017) Field evaluation of *Streptomyces rubrogriseus* HDZ-9-47 for biocontrol of *Meloidogyne incognita* on tomato. J Integr Agric 16(6):1347–1357

Jing T, Zhou D, Zhang M, Yun T, Qi D, Wei Y, Chen Y, Zang X, Wang W, Xie J (2020) Newly isolated *Streptomyces* sp. JBS5-6 as a potential biocontrol agent to control banana fusarium wilt: genome sequencing and secondary metabolite cluster profiles. Front Microbiol 11:602591

Jose PA, Jebakumar SR (2014) Unexplored hypersaline habitats are sources of novel actinomycetes. Front Microbiol 5:242

Kaari M, Joseph J, Manikkam R, Sreenivasan A, Venugopal G, Alexander B, Krishnan S (2022) Anti-biofilm activity and biocontrol potential of *Streptomyces* cultures against *Ralstonia solanacearum* on tomato plants. Indian J Microbiol 1:1–8

Kabaluk JT, Svircev AM, Goettel MS, Woo SG (eds) (2010) The use and regulation of microbial pesticides in representative jurisdictions worldwide. International Organization for Biological Control of Noxious Animals and Plants (IOBC), St. Paul

Kang MK, Kim MH, Liu MJ, Jin CZ, Park SH, Lee JM, Kim J, Park DJ, Park HR, Kim YH, Kim CJ (2021) Nematicidal activity of teleocidin B4 isolated from *Streptomyces* sp. against pine wood nematode, *Bursaphelenchus xylophilus*. Pest Manag Sci 77(4):1607–1615

Kang MK, Kim JH, Liu MJ, Jin CZ, Park DJ, Kim J, Sung BH, Kim CJ, Son KH (2022) New discovery on the nematode activity of aureothin and alloaureothin isolated from endophytic bacteria *Streptomyces* sp AE170020. Sci Rep 12(1):3947

Karimi R, Noori A (2022) *Streptomyces rimosus* rhizobacteria and *Glomus mosseae* mycorrhizal fungus inoculation alleviate salinity stress in grapevine through morphophysiological changes and nutritional balance. Sci Hort 305:111433

Kaur T, Jasrotia S, Ohri P, Manhas RK (2016) Evaluation of in vitro and in vivo nematicidal potential of a multifunctional streptomycete, *Streptomyces hydrogenans* strain DH16 against *Meloidogyne incognita*. Microbiol Res 192:247–252

Kaur T, Khanna K, Sharma S, Manhas RK (2023) Mechanistic insights into the role of actinobacteria as potential biocontrol candidates against fungal phytopathogens. JBM 63(11):1196–1218

Khan S, Srivastava S, Karnwal A, Malik T (2023) *Streptomyces* as a promising biological control agents for plant pathogens. Front Microbiol 14:1285543

Khanna K, Jamwal VL, Kohli SK, Gandhi SG, Ohri P, Bhardwaj R, Wijaya L, Alyemeni MN, Ahmad P (2019a) Role of plant growth promoting bacteria (PGPRs) as biocontrol agents of Meloidogyne incognita through improved plant defense of *Lycopersicon esculentum*. Plant Soil 436:325–345

Khanna K, Sharma A, Ohri P et al (2019b) Impact of plant growth promoting rhizobacteria in the orchestration of *Lycopersicon esculentum* mill. Resistance to plant parasitic nematodes: a metabolomic approach to evaluate defense responses under field conditions. Biomol Ther 9(11):676

Khilyas IV, Markelova MI, Valeeva LR, Ivoilova TM, Shagimardanova E, Laikov AV, Elistratova AA, Berkutova ES, Lochnit G, Sharipova MR (2024) Genomic insights and anti-phytopathogenic potential of siderophore metabolome of endolithic *Nocardia mangyaensis* NH1. Sci Rep 14(1):5676

Kim YH, Park BS, Bhatia SK, Seo HM, Jeon JM, Kim HJ, Yi DH, Lee JH, Choi KY, Park HY, Kim YG (2014) Production of rapamycin in *Streptomyces hygroscopicus* from glycerol-based media optimized by systemic methodology. J Microbiol Biotechnol 24(10):1319–1326

Kim JH, Choi JY, Park DH, Park DJ, Park MG, Kim SY, Ju YJ, Kim JY, Wang M, Kim CJ, Je YH (2020) Isolation and characterization of the insect growth regulatory substances from actinomycetes. Comp Biochem Physiol C Toxicol Pharmacol 228:108651

Kim YS, Umurzokov M, Cho KM, Choi JS, Park KW (2022) Insecticidal characteristics and structural identification of the potential active compounds from *Streptomyces* sp. KR0006: strain improvement through mutagenesis. PLoS One 17(9):e0274766

Kishimoto K, Park YS, Okabe M et al (1996) Effect of ferrous ion on mildiomycin production by *Streptoverticillium rimofaciens*. J Antibiot 49(8):770–774

Kobinata K, Koshino H, Kudo T, Isono K, Osada H (1993) RK-397, a new oxo pentaene antibiotic. J Antibiot 46(10):1616–1618

Komaki H, Ichikawa N, Hosoyama A, Takahashi-Nakaguchi A, Matsuzawa T, Suzuki KI, Fujita N, Gonoi T (2014) Genome based analysis of type-I polyketide synthase and nonribosomal peptide synthetase gene clusters in seven strains of five representative Nocardia species. BMC Genomics 15:1–1

Kulkarni M, Gorthi S, Banerjee G, Chattopadhyay P (2017) Production, characterization and optimization of actinomycin D from *Streptomyces hydrogenans* IB310, a (n antagonistic bacterium against phytopathogens). Biocatal Agric Biotechnol 10:69–74

Kurth F, Mailänder S, Bönn M, Feldhahn L, Herrmann S, Große I, Buscot F, Schrey SD, Tarkka MT (2014) Streptomyces-induced resistance against oak powdery mildew involves host plant responses in defense, photosynthesis, and secondary metabolism pathways. Mol Plant-Microbe Interact 27(9):891–900

Latake SB, Borkar SG (2017) Characterization of marine actinomycete having antiviral activity against cucumber mosaic virus. Curr Sci 10:1442–1447

Le KD, Kim J, Nguyen HT, Yu NH, Park AR, Lee CW, Kim JC (2021) *Streptomyces* sp. JCK-6131 protects plants against bacterial and fungal diseases via two mechanisms. Front Plant Sci 12:726266

Le KD, Yu NH, Park AR, Park DJ, Kim CJ, Kim JC (2022) *Streptomyces* sp. AN090126 as a biocontrol agent against bacterial and fungal plant diseases. Microorganisms 10(4):791

Lee SY, Tindwa H, Lee YS, Naing KW, Hong SH, Nam Y, Kim KY (2012) Biocontrol of anthracnose in pepper using Chitinase, β-1, 3 Glucanase, and 2-Furancarboxaldehyde produced by *Streptomyces cavourensis* SY224. J Microbiol Biotechnol 22(10):1359–1366

Li W, Csukai M, Corran A, Crowley P, Solomon PS, Oliver RP (2008) Malayamycin, a new streptomycete antifungal compound, specifically inhibits sporulation of *Stagonospora nodorum* (Berk) castell and Germano, the cause of wheat glume blotch disease. Pest Manag Sci 64(12):1294–1302

Li Y, He F, Lai H, Xue Q (2017) Mechanism of in vitro antagonism of phytopathogenic *Scelrotium rolfsii* by actinomycetes. Eur J Plant Pathol 149:299–311

Li Y, Guo Q, Li Y, Sun Y, Xue Q, Lai H (2019a) *Streptomyces pactum* Act12 controls tomato yellow leaf curl virus disease and alters rhizosphere microbial communities. Biol Fertil Soils 55:149–169

Li Y, Hu XL, Trigiano RN, Aldwinckle H, Cheng ZM (2019b) Evaluation of 110 apple cultivars for resistance to *Alternaria* blotch caused by *Alternaria alternata* apple pathotype. HortScience 54(8):1268–1274

Li JS, Qi H, Zhang SY, Xiang WS, Zhang LQ, Wang HY, Hao ZK, Wang JD (2021a) Two new milbemycin derivatives from a genetically engineered strain *Streptomyces bingchenggensis*. J Asian Nat Prod Res 23(7):660–665

Li X, Jing T, Zhou D, Zhang M, Qi D, Zang X, Zhao Y, Li K, Tang W, Chen Y, Qi C (2021b) Biocontrol efficacy and possible mechanism of *Streptomyces* sp. H4 against postharvest anthracnose caused by *Colletotrichum fragariae* on strawberry fruit. Postharvest Biol Technol 175:111401

Liu MJ, Hwang BS, Jin CZ, Li WJ, Park DJ, Seo ST, Kim CJ (2019) Screening, isolation and evaluation of a nematicidal compound from actinomycetes against the pine wood nematode, *Bursaphelenchus xylophilus*. Pest Manag Sci 75(6):1585–1593

Liu W, Wang J, Li S, Zhang H, Meng L, Liu L, Ping W, Du C (2022) Genomic and biocontrol potential of the crude lipopeptide by *Streptomyces bikiniensis* HD-087 against *Magnaporthe oryzae*. Front Microbiol 13:888645

Lyu A, Yang L, Wu M, Zhang J, Li G (2020) High efficacy of the volatile organic compounds of *Streptomyces yanglinensis* 3-10 in suppression of *Aspergillus* contamination on peanut kernels. Front Microbiol 11:142

Manjunatha L, Rajashekara H, Uppala LS, Ambika DS, Patil B, Shankarappa KS, Nath VS, Kavitha TR, Mishra AK (2022) Mechanisms of microbial plant protection and control of plant viruses. Plan Theory 11(24):3449

Meidani C, Savvidis A, Lampropoulou E, Sagia A, Katsifas E, Monokrousos N, Hatzinikolaou DG, Karagouni AD, Giannoutsou E, Adamakis ID, Ntalli NG (2020) The Nematicidal potential of bioactive *Streptomyces* strains isolated from Greek rhizosphere soils tested on Arabidopsis plants of varying susceptibility to Meloidogyne spp. Plan Theory 9(6):69

Migunova VD, Sasanelli N (2021) Bacteria as biocontrol tool against phytoparasitic nematodes. Plan Theory 10(2):389

Minuto A, Spadaro D, Garibaldi A, Gullino ML (2006) Control of soilborne pathogens of tomato using a commercial formulation of *Streptomyces griseoviridis* and solarization. Crop Prot 25(5):468–475

Mitra D, Mondal R, Khoshru B, Senapati A, Radha TK, Mahakur B, Uniyal N, Myo EM, Boutaj H, Sierra BE, Panneerselvam P (2022) Actinobacteria-enhanced plant growth, nutrient acquisition, and crop protection: advances in soil, plant, and microbial multifactorial interactions. Pedosphere 32(1):149–170

Mohamed SH, Galal AM (2005) Identification and antiviral activities of some halotolerant *Streptomycetes* isolated from Qaroon Lake. Int J Agric Biol 7(5):747–753

Muangham S, Pathom-Aree W, Duangmal K (2015) Melanogenic actinomycetes from rhizosphere soil – antagonistic activity against *Xanthomonas oryzae* and plant-growth-promoting traits. Can J Microbiol 61(2):164–170

Nazari MT, Schommer VA, Braun JC, dos Santos LF, Lopes ST, Simon V, Machado BS, Ferrari V, Colla LM, Piccin JS (2023) Using *Streptomyces* spp. as plant growth promoters and biocontrol agents. Rhizosphere 30:100741

Ngamcharungchit C, Chaimusik N, Panbangred W, Euanorasetr J, Intra B (2023) Bioactive metabolites from terrestrial and marine actinomycetes. Molecules 28(15):5915

Nimnoi P, Ruanpanun P (2020) Suppression of root-knot nematode and plant growth promotion of chili (*Capsicum flutescens* L.) using co-inoculation of *Streptomyces* spp. Biol Control 145:104244

Niu S, Gao Y, Zi H, Liu Y, Liu X, Xiong X, Yao Q, Qin Z, Chen N, Guo L, Yang Y (2022) The osmolyte-producing endophyte *Streptomyces albidoflavus* OsiLf-2 induces drought and salt tolerance in rice via a multi-level mechanism. Crop J 10(2):375–386

Nozari RM, Ramos LM, da Luz LA, Almeida RN, Lucas AM, Cassel E, de Oliveira SD, Astarita LV, Santarém ER (2022) Halotolerant *Streptomyces* spp. induce salt tolerance in maize through systemic induction of the antioxidant system and accumulation of proline. Rhizosphere 24:100623

Ouchari L, Boukeskasse A, Bouizgarne B, Ouhdouch Y (2019) Antimicrobial potential of actinomycetes isolated from the unexplored hot Merzouga desert and their taxonomic diversity. Biol Open 8(2):bio035410

Oyedoh OP, Yang W, Dhanasekaran D, Santoyo G, Glick BR, Babalola OO (2023) Rare rhizo-Actinomycetes: a new source of agroactive metabolites. Biotechnol Adv 24:108205

Padilla-Gálvez N, Luengo-Uribe P, Mancilla S, Maurin A, Torres C, Ruiz P, France A, Acuña I, Urrutia H (2021) Antagonistic activity of endophytic actinobacteria from native potatoes (*Solanum tuberosum* subsp. *tuberosum* L.) against *Pectobacterium carotovorum* subsp. carotovorum and *Pectobacterium atrosepticum*. BMC Microbiol 21:1–7

Pal KK, Gardener BM (2006) Biological control of plant pathogens. Plant Health Instr 2(5):1117–1142

Palaniyandi SA, Yang SH, Suh JW (2013) Extracellular proteases from *Streptomyces phaeopurpureus* ExPro138 inhibit spore adhesion, germination and appressorium formation in Colletotrichum coccodes. J Appl Microbiol 115(1):207–217

Patil HJ, Srivastava AK, Singh DP, Chaudhari BL, Arora DK (2011) Actinomycetes mediated biochemical responses in tomato (*Solanum lycopersicum*) enhances bioprotection against *Rhizoctonia solani*. Crop Prot 30(10):1269–1273

Pérez-Corral DA, Ornelas-Paz JD, Olivas-Orozco GI, Acosta-Muñiz CH, Salas-Marina MÁ, Berlanga-Reyes DI, Ruiz-Cisneros MF, Rios-Velasco C (2022) Molecular, morphological and biochemical characterization of actinomycetes and their antagonistic activity against phytopathogenic fungi. Rev Fitotec Mex 45(1):103–115

Pires D, Vicente CS, Menéndez E, Faria JM, Rusinque L, Camacho MJ, Inácio ML (2022) The fight against plant-parasitic nematodes: current status of bacterial and fungal biocontrol agents. Pathogens 11(10):1178

Priyadharsini P, Dhanasekaran D, Kanimozhi B (2013) Isolation, structural identification and herbicidal activity of N-phenylpropanamide from *Streptomyces* sp. KA1-3. Arch Phytopathol 46(3):364–373

Priyadharsini P, Dhanasekaran D, Gopinath PM, Ramanathan K, Shanthi V, Chandraleka S, Biswas B (2017) Spectroscopic identification and molecular modeling of diethyl 7-hydroxytrideca-2, 5, 8, 11-tetraenedioate: a herbicidal compound from *Streptomyces* sp. Arab J Sci Eng 42:2217–2227

Quach NT, Vu TH, Nguyen TT, Le PC, Do HG, Nguyen TD, Thao PT, Nguyen TT, Chu HH, Phi QT (2023) Metabolic and genomic analysis deciphering biocontrol potential of endophytic Streptomyces albus RC2 against crop pathogenic fungi. Braz J Microbiol 54(4):2617–2626

Quintana-Jones TA (2011) Evaluation of drip applications and foliar sprays of the biocontrol product Actinovate on powdery mildew and other fungal diseases of tomato. California Polytechnic State University

Raja VV, Dhanasekaran D (2022) An in vitro Nematicidal activity of Actinobacteria: juvenile mortality and egg hatching inhibition. Methods Actinobacteriol:615–619

Rajendran K, Krishnamoorthy M, Karuppiah K, Ethiraj K, Sekar S (2024) Chitinase from *Streptomyces mutabilis* as an effective eco-friendly biocontrol agent. Appl Biochem Biotechnol 196(1):18–31

Renuka R, Prabakar K, Anandham R, Pugalendhi L, Rajendran L, Raguchander T, Karthikeyan G (2023) Exploring the potentiality of native actinobacteria to combat the chilli fruit rot pathogens under post-harvest pathosystem. Life 13(2):426

Ruanpanun P, Dame ZT, Laatsch H, Lumyong S (2011a) 3-Methoxy-2-methyl-carbazole-1, 4-quinone, carbazomycins D and F from *Streptomyces* sp. CMU-JT005. FEMS Microbiol Lett 322(1):77–81

Ruanpanun P, Laatsch H, Tangchitsomkid N, Lumyong S (2011b) Nematicidal activity of fervenulin isolated from a nematicidal actinomycete, *Streptomyces* sp. CMU-MH021, on *Meloidogyne incognita*. World J Microbiol Biotechnol 27:1373–1380

Sabaratnam S, Traquair JA (2002) Formulation of a *Streptomyces* biocontrol agent for the suppression of *Rhizoctonia* damping-off in tomato transplants. Biol Control 23(3):245–253

Saberi-Riseh R, Moradi-Pour M (2021) A novel encapsulation of *Streptomyces fulvissimus* Uts22 by spray drying and its biocontrol efficiency against *Gaeumannomyces graminis*, the causal agent of take-all disease in wheat. Pest Manag Sci 77(10):4357–4364

Saberi-Riseh R, Moradi Pour M, Ait Barka E (2022) A novel route for double-layered encapsulation of *Streptomyces fulvissimus* Uts22 by alginate–Arabic gum for controlling of *Pythium aphanidermatum* in cucumber. Agronomy 12(3):655

Saxena S, Pandey AK (2001) Microbial metabolites as eco-friendly agrochemicals for the next millennium. Appl Microbiol Biotechnol 55:395–403

Shafie RM, Hamed A, El-Sharkawy HH (2016) Inducing systemic resistance against cucumber mosaic cucumovirus using *Streptomyces* spp. Egypt J Phytopathol 44(1):127–142

Shariffah-Muzaimah SA, Idris AS, Nur-Rashyeda R, Naidu Y, ZainolHilmi NH, Norman K (2020) Impact of pre-inoculating soil with *Streptomyces* sp. GanoSA1 on oil palm growth and Ganoderma disease development. Biocatal Agric Biotechnol 29:101814

Sharma M, Manhas RK (2020) Purification and characterization of salvianolic acid B from *Streptomyces* sp. M4 possessing antifungal activity against fungal phytopathogens. Microbiol Res 237:126478

Sharma M, Jasrotia S, Ohri P, Manhas RK (2019) Nematicidal potential of *Streptomyces antibioticus* strain M7 against *Meloidogyne incognita*. AMB Express 9:1–8

Sharma P, Aswini K, Sai Prasad J, Kumar N, Pathak D, Gond S, Venkadasamy G, Suman A (2023) Characterization of actinobacteria from wheat seeds for plant growth promoting traits and protection against fungal pathogens. J Basic Microbiol 63(3–4):439–453

Shekhar N, Bhattacharya D, Kumar D, Gupta RK (2006) Biocontrol of wood-rotting fungi with *Streptomyces violaceusniger* XL-2. Can J Microbiol 52(9):805–808

Sheng Y, Lam PW, Shahab S, Santosa DA, Proteau PJ, Zabriskie TM, Mahmud T (2015) Identification of elaiophylin skeletal variants from the Indonesian *Streptomyces* sp. ICBB 9297. J Nat Prod 78(11):2768–2775

Sholkamy EN, Muthukrishnan P, Abdel-Raouf N, Nandhini X, Ibraheem IB, Mostafa AA (2020) Antimicrobial and antinematicidal metabolites from *Streptomyces cuspidosporus* strain SA4 against selected pathogenic bacteria, fungi and nematode. Saudi J Biol Sci 27(12):3208–3220

Shu J, Cao K, Fei C, Dai H, Li Y, Cao Y, Zhou T, Yu M, Xia Z, An M, Wu Y (2024) Antiviral mechanisms of Anisomycin produced by *Streptomyces albulus* SN40 on potato virus Y. Agric Food Chem 72(7):3506–3519

Silambarasan S, Logeswari P, Sivaramakrishnan R, Cornejo P, Sipahutar MK, Pugazhendhi A (2022) Amelioration of aluminum phytotoxicity in *Solanum lycopersicum* by co-inoculation of plant growth promoting *Kosakonia radicincitans* strain CABV2 and *Streptomyces corchorusii* strain CASL5. Sci Total Environ 832:154935

Silva GD, Kitano IT, Ribeiro IA, Lacava PT (2022) The potential use of actinomycetes as microbial inoculants and biopesticides in agriculture. Front Soil Sci 21(2):833181

Singh AK, Chhatpar HS (2011) Purification, characterization and thermodynamics of antifungal protease from *Streptomyces* sp. A6. J Basic Microbiol 51:424–432

Singh SP, Gaur R (2017) Endophytic *Streptomyces* spp. underscore induction of defense regulatory genes and confers resistance against *Sclerotium rolfsii* in chickpea. Biol Control 104:44–56

Singh SP, Gupta R, Gaur R, Srivastava AK (2017) Antagonistic actinomycetes mediated resistance in *Solanum lycopersicon* mill. Against *Rhizoctonia solani* Kühn. Proc Natl Acad Sci India B – Biol Sci 87:789–798

Solanki MK, Malviya MK, Wang Z (2016) Actinomycetes bio-inoculants: a modern prospectus for plant disease management. In: Plant growth promoting Actinobacteria: a new avenue for enhancing the productivity and soil fertility of grain legumes, pp 63–81

Srividya S, Thapa A, Bhat DV, Golmei K, Dey N (2012) *Streptomyces* sp. 9p as effective biocontrol against chilli soilborne fungal phytopathogens. Eur J Exp Biol 2(1):163–173

Sutherland ED, Baker KK, Lockwood JL (1984) Ultrastructure of *Phytophthora megasperma* f.sp. glycinea oospores parasitised by *Actinoplanes missouriens* and *Humicola fuscoatra*. Trans Brit Mycol Soc 82:726–728

Taha M, Ghaly M, Atwa H, Askoura M (2021) Evaluation of the effectiveness of soil *Streptomyces* isolates for induction of plant resistance against tomato mosaic virus (ToMV). Curr Microbiol 78(8):3032–3043

Tapio E, Pohto-Lahdenperä A (1991) Scanning electron microscopy of hyphal interaction between *Streptomyces griseoviridis* and some plant pathogenic fungi. Agric Food Sci 63(5):435–441

Trejo-Estrada SR, Sepulveda IR, Crawford DL (1998) In vitro and in vivo antagonism of *Streptomyces violaceusniger* YCED9 against fungal pathogens of turfgrass. World J Microbiol Biotechnol 14:865–872

Trojak-Goluch A (2024) The use of bacteria, Actinomycetes and fungi in the bioprotection of Solanaceous crops against tobacco mosaic virus (TMV). Agriculture 14(8)

Tu JC (1988) Antibiosis of *Streptomyces griseus* against *Colletotrichum lindemuthianum*. J Phytopathol 121(2):97–102

Umurzokov M, Lee YM, Kim HJ, Cho KM, Kim YS, Choi JS, Park KW (2022) Herbicidal characteristics and structural identification of a potential active compound produced by *Streptomyces* sp. KRA18–249. Pestic Biochem Physiol 187:105213

Van Elsas JD, Chiurazzi M, Mallon CA, Elhottovā D, Krištůfek V, Salles JF (2012) Microbial diversity determines the invasion of soil by a bacterial pathogen. Proc Natl Acad Sci 109(4):1159–1164

Vatsa-Portugal P, Aziz A, Rondeau M, Villaume S, Morjani H, Clément C, Ait Barka E (2017) How *Streptomyces anulatus* primes grapevine defenses to cope with gray mold: a study of the early responses of cell suspensions. Front Plant Sci 8:1043

Vilasinee S, Toanuna C, McGovern RJ, Nalumpang S (2019) Expression of pathogenesis-related (PR) genes in tomato against fusarium wilt by challenge inoculation with *Streptomyces* NSP3. Int J Agric Technol 15(1):157–170

Wan M, Li G, Zhang J, Jiang D, Huang HC (2008) Effect of volatile substances of *Streptomyces platensis* F-1 on control of plant fungal diseases. Biol Control 46(3):552–559

Wang W, Qiu Z, Tan H, Cao L (2014) Siderophore production by actinobacteria. Biometals 27:623–631

Wang R, Zhang H, Sun L, Qi G, Chen S, Zhao X (2017) Microbial community composition is related to soil biological and chemical properties and bacterial wilt outbreak. Sci Rep 7(1):343

Wattana-Amorn P, Charoenwongsa W, Williams C, Crump MP, Apichaisataienchote B (2016) Antibacterial activity of cyclo (L-Pro-L-Tyr) and cyclo (D-Pro-L-Tyr) from *Streptomyces* sp. strain 22-4 against phytopathogenic bacteria. Nat Prod Res 30(17):1980–1983

Wei Y, Zhao Y, Zhou D, Qi D, Li K, Tang W, Chen Y, Jing T, Zang X, Xie J, Wang W (2020) Newly isolated *Streptomyces* sp. YYS-7 with a broad-spectrum antifungal activity improves the banana plant resistance to *Fusarium oxysporum* f. sp. cubense tropical race 4. Front Microbiol 11:1712

Widada J, Damayanti E, Alhakim MR, Yuwono T, Mustofa M (2021) Two strains of airborne *Nocardiopsis alba* producing different volatile organic compounds (VOCs) as biofungicide for *Ganoderma boninense*. FEMS Microbiol Let 368(20):fnab138

Won OJ, Kim YT, Choi JS, Oh TK, Shinogi Y, Park KW (2016) Herbicidal activity and mode of action of *Streptomyces scopuliridis* metabolites. J Fac Agric Kyushu 61:47–51

Xiang WS, Wang JD, Wang M, Wang XJ (2010) New nemadectin congener from *Streptomyces microflavus* neau3: fermentation, isolation, structure elucidation and biological activities. J Antibiot 63(4):171–175

Xu W, Wang F, Zhang M, Ou T, Wang R, Strobel G, Xiang Z, Zhou Z, Xie J (2019) Diversity of cultivable endophytic bacteria in mulberry and their potential for antimicrobial and plant growth-promoting activities. Microbiol Res 229:126328

Yandigeri MS (2021) Role of actinomycetes in insect pest and plant disease management. Biopestici Horticul Crops 17:152–172

Yandigeri MS, Malviya N, Solanki MK, Shrivastava P, Sivakumar G (2015) Chitinolytic *Streptomyces vinaceusdrappus* S5MW2 isolated from Chilika lake, India enhances plant growth and biocontrol efficacy through chitin supplementation against *Rhizoctonia solani*. World J Microbiol Biotechnol 31:1217–1225

Yang H, Li J, Xiao Y, Gu Y, Liu H, Liang Y, Liu X, Hu J, Meng D, Yin H (2017) An integrated insight into the relationship between soil microbial community and tobacco bacterial wilt disease. Front Microbiol 8:2179

Yang M, Huang C, Xue Y, Li S, Lu L, Wang C (2018) Biofumigation with volatile organic compounds from *Streptomyces alboflavus* TD-1 and pure chemicals to control *Aspergillus ochraceus*. Ann Appl Biol 3(3):313–322

Yoon GY, Lee YS, Lee SY, Park RD, Hyun HN, Nam Y, Kim KY (2012) Effects on *Meloidogyne incognita* of chitinase, glucanase and a secondary metabolite from *Streptomyces cacaoi* GY525. Nematology 14(2):175–184

Zakaria L (2023) Fusarium species associated with diseases of major tropical fruit crops. Horticulturae 9(3):322

Zambrano EC, Parra AS, Ortiz ÁM (2021) Biocontrol of rice sheath blight with microorganisms obtained in rice cultivated soils. Bragantia 80:e0921

Zeng Q, Huang H, Zhu J, Fang Z, Sun Q, Bao S (2013) A new nematicidal compound produced by *Streptomyces albogriseolus* HA10002. Antonie Van Leeuwenhoek 103:1107–1111

Zhang G, Han L, Zhang G, Zhang X, Feng J (2015) Purification and characterization of a novel glycoprotein from *Streptomyces* sp. ZX01. Int J Biol Macromol 78:195–201

Zhang D, Lu Y, Chen H, Wu C, Zhang H, Chen L, Chen X (2020) Antifungal peptides produced by actinomycetes and their biological activities against plant diseases. J Antibiot 73(5):265–282

Zhang L, Liu Z, Wang Y, Zhang J, Wan S, Huang Y, Yun T, Xie J, Wang W (2022) Biocontrol potential of endophytic *Streptomyces malaysiensis* 8ZJF-21 from medicinal plant against banana fusarium wilt caused by *Fusarium oxysporum* f. sp. cubense tropical race 4. Front Plant Sci 13:874819

Zhao S, Du CM, Tian CY (2012) Suppression of fusarium oxysporum and induced resistance of plants involved in the biocontrol of cucumber fusarium wilt by *Streptomyces bikiniensis* HD-087. World J Microbiol Biotechnol 28:2919–2927

Zheng X, Wang J, Chen Z, Zhang H, Wang Z, Zhu Y, Liu B (2019) A *Streptomyces* sp. strain: isolation, identification, and potential as a biocontrol agent against soilborne diseases of tomato plants. Biol Control 136:104004

Zhou D, Jing T, Chen Y, Yun T, Qi D, Zang X, Zhang M, Wei Y, Li K, Zhao Y, Wang W (2022) Biocontrol potential of a newly isolated *Streptomyces* sp. HSL-9B from mangrove forest on postharvest anthracnose of mango fruit caused by *Colletotrichum gloeosporioides*. Food Control 135:108836

Zhou S, Zhou Y, Li C, Wu W, Xu Y, Xia W, Huang D, Huang X (2023) Identification and genomic analyses of a novel endophytic actinobacterium *Streptomyces endophytica* sp. nov. with potential for biocontrol of yam anthracnose. Front Microbiol 14:1139456

Chapter 4
Commercial Applications of Elicitors in Crop Protection Products, Seed Treatments, Biostimulants, Sustainable Agriculture, and Precision Farming

Alia Telli, Heba I. Mohamed, and Murad Muhammed

Abstract Recently, researchers have looked for a comprehensive plan to lessen the negative impacts of synthetic chemicals in agricultural production in order to increase quality and output. By preventing the negative environmental consequences of pesticides and chemical fertilizers and safeguarding soil and water resources, biostimulants help plants. Plant biostimulants, also known as bioactivators, are gaining popularity in the agricultural industry because of their ability to accelerate a plant's growth and strengthen its tolerance to stress. Since biostimulants can improve crop quality, nutrient assimilation, growth rate, and stress tolerance, they are widely employed. This chapter offers a comprehensive analysis of the function of elicitors in agriculture, emphasizing their capacity to improve crop protection and foster sustainable agricultural practices. Elicitors, chemicals that activate the innate defense systems of plants, have garnered considerable interest in recent years for their capacity to enhance crop resilience against biotic and abiotic stresses. This chapter examines the progress and utilization of elicitors in seed treatments, biostimulants, and precision agriculture, emphasizing their economic and ecological advantages. Despite existing hurdles and constraints, including regulatory factors,

A. Telli (✉)
Laboratory of Preservation of Ecosystems in Arid and Semi-Arid Areas, Faculty of Nature and Life Sciences, University of KASDI Merbah-Ouargla, Ouargla, Algeria
e-mail: telli.alia@univ-ouargla.dz

H. I. Mohamed (✉)
Biological and Geological Sciences, Faculty of Education, Ain Shams University, Cairo, Egypt

M. Muhammed
State Key Laboratory of Desert and Oasis Ecology, Key Laboratory of Ecological Safety and Sustainable Development in Arid Lands, Xinjiang Institute of Ecology and Geography, Chinese Academy of Sciences, Xinjiang, China

University of Chinese Academy of Sciences, Beijing, People's Republic of China

K. A. Abd-Elsalam, H. I. Mohamed (eds.), *Elicitors for Sustainable Crop Production*, Sustainability Sciences in Asia and Africa,
https://doi.org/10.1007/978-981-96-6569-3_4

the application of elicitors in agriculture has surfaced as a viable strategy for fostering a more sustainable and resilient agricultural system.

Keywords Elicitors · Crop protection · Sustainable agriculture · Seed treatments · Biostimulants · Biotic and abiotic stresses

4.1 Introduction

The main goal of integrated agriculture is to maximize yield while minimizing treatments and preventing environmental impact, coupling socially and economically based strategies. Emphasizing the need for greener agriculture, significant international resources are currently being allocated to bio-based products, and related market forecasts are continuously updated. The current trend in innovative agriculture intends to address the main challenges of sustainable world food security. Many low-risk crop protection tools should be developed to successfully replace the tools that are more prone to generating resistance, and unwanted environmental and human health impacts (Suwanchaikasem et al. 2022; Jeyasri et al. 2023).

Among the promising approaches for inducing plant disease resistance reducing stress damage and stimulating plant growth is the application of different elicitors. Elicitors are recognized as compounds that trigger or amplify plant defense responses, enabling them to respond more effectively to biotic and abiotic stresses. They can mimic or interact with molecules produced during pathogen invasion to stimulate the plant's defense mechanisms. Elicitors, also known as resistance inducers, are affordable tools that have the potential to overcome some of the current limitations in agriculture (Tütüncü et al. 2024). In addition to significant environmental benefits, the regular use of these sustainable tools in crop protection and sustainable agriculture also serves as a viable solution to feed the ever-increasing world population. Commercial products based on elicitors are available as standalone treatments, in mixture with other elicitors and fertilizers, or combined with phyto-stimulating agents (Mejia-Teniente et al 2010; Bektas and Eulgem 2015; Guru et al. 2022; Gowtham et al. 2024).

This review seeks to elucidate the progress and applications of elicitors in agriculture, examining their commercial utilization in crop protection and seed treatments, their role as biostimulants for sustainable agriculture, and their economic and environmental advantages in precision farming. The discussion will also encompass the problems and constraints of integrating elicitors into crop protection solutions, as well as regulatory issues and prospects.

4.2 Most Important Concepts

Disease refers to the disruption of normal physiological functions in plants induced by microbes or an abiotic agent. A pathogen is an agent that induces infection or sickness, particularly a microbe such as a bacteria, protozoa, or virus (Lahlali et al. 2022). A disease-causing pathogen is referred to as virulent. A pathogen that does not induce disease is referred to as avirulent. Plant pathogens are categorized into three kinds, which are necrotrophic pathogen (necrotrophic pathogens invade and destroy host tissue, subsequently deriving nourishment from the deceased host cells, e.g., *Botrytis cinerea* and *Erwinia caratovora*), biotrophic pathogens (biotrophic infections infiltrate living plant tissue and derive nourishment from viable host cells, e.g., *Erysiphe cruciferarum*), and hemibiotrophic pathogens (hemibiotrophic pathogens are effective groups of plant diseases that necessitate living plant tissue for survival and the completion of their life cycle, e.g., *Phytophtora infestans* and *Pseudomonas syringae*) (Gowthami 2018).

Plant defense mechanisms refer to the complex and multifaceted strategies employed by plants to protect themselves against biotic and abiotic stresses, including pathogens, pests, and environmental stressors, by triggering physical and biochemical responses that enhance their resilience and resistance to damage (Doughari 2015; Kaur et al. 2022). Abiotic stress denotes the nonbiological variables that adversely affect a plant's capacity to survive, develop, and reproduce, including severe temperatures, drought, flooding, salinity, and excessive light or radiation exposure (Onaga and Wydra 2016). Biotic stress implies the living forces that adversely affect a plant's capacity to survive, grow, and reproduce, including diseases, pests, insects, and other organisms that may induce disease or harm to the plant (Boyd et al. 2013).

Sustainable agriculture means the cultivation of crops and livestock management that emphasizes enduring environmental stewardship, social accountability, and economic sustainability, striving to fulfill current needs without jeopardizing future generations' capacity to satisfy their own requirements (Peigné et al. 2016). This strategy prioritizes the amalgamation of three primary objectives: ecological well-being, financial viability, and social and economic equality. Elicitors advance the attainment of these objectives by enhancing crop resilience, decreasing chemical inputs, and mitigating the environmental impact of agricultural practices through the promotion of sustainable methods (Velten et al. 2015).

4.3 Definition and Types of Elicitors

The term elicitor was first designated for chemicals that induce phytoalexin synthesis, but it is now broadly applied to substances that stimulate any form of plant defense (Thakur and Sohal 2013). Elicitors are molecules or compounds that, when applied to plants, induce significant physiological modifications akin to the mechanisms observed in plant responses. Pathogen or environmental stress responses are initiated, influencing plant metabolism and increasing the biosynthesis of secondary metabolites (Moreno-Pérez et al. 2020). The use of elicitors offers several advantages compared to alternative methods; they are nontoxic, environmentally friendly, and, importantly, require only small quantities to provide long-term protection to plants against a broad spectrum of pathogens (Zheng et al. 2020).

Elicitors can be classified in several ways. According to Radman et al. (2003), they can be categorized as physical or chemical, biotic or abiotic, and complicated or well-defined, based on their nature, origin, or molecular structure. Vasconsuelo and Boland (2007), in their investigation of elicitation to enhance the production of secondary metabolites in vitro, along with Baenas et al. (2014), in their review of elicitation as a means to augment the concentration of bioactive compounds, underscore the significance of origin and categorize elicitors into biotic (exogenous, derived from biological sources due to fungal, bacterial, or viral infections, or endogenous, generated by the plant in response to pathogen assault) and abiotic (physical or chemical) factors, as well as plant hormones. A third classification approach involves examining the relationship between the plant and the elicitor, categorizing general elicitors as those that trigger nonspecific defensive mechanisms across various crops, while specific elicitors exclusively influence a particular host (Vasconsuelo and Boland 2007; Thakur and Sohal 2013; Baenas et al. 2014; Moreno-Pérez et al. 2020). Figure 4.1 presents a classification overview of elicitors. Recent investigations have revealed significant parallels between the defense systems activated by general elicitors and the innate immunity of animals, suggesting that the detection of general elicitors may subsequently initiate plant innate immunity (Thakur and Sohal 2013).

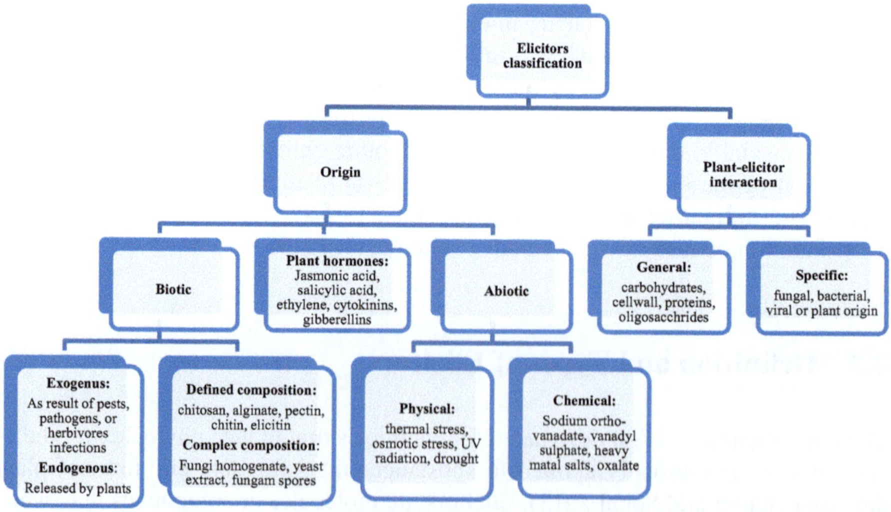

Fig. 4.1 Types of elicitors based on their origin or their interaction with plants

4.4 Elicitors' Mechanisms of Action

The concept of elicitors is based on the idea that plants have an innate ability to defend themselves against pathogens and pests and that certain substances can trigger this defense response. Elicitors can be derived from various sources, including plants, fungi, bacteria, and synthetic compounds (Jalota et al. 2024). They work by binding to specific receptors on the plant cell surface, triggering a signaling cascade that activates the plant's defense genes, leading to the production of defense-related proteins and compounds. This enhanced defense response enables the plant to better resist pathogen and pest attacks, reducing the need for pesticides and other chemicals (Jalota et al. 2024). Elicitors are typically classified into two main categories: pathogen-associated molecular patterns (PAMPs) and damage-associated molecular patterns (DAMPs). PAMPs are conserved molecules associated with pathogens, such as bacterial flagellin and fungal chitin, which are recognized by the plant's immune system (Ramirez-Estrada et al. 2016).

DAMPs, on the other hand, are endogenous molecules released by damaged plant cells that signal the presence of pathogens or pests. These molecules trigger an immune response in neighboring cells, preparing them to defend against potential threats. Despite their different origins, both PAMPs and DAMPs ultimately activate similar defense pathways in plants. This activation process involves a series of biochemical reactions that culminate in the expression of defense-related genes and the production of antimicrobial compounds. Elicitors can also induce systemic acquired resistance (SAR), a long-lasting and broad-spectrum defense mechanism that protects plants against a wide range of pathogens and pests. This phenomenon involves the production of mobile signals that travel throughout the plant, priming distant tissues for enhanced defense responses. Elicitors, therefore, represent a promising tool for sustainable crop protection, harnessing the plant's natural defense mechanisms to reduce reliance on chemical inputs (Tripathi et al. 2019; Kamle et al. 2020; Meena et al. 2022).

Different receptors in various plant species interact with elicitors to further trigger signals inside the cell. There is reversible phosphorylation and dephosphorylation of the cytosolic and plasma membrane proteins. Intracellular acidification and extracellular alkalization are caused by the passage of hydrogen ions and the efflux of potassium and chloride ions from the cell. Mitogen-activated protein kinase (MAPK) cascade activation by G-protein in conjunction with membrane receptors that are phosphorylated in turn. Moreover, calcium ions enter the plant cell's cytoplasm by infiltration. Calcium ion-induced production of reactive oxygen species, reactive nitrogen species, and NADPH oxidase activation. The calcium-dependent protein kinases (CDPK), which are involved in controlling the biosynthetic pathways, are likewise stimulated by calcium ions (Jalota et al. 2024).

Expression of functionally active genes results in the production of bioactive chemicals from plants. Phospholipases create signaling molecules like salicylic acid, jasmonate, and others that trigger the de novo biosynthesis of transcriptional factors. These factors regulate the expression of genes involved in the creation of

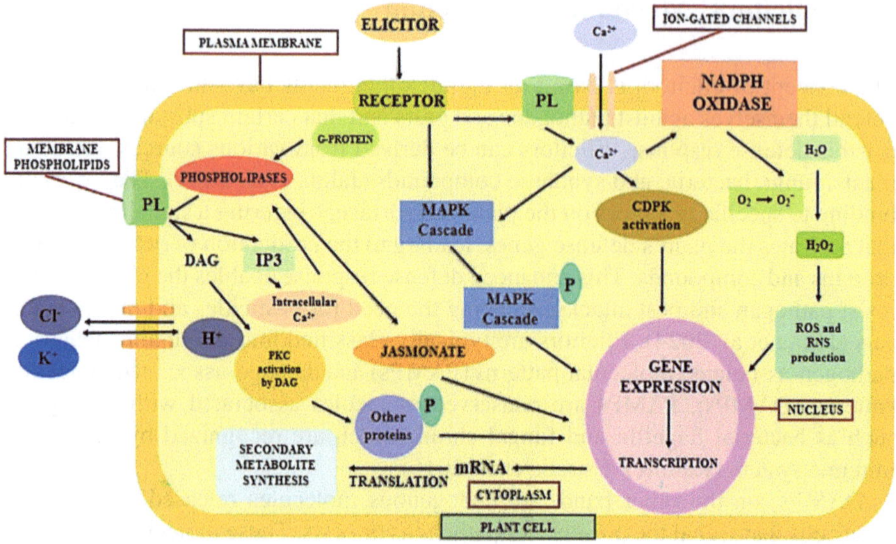

Fig. 4.2 The ubiquitous mode of action of elicitors within a plant cell (Jalota et al. 2024). *CDPK* calcium-dependent protein kinases, *MAPK* mitogen-activated protein kinase, *DAG* diacylglycerol, *IP3* inositol triphosphate, *PKC* protein kinase C

enzymes and, eventually, the manufacture of secondary metabolites (Ramirez-Estrada et al. 2016; Narayani and Srivastava 2017). The elicitors' widespread mode of action within a plant cell is shown in Fig. 4.2.

4.5 Factors Affecting the Elicitation Process

Elicitation has been widely employed in vitro culture settings to improve synthesis or initiate the production of phytochemicals. Numerous aspects are crucial in determining the productivity of plant secondary metabolites, including the amount and specificity of the elicitor, exposure duration, growth regulation, nutritional composition, and the quality of the chemicals being employed. Below, a specific number of the factors are covered.

4.5.1 Elicitor Concentration

The elicitor's concentration is a critical component of the elicitation process. When a dose is administered over the optimal level, a hypersensitive reaction is triggered in the plants, resulting in cell death; however, an optimal concentration of elicitors is necessary to induce secondary metabolite synthesis (Patel and Krishnamurthy 2013).

4.5.2 Exposure Time to Elicitor

The time of exposure of elicitor effect on the plant and production of secondary metabolites. The maximum biosynthesis of gymnemic acid (135.41 ± 0.43 mg/g dry cell weight) was observed at the 72-hour mark after treating *Gymnema sylvestre* cell suspension culture with methyl jasmonate at a concentration of 150 μM for varying lengths of time (24, 48, and 72 h). When compared to cultures without elicitors, the concentration of gymnemic acid in evoked cultures was 15.4 times higher (Chodisetti et al. 2015; Narayani and Srivastava 2017). Various fungal and bacterial preparations (*Agrobacterium rhizogenes, Bacillus subtilis, Escherichia coli, Aspergillus niger*, and *Saccharomyces cerevisiae*) were used to increase the amount of gymnemic acid in the cell suspension culture of *Gymnema sylvestre* at different times (24, 48, 48, 72, and 72 h, respectively). The highest amount was recorded with *Aspergillus niger* after 72 h (Chodisetti et al. 2013). After 2 days, the *Bacopa monnieri* shoot culture treated with methyl jasmonate produced the highest concentration of bacoside compounds, 1.5 times that of the control cultures (Sharma et al. 2013).

For varying lengths of time (24, 48, 72, and 96 hours), the *Catharanthus roseus* cell culture was exposed to 5% filtrates of fungi such as *Trichoderma viride, Aspergillus niger*, and *Fusarium moniliforme*, respectively. Cell cultures induced with *Trichoderma viride* for 48 hours showed a threefold increase in ajmalicine production, whereas cultures to which *Aspergillus niger* and *Fusarium moniliforme* biotic elicitors were added showed a twofold increase (Namdeo et al. 2002) In *Andrographis paniculate* cell culture, the hormonal elicitor methyl jasmonate (5 μM) promoted the maximum storage of andrographolide compounds at 24 hours as opposed to 48 and 72 hours (Sharma et al. 2015).

4.5.3 Culture Age

Culture age is another important determinant in the production of secondary metabolites in plants. After 40 days of culture, the *Withania somnifera* hairy root culture elicited salicylic acid, which resulted in the highest storage of withanolide A, withanone, and withaferin A compounds (Sivanandhan et al. 2013). Ajmalicine and serpentine were synthesized separately when methyl jasmonate was added to suspension cultures of *Catharanthus roseus* (6 days old) at concentration levels of 10 μM and 100 μM. Re-elicitation, however, showed an antagonistic effect on growth and alkaloid formation (Sumit et al. 2011). *Catharanthus roseus* cultures that were stimulated 20 days ago produced more bioactive ajmalicine. When treated with fungal extracts of *Trichoderma viride*, the 20-day-old culture accumulated the most ajmalicine (166 μg/g of dry weight); in contrast, the cell culture obtained with exogenous extracts of *Aspergillus niger* and *Fusarium moniliforme* accumulated 90 μg/g and 88 μg/g of dry weight of ajmalicine compounds, respectively (Namdeo et al. 2002; Patel and Krishnamurthy 2013).

4.5.4 Nutritional Makeup

Another crucial element in the elicitation process is the medium's composition or media selection. The components of the culture medium of *Erythroxylum coca* callus culture had a notable impact on the amounts of various secondary metabolites, including cocaine, cinnamoyl cocaine, chlorogenic acid (CGA), and 4-coumaroyl quinate (CQA) (Docimo et al. 2015). According to reports, the synthesis of cocaine on culture media like Gamborg B5 and Murashige–Tucker medium is nearly identical. The medium differed greatly from one another in terms of the generation of chlorogenic acid, with ARM showing the highest synthesis and MMT showing the lowest. The composition of the nutritional medium was the primary factor in charge of monitoring the buildup of tropane, an alkaloid. Due to the existence of distinct components in ARM media as opposed to other media, it was found that employing ARM as the culture medium increased the content of tropane synthesis (Docimo et al. 2015). In addition to these features, the specificity of the elicitor, the presence of PGRs, the medium composition, the cell lines or clones of microbial elicitors, and the ambient conditions supplied all have an impact on elicitation efficiency (Jalota et al. 2024).

4.6 Role of Elicitors in Agriculture

Elicitors are a unique group of plant protection agents. Elicitors represent a wide range of compounds, including microbial preparations or cell wall components of host and nonhost plants, which are applied via seed treatment, foliar sprays, soil drenches, or in other ways that potentially provide a broad spectrum of effects on their activity. The principles of action of elicitors are generally based on the boosting of endogenous defense mechanisms of plants that are triggered hypersensitively by pathogen-associated elicitors. The application of elicitors provides long-term protection of plants. It is important to note that elicitors never induce a direct toxic effect on pathogenic organisms that might lead to resistance issues (Malook et al. 2022; Altun and Orcan 2024).

4.7 Commercial Applications of Elicitors in Crop Protection Products

The rapidly evolving landscape of crop protection necessitates innovative approaches to ensure sustainable agricultural practices. One promising strategy involves the utilization of elicitors, substances that trigger plant defense mechanisms against pathogens and pests. Elicitors have a wide range of applications in crop protection, including disease management, pest control, and abiotic stress tolerance. In the following sections, we will explore the applications of elicitors in disease management, pest control, and abiotic stress tolerance.

4.7.1 Diseases Management

Elicitors can help protect crops against a wide range of plant diseases by activating defense mechanisms that inhibit pathogen growth and spread. These compounds can be applied preventively or in response to disease outbreaks, providing growers with flexible options for disease management. By priming plants for enhanced defense responses, elicitors can help reduce the incidence and severity of diseases, minimizing yield losses and ensuring crop quality (Poveda and Díez 2023; Gowthami 2018).

4.7.2 Pests' Control

Elicitors can also help protect crops against a variety of pests by activating defense mechanisms that deter pest feeding and oviposition. These compounds can be applied preventively or in response to pest infestations, providing growers with effective tools for pest control. By deterring pest activities, elicitors can help minimize yield losses and ensure crop quality. (Thakur and Sohal 2013; Malik et al. 2020).

4.8 Commercial Applications of Elicitors in Biostimulants

The agricultural industry is facing significant challenges in meeting the increasing global demand for food while minimizing its environmental footprint. In recent years, biostimulants have emerged as a promising solution to enhance crop growth and yield, improve resource use efficiency, and reduce the environmental impact of agricultural practices. Biostimulants are a class of substances that, when applied to plants or the soil, stimulate natural processes to enhance plant growth and development (Backer et al. 2018).

4.8.1 Definition and Classification of Biostimulants

Biostimulants can be defined as substances or microorganisms that, when applied to plants or the soil, stimulate natural processes to enhance plant growth and development. The term "biostimulant" was first introduced in the 1990s to describe a new category of agricultural products that did not fit into the traditional categories of fertilizers or pesticides (Singh 2013). Biostimulants can be classified into several categories based on their composition, mode of action, and application method. Some of the main categories of biostimulants include (Bulgari et al. 2015; Caradonia et al. 2019; Sible et al. 2021; Soudani et al. 2023).

(a) **Plant Growth Regulators (PGRs):** These are substances that regulate plant growth and development by influencing hormone activity (Gupta et al. 2023).
(b) **Plant Extracts:** These are concentrated plant extracts that contain bioactive compounds that stimulate plant growth (Yadav et al. 2024).
(c) **Microbial Biostimulants:** These are microorganisms such as bacteria, fungi, and algae that stimulate plant growth and development (Calvo et al. 2014; González-González et al. 2020).
(d) **Amino Acid Biostimulants:** These are amino acid-based products that stimulate plant growth and development (Sun et al. 2024).
(e) **Humic and Fulvic Acid Biostimulants:** These are products derived from humic and fulvic acids that stimulate plant growth and development (Calvo et al. 2014).

4.8.2 Biostimulants' Mode of Action

Overall, biostimulants can be classified into different categories based on their composition, mode of action, and application method. Understanding the different types of biostimulants is essential for selecting the right product for specific agricultural applications. In the following sections, we will explore the mechanisms of action of biostimulants and their impact on crop growth and yield (Jardin 2015; Baltazar et al. 2021; Franzoni et al. 2022; Soudani et al. 2023). These mechanisms include

(a) **Hormonal Regulation:** Biostimulants can regulate plant growth and development by influencing hormone activity. Plant hormones such as auxins, cytokinins, gibberellins, abscisic acid, and ethylene play crucial roles in controlling various aspects of plant growth and development. Biostimulants can influence hormone activity by mimicking the effects of natural hormones, altering hormone metabolism, or affecting hormone transport within the plant (Gupta et al. 2023).
(b) **Antioxidant Activity:** Biostimulants can enhance plant stress tolerance by scavenging reactive oxygen species (ROS) and reducing oxidative damage in plants. ROS are natural byproducts of plant metabolism that can accumulate to toxic levels under abiotic and biotic stress conditions. Biostimulants with antioxidant activity can help plants cope with stress by neutralizing ROS and protecting plant cells from oxidative damage (Ali et al. 2024).
(c) **Nutrient Uptake and Use Efficiency:** Biostimulants can improve nutrient uptake and use efficiency in plants by enhancing root development, increasing nutrient availability in the rhizosphere, and promoting nutrient translocation within the plant (Mohamed et al. 2021).

(d) **Enhanced Metabolic Activity:** Biostimulants can stimulate plant metabolism and enhance various biochemical and physiological processes in plants. These processes include photosynthesis, respiration, protein synthesis, and enzyme activity. Biostimulant components may be linked to the creation of molecules or precursors involved in metabolic activities, even though their exact impact on these processes has not yet been fully understood. For example, because they contain short-chain peptides, protein hydrolisates may regulate plant development, and some amino acids, such phenylalanine, have been shown to act as signaling molecules that enhance the synthesis of endogenous auxin (Raguraj et al. 2022). Furthermore, the synthesis of secondary metabolites that improve plant defense and stress tolerance may be influenced by biostimulants like phenolic chemicals (Nephali et al. 2021).

4.8.3 Agricultural, Economic, and Environmental Benefits of Biostimulants

The application of biostimulants has been shown to have a significant impact on crop growth and yield. Biostimulants can enhance plant growth and development, leading to increased crop yields and improved crop quality (Hamid et al. 2021). The effects of biostimulants on crop growth and yield can be attributed to their ability to stimulate plant metabolism, enhance nutrient uptake and use efficiency, and promote plant stress tolerance (Rouphael and Colla 2020). Studies have shown that biostimulants can increase crop yields by up to 20% or more, depending on the type of crop, application method, and environmental conditions (Li et al. 2022). In addition to increasing crop yields, biostimulants can also improve crop quality. For example, biostimulants can enhance fruit size, color, and texture, leading to improved marketability and consumer acceptance. Biostimulants can also reduce the incidence of crop diseases and pests, leading to reduced pesticide use and improved environmental sustainability (Hamid et al. 2021; Miranda et al. 2024). Biostimulants provide several environmental and economic benefits that make them an attractive option for sustainable agriculture. These benefits include

(a) **Reduced Chemical Use:** Biostimulants can reduce the need for chemical fertilizers and pesticides, leading to a decrease in environmental pollution and the promotion of sustainable agriculture practices (Miranda et al. 2024).
(b) **Improved Water Quality:** Biostimulants can improve water quality by reducing the amount of nutrients and pollutants that enter waterways through agricultural runoff (de Almeida Silva et al. 2024).
(c) **Soil Conservation:** Biostimulants can help conserve soil health by promoting soil biota and improving soil structure, reducing the risk of soil erosion and degradation (Rajesaheb et al. 2024).
(d) **Climate Change Mitigation:** Biostimulants can help mitigate climate change by promoting carbon sequestration in soils, reducing greenhouse gas emissions,

and enhancing plant stress tolerance to extreme weather conditions (Zulfiqar et al. 2024).

(e) **Cost Savings:** Biostimulants can provide cost savings to farmers by reducing the need for chemical inputs, improving crop yields, and enhancing crop quality.
(f) **Increased Crop Value:** Biostimulants can increase crop value by improving crop quality, texture, and appearance, leading to higher market prices and increased profitability for farmers (Matthews et al. 2024).
(g) **Job Creation:** The biostimulant industry can create new job opportunities in manufacturing, sales, and application, contributing to local economic growth and development.
(h) **Improved Food Safety:** Biostimulants can improve food safety by reducing the use of chemical pesticides and fertilizers, leading to lower residue levels on crops and a reduced risk of chemical contamination (Uddain 2024).

4.8.4 Commercial Applications of Biostimulants in Agriculture

Biostimulants have various commercial applications in agriculture (Singh et al. 2024) including

(a) **Crop Yield Enhancement:** Biostimulants can enhance crop yields by stimulating plant growth and development.
(b) **Stress Tolerance:** Biostimulants can help plants cope with abiotic and biotic stress, such as drought, salinity, and extreme temperatures (Sangiorgio et al. 2020).
(c) **Soil Health Improvement:** Biostimulants can improve soil health by promoting beneficial microorganisms and increasing nutrient availability.
(d) **Organic Farming:** Biostimulants can be used in organic farming to promote sustainable agriculture practices (Sani and Yong 2021).
(e) **Food and Agricultural Wastes Management:** Developing biostimulants from by-products paves the path to waste recycling and reduction, generating benefits for growers, food industry, registration and distribution companies, as well as consumers (Xu and Geelen 2018).
(f) **Greenhouse Production:** Biostimulants can be used in greenhouse production to enhance crop growth and yield.
(g) **Turf and Ornamental Care:** Biostimulants can be used in turf and ornamental care to promote healthy plant growth and appearance.
(h) **Seed Treatment:** Biostimulants can be used as seed treatments to enhance seed germination and seedling establishment.
(i) **Biofertilizers:** Biostimulants can be used as biofertilizers to promote plant growth and reduce chemical fertilizer use (Ahmed et al. 2023).
(j) **Bioremediation:** Biostimulants can be used for bioremediation to clean pollutants from contaminated soil and water (Bartucca et al. 2022).

(k) **Aquaculture:** Biostimulants can be used in aquaculture to promote healthy fish and plant growth. The commercial applications of biostimulants in agriculture are diverse and can contribute to sustainable agriculture practices, improved crop yields, and reduced environmental impacts (Domínguez et al. 2024).

4.8.5 Challenges and Limitations of Biostimulants

Despite the potential benefits of biostimulants, several challenges and limitations need to be addressed (Bartucca et al. 2022; Li et al. 2022; Bashir et al. 2021; Kapoore et al. 2021). These include

(a) **Lack of Standardization:** There is currently a lack of standardization in the biostimulant industry, which can make it difficult to compare products and ensure consistency in quality.
(b) **Limited Understanding of Mechanisms:** While biostimulants are effective in enhancing plant growth and development, there is still a limited understanding of their mechanisms of action, which can make it difficult to optimize their use.
(c) **Variable Efficacy:** The efficacy of biostimulants can vary depending on factors such as soil type, climate, and crop species, which can make it challenging to predict their performance in different environments.
(d) **High Cost:** Biostimulants can be more expensive than traditional fertilizers and pesticides, which can be a barrier to adoption for some farmers.
(e) **Limited Availability:** Biostimulants may not be widely available in all regions, which can limit their adoption and use.
(f) **Regulatory Framework:** The regulatory framework for biostimulants is still evolving and can vary from country to country, which can create uncertainty and confusion for manufacturers and users (Therond et al. 2017).
(g) **Public Perception:** There may be public perception issues related to the use of biostimulants, particularly if they are perceived as being similar to genetically modified organisms (GMOs) or other biotechnology products.

4.9 Commercial Applications of Elicitors in Seed Treatment

Seed treatments promote seed germination and prevent yield reduction due to disease and biotic stress attacks. Successful germination and healthy establishment of the seedlings are essential for achieving good crop yield. However, plants are more susceptible to numerous diseases during germination or in the seedling or early growth stage. Nearly 70% of crop products depend on seeds or any of their derivatives to establish and produce the next generation of crops in the field. Seed quality, defined as the seed's potential to produce a normal seedling under a wide range of environmental conditions, is critical for plant establishment. This is directly related to several seed components like seed size, less structural damage, high germination and vigor indices, low disease levels, essential nutrients, and microbial control. It precludes poor yields, and managers have received a high return on seed investment (Meena et al. 2022; Abdul et al. 2020).

Seed treatment has emerged as a crucial aspect of agricultural practices, enabling farmers to optimize crop yields and reduce losses due to pests and diseases. Recently, the application of elicitors in seed treatment has gained significant attention for its potential to stimulate plant defense mechanisms, promoting healthy growth and development (Samota et al. 2024). Elicitors can serve as biomolecules that stimulate the plant immune system, activating the synthesis of numerous defense proteins and metabolites in plant tissues, making them more resistant to subsequent stress and preventing damage caused by both internal and exogenous factors. This mechanism works at the onset of disease and in a multifunctional approach. They exhibit antimicrobial activity, stimulate the activity of plant defense enzymes, perform signal transduction, improve tolerance to agents such as drought or soil contamination, promote plant growth, and interact with beneficial microbiota (Gowtham et al. 2024).

Products containing these biomolecules are becoming more common in the market for protection against pests and diseases. Elicitors such as chitosan, extracts of group 4 thiazolyl tetrazole, and phosphite salts containing ions are increasingly demanded for improving plant defense and remediation (Gowtham et al. 2024). Using elicitors as components of seed treatment in the production of seeds additionally stimulates the immunity associated with increasing resistance to pathogens in soil or crops that are transmitted only through seeds; the development of crop stands; the formation of amicable plant microflora; diseases of crops that persist in the soil; and pathogens carried on the seeds (Malook et al. 2022).

Elicitors have shown considerable promise as seed treatment agents, offering a sustainable and environmentally friendly alternative to traditional chemical treatments. These molecules can stimulate plant defense responses, enabling the plant to resist pests and diseases and promoting healthy growth and development. As a result, elicitors have the potential to enhance agricultural productivity and reduce reliance on chemical inputs, making them an attractive option for modern agricultural practices (Selwal et al. 2024). Several elicitors have recently been commercialized for use in seed treatment, providing farmers with various options to protect their crops and improve yields. These commercial elicitors can be categorized based on their origins and modes of action (Chalfoun et al. 2018).

Microbial elicitors are derived from pathogens and recognized by plant pattern recognition receptors (PRRs), triggering defense responses in the plant. These elicitors can stimulate systemic acquired resistance (SAR) and induce the expression of defense-related genes, enhancing plant resistance to numerous pests and diseases (Gowtham et al. 2024).

Plant-derived elicitors are endogenous molecules produced by plants in response to stress. These elicitors can enhance plant defense mechanisms and promote healthy growth and development, making them valuable tools for seed treatment.

Synthetic elicitors are artificially synthesized compounds that mimic the effects of natural elicitors. These compounds can activate plant defense responses and protect the plant from pests and diseases, offering a sustainable alternative to traditional chemical treatments (Bektas and Eulgem 2015).

Alternaria leaf blight on mustard is a critical problem, and chemical pesticides are the sole remedy. Induced resistance, which enhances defense-related enzymes and antioxidants, is a favored approach. A study on three mustard varieties revealed that pre-sowing elicitor treatments enhanced seed vigor and resilience. This indicates that pre-sowing elicitor therapy is an effective and economical approach for enhancing plant defense mechanisms and seedling characteristics, rendering it a practical solution (Rayanootha et al. 2022).

4.10 Commercial Applications of Elicitors in Sustainable Agriculture and Precision Farming

The phyto-industry is showing an increased interest in finding new biologically active compounds that help to establish sustainable agricultural practices. Climate change emphasizes the need to establish a set of values that meet the need for new and diversified plant protection strategies, acting as a complement to the restricted range of modes of action accessible through synthetic molecules. In this context, the concept of sustainable agriculture arises, a philosophy that advocates the joint development of economic performance, respect for the environment, and social cohesion. The current agricultural directives aim at promoting agricultural systems with low dependency on chemical products. Moreover, there is a perspective of a substantial reduction in synthetic pesticide use. At a global level, future research and innovation activities, as well as the policies, measures, and funding introduced, also aim at the use of innovative biotechnological tools for sustainable agriculture, reducing the chemical load to guarantee the preservation of human health and environmental quality, while ensuring the required plant protection level (Asghar et al. 2024).

4.11 Regulatory Approval of Elicitors

For the commercial registration of a new crop protection product, plant biostimulant, or seed treatment, the following safety requirements are demanded regulatory control: safety of exposed workers; operators involved in mixing, loading, application, and disposal; safety for consumers; residues in food, feed, and water; operator and bystander exposure; worker exposure to chemical and biological agents; environmental risk assessment of the active substance and plant protection product; protection of aquatic organisms; birds and mammals; nontarget arthropods; and honeybees; ecotoxicologically effective microdose for bees; earthworms and soil microorganisms. These necessary safety requirements can be confirmed by the registration process, comprising: (i) laboratory bench testing and scaled field trials for bio- and agro-ecological efficacy; (ii) bioavailability and metabolism, degradation, and mobility of the molecules; and (iii) extensive toxicology, metabolism, and exposure studies across all clauses of chemistry, manufacturing, and control for agrochemicals. Any change in the commercial name, manufacturing process, manufacturing method, or formulation must be approved by the regulatory authority. The assignment of a unique agricultural active substance name and structure enables the industry and regulatory authorities to globally characterize chemical data and to link physical, chemical, and biological properties via the identifier (Collinge et al. 2022).

In general, regulatory guidelines refer to substances that serve as adjuvants for plant protection products: fertilizer products: The national consultants' guidelines suggest a national regulatory notification system applicable to relevant authorities for the major categories of fertilizers that can serve as adjuvants. These guidelines also regulate plant protection products regarding their classification, bringing to account whether the product is placed on the market as a tank mix adjuvant or as a stand-alone adjuvant and if the active ingredient is already assessed for plant protection purposes. As of today, there are standards for various crops that require an assessment of the potential effects of tank-mix adjuvants. In particular, the determination of the biological effects of adjuvants must be performed in accordance with the respective legislative requirements. The Biological Efficacy Regulating Authority implements regulations that establish specific standards applicable to adjuvants incorporated in the legislative framework supporting plant protection products using elastomeric capsule technology.

The commercialization of elicitors is subject to regulatory approval by government agencies responsible for pesticide registration and crop protection. Regulatory approval ensures that elicitors meet stringent safety and efficacy standards before they can be marketed and used in agriculture. In the United States, the Environmental Protection Agency (EPA) is responsible for regulating elicitors under the Federal Insecticide, Fungicide, and Rodenticide Act (FIFRA). Under FIFRA, elicitors are classified as biopesticides, a category that includes microbial pesticides, biochemical pesticides, and plant-incorporated protectants. Biopesticides are derived from natural materials, such as animals, plants, bacteria, and certain minerals, that are used to control pests and diseases. Biopesticides are often considered lower-risk alternatives to conventional pesticides as they typically pose fewer risks to human health and the environment. In addition to the EPA, the U.S. Department of Agriculture (USDA) plays a role in regulating elicitors under the Plant Protection Act. The USDA's Animal and Plant Health Inspection Service (APHIS) is responsible for safeguarding US agriculture from pests and diseases by regulating the import, transit, and release of certain organisms, including genetically engineered plants and plant pests. In the European Union, elicitors are regulated under Regulation (EC) No. 1107/2009 concerning the placing of plant protection products on the market. Under this regulation, plant protection products must undergo a thorough evaluation to ensure their safety and efficacy before they can be authorized for marketing and use. In Canada, elicitors are regulated under the Pest Control Products Act, which governs the registration, sale, and use of pest control products in the country (Karamaouna et al. 2023; Pujari 2023; Akutse et al. 2020; Regnault-Roger 2020; Regnault 2011).

4.12 Economic and Social Implications of Elicitors

The adoption of elicitors in crop protection has significant economic implications for growers, input suppliers, and other stakeholders in the agricultural value chain. Elicitors can help reduce the costs associated with chemical inputs, such as pesticides and fertilizers, by providing effective alternatives for disease and pest management. These compounds can also help minimize yield losses and ensure crop quality, thereby enhancing growers' profitability and competitiveness. The use of elicitors in crop protection has social implications for growers, consumers, and society at large. Elicitors can help enhance food safety by reducing the residues of synthetic pesticides in agricultural products, ensuring that consumers have access to safe and high-quality food. These compounds can also help minimize the risks associated with pesticide exposure for farmworkers and rural communities, promoting a safer working environment and healthier living conditions (Liu et al. 2022; Caicedo-López et al. 2021; Chandler et al. 2011).

4.13 Conclusion

The use of elicitors in agriculture has emerged as a promising approach to enhance crop protection and promote sustainable agricultural practices. Through their ability to stimulate the natural defense mechanisms of plants, elicitors have shown significant potential in improving crop resilience to biotic and abiotic stresses. The advancements and applications of elicitors in seed treatments, biostimulants, and precision farming have demonstrated economic and environmental benefits, aligning with the growing demand for environmentally friendly and sustainable agricultural solutions. However, challenges and limitations remain, including regulatory considerations and the need for further research to fully explore the potential of elicitors in crop protection. Future work should focus on addressing these challenges and harnessing the potential of elicitors to contribute to a more sustainable and resilient agricultural system in the climate change era.

References

Abdul Malik NA, Kumar IS, Nadarajah K (2020) Elicitor and receptor molecules: orchestrators of plant defense and immunity. Inter J Mol Sci 21(3):963

Ahmed W, Shah AN, Abbas A, Nawaz M, Qayyum A, Hassan MU, Qaisrani MM, Khan J (2023) Role of plant bio-stimulants and their classification. In: Biofertilizers for sustainable soil management. CRC Press, pp 65–88

Akutse KS, Subramanian S, Maniania NK et al (2020) Biopesticide research and product development in Africa for sustainable agriculture and food security–experiences from the International Centre of Insect Physiology. Front Sustainable Food Syst 4:563016

Ali S, Akhtar MS, Siraj M, Zaman W (2024) Molecular communication of microbial plant biostimulants in the rhizosphere under abiotic stress conditions. Inter J Mol Sci 25(22):12424

Altun H, Orcan P (2024) Responses to exogenous elicitor treatment in lead-stressed *Oryza sativa* L. BMC Plant Biol 24:897

Asghar W, Craven KD, Kataoka R, Mahmood A, Asghar N, Raza T, Iftikhar F (2024) The application of Trichoderma spp., an old but new useful fungus, in sustainable soil health intensification: a comprehensive strategy for addressing challenges. Plant. Stress 12:100455

Backer R, Rokem JS, Ilangumaran G et al (2018) Plant growth-promoting rhizobacteria: context, mechanisms of action, and roadmap to commercialization of biostimulants for sustainable agriculture. Front Plant Sci 9:1473

Baenas N, García-Viguera C, Moreno DA (2014) Elicitation: a tool for enriching the bioactive composition of foods. Molecules 19(9):13541–13563

Bartucca ML, Cerri M, Del Buono D, Forni C (2022) Use of biostimulants as a new approach for the improvement of phytoremediation performance—a review. Plan Theory 11(15):1946

Bektas Y, Eulgem T (2015) Synthetic plant defense elicitors. *Front Plant Sci, Sec Plant Physiology 5*: https://doi.org/10.3389/fpls.2014.00804

Bulgari R, Cocetta G, Trivellini A et al (2015) Biostimulants and crop responses: a review. Biol Agric Hortic 10(1):1–18

Caicedo-López LH, Aranda ALV, Sáenz de la OD, Gómez CEZ, Márquez EE, Zepeda HR (2021) Elicitors: bioethical implications for agriculture and human health. Revista Bioética 29:76–86

Calvo P, Nelson L, Kloepper JW (2014) Agricultural uses of plant biostimulants. Plant Soil 383:3–41

Caradonia F, Battaglia V, Righi L, Pascali G, La Torre A (2019) Plant biostimulant regulatory framework: prospects in Europe and current situation at international level. J Plant Growth Regul 38:438–448

Chalfoun NR, Durman SB, González-Montaner J, Reznikov S, De Lisi V, González V, Moretti ER, Devani MR, Ploper LD, Castagnaro AP, Welin B (2018) Elicitor-based biostimulant PSP1 protects soybean against late season diseases in field trials. FrontPlant Sci 9:763

Chandler D, Bailey AS, Tatchell GM, Davidson G, Greaves J, Grant WP (2011) The development, regulation and use of biopesticides for integrated pest management. Philosophical Trans Royal Soc B: Biol Sci 366(1573):1987–1998

Chodisetti B, Rao K, Gandi S, Giri A (2013) Improved gymnemic acid production in the suspension cultures of *Gymnema sylvestre* through biotic elicitation. Plant Biotechnol Rep 7:519–525

Chodisetti B, Rao K, Gandi S, Giri A (2015) Gymnemic acid enhancement in the suspension cultures of *Gymnema sylvestre* by using the signaling molecules—methyl jasmonate and salicylic acid. Vitro Cell Dev Biology-Plant 51:88–92

Collinge DB, Jensen DF, Rabiey M, Sarrocco S, Shaw MW, Shaw RH (2022 Jun) Biological control of plant diseases–what has been achieved and what is the direction? Plant Pathol 71(5):1024–1047

de Almeida Silva M, Mendes CR, de Sousa FL, Santos HL, Almeida LC, Carnietto MR (2024) The composition of biostimulants applied in seed treatment interferes with the soybean emergence under water deficiency. Nativa 12(4):819–833

Docimo T, Davis AJ, Luck K, Fellenberg C, Reichelt M, Phillips M, Gershenzon J, D'Auria JC (2015) Influence of medium and elicitors on the production of cocaine, amino acids and phytohormones by *Erythroxylum coca* calli. Plant Cell Tissue Organ Cult 120:1061–1075

Domínguez H, Iñarra B, Labidi J, Bald C (2024) Fish viscera hydrolysates and their use as biostimulants for plants as an approach towards a circular economy in Europe: a review. Sustainability 16(20):8779

Doughari J (2015) An overview of plant immunity. J Plant Pathol Microbiol 6(11):10–4172

González-González MF, Ocampo-Alvarez H, Santacruz-Ruvalcaba F, Sánchez-Hernández CV, Casarrubias-Castillo K, Becerril-Espinosa A, Castañeda-Nava JJ, Hernández-Herrera RM (2020) Physiological, ecological, and biochemical implications in tomato plants of two plant biostimulants: Arbuscular mycorrhizal fungi and seaweed extract. Front Plant Sci 11:999

Gowtham HG, Murali M, Shilpa N, Amruthesh KN, Gafur A, Antonius S, Sayyed RZ (2024) Harnessing abiotic elicitors to bolster plant's resistance against bacterial pathogens. Plant Stress 24:100371

Gowthami L (2018) Role of elicitors in plant defense mechanism. J Pharmacog Phytochem 7(6):2806–2812

Gowtham HG, Hema P, Murali M, Shilpa N, Nataraj K, Basavaraj GL, Singh SB, Aiyaz M, Udayashankar AC, Amruthesh KN (2024) Fungal Endophytes as Mitigators against Biotic and Abiotic Stresses in Crop Plants. J Fungi (Basel)

Gupta S, Bhattacharyya P, Kulkarni MG, Doležal K (2023) Growth regulators and biostimulants: upcoming opportunities. Front Plant Sci 14:1209499

Guru PN, Mridula D, Dukare AS, Ghodki BM, Paschapur AU, Samal I, Nikhil Raj M, Padala VK, Rajashekhar M and Subbanna ARNS (2022) A comprehensive review on advances in storage pest management: Current scenario and future prospects. Front. Sustain. Food Syst. 6:993341. doi: 10.3389/fsufs.2022.993341

Hamid B, Zaman M, Farooq S, Fatima S, Sayyed RZ, Baba ZA, Sheikh TA, Reddy MS, El Enshasy H, Gafur A, Suriani NL (2021) Bacterial plant biostimulants: a sustainable way towards improving growth, productivity, and health of crops. Sustain For 13(5):2856

Jalota K, Sharma V, Agarwal C, Jindal S (2024) Eco-friendly approaches to phytochemical production: elicitation and beyond. Natural Prod Bioprosp 14(1):5

Jeyasri R, Muthuramalingam P, Karthick K, Shin H, Choi SH, Ramesh M (2023) Methyl jasmonate and salicylic acid as powerful elicitors for enhancing the production of secondary metabolites in medicinal plants: an updated review. Plant Cell Tissue Organ Cult 153(3):447–458

Kamle M, Borah R, Bora H, Jaiswal AK, Singh RK, Kumar P (2020) Systemic acquired resistance (SAR) and induced systemic resistance (ISR): role and mechanism of action against phytopathogens. Fungal. Biotechnol Bioeng:457–470

Karamaouna F, Economou LP, Lykogianni M, Mantzoukas S, Eliopoulos PA (2023) Biopesticides in the EU: state of play and perspectives after the green Deal for agriculture. In: Development and commercialization of biopesticides. Academic Press, pp 213–239

Kaur S, Samota MK, Choudhary M, Choudhary M, Pandey AK, Sharma A, Thakur J (2022) How do plants defend themselves against pathogens-biochemical mechanisms and genetic interventions. Physiol Mol Biol Plants 28(2):485–504

Lahlali R, Ezrari S, Radouane N, Kenfaoui J, Esmaeel Q, El Hamss H, Belabess Z, Barka EA (2022) Biological control of plant pathogens: a global perspective. Microorganisms 10(3):596

Li J, Van Gerrewey T, Geelen D, Li et al (2022) A meta-analysis of biostimulant yield effectiveness in field trials. Front Plant S13:836702

Liu Y, Shi R, Peng Y, Wang W, Fu X (2022) Impacts of technology training provided by agricultural cooperatives on farmers' adoption of biopesticides in China. Agriculture 12(3):316

Malook SU, Maqbool S, Hafeez M, Karunarathna SC, Suwannarach N (2022) Molecular and biochemical mechanisms of elicitors in pest resistance. Life 6:844

Matthews S, Siddiqui Y, Ali A (2024) Unleashing the power of bio-stimulants for enhanced crop growth, productivity, and quality: a comprehensive review. J Plant Nutrit 3:1–23

Meena M, Yadav G, Sonigra P, Nagda A, Mehta T, Swapnil P, Marwal A (2022) Role of elicitors to initiate the induction of systemic resistance in plants to biotic stress. Plant Stress 5:100103

Mejía-Teniente L, Torres-Pacheco I, González-Chavira MM, Ocampo-Velazquez RV, Herrera-Ruiz G, Chapa-Oliver AM and Guevara-González RG (2010) Use of elicitors as an approach for sustainable agriculture. *Afr J Biotechnol 9* (54): 9155–9162 http://www.academicjournals.org

Miranda AM, Hernandez-Tenorio F, Villalta F, Vargas GJ, Sáez AA (2024) Advances in the development of biofertilizers and biostimulants from microalgae. Biology 13(3):199

Mohamed M, Sami R, Almushhin A, Ali M, El Desouky HS, Ismail K, Khalil RR, Zewail R (2021) Impacts of effective microorganisms, compost tea, fulvic acid, yeast extract, and foliar spray with seaweed extract on sweet pepper plants under greenhouse conditions. Plan Theory 9(10):1927

Moreno-Pérez A, Martínez-Ferri E, Pliego-Alfaro F, Pliego C (2020) Elicitors and plant Defence. JOJ Hortic Arboric 2(5):555600

Namdeo A, Patil S, Fulzele DP (2002) Influence of fungal elicitors on production of ajmalicine by cell cultures of Catharanthus roseus. Biotechnol Prog 18(1):159–162

Narayani M, Srivastava S (2017) Elicitation: a stimulation of stress in in vitro plant cell/tissue cultures for enhancement of secondary metabolite production. Phytochem Rev 16:1227–1252

Nephali L, Moodley V, Piater L, Steenkamp P, Buthelezi N, Dubery I, Burgess K, Huyser J, Tugizimana F (2021) A metabolomic landscape of maize plants treated with a microbial biostimulant under well-watered and drought conditions. Front Plant Sci 12:676632

Onaga G, Wydra K (2016) Advances in plant tolerance to abiotic stresses. Plant Genom 10(9):229–272

Patel H, Krishnamurthy R (2013) Elicitors in plant tissue culture. J Pharmacogn Phytochem 2(2):60–65

Peigné J, Casagrande M, Payet V, David C, Sans FX, Blanco-Moreno JM, Cooper J, Gascoyne K, Antichi D, Bàrberi P, Bigongiali F (2016) How organic farmers practice conservation agriculture in Europe. Renewable Agric Food Syst 31(1):72–85

Poveda J, Díez-Méndez A (2023) Use of elicitors from macroalgae and microalgae in the management of pests and diseases in agriculture. Phytoparasitica 51(4):667–701

Pujari AA (2023) Regulatory frameworks for biopesticides and biostimulants: a comparative analysis of the EU, US, India and Japan. Inter J Analysis of Basic Appl Sci 7(4):12–17

Radman R, Saez T, Bucke C, Keshavarz T (2003) Elicitation of plants and microbial cell systems. Biotechnol Appl Biochem 37(1):91–102

Raguraj S, Kasim S, Md Jaafar N, Nazli MH (2022) Growth of tea nursery plants as influenced by different rates of protein hydrolysate derived from chicken feathers. Agronomy 12(2):299

Rajesaheb KS, Subramanian S, Boominathan P, Thenmozhi S, Gnanachitra M (2024) Biostimulant in improving crop yield and soil health. Comm Soil Sci Plant Anal 56(3):458–493

Ramirez-Estrada K, Vidal-Limon H, Hidalgo D, Moyano E, Golenioswki M, Cusidó RM, Palazon J (2016) Elicitation, an effective strategy for the biotechnological production of bioactive high-added value compounds in plant cell factories. Molecules 21(2):182

Rayanootha PS, Mahapatra S, Das S (2022) Effect of several pre-sowing elicitor seed treatments on mustard (*Barassica juncea* L.) vogour index against Alternaria blight disease in. India J Crop and Weed 18(1):254–257

Regnault C (2011) Trends for commercialization of biocontrol agent (biopesticide) products. Springer, pp 1–18. https://link.springer.com/chapter/10.1007/978-94-007-1933-0_6

Regnault-Roger C (2020) Trends for commercialization of biocontrol agents (biopesticide). In: Mérillon JM, Ramawat KG (eds) Plant Defence: biological control. Progress in biological control, vol 22. Springer, Cham, pp 445–471. https://doi.org/10.1007/978-3-030-51034-3_17

Rouphael Y, Colla G (2020) Biostimulants in agriculture. Front. Plant Sci 11:40

Samota MK, Awana M, Krishnan V, Kumar S, Tyagi A, Pandey R, Mithra SA, Singh A (2024) A novel micronutrients and methyl jasmonate cocktail of elicitors via seed priming improves drought tolerance by mitigating oxidative stress in rice (*Oryza sativa* L.). Protoplasma 261(3):553–570

Sangiorgio D, Cellini A, Donati I, Pastore C, Onofrietti C, Spinelli F (2020) Facing climate change: application of microbial biostimulants to mitigate stress in horticultural crops. Agronomy 10(6):794

Sani MN, Yong JW (2021) Harnessing synergistic biostimulatory processes: a plausible approach for enhanced crop growth and resilience in organic farming. Biology 11(1):41

Selwal N, Goutam U, Akhtar N, Sood M, Kukreja S (2024) Elicitation: "a trump card" for enhancing secondary metabolites in plants. J Plant Growth Regul 8:1–21

Sharma SN, Jha Z, Sinha RK, Geda AK (2015) Jasmonate-induced biosynthesis of andrographolide in *Andrographis paniculata*. Physiol Plant 153(2):221–229

Sible CN, Seebauer JR, Below FE (2021) Plant biostimulants: a categorical review, their implications for row crop production, and relation to soil health indicators. Agronomy 11(7):1297

Singh JS (2013) Plant growth promoting rhizobacteria: potential microbes for sustainable agriculture. Resonance 18(3):275–281

Singh R, Kaur S, Bhullar SS, Singh H, Sharma LK (2024) Bacterial biostimulants for climate smart agriculture practices: mode of action, effect on plant growth and roadmap for commercial products. J Sustain Agric Environ 3(1):e12085

Sivanandhan G, Kapil Dev G, Jeyaraj M, Rajesh M, Arjunan A, Muthuselvam M, Manickavasagam M, Selvaraj N, Ganapathi A (2013) Increased production of withanolide a, withanone, and withaferin a in hairy root cultures of *Withania somnifera* (L.) Dunal elicited with methyl jasmonate and salicylic acid. Plant Cell Tissue Org Cult 114:121–129

Soudani N, Toumi K, El-Jarroudi M, Fattouch S (2023) Microbial consortia as plant biostimulants: definition, concept, categories, and regulation. In: Microbial Biostimulants. Apple Academic Press, pp 365–380

Sumit G, Amit M, Archana T (2011) Catharanthus alkaloids and their enhanced production using elicitors: a review. Int J Pharm Technol 3(1):713–724

Sun W, Shahrajabian MH, Kuang Y, Wang N (2024) Amino acids biostimulants and protein hydrolysates in agricultural sciences. Plan Theory 13(2):210

Suwanchaikasem P, Idnurm A, Selby-Pham J, Walker R, Boughton BA (2022) Root-TRAPR: a modular plant growth device to visualize root development and monitor growth parameters, as applied to an elicitor response of *Cannabis sativa*. Plant Methods 18(1):46

Thakur M, Sohal BS (2013) Role of elicitors in inducing resistance in plants against pathogen infection: a review. Int Scholarly Res Notices 1:762412

Therond O, Duru M, Roger-Estrade J, Richard G (2017) A new analytical framework of farming system and agriculture model diversities. Review Agronomy Sustainable Devel 37:1–24

Tripathi D, Raikhy G, Kumar D (2019) Chemical elicitors of systemic acquired resistance—salicylic acid and its functional analogs. Current Plant Biol 17:48–59

Tütüncü M, Şekerci AD, Dönmez D, İzgü T, Isak MA, Şimşek Ö (2024) Plant biostimulants in ornamentals: enhancing growth and stress tolerance. Adv Hort Sci 38(2):211222

Uddain J (2024) Enhancing food safety and security through organic agriculture and innovative fertilizer management. Asian-Australasian J Food Safety Sec 8(2):27–31

Vasconsuelo A, Boland R (2007) Molecular aspects of the early stages of elicitation of secondary metabolites in plants. Plant Sci 172(5):861–875

Velten S, Leventon J, Jager N, Newig J (2015) What is sustainable agriculture? Systematic Review Sustain 7(6):7833–7865

Xu L, Geelen D (2018) Developing biostimulants from agro-food and industrial by-products. Front Plant Sci 9:1567

Yadav A, Singh S, Yadav V (2024) Screening herbal extracts as biostimulant to increase germination, plant growth and secondary metabolite production in wheatgrass. Sci Rep 14(1):607

Zheng F, Chen L, Zhang P, Zhou J, Lu X et al (2020) Carbohydrate polymers exhibit great potential as effective elicitors in organic agriculture: a review. Carbohydr Polym 230(15):115637

Zulfiqar F, Moosa A, Ali HM, Bermejo NF, Munné-Bosch S (2024) Biostimulants: a sufficiently effective tool for sustainable agriculture in the era of climate change? Plant Physiol Biochem 9:108699

Chapter 5
Nanoelicitors: A Promising Strategy for Sustainable Crop Production Under Biotic Stress Conditions

Anu Kalia and M. V. Sreelakshmi

Abstract Biotic stress refers to the altered metabolic status of the plant in response to infestation or infection by a variety of microorganisms, including bacteria, fungi, viruses, pests, and weeds. Biotic stress directly affects the growth and overall health of the host plant such that the host plant is deprived of nutrient access or uptake, leading to loss of vigor and culminating to the death of the plant. Therefore, biotic stress agents jeopardize global food security. The conventional biotic stress agent management strategies include the application of chemical pesticides. However, pesticides seriously harm both human health and the environment. The recent decade witnessed the application of nanoparticles as a promising substitute for chemical pesticides. A variety of nanomaterials have direct pesticidal effects while these can also stimulate localized systemic immune responses in the plant tissues. The latter trait earmarked the role of nanoscale materials as elicitors of immune responses for improved plant health management. This chapter presents the published literature on recent advancements in the application of nanoelicitors for the management of biotic stress agents across various crop plants. Furthermore, different mechanisms of action of the nanoelicitors to enhance crop tolerance and/or resistance to biotic stress have been discussed. The prospects, highlighting the scope of further research to explore the full potential of nanoelicitors in crop health management, have also been discussed.

Keywords Biostimulants · Elicitor · Pathogen · Pest · Plant immunity

---

A. Kalia (✉)
Electron Microscopy and Nanoscience Laboratory, Punjab Agricultural University, Ludhiana, Punjab, India
e-mail: kaliaanu@pau.edu

M. V. Sreelakshmi
Department of Microbiology, College of Basic Sciences and Humanities, Punjab Agricultural University, Ludhiana, Punjab, India

K. A. Abd-Elsalam, H. I. Mohamed (eds.), *Elicitors for Sustainable Crop Production*, Sustainability Sciences in Asia and Africa,
https://doi.org/10.1007/978-981-96-6569-3_5

5.1 Introduction

The recent decade has witnessed several global challenges such as climate change, decreasing nonrenewable resources, environmental pollution, and the population big bang. These factors are compromising or presenting a negative impact on the land resources for cultivation besides crop health that are the critical factors for achieving higher agricultural productivity. Various biotic and abiotic stresses pose a great challenge on the crop health and productivity (Abdullah et al. 2017; Chaudhry et al. 2021). Biotic stress arising from living organisms such as bacteria, viruses, fungi, pests, nematodes, and weeds disrupts the normal metabolism of plants, thereby negatively impacting crop growth, vigor, and productivity (Mishra et al. 2021; Saed-Moucheshi et al. 2024). The plants exhibit physiological and morpho-anatomical alterations in response to biotic stress. The phenotypic variations may include the formation of trichomes, spines, waxy cuticles, and setae. While bio-chemical changes involve the production of various secondary metabolites post-exposure to biotic stress. These secondary metabolites can elicit different defense-related pathways in plants through a cascade of signal transduction events (Saddique et al. 2018). Biotic stress agents, that is, phytopathogens and pests, have traditionally been managed using synthetic or natural organic pesticides, which pose harmful effects on the ecosystem, and human and animal health. Therefore, several novel molecules/compounds have been evaluated for managing biotic stress in crop plants.

Elicitors are a type of substances that activate plant chemical defenses. These molecules hold great promise for improving crop resistance against biotic stress agents. The elicitor molecules can trigger plant defense responses unlike the effector molecules that weaken the plant defenses (Chen and Mao 2020). A variety of molecules could function as elicitors such as chitosan (Chun and Chandrasekaran 2019), methyl salicylate (Zehra et al. 2021), salicylic acid (Lukan and Coll 2022), benzoic acid (Meena et al. 2022), benzothiadiazole (Thakur and Sohal 2013), phytohormones (auxin, gibberellin, and cytokinins), and different types of compounds produced and secreted by the plant beneficial microbes including mycorrhizae. The chemical nature and the type of elicitor affect the biosynthetic pathways triggered in plants on application (Kalia et al. 2024). The conventional chemical elicitors influence the phenolic compound production and lead to the stimulation of various defense-related enzymes in plants. However, negative effects of the application of various chemical elicitors are manifested as stunted vegetative growth of the plant besides a decrease in the photosynthetic pigments and photosynthesis rate. Therefore, improved elicitor formulations are desirable to address the negative issues of application of the conventional elicitors.

Nanotechnology-inspired elicitors could be a potential alternative (Kaur et al. 2023). Nanoelicitors are diverse in origin and encompass organic (polymeric, carbon nanomaterial) as well as inorganic (metal/metal oxide nanoparticles, nonmetal nanoparticles particularly SiO_2NPs) nanomaterials that can offer targeted and more effective activities compared to conventional elicitors (Yıldız et al. 2024). These molecules represent a sustainable alternative to genetically modified crops and hold potential for widespread agricultural use. This chapter delves into the mechanisms of action of nanoelicitors, such as their role in enhancing cell membrane permeability via reactive oxygen species (ROS) and activating signaling pathways that contribute to induced systemic resistance (ISR) and phytohormone synthesis. Additionally, nanoelicitors exhibit secondary passive mechanisms to alter the plant physiology and metabolism through modulation of the plant-associated microbiomes, and activation of the antibiotic production by microbes (actinobacteria-*Streptomyces*) (Zong et al. 2022). The results of the existing published reports on the potential use of nanoelicitors for biotic stress tolerance have been discussed, highlighting their ability to reduce the damage caused by pests and pathogens.

Furthermore, the mitigation of the environmental hazards associated with the application of synthetic pesticides to manage pests and pathogens has also been presented. Also, the advancements and prospects emphasizing the need for more focused research on nanoelicitors have been delved, which could help to harness the full potential of nanoelicitors in crop protection strategies. The application of elicitors has various advantages over the conventional techniques of biotic stress control. These include reduced environmental hazards, reduced toxicity to other organisms, function as a protective agrochemical, etc. This chapter aims to identify only the elicitor action of the nanomaterials that could be harnessed for stimulating the immune response against the biotic stress agents affecting the health of crop plants. Therefore, this elicitor action has been showcased and dissected from the conventional antimicrobial potential of the nanomaterials. The nanoelicitors have been contrasted with the conventional elicitors for the extent and efficacy of stimulation imparted to the crop plants on application.

5.2 Conventional Elicitors

The conventional elicitors include chemicals like chitosan, salicylic acid, benzoic acid, and benzothiadiazole, biological elicitors (mycorrhizae and other beneficial microbes), and growth regulators (auxins, gibberellins, cytokinins). These elicitors exhibit variable routes to accomplish elicitation though sometimes single to multiple and shared pathways are elicited by the elicitor molecule. Among the conventional elicitors, the research reports advocating agri-applications of chitosan or its derivatives predominate because of its promising eliciting properties. The availability of chitosan (high molecular weight: >700 kDa, medium molecular weight: 150–700 kDa, and low molecular weight: <150 kDa) provides the choice to the researchers to articulate the development of formulations and their efficacy. This oligosaccharide molecule is involved in the inhibition of the proteinase enzyme, promotes the formation of callose, and aids in phytoalexin synthesis. The callose deposition in the cell wall and plasma membrane helps in withstanding wound stress caused by various biotic stresses. Chitosan promotes the production of phytoalexins, which are low-molecular-weight antimicrobial compounds synthesized by plants in response to biotic stress (Riseh et al. 2022). Chitosan can elicit the formation of lytic or degrading enzymes in plants such as chitinase and glucanase enzymes, which help in the degradation of the cell wall components of the fungal pathogens.

Salicylic acid is another important elicitor that aids in biotic stress tolerance in plants. It functions as a signaling molecule and leads to induction of antioxidant enzyme machinery. Different derivatives of salicylic acid like methyl salicylate, and benzyl ester of salicylic acid, have been used as elicitors. Salicylic acid and its derivatives activate the formation of phytoalexins (Mishra et al. 2024), callose depositions (Zhao et al. 2024), cell wall thickening (Yin et al. 2024a, b), secondary metabolite production (Hou et al. 2024), and PR protein synthesis (Enebe and Babalola 2019).

The precursor compounds, functional analogs as well as small molecules derived from salicylic acid, also exhibit elicitor function. Benzoic acid induces the synthesis of chitinase, peroxidase, and polyphenol oxidase enzymes. It is the precursor of SA, which plays the role of an important signaling molecule in systemic acquired resistance (Klessig et al. 2018). A functional analog of SA, benzothiadiazole is another synthetic inducer of plant defense against pathogens that inhibits catalase enzyme activity. It also inhibits peroxidase and induces the synthesis of a variety of PR proteins. The inhibition of the two important H_2O_2 scavenging enzymes by this molecule aids in the accumulation of H_2O_2, which acts as an inducer of the expression of different PR proteins. Furthermore, it helps in the prevention of pathogens through the elicitation of the hypersensitive response.

The biological elicitors comprise different microbial elicitors produced and secreted by beneficial microbes. The obligatory plant–fungal symbiotic association i.e. the mycorrhiza exhibits defense priming, which helps the AM plants to withstand the stress more effectively. The mechanism includes the production of PR proteins, phenolic compounds, and callose-rich cell wall appositions at pathogen entry sites to protect against root pathogens (Sanmartín et al. 2020). AMF can also improve plant tolerance to pathogen attack through enhanced synthesis of numerous secondary metabolites (Zhao et al. 2022). Other PGPR microbes and endophytes can also trigger systemic acquired resistance through the activation of various defense-related pathways and genes. The activated genes lead to enhanced synthesis of PR proteins and antioxidant enzymes (Oukala et al. 2021).

Plant hormones or phytohormones regulate several developmental processes in plants. Phytohormones, including auxins, gibberellins, cytokinins, abscisic acid, ethylene, jasmonic acid, salicylic acid, and brassinosteroids, are key regulators of plant growth and metabolism. Among them, ethylene, jasmonic acid, and salicylic acid are crucial for plant defense, while auxins, gibberellins, abscisic acid, cytokinins, and brassinosteroids modulate their effects. For an effective response to biotic stress, the signaling pathways of these hormones are interconnected, enabling crosstalk at various levels (Rasool 2022). The jasmonic and salicylic acids have key roles in the regulation of plant defense systems during stress (Meena et al. 2022). Additionally, JA and SA are the main regulators of non-expressers of pathogenesis-related protein 1, the transcription factor responsible for activating PR gene

(Boubakri 2020). Application of these phytohormonal elicitors can be adopted in plants to overcome biotic stress (Rasool 2022). Auxin is a well-studied stress hormone that plays an essential role in plant growth and development. Auxin is also regarded as an elicitor of biotic stress tolerance (Singh et al. 2022).

5.3 Nanoelicitors: The Nanotechnology-Inspired Elicitors

Nanomaterials (NMs) are natural or synthetic substances with size dimensions between 1 and 100 nm (Palmqvist 2017). In recent years, NMs have gained attention for their unique morphological, chemical, optical, and competitive binding site-related properties. The distinct properties of nanoparticles have generated significant interest in agriculture and related fields (Fig. 5.1). One of the most valuable applications of nanomaterials in crop production is reducing agrochemical use while enhancing yield by managing both biotic and abiotic stresses (Table 5.1). The latest findings have indicated the possible uses of nanotechnology in agriculture for biotic stress management (Wasule et al. 2024).

Nanoelicitors enhance the stress resistance in crop plants, particularly by playing a significant role in improving the biotic stress tolerance potentials of the plants (Fig. 5.2). Diverse types of nanoparticles could function as novel and effective elicitors for biotic stress management (Gao and Tuda 2024). These can be classified as organic and inorganic based on their chemical origin (Tortella et al. 2023). Inorganic nanoparticles, including the metal/metal oxides (Banik and Pérez-de-Luque 2017)/sulfides/nitrides which themselves possess antimicrobial potentials against phytopathogens (El-Shetehy et al. 2021; Khan et al. 2021).

While the organic polymeric NPs such as chitosan (Chun and Chandrasekaran 2019, Iber et al. 2022), polylactic-co-glycolic acid (PLGA), and carbon-based NMs are primarily employed as nanodelivery vehicles or carriers, facilitating the effective release of the active ingredient (Shakiba et al. 2020).

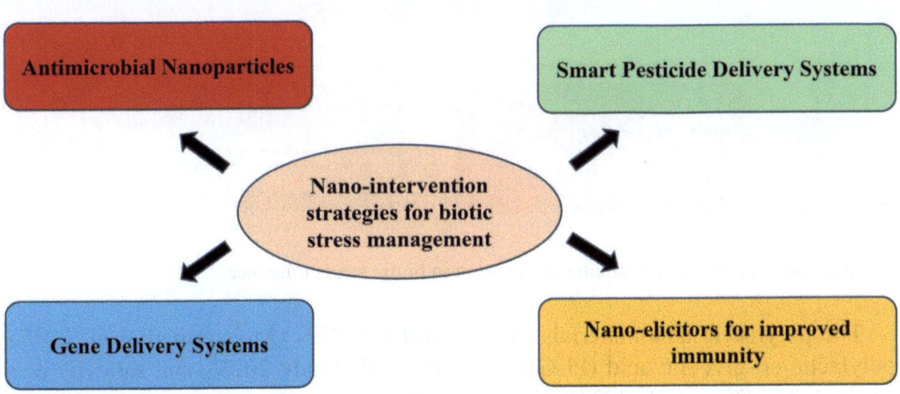

Fig. 5.1 Application of nanotechnology for biotic stress management

Table 5.1 Direct antimicrobial effects of different nanomaterials on a variety of phytopathogens

Nanomaterial type	Concentration used	Application method	Pathogen controlled	Host plant	Reference
Chitosan nanoparticles	0.1% w/v	Seed coating	*Colletotrichum gloeosporioides*	Mango	Lee et al. (2019)
Chitosan nanoparticles	100 mg/L	Seed treatment	*Pseudomonas syringae*	Rice	Smith et al. (2019)
Copper nanoparticles	200 ppm	Root dipping	*Phytophthora infestans*	Potato	Kim et al. (2023)
Nickel–silica nanoparticles	200 µg/mL	Foliar spray	*Xanthomonas oryzae pv. Oryzae (BLB)*	Rice	Abdallah et al. (2023)
Silicon nanoparticles	100 ppm	Foliar spray	*Xanthomonas campestris*	Rice	Chen et al. (2022)
Silicon nanoparticles	1 mM	Soil amendment	*Blumeria graminis*	Wheat	Lee et al. (2021)
Silver nanoparticles	10 ppm	Foliar spray	*Fusarium oxysporum*	Tomato	Doe et al. (2020)
Zinc oxide nanoparticles	50 ppm	Soil application	*Alternaria solani*	Potato	Smith et al. (2021)
Zinc oxide nanoparticles	50 ppm	Foliar spray	*Botrytis cinerea*	Strawberry	Brown et al. (2022)

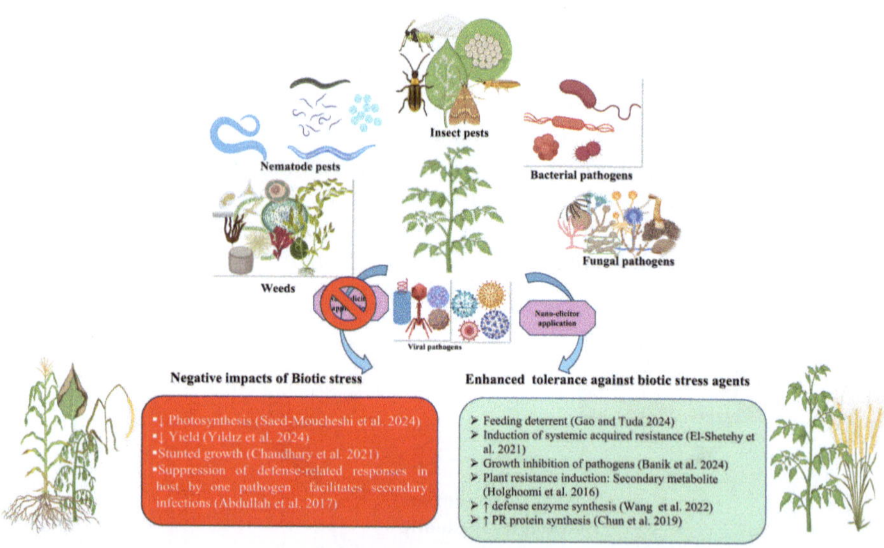

Fig. 5.2 Advantages of nanoelicitor application in biotic stress tolerance

The polymeric nanomaterials include chitosan NPs (Mohammed et al. 2017), polylactic-co-glycolic acid (PLGA) (Vemula and Reddy 2023), and salicylic acid nanospheres (Yin et al. 2024a, b). Among the polymeric nanomaterials, chitosan

NPs are the most widely studied. These have been evaluated alone and as nanocarrier formulations for other chemicals for induction of biotic stress resistance in plants. The cationic, biodegradable, and biocompatible nature and the associated antibacterial properties make it an ideal alternative for use in agriculture. Chitosan nanoparticles trigger antioxidants and defense enzymes in crop plants. Transcriptome analysis revealed increased levels of defense responses caused by greater expression of defense-related genes in chitosan NPs-treated plants (Selvaraj et al. 2022). Various plant defense mechanisms such as antioxidants and defense enzymes get induced by chitosan NPs besides up-regulating defense-related gene expression. Compared to the bulk form of chitosan, the nanoform exhibits more effective plant defense elicitation because of the nanoscale-related large surface area and charge density properties (Chun and Chandrasekaran 2019). The slow release of the active compounds, that is, phytohormones, thiamine, $Harpin_{Pss}$, metal oxide nanoparticles, and histidine from the nano-chitosan-based formulation and the role of nano chitosan as nanoreactor is presented in Table 5.2.

PLGA nanoparticles alone do not exhibit any elicitor action in plants on the biotic stress. However, its role as a phytohormone carrier has been well explored. The nano PLGA particles were utilized for entrapment of methyl jasmonate (MeJA) (Chronopoulou et al. 2019). On delivering MeJA, plant defense responses were elicited, presenting an alternative and sustainable pest control technique. PLGA NPs held a substantial quantity of MeJA and released it gradually over 5 days (Chronopoulou et al. 2019). The stilbene synthesis in *V. vinifera* cells exposed to free MeJA or MeJA-loaded PLGA NPs was evaluated. The results showcased the promising future applications of this formulation as nanopesticide. The introduction of PLGA NPs enabled the synthesis of stilbenes earlier and more reliably than free MeJA. These stilbenes are found to be toxic against several pathogenic microorganisms, thereby making the plants more resistant to pathogen attack. Likewise, PLGA NPs have also been utilized to entrap harpins (Liang et al. 2009).

Pathogenic microorganisms release proteins called hairpins that cause diseases, which trigger a hypersensitive reaction (HR) as a defense mechanism in resistant plants (Atkinson and Urwin 2012). Harpins can activate the SAR pathway in non-host plants shielding them from several diseases (Abdul Malik et al. 2020). The impact of recombinant HrpZ and recombinant HrpZ PLGA NPs on tobacco leaf phenylalanine ammonia lyase activity and pathogenesis-related protein-5 dB gene expression was studied (Liang et al. 2009). The recombinant HrpZ resulted in a quick and transitory increase in both phenyl alanine ammonia lyase activity and pathogenesis-related protein-5 dB expression, although the impact faded after many applications. Spraying recombinant HrpZ PLGA NPs on leaves resulted in consistent and long-lasting changes in the PAL activity and PR-5 dB expression. PLGA NP role in this study provided insights to resolve poor permeability and transient occurrence hurdles of HrpZ in plant organs. Another study showcased the specific role of RPP1-like gene cluster (Recognition of Peronospora Parasitica1) in *Arabidopsis thaliana* to alter the phyllosphere microbiome (Runge et al. 2024). Such genes could be useful to create a differential response to pathogenic attack.

Table 5.2 Chitosan NP application in biotic stress management

Combination used	Agrochemical loaded	Crop	Effects	Biotic stress	Reference
Chitosan NPs	–	Tomato	↑ β-1,3-glucanase expression ↑ chitinase expression ↑ PR-1 and PR-10 genes expression SOD gene upregulation	Wilt disease (*Fusarium andiyazi*)	Chun and Chandrasekaran (2019)
Chitosan NPs	–	Leaves of tea	Defense enzyme activation Defense-related genes upregulation ↑ total phenolics	–	Chandra et al. (2015)
Chitosan NPs	–	Wheat	Direct antimicrobial activity ↑ defense-related genes Restricts proliferation of pathogen	Fusarium head blight	Kheiri et al. (2017)
Chitosan thiamine NPs	Thiamine	Chickpea	↑ enzymatic and nonenzymatic antioxidants ↑ lignin deposition in vascular bundles of stem ↑ accumulation of antioxidants ↑ accumulation of phenylpropanoids Stress response elicitation	*Fusarium oxysporum* f. sp. *ciceri*	Sathiyabama and Indhumathi (2022)
Chitosan NPs	Salicylic acid (SA)	Wheat	↑ blade thickness ↑ mesophyll tissue thickness ↑ lower and upper epidermis thickness ↑ in the bundle length and width in midrib	Wheat leaf rust (*Puccinia triticina*)	Elsharkawy et al. (2022)
Chitosan NPs	Harpin_{Pss}	Tomato	↑ expression of JA metabolism genes H-CSNPs effectively enabled harpin_{Pss} sustained discharge Ensure long lasting resistance	*Rhizoctonia solani*	Nadendla et al. (2018)
Chitosan NPs	Salicylic acid Metal oxide Endophytic *Bacillus subtilis*	Citrus	SA activates PR proteins *B. subtilis* + SA: stress mitigation via well-established roots Enhances root exploration of soil ↑ capacity of nutrient and water absorption SACS NPs + B-CS NPs best for other plant growth traits	Citrus decline (*Candidatus Liberibacter* asiaticus)	Tadayon et al. (2023)

Chitosan NPs	Salicylic acid	Wheat	↑ of PR genes ↑ activity of defense enzymes ↑ phenylalanine ammonia lyase ↑ tyrosine ammonia lyase ↑ polyphenol oxidase	Stripe rust	Hembade et al. (2022)
Chitosan–histidine nanoformulation (C–H NF)	Chitosan has been functionalized via histidine	Tomato	Enhance growth parameters Enhance antioxidant- enzyme Positively impacts ROS level	–	Meena et al. (2024)
Daphnetin-carboxymethyl chitosan NPs	Daphnetin Carboxymethyl chitosan (biopolymer)	Tobacco	Improved systemic resistance CAT, PPO, and SOD ↑ PR1 and NPR1 ↑ JAZ3 ↑ PR proteins overexpression	*R. solanacearum*	Wang et al. (2023)

Liposomes are spherical structures of lipid bilayers containing an internal aqueous cavity (Nsairat et al. 2022). These polymeric organic NPs have been commercialized as a successful drug nanodelivery system and can carry hydrophilic drugs in the internal aqueous core and hydrophobic drugs on an external lipid bilayer (Khan et al. 2024). Agricultural applications involve their use to deliver various biostimulant elicitor molecules to enhance the immunity of crop plants against pathogenic organisms. A novel plant biostimulant, ELICE16INDURES®, has been developed by liposomal encapsulation of botanical extracts for application on barley plants (Hegedűs et al. 2022). The application of this liposomal formulation in barley led to an increase in the expression of PR and other associated genes with induced resistance as analyzed through a variety of molecular characterization tools such as KEGG analysis, Illumina sequencing, and gene expression profiling. The garlic extract has been encapsulated in liposomes and application of this formulation on wheat plants improved the biotic stress management through an increase in the expression of ABA pathways and PR protein-related genes (Kutasy et al. 2023).

The other category of organic nanomaterials includes carbon-based nanomaterials such as carbon nanotubes, graphene oxide, fullerenes, and biochar nanoparticles (Verma and Singh 2023). Biochar nanoparticles help improve biotic stress tolerance. Studies performed on *Nicotiana benthamiana* and *Phytophthora nicotiana* interactions revealed a ten-fold increased SA-induced immunity on application of nanobiochar. Compared to bulk biochar, nanobiochar significantly increased the amount of PR-1 protein in plants (Kong et al. 2022). Nanobiochar can also be used as a carrier platform for delivery of other elicitors. Nanobiochar as a carrier has been utilized for the elicitor molecule methoxyindole (Kong et al. 2022). Application of this formulation helped in improving the plant growth parameters during biotic stress. The impact of rice straw biochar nanoparticles on pepper leaf spots induced by *Xanthomonas campestris* was studied under both in vivo and in vitro conditions (Aftab et al. 2022). Use of nanoparticles as a soil additive significantly boosted chili pepper plant development. However, rice straw biochar nanoparticles were found to be most suitable in increasing plant biomass output, both below and above ground (Aftab et al. 2022). In vitro studies revealed that the rice straw biochar nanoparticles inhibited growth of *X. campestris.* The X-ray diffraction spectroscopy analysis in rice straw biochar nanoparticles established their distinct nanoscale dimensions, which may help induce defense against pathogens in plants (Aftab et al. 2022).

The inorganic NPs include metal/metal oxide nanoparticles (Fe, Fe_3O_4, TiO_2, CuO, ZnO, MgO, MnO_2) and nonmetal oxide NPs (SiO_2). These NPs possess established direct antimicrobial properties, particularly silver and selenium NPs, and can improve the resistance to biotic stress in plants by induction of SAR, thereby triggering the defense pathway (Khosropour et al. 2023). Inorganic nanoparticles improve plant health during disease infestation by enhancing antioxidant defense enzyme content and by modulation of defense gene expression related to stress response (Abdelrhim et al. 2021).

Several published reports have indicated the use of metal and metal oxides NPs as nanoelicitors. These materials have the potential to elicit secondary metabolite production (Holghoomi and Colagar 2024). The small size and high surface area of

NMs lead to complex and intricate interaction of the NMs at cellular and molecular levels (Prasad et al. 2024). These interactions help in triggering the plant defense response against various biotic stresses. The NMs on application (as seed priming, foliar spray) exhibit a priming effect on the defense system of the plant, thereby improving the response of the plants and lead to effective management of the pathogen (Yadav et al. 2024). The defense priming may be attributed to maintenance of the highest state of alert through activation of defense genes on exposure to biotic stress. Since mineral elements immediately contribute to plant immunity and protection as structural elements and metabolic regulators, the application of these NMs improves plant nutrition (complete and balanced plant nutrition) and has traditionally been the first line of defense for plants. Therefore, a plant's susceptibility to illnesses caused by different types of biotic stresses can be influenced by its nutritional state. Different metal nanoparticle priming helps in improving the immunity of the crop plants (Shams et al. 2023).

The CaNP and FeNP effect was studied on foliar spray application on strawberry fruits against *Botrytis cinerea* has been evaluated (Mogazy et al. 2022). The application of these NPs increased the vitamin content, and enzyme activities, including the peroxidase, β-1, 3-glucanase, and PAL, cell wall fractions, mineral content, phenolics, and flavonoids. Though nanoparticles can act as effective control agents against infections caused by a variety of fungal and bacterial pathogens, limited studies have been performed on control of viral pathogens. A study reported the use of green synthesized ZnO NPs for the commencement of plant SAR against TMV (Abdelkhalek and Al-Askar 2020). Both protective and curative effects of zinc oxide NPs have been reported. All treated plants exhibited an increase in the expression of the PAL, PR-1, and POD genes (Abdelkhalek and Al-Askar 2020). Activation of plant immunity in tobacco against TMV has been recorded in the NP treatment, leading to increased levels of reactive oxygen species, peroxidase and catalase activities, besides genes linked to systemic resistance; PR1 and PR2 (Cai et al. 2019). Subsequent research revealed that ZnONPs increased the levels of the phytohormones SA and ABA by 162% and 517%, respectively. The NPs also improved the fresh weight and dry weight of tobacco plants (Cai et al. 2019).

The SiO_2 NPs have been quite extensively evaluated for agri-applications as these NPs have been proposed as a safer agrochemical (Wang et al. 2022). The SiO_2-NPs offer a cost-effective, efficient, safe (even at ten-fold higher dose needed for plant protection), and sustainable solution for protecting plants from pathogen infection (Wang et al. 2022). The SiO_2 NPs have been evaluated for imparting resistance to infection caused by *Pseudomonas* in *Arabidopsis* (El-Shetehy et al. 2021). The salicylic acid-mediated immunity response is the major pathway governing the protective effect of SiO_2-NPs. The SiO_2-NPs in combination or as alone can safeguard the plants from pathogenic attacks and the progression of infections through the controlled induction of SAR.

SiO_2 NPs also promise to provide a more affordable, consumer-safe, and sustainable option than currently used treatments. This is because the controlled induction of SAR will not negatively impact yield or nontarget organisms, which was previously linked to the action of plant biostimulants or pesticides, previously described.

Direct antifungal activity and elicitation of defense mechanism in wheat against *Rhizoctonia solani* by SiO_2NPs application has been reported (Abdelrhim et al. 2021). The researchers observed that the SiO_2NPs increased defense-related chemicals, including SA, and decreased oxidative stress by initiating both enzymatic and nonenzymatic antioxidant defense mechanisms and functioned as a plant therapeutic agent for a variety of phytopathogens. The development of controlled release formulations of silicon dioxide nanospores can help reduce the environmental footprints on application. The mesoporous silica loaded with SA has also been tested to promote long-term disease resistance in plants. SA nanospheres can activate polyphenol oxidase, phenylalanine ammonia-lyase, peroxidase, and chitinase to boost plant defense in opposition to *Phytophthora nicotianae*. The SA nanospheres can provide long-term fungistatic effect and plant protection due to their continuous release mechanism (Feng et al. 2020).

5.4 Mechanisms Governing the Effect of Nanoelicitors for Biotic Stress Tolerance in Plants

Nanomaterials exhibit a dual impact on the biotic stress of plants (Fig. 5.3). The direct effects of the application of NMs have been extensively studied and commercial products have been launched, particularly silver NPs-based products. However, the indirect effects of NMs as elicitors have been sparingly explored with a few published reports on the aspect in recent years. The indirect effect of application of NMs involves improved production of defense-related secondary metabolites like phenols, terpenoids, phytoalexins, and polyketides, mediated through activation of the plant immune response, and strengthening of the overall resilience of the plant.

Nanoseed priming with an appropriate dosage of nanoparticles makes the plant ready to combat/tolerate a range of biotic stress agents. This stress-readiness is

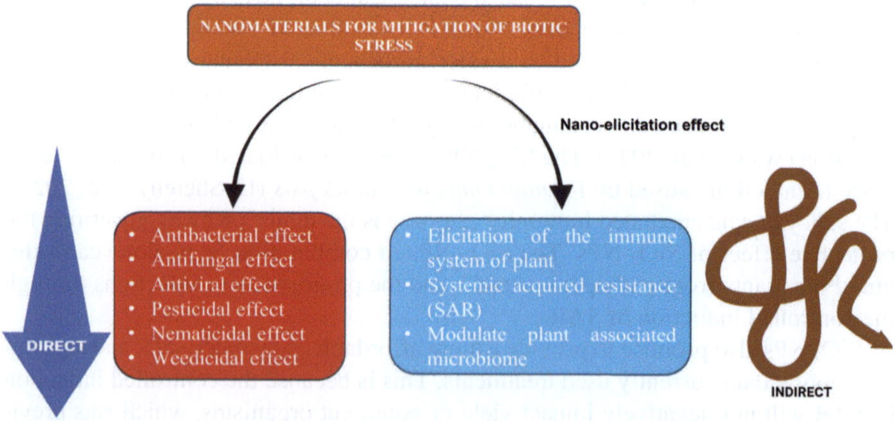

Fig. 5.3 Nanointerventions for effective mitigation of biotic stress in plant crops

ushered through NP exposure of the plant mimicking a controlled level of stress, which in turn improves the response to subsequent stressors. The secondary responses of the plant against biotic stress get improved. A prior optimization of the size dimensions, time duration of exposure, and concentration of the NPs must be performed as these critical factors alter the extent of impact on the crop plant post-exposure. Therefore, an optimized adequate concentration imparts a promising innate immune stimulation response in plants. Nanoseed priming aids in the development of augmented defense capacity in the plants in an early and durable manner. This leads to resistance in the new generation of plants through epigenetic modulation (Yadav et al. 2024).

The primary mechanisms include induction of ROS, PR proteins, phytoalexins production, cell wall strengthening, callose deposition, and stimulation of signaling pathways; ISR, improved phytohormone synthesis, and modulation of plant-associated microbiomes (Verma et al. 2024). Especially metal nanoparticles are found to contribute immediately to plant immunity because of their role in different signaling pathways, activation of defense response, and structural reinforcement (Alikhani et al. 2021; Mogazy et al. 2022).

5.5 Conclusion and Future Prospects

Nanoelicitors offer an innovative and sustainable method for enhancing crop resilience to biotic stresses. They trigger various plant defense mechanisms, enabling crops to withstand pests and diseases better while decreasing the need for harmful agrochemicals like pesticides and herbicides. However, there are still challenges which must be resolved before their widespread use. Future research should focus on understanding the long-term impacts of nanoparticles on plant health, soil quality, and nontarget organisms. Advancements in nanotechnology could also enhance the precision of targeted nanoelicitation and improve the delivery of other agrochemicals, contributing to a transition toward sustainable agriculture. This aligns with the increasing emphasis on developing resilient and sustainable food production systems.

References

Abdallah Y, Nehela Y, Ogunyemi SO, Ijaz M, Ahmed T, Elashmony R, Alkhalifah DHM, Hozzein WN, Xu L, Yan C, Chen J, Li B (2023) Bio-functionalized nickel-silica nanoparticles suppress bacterial leaf blight disease in rice (*Oryza sativa* L.). Front. Plant Sci 14:1216782

Abdelkhalek A, Al-Askar AA (2020) Green synthesized ZnO nanoparticles mediated by *Mentha spicata* extract induce plant systemic resistance against tobacco mosaic virus. Appl Scis 10(15):5054

Abdelrhim AS, Mazrou YS, Nehela Y, Atallah OO, El-Ashmony RM, Dawood MF (2021) Silicon dioxide nanoparticles induce innate immune responses and activate antioxidant machinery in wheat against *Rhizoctonia solani*. Plan Theory 10(12):2758

Abdullah AS, Moffat CS, Lopez-Ruiz FJ, Gibberd MR, Hamblin J, Zerihun A (2017) Host–multi-pathogen warfare: pathogen interactions in co-infected plants. Front Plant Sci 8:1806

Abdul Malik NA, Kumar IS, Nadarajah K (2020) Elicitor and Receptor Molecules: Orchestrators of Plant Defense and Immunity. Int J Mol Sci 21:963. https://doi.org/10.3390/ijms21030963

Aftab Z-H, Aslam W, Aftab A, Shah AN, Akhter A, Fakhar U, Siddiqui I, Ahmed W, Majid F, Wróbel J, Ali MD, Aftab M, Ahmed MAA, Kalaji HM, Abbas A, Khalid U (2022) Incorporation of engineered nanoparticles of biochar and fly ash against bacterial leaf spot of pepper. Sci Rep 12(1):8561

Alikhani TT, Tabatabaei SJ, Torkashvand AM, Talei D (2021) Silica nanoparticles and calcium on the histological characteristics and stem bending in gerbera cut flower. Ornamental Horticul 27(03):334–343

Atkinson NJ, Urwin PE (2012) The interaction of plant biotic and abiotic stresses: from genes to the field. J Exp Bot 63:3523–3543. https://doi.org/10.1093/jxb/ers100

Banik S, Pérez-de-Luque A (2017) In vitro effects of copper nanoparticles on plant pathogens, beneficial microbes and crop plants. Spanish J Agricul Res 15(2):e1005–e1005

Boubakri H (2020) Induced resistance to biotic stress in plants by natural compounds: possible mechanisms. In: Hossain MA, Liu F, Burritt DJ, Fujita M, Huang B (eds) Priming-mediated stress and cross-stress tolerance in crop plants. Academic Press, pp 79–99

Brown T, Davis L, Wilson K (2022) Zinc oxide nanoparticles as a foliar spray to manage *Botrytis cinerea* in strawberries. Nanomed Plant Sci 18(7):567–574

Cai L, Liu C, Fan G, Liu C, Sun X (2019) Preventing viral disease by ZnONPs through directly deactivating TMV and activating plant immunity in *Nicotiana benthamiana*. Environ Sci Nano 6(12):3653–3669

Chandra S, Chakraborty N, Dasgupta A, Sarkar J, Panda K, Acharya K (2015) Chitosan nanoparticles: a positive modulator of innate immune responses in plants. Sci Rep 5(1):15195

Chaudhry V, Runge P, Sengupta P, Doehlemann G, Parker JE, Kemen E (2021) Shaping the leaf microbiota: plant–microbe–microbe interactions. J Exp Bot 72:36–56. https://doi.org/10.1093/jxb/eraa417

Chen C-Y, Mao Y-B (2020) Research advances in plant–insect molecular interaction. F1000Research 9:198. https://doi.org/10.12688/f1000research.21502.1

Chen L, Zhang Y, Li X (2022) Silicon nanoparticles for controlling *Xanthomonas campestris* in rice through foliar spray. J Plant Pathol 30(2):211–218

Chronopoulou L, Donati L, Bramosanti M, Rosciani R, Palocci C, Pasqua G, Valletta A (2019) Microfluidic synthesis of methyl jasmonate-loaded PLGA nanocarriers as a new strategy to improve natural defenses in *Vitis vinifera*. Sci Rep 9(1):18322

Chun SC, Chandrasekaran M (2019) Chitosan and chitosan nanoparticles induced expression of pathogenesis-related proteins genes enhances biotic stress tolerance in tomato. Intl J Biol Macromol 125:948–954

Doe J, Smith A, Johnson B (2020) Effect of silver nanoparticles on *Fusarium oxysporum* in tomato. Nanotech Agric 15(3):123–130

Elsharkawy MM, Omara RI, Mostafa YS, Alamri SA, Hashem M, Alrumman SA, Ahmad AA (2022) Mechanism of wheat leaf rust control using chitosan nanoparticles and salicylic acid. J Fungi 8(3):304

El-Shetehy M, Moradi A, Maceroni M, Reinhardt D, Petri-Fink A, Rothen-Rutishauser B, Mauch F, Schwab F (2021) Silica nanoparticles enhance disease resistance in *Arabidopsis* plants. Nature Nanotechnol 16(3):344–353

Enebe MC, Babalola OO (2019) The impact of microbes in the orchestration of plants' resistance to biotic stress: a disease management approach. Appl Microbiol Biotechnol 103(1):9–25

Feng C, Tian X, Wang X, Cui M, Xu C, Wang W, Wang W (2020) Fabrication of salicylic acid nanosphere for long-term induced immunity performance. RSC Adv 10(48):28576–28584

Gao S, Tuda M (2024) Silica and selenium nanoparticles attract or repel scale insects by altering physicochemical leaf traits. Plan Theory 13(7):952

Hegedűs G, Kutasy B, Kiniczky M, Decsi K, Juhász Á, Nagy Á, Pallos JP, Virág E (2022) Liposomal formulation of botanical extracts may enhance yield triggering PR genes and phenylpropanoid pathway in barley (Hordeum vulgare). Plan Theory 11(21):2969

Hembade VL, Yashveer S, Taunk J, Sangwan S, Tokas J, Singh V, Redhu NS, Grewal S, Malhotra S, Kumar M (2022) Chitosan-salicylic acid and zinc sulphate nano-formulations defend against yellow rust in wheat by activating pathogenesis-related genes and enzymes. Plant Physiol Biochem 192:129–140

Holghoomi R, Colagar AH (2024) Applications of biocompatible nanoparticles in plant biotechnology for enhanced secondary metabolite biosynthesis. Inorg Chem Commun 167:112753

Hou J, Ai M, Li J, Cui X, Liu Y, Yang Q (2024) Exogenous salicylic acid treatment enhances the disease resistance of *Panax vietnamensis* by regulating secondary metabolite production. Fron Plant Sci 15:1428272

Iber BT, Kasan NA, Torsabo D, Omuwa JW (2022) A review of various sources of chitin and chitosan in nature. J Renew Mater 10(4):1097

Kalia A, Sreelakshmi MV, Kaur S (2024) Inorganic nanomaterials for improved abiotic stress tolerance in crop plants. In: Avila-Quezada G (ed) Rai M. Nanotechnology in Plant Health, CRC Press, pp 236–255

Kaur J, Dhiman D, Kalia A (2023) Antioxidant activity of Mycosynthesized nanoparticles. In: Rai M, Golinska P (eds) Myconanotechnology: emerging trends and applications, 1st edn. CRC Press, pp 108–127

Khan M, Khan AU, Bogdanchikova N, Garibo D (2021) Antibacterial and antifungal studies of biosynthesized silver nanoparticles against plant parasitic nematode *Meloidogyne incognita*, plant pathogens *Ralstonia solanacearum* and *Fusarium oxysporum*. Molecules 26(9):2462

Khan MS, Gupta G, Alsayari A, Wahab S, Sahebkar A, Kesharwani P (2024) Advancements in liposomal formulations: A comprehensive exploration of industrial production techniques. Int J Pharm 658:124212. https://doi.org/10.1016/j.ijpharm.2024.124212

Kheiri A, Jorf SM, Malihipour A, Saremi H, Nikkhah M (2017) Synthesis and characterization of chitosan nanoparticles and their effect on Fusarium head blight and oxidative activity in wheat. Intl J Biol Macromol 102:526–538

Khosropour E, Hakimi L, Weisany W (2023) Role of engineered nanomaterials in biotic stress managements. In: Husen A (ed) Engineered nanomaterials for sustainable agricultural production, soil improvement and stress management, 1st edn. Academic Press, pp 257–272

Kim J, Park H, Lee Y (2023) Copper nanoparticles for managing *Phytophthora infestans* in potatoes. J Nano Agric 25(2):89–96

Klessig DF, Choi HW, Dempsey DMA (2018) Systemic acquired resistance and salicylic acid: past, present, and future. Mol Plant-Microbe Interac 31(9):871–888

Kong M, Liang J, White JC, Elmer WH, Wang Y, Xu H, He W, Shen Y, Gao X (2022) Biochar nanoparticle-induced plant immunity and its application with the elicitor methoxyindole in *Nicotiana benthamiana*. Environ Sci Nano 9(9):3514–3524

Kutasy B, Kiniczky M, Decsi K, Kálmán N, Hegedűs G, Alföldi ZP, Virág E (2023) 'Garlic-lipo'4Plants: liposome-encapsulated garlic extract stimulates ABA pathway and PR genes in wheat (*Triticum aestivum*). Plan Theory 12(4):743

Lee H, Kim J, Park S (2019) Chitosan nanoparticles for managing *Colletotrichum gloeosporioides* in mango. J Plant Dis Prot 21(3):189–196

Lee C, Park J, Choi M (2021) Role of silicon nanoparticles in reducing *Blumeria graminis* infection in wheat. Plant Nanobiotechnol 30(4):345–352

Liang Z, Xu JP, Meng XL, Lu W, Wang J, Xia H (2009) Improve bioavailability of Harpin protein on plant use PLGA based nanoparticle. J Biotechnol 143(4):296–301

Lukan T, Coll A (2022) Intertwined roles of reactive oxygen species and salicylic acid signaling are crucial for the plant response to biotic stress. Intl J Mol Scis 23(10):5568

Meena M, Yadav G, Sonigra P, Nagda A, Mehta T, Swapnil P, Marwal A (2022) Role of elicitors to initiate the induction of systemic resistance in plants to biotic stress. Plant Stress 5:100103
Meena M, Saharan V, Meena KK, Singh B, Pilania S, Gupta NK, Pal A, Garhwal OP, Sharma YK, Singh U, Bagri R, Sharma MK, Sharma R, Jakhar BL, Chandel P, Prajapati D, Mondal K, Mahala M, Bairwa DK, Meena MB (2024) Synthesis and characterization of novel histidine functionalized chitosan nanoformulations and its bioactivity in tomato plant. Sci Rep 14(1):15118
Mishra D, Kumar A, Tripathi S, Chitara MK, Chaturvedi P (2021) Endophytic fungi as biostimulants: an efficient tool for plant growth promotion under biotic and abiotic stress conditions. In: Gupta S, Van Staden J (eds) Biostimulants for crops from seed germination to plant development, Academic Press, pp 365–391
Mishra S, Roychowdhury R, Ray S, Hada A, Kumar A, Sarker U, Aftab T, Das R (2024) Salicylic acid (SA)-mediated plant immunity against biotic stresses: an insight on molecular components and signaling mechanism. Plant Stress 11:100427
Mogazy AM, Mohamed HI, El-Mahdy OM (2022) Calcium and iron nanoparticles: a positive modulator of innate immune responses in strawberry against *Botrytis cinerea*. Process Biochem 115:128–145
Mohammed MA, Syeda JT, Wasan KM, Wasan EK (2017) An overview of chitosan nanoparticles and its application in non-parenteral drug delivery. Pharmaceutics 9(4):53
Nadendla SR, Rani TS, Vaikuntapu PR, Maddu RR, Podile AR (2018) HarpinPss encapsulation in chitosan nanoparticles for improved bioavailability and disease resistance in tomato. Carbohydr Polym 199:11–19
Nsairat H, Khater D, Sayed U, Odeh F, Al Bawab A, Alshaer W (2022) Liposomes: structure, composition, types, and clinical applications. Heliyon 8:e09394. https://doi.org/10.1016/j.heliyon.2022.e09394
Oukala N, Aissat K, Pastor V (2021) Bacterial endophytes: the hidden actor in plant immune responses against biotic stress. Plan Theory 10(5):1012
Palmqvist NGM (2017) Nanoparticles: Case studies of their synthesis, properties and biological interaction. Acta Universitatis Agriculturae Sueciae, 2017:105. https://res.slu.se/id/publ/104146
Prasad A, Sidhic J, Sarbadhikary P, Narayanankutty A, George S, George BP, Abrahamse H (2024) Role of metal nanoparticles in organogenesis, secondary metabolite production and genetic transformation of plants under in vitro condition: a comprehensive review. PCTOC 158(2):33
Rasool N (2022) Plant hormones: role in alleviating biotic stress. In: Hano C (ed) Plant hormones: recent advances and new perspectives and applications. IntechOpen. Available at: https://doi.org/10.5772/intechopen.102689
Riseh RS, Hassanisaadi M, Vatankhah M, Babaki SA, Barka EA (2022) Chitosan as a potential natural compound to manage plant diseases. Intl J Biol Macromols 220:998–1009
Runge P (2024) Structural variation of a polymorphic RPP1-like cluster in local Arabidopsis thaliana populations alters phyllosphere microbiome in Gorzów Wielkopolski. Doctoral dissertation, Universität Tübingen. https://tobias-lib.ub.uni-tuebingen.de/xmlui/handle/10900/154369
Saddique M, Kamran M, Shahbaz M (2018) Differential responses of plants to biotic stress and the role of metabolites. In: Ahmad P, Ahanger MA, Singh VP, Tripathi DK, Alam P, Alyemeni MN (eds) Plant metabolites and regulation under environmental stress, Academic Press, pp 69–87
Saed-Moucheshi A, Marvasti FB, Moeinzadeh A, Ansarshourijeh F, Tabarzad A, Pessarakli M, Hossain MA (2024) Photosynthesis efficiency and status of photosynthetic components under biotic stresses. In: Pessarakli M (ed) Handbook of photosynthesis, 4th edn. CRC Press, pp 694–713
Sanmartín N, Pastor V, Pastor-Fernández J, Flors V, Pozo MJ, Sánchez-Bel P (2020) Role and mechanisms of callose priming in mycorrhiza-induced resistance. J Exptl Bot 71(9):2769–2781
Sathiyabama M, Indhumathi M (2022) Chitosan thiamine nanoparticles intervene innate immunomodulation during chickpea-Fusarium interaction. Intl J Biol Macromol 198:11–17

Selvaraj M, Ramasami N, Jagan EG (2022) The effect of chitosan nanoparticles on immune responses in plants. In: Kumar S, Madihally SV (eds) Role of chitosan and chitosan-based nanomaterials in plant sciences. Academic Press, pp 185–196

Shakiba S, Astete CE, Paudel S, Sabliov CM, Rodrigues DF, Louie SM (2020) Emerging investigator series: polymèric nanocarriers for agricultural applications: synthesis, characterization, and environmental and biological interactions. Environ Sci Nano 7(1):37–67

Shams AH, Helaly AA, Algeblawi AM, Awad-Allah EF (2023) Efficacy of seed-biopriming with *Trichoderma* spp. and foliar spraying of ZnO-nanoparticles induce cherry tomato growth and resistance to Fusarium wilt disease. Plan Theory 12(17):3117

Singh D, Dhiman VK, Pandey H, Dhiman VK, Pandey D (2022) Crosstalk between salicylic acid and auxins, cytokinins and gibberellins under biotic stress. In: Aftab T (ed) Auxins, cytokinins and gibberellins signaling in plants. Springer, pp 249–262

Smith R, Lee H, Kim S (2019) Chitosan nanoparticles for controlling *pseudomonas syringae* in rice. J Plant Pathol 22(5):245–250

Smith R, Johnson T, Lee H (2021) Zinc oxide nanoparticles for controlling *Alternaria solani* in potato through soil application. J Plant Nanobiotechnol 28(4):345–352

Tadayon MS, Safaiefarahani B, Sadeghi S (2023) Evaluation of chitosan encapsulated salicylic acid, metal oxide, and endophytic *Bacillus subtilis* in amelioration of citrus decline. Physiol Mol Plant Pathol 127:102104

Thakur M, Sohal BS (2013) Role of elicitors in inducing resistance in plants against pathogen infection: a review. Intl Scholarly Res Notices (1): 762412,

Tortella G, Rubilar O, Pieretti JC, Fincheira P, de Melo SB, Fernández-Baldo MA, Benavides-Mendoza A, Seabra AB (2023) Nanoparticles as a promising strategy to mitigate biotic stress in agriculture. Antibiotics 12(2):338

Vemula M, Reddy AVB (2023) Polymeric nanoparticles as effective delivery systems in agriculture sustainability. Nanotechnol Environ Engin 8(3):805–814

Verma L, Singh J (2023) Production of biochar-based nanocomposites from chemical and biological methods. In: Mishra D, Singh R, Khare P (eds) Biochar-based nanocomposites for contaminant management: synthesis, contaminants removal, and environmental sustainability. Springer, pp 47–56

Verma KK, Joshi A, Song XP, Singh S, Kumari A, Arora J, Singh SK, Solanki MK, Seth CS, Li YR (2024) Synergistic interactions of nanoparticles and plant growth promoting rhizobacteria enhancing soil-plant systems: a multigenerational perspective. Front Plant Sci 15:1376214

Wang L, Ning C, Pan T, Cai K (2022) Role of silica nanoparticles in abiotic and biotic stress tolerance in plants: a review. Intl J Mol Scis 23(4):1947

Wang Y, Yang L, Zhou X, Wang Y, Liang Y, Luo B, Dai Y, Wei Z, Li S, He R, Ding W (2023) Molecular mechanism of plant elicitor daphnetin-carboxymethyl chitosan nanoparticles against *Ralstonia solanacearum* by activating plant system resistance. Intl J Biol Macromol 241:124580

Wasule DL, Shingote PR, Saxena S (2024) Exploitation of functionalized green nanomaterials for plant disease management. Discover Nano 19(1):118

Yadav N, Bora S, Devi B, Upadhyay C, Singh P (2024) Nanoparticle-mediated defense priming: a review of strategies for enhancing plant resilience against biotic and abiotic stresses. Plant Physiol Biochem 213:108796

Yildiz Ş, Küçükrecep A, Tekdal D (2024) Physiological and molecular responses of nanoparticle application to plants grown under abiotic and biotic stress. In: Rahman A (ed) Studies in natural products chemistry, vol 80. Elsevier, pp 113–145

Yin J, Li K, Liu X, Han S, Han X, Liu W, Li Y, Chen Y, Zhu Y (2024a) Silica nanoparticles enhance wheat resistance to fusarium head blight through modulating antioxidant enzyme activities and salicylic acid accumulation. Environ Sci Nano 11:4765–4778

Yin J, Zhao J, Wang Z, Xue F, Wang Q, Guo H, Cheng H, Li J, Shen J, Yin M, Su Y, Yan S (2024b) Preparation of salicylic acid nano-protectant with dual synergistic mechanism: high direct fungicidal activity and plant defence toward cotton *Verticillium* wilt. Chem Engin J 496:154036

Zehra A, Raytekar NA, Meena M, Swapnil P (2021) Efficiency of microbial bio-agents as elicitors in plant defense mechanism under biotic stress: a review. Curr Res Microbial Scis 2:100054

Zhao Y, Cartabia A, Lalaymia I, Declerck S (2022) Arbuscular mycorrhizal fungi and production of secondary metabolites in medicinal plants. Mycorrhiza 32(3):221–256

Zhao X, Liu Q, Tan L (2024) Callose and salicylic acid are key determinants of Strigolactone-mediated disease resistance in *Arabidopsis*. Plan Theory 13(19):2766

Zong G, Fu J, Zhang P, Zhang W, Xu Y, Cao G, Zhang R (2022) Use of elicitors to enhance or activate the antibiotic production in streptomyces. Crit Rev Biotechnol 42:1260–1283. https://doi.org/10.1080/07388551.2021.1987856

Chapter 6
Role of Carbohydrates in Enhancing Plant Immunity and Defense

Fatima Zahra Chakou, Karima Guehaz, Zakaria Boual, and Alia Telli

Abstract This study examines the essential role of carbohydrates in plant immunity and protection, highlighting their dual function as structural and signaling molecules vital for plant defense mechanisms. Carbohydrates, such as polymers containing cellulose, hemicellulose, and pectin, form the principal defensive barrier in plant cell walls against microbial invasion. The research investigates the role of oligosaccharide fragments such as microbe-associated molecular patterns (MAMPs), pathogen-associated molecular patterns (PAMPs), and damage-associated molecular patterns (DAMPs) in eliciting innate immune responses. Furthermore, it explores the importance of glycoproteins and glycolipids in pathogen identification and signal transmission, emphasizing the role of oligogalacturonides (OGs) in activating defense systems. Recent results highlight the role of carbohydrates in systemic acquired resistance (SAR), emphasizing their signaling functions beyond simple energy supply. The present study will categorize plant carbohydrates, clarify carbohydrate-mediated immunological processes, and evaluate their effects on disease resistance while examining current research trends and future directions in this field. Research on enhancing plant defense through carbohydrate tailoring presents a new avenue for sustainable strategies, contributing to global food security and ecosystem stability. Future studies should explore these connections.

Keywords Carbohydrates · Plant immunity · Biotic stress · Food security · Elicitors · Plant diseases

F. Z. Chakou · K. Guehaz · Z. Boual · A. Telli (✉)
Laboratory of preservation of ecosystems in arid and semi-aride areas, Faculty of Nature and Life Sciences, University of KASDI Merbah-Ouargla, Ouargla, Algeria
e-mail: telli.alia@univ-ouargla.dz

K. A. Abd-Elsalam, H. I. Mohamed (eds.), *Elicitors for Sustainable Crop Production*, Sustainability Sciences in Asia and Africa,
https://doi.org/10.1007/978-981-96-6569-3_6

6.1 Introduction

Plant immunity and protection is a multifaceted and highly regulated process that involves the coordinated action of various mechanisms to defend against pathogens and other foreign invaders. As the development of effective crop protection and improvement strategies is contingent upon a comprehensive understanding of the mechanisms of plant immunity, elucidating the role of carbohydrates in this complex process is of paramount importance. Carbohydrates have been found to play a pivotal role in plant defense mechanisms, serving as key components in the recognition, signaling, and activation of immune responses against pathogens and other foreign invaders (Su et al. 2024).

Plants possess immune systems akin to those of animals, which safeguard them against diseases. Pathogen recognition constitutes a crucial element of plant immunity. Plants employ two distinct immune responses to counteract diverse threats: pathogen/microbe-associated molecular patterns and effector-triggered immunity. Developing plants that enhance one or both of these defense mechanisms, either by establishing new systems or augmenting existing ones, presents intriguing opportunities. The pathway to attaining this can be effectively investigated by elucidating the immunological disease profiles of plants and enhancing the comprehension of these responses. Consequently, a thorough understanding of plant immune responses and the formulation of resistance profiles can yield plants that are more robust and less vulnerable to disease (Chiquito-Contreras et al. 2024; Gai and Wang 2024; Ji et al. 2021).

Carbohydrates are essential for plant immunity and protection, functioning as both structural and signaling molecules crucial for plant health and defense processes. In the plant kingdom, carbohydrates are crucial for energy storage and structural integrity, and they also significantly influence immune responses to various pathogens. Complex carbohydrate polymers in cell walls, including cellulose, hemicellulose, and pectin, constitute a principal defense barrier against microbial invasion. Furthermore, particular oligosaccharide fragments derived from these polysaccharides may function as damage-associated molecular patterns (DAMPs), eliciting innate immune responses (Molina et al. 2024; Bacete et al. 2018).

Glycoproteins and glycolipids, composed of carbohydrate components attached to proteins and lipids, are essential for cell surface interactions related to pathogen identification and signal transduction. Oligogalacturonides (OGs), resulting from pectin degradation, can stimulate defense mechanisms by activating pathogenesis-related (PR) proteins and inducing phytoalexin synthesis. Moreover, the significance of raffinose family oligosaccharides in abiotic stress tolerance highlights the diverse protective roles that carbohydrates provide to plants (Hassan et al. 2020; Lin et al. 2020; Egert et al. 2015).

Recent studies have also highlighted the critical involvement of carbohydrates in systemic acquired resistance (SAR), a "whole-plant" resistance response that provides long-term protection against a broad spectrum of pathogens. The initiation and propagation of SAR involve complex carbohydrate signaling networks, emphasizing the importance of sugars not only as energy sources but also as key regulators of plant defensive strategies (Gao et al. 2020; Singh et al. 2017).

Numerous carbohydrates and their derivatives identified as elicitors can stimulate plant defense mechanisms and prompt the synthesis of disease-resistant chemicals. Particularly, carbohydrates from various sources can be used by distinct infections, resulting in a range of carbohydrate-based insecticides that can be formulated for many illnesses, addressing the challenge of ecological heterogeneity among pathogen races, an issue that is difficult to resolve by alternative means. As a result, carbohydrate polymers exhibiting elicitor activity may be optimal for biopesticides in sustainable agriculture. In recent years, the structures of several polysaccharides and oligosaccharides have been elucidated. Extensive research has been conducted on defense signaling triggered by various carbohydrate elicitors. The structural and activity connections of carbohydrate elicitors remain poorly comprehended. The comprehension of the molecular pathways associated with carbohydrate-elicitor-induced downstream signaling remains incomplete (Zheng et al. 2020).

This work aims to explore carbohydrates' significance in plant immunity and protection. It examines their structural and functional roles in plant defense mechanisms, their interactions with other defense pathways, and potential applications in agriculture and biotechnology.

6.2 Overview of Plant Immune System Components and Responses

Plants possess immunity to most microorganisms they encounter, due to a variety of constitutively produced physical and chemical barriers that serve as their primary line of defense. The plant cell wall is a significant physical barrier. The cell wall consists of many basic macromolecules, such as pectic polysaccharides, hemicelluloses, cellulose, and structural glycoproteins. All these polymers can be changed post-synthesis, for instance, by glycosylation, and may play a significant role in plant-microbe interactions (Ha et al. 2021; Kozieł et al. 2021; Wan et al. 2021).

The plant immune system consists of numerous interacting elements, including cell surface and intracellular receptors, signaling proteins, transcription factors, and diverse regulatory proteins including kinases and protein phosphatases. Both cell surface and internal receptors are implicated, similar to their role in mammals. A significant distinction, however, is the absence of an adaptive immune system in plants, as they do not possess immune cells or somatic gene rearrangements; instead, they depend on a static array of germline-encoded receptors and immunological signals to react to various invading invaders. Plants are adept at swiftly responding to and recuperating from pathogen-induced immunity and tissue damage, partly due to sophisticated metabolic systems that optimize resource allocation inside cells (Delplace et al. 2022; Tiwari et al. 2022).

The plant immune system comprises systems that defend against several forms of biotic assault. Immune responses are categorized as "innate" or "adaptive." Innate immunity constitutes the primary defense mechanism and is operational in all multicellular organisms, encompassing plants, invertebrates, and vertebrates. The identification of patterns regarded as alien is chiefly facilitated by specialized receptors known as pattern recognition receptors. These receptors are situated on the surface of animal cells and on the cell wall and plasma membrane of plant cells. Upon detecting these foreign ligands, the receptors activate signal transduction pathways that trigger defense responses, including the synthesis of reactive oxygen species and phytoalexins, as well as the upregulation of proteins with defensive functions. Innate responses are strong yet generalized, facilitating resistance to diverse categories of infections. Plants subsequently establish supplementary snowballing mechanisms that result in adaptive resistance, extending beyond pathogenic resistance to include resistance against predators. Furthermore, these cascades significantly impact plants, affecting metabolism and growth that are induced by pathogens or wounds (Appu et al. 2021; Iqbal et al. 2021).

The hypersensitive response is frequently linked to increased resistance against subsequent attacks by the same pathogen and is also typically associated with the establishment of systemic acquired resistance, which is a prolonged accumulation of defense molecules that confers resistance to a wide array of pathogens. The initiation of diverse immune responses and the cascading effects of immune signaling constitute intricate networks facilitated by numerous molecules that are synthesized, degraded, and altered through various regulatory processes, including the covalent attachment of pharmacological agents or the assembly of protein complexes. In addition to disease resistance, most plants possess excellent strategies to repel herbivores. Systemic mechanisms may be activated via systemic acquired resistance, exhibiting durability and extensive effects, generally characterized by necrosis at the assault site and the release of particular volatile organic chemicals that inhibit herbivore colonization (Bhavani et al. 2022; Chen et al. 2023).

6.3 Types and Impact of Biotic Stress on Plants

Biotic stress can influence numerous facets of plant health. Pathogens and pests are two principal sources of biotic stress that impact millions globally, leading to substantial reductions in yield production and overall productivity. Diseases caused by pathogens and pests adversely influence land production and human health, resulting in annual costs amounting to billions of dollars and influencing multiple facets of human life. Land biodiversity has diminished, a phenomenon linked to soil degradation and incorrect land use resulting from inadequate agricultural practices, crop harvesting, and waste disposal. The term "biotic" denotes something that is "alive or derived from living organisms." Biotic stress is defined as any adversity induced by living organisms. Plants endure stress due to such pressures. Evolution has influenced plant responses, and several systems exist to assist plants in adapting to them. Plants possess the ability to detect impact or stress and subsequently coordinate an appropriate immunological and/or adaptive response. Ultimately, a plant must determine which of these responses—each with its associated costs—will best facilitate its survival. In this context, plant stress can be regarded as a shift in the allocation of resources inside a plant, between those utilized for fundamental growth and development and those employed to combat biotic threats (Gallusci et al. 2023; Mittler et al. 2022).

Biotic stress includes the various viruses and pests that threaten agricultural sustainability. These primarily encompass bacteria, fungi, viruses, and herbivores that infect or reside within plant tissues throughout their life cycle. Numerous examples demonstrate the influence of pathogenic bacteria and herbivorous insects on the physiology and economic significance of agriculturally significant plants. A prevalent necrotrophic fungus diminishes grain production by decreasing seed and straw quantity. A bacterium is recognized as the causative agent of bacterial wilt disease. A viral infection frequently leads to a deterioration in crop flavor and quality. Environmental stress frequently exacerbates biotic stress in plants by eliciting several responses that increase vulnerability to an additional stressor. The environment typically engages with consistent challenges including temperature, drought, salinity, and nutrient availability in plants. These stressors exert diverse effects on different developmental stages of the plant, resulting in a notable reduction in plant or grain-to-seed yield. In contrast to abiotic stresses, biotic stress results in the death of the entire plant or significant portions of it. A significant consequence of herbivores is the diminishment of photosynthesis resulting from their consumption and nutrient utilization, alongside a decrease in the capacity for further growth due to the ingestion of both assimilates and roots. Biotic stress involves the infiltration and progression within the plant, resulting in a sequence of detrimental events that are either directly linked to autotrophic growth mechanisms, such as photosynthesis, respiration, and nutrient utilization, or indirectly related to plant development (Adetunji et al. 2021; Kashyap and Kumar 2021).

6.4 Carbohydrates in Plant Biology

Carbohydrates represent a significant category of biomolecules typically present in plants. These compounds consist of carbon, hydrogen, and oxygen atoms, following the general formula $(CH_2O)_n$. Carbohydrates can be classified into three distinct categories based on their size and structure: monosaccharides, oligosaccharides, and polysaccharides. Carbohydrates serve critical physiological functions by providing a primary energy source through sugar metabolism. Carbohydrates also constitute structural components of various tissues, existing as complex carbohydrates including cellulose, hemicellulose, and pectin. Additionally, carbohydrates participate alongside proteins and lipids in several critical cellular functions. Glycolipids and glycoproteins serve essential functions in both prokaryotic and eukaryotic systems. Oligosaccharides and polysaccharides in plants play critical roles in various physiological and developmental processes, including light reception, pollen growth, meristem development, seed germination, thigmomorphogenesis, and responses to biotic and abiotic stress. Understanding the functions of carbohydrates in physiological phenomena is advancing swiftly due to discoveries regarding their chemical nature and improvements in structural, biochemical, and genetic investigative methods. Polysaccharides are diverse homopolymers and heteropolymers of sugars that primarily provide structural support through polymers such as cellulose and hemicelluloses (Shah et al. 2022; Naithani et al. 2021).

The effective absorption, transport, and breakdown of carbohydrates are essential for sustaining plant biology. Carbohydrates serve dual roles as structural components and as compounds for long- or short-term storage. They are categorized as monosaccharides, disaccharides, or polysaccharides. Monosaccharides and disaccharides serve as the primary energy source for many organisms, act as direct signaling molecules, and participate in redox reactions. Polysaccharide degradation yields small sugars that serve as energy sources and building blocks for various metabolites. Hemicellulose, cellulose, and callose polysaccharides constitute the cell wall of terrestrial plants, with their precursors being highly conserved among all organisms, regardless of cell wall composition (Costa et al. 2024; Ellouzi et al. 2024; Iqbal et al. 2023).

Energy is stored as disaccharides and polysaccharides in organs like tubers or mature seeds, or in specific cell types, and can be remobilized when carbohydrates are transported to the organs requiring them. Storage compounds may be organized into structural polysaccharides throughout growth and development. A significant amount of starch stored in leaves during light is degraded in the dark to supply energy for nocturnal activities. Metabolism and growth, particularly in plants, are contingent upon the utilization and regulation of carbohydrates. Storage carbohydrates represent a substantial portion of annual plant biomass and serve as intermediaries for long-term storage, containing information pertinent to biological processes such as growth and development. They also function as signaling compounds and protective barriers that interact with the environment. Many plant species have developed defense systems to assess and, if required, metabolically dismantle carbohydrate-based signaling mechanisms and wall polymers encountered at wound sites. This text provides an overview of the recent understanding of the roles of cellular carbohydrates as social messengers and structural polymers in energy provision and plant immunity (Blumstein et al. 2022; El Omari 2022; Hiltbrunner et al. 2021; Lubbe et al. 2021).

The diversity of PAMPs reflects the extensive variety of glycans that can adorn a pathogen's surface. Carbohydrate-mediated signaling provides an additional mechanism for plants to adjust their responses according to infection kinetics. This is categorized into early, middle, and late phases of immune activation due to the differing kinetics of molecular immune events. Temporal signaling events are as crucial as spatial factors in carbohydrate recognition for the effectiveness of an immune response. The interaction between pathogens, particularly those that produce a variety of carbohydrate-active enzymes, and various effectors introduces further complexity. This represents a redundancy likely arising from the need to ensure plant fitness, despite the abundance of carbohydrates in the form of PAMPs present in the environment (Galindo-Trigo et al. 2024; Nance 2023; Henao 2022).

6.5 Principal Carbohydrates Categories Associated with Plant Immunity and Recognition by Plant Receptors

Carbohydrates play an emerging role in plant immunity via signaling pathways. Carbohydrates influence gene expression and initiate plant defense mechanisms. The roles of carbohydrates in immunity are often characterized by the events that occur following the detection of potentially harmful molecules by the plant's recognition machinery. This category of plant-derived immune signals comprises two groups of molecules "damage-associated molecular patterns" (DAMPs) and "endogenous elicitors." Distinct plant cell-surface pattern recognition receptors (PRRs), referred to as "sentinels," perceive them. Endogenous elicitors are defined as carbohydrates derived from the cell wall, primarily consisting of monosaccharides. Recent evidence indicates that these hexoses function as immune signals. The activation of systemic immunity by carbohydrates is particularly noteworthy. This contrasts with certain other biological effects, including their rapid synergistic interaction with jasmonic acid and the necessity of PRR-activated ROS for optimal functionality (Meresa et al. 2024; Kaur et al. 2022; Ochoa-Meza et al. 2021).

Several carbohydrates derived from various sources—plants, animals, and microorganisms—are presently being studied for their possible roles as inducers/elicitors of plant immune system from invading pathogens, such as oligogalacturonoids (Howlader et al. 2020), hemicellulose- and cellulose-derived oligomers (Meresa et al. 2024), xyloglucans (Kongala and Kondreddy 2023), chitin and chitosan oligomers, beta-glucans, fucans, alginates, ulvans, and carrageenans (Brown 2022).

The systemic effects mediated by carbohydrates can be biochemically characterized; hexoses enhance the synthesis of salicylic acid, a crucial phenolic immune signal in systemic immunity. The hormonal phenolic signals subsequently trigger the expression of defense genes, which encode proteins that strengthen cell walls, detoxify reactive oxygen species (ROS) and their derivatives, produce phytoalexins, and facilitate regenerative enzymes. The examples of observed functions of carbohydrate triggers are expanding. Plant cells respond to oligogalacturonides that are derived from pectin, a key component of the primary cell wall. Oligogalacturonide (OG) perception receptors are distributed across various subcellular compartments, and the accumulation of secreted and apoplastic oligogalacturonides is triggered by distinct immune responses. The spatial and temporal specificity of this extensive compartmentalization is accomplished through the synthesis of OG in reaction to fungal and bacterial pathogens, alongside the concurrent secretion of OG-digestive enzymes (Su et al. 2024; Mijailovic et al. 2022; Bauer et al. 2021).

The recognition of carbohydrate signals by their respective receptors is crucial in the initial processes that contribute to plant immunity. The complexity of these interactions is notable and underscores the significance of carbohydrates in immune function. Upon pathogen invasion, both specific and nonspecific receptors present at the cellular level and within the cytoplasm can identify largely invariant molecular patterns, often structural, that are typically associated with microorganisms. Structural analysis of numerous receptors indicates that soluble receptor proteins possess distinct carbohydrate-binding pockets capable of binding specific carbohydrate ligands, exhibiting varying affinities. Some receptors necessitate classical carbohydrate moieties for recognition, while others do not (Li and Wu 2021).

Pattern recognition receptors (PRRs) play a crucial role in initiating responses to pathogens. They encompass various receptor families, including NLR-like receptors, which function as components in numerous PRR pathways or as intracellular proteins that enhance PRR-mediated immune responses. Additionally, small receptor-like kinases (RLKs) are located on the surface of plant cells. Small leucine-rich repeat kinases (LRRKs) are present on the surface of plant cells. Although this topic is not the primary focus of this review, there is a notable lack of information regarding the binding profile and targets for ligands associated with this class of receptor. Each PRR typically binds specific microbial structures at any given time. Pathogen carbohydrates differ significantly from those of plant cells, making early carbohydrate ligand recognition a key indicator for plants of invasion. This represents a distinct form of exclusion signal, separate from the recognition of effectors and stresses, serving as a unique cue for pathogens. The specificity of these interactions enables the utilization of these receptors for targeted immunological memory. Interactions influence pathogen behavior, with many having evolved the capacity to mimic carbohydrate signals to circumvent plant immune responses. In fungal ecology, plant surface sugars have been observed to function as fungicides by binding to and inhibiting the virulence activity of enzymes. From an evolutionary perspective, utilizing carbohydrates to distinguish self from nonself would have been relatively straightforward, as these molecules are found in essential resources and are observed in more primitive organisms like bacteria. The existence of these conserved receptors aligns with the hypothesis regarding their function (Molina et al. 2024; Kongala and Kondreddy 2023; Breia et al. 2021; Lorrai and Ferrari 2021; Wan et al. 2021).

6.5.1 Polysaccharides

Carbohydrates, including cellulose, hemicellulose, and pectin, constitute the cell wall matrix, providing crucial physical barriers against pathogens. These structural carbohydrates confer mechanical stability to plant cells and serve as substrates for the synthesis of many defense-related chemicals. The accumulation of polysaccharides in the cell wall might enhance the structural integrity of the plant, hence increasing its overall resilience against pathogen invasion (Zaynab et al. 2019).

(a) **Cellulose**

Cellulose is a major component of the plant cell walls, resulting from the linear assemblage of β-(1,4)-D-glucopyranoses residues (Garon 2014), and plays an important role as a physical protection against pathogens (Kongala and Kondreddy 2023). Enzymatic derivatives of cellulose named cyclodextrins (CD) or cello-oligomers are known to induce a variety of defense responses in plants including the rise in H_2O_2 levels, calcium, and induction of phenyl proposed pathway genes (Aziz et al. 2007). The degree of polymerization (DP), that is, the length of the polymer regularly influences the intensity and the type of defense responses induced by the oligosaccharides (Trouvelot et al. 2014). In grapevine, cellodextrin with DP 7 was more active than those with DP 8 and 9 in eliciting H_2O_2 production (Aziz et al. 2007). As well, cellotriose (DP 2) is regarded as the most effectual elicitor as it induces reactive oxygen species (ROS) and Ca^{+2} influx in *Arabidopsis* (De Azevedo Souza et al. 2017). Similarly, C1-oxidized cellobionic acid demonstrated the highest defense-eliciting activity among cellodextrins at DP 2–4 (Zarattini et al. 2021), while it was null in cellobiose-treated *Arabidopsis* plants (Locci et al. 2019). Authors deduce that the oxidation of the CDs makes them inaccessible for the pathogen as a carbon source (Locci et al. 2019).

(b) **Hemicelluloses**

Hemicelluloses are complex polysaccharides composed of linear chains of β-(1,4)-D-pyran residues, more or less branched, combined with cellulose microfibrils to form a cross-linked matrix (Carpita 2000). Hemicellulose-derived oligomers are reported for their elicitor effect in plant immunity. Heptamaloxyloglucan (DP 7) mixed with a low concentration of higher DP 8–9 efficiently elicited defense responses in Arabidopsis and grapevine (Claverie et al. 2018). In Arabidopsis, the strongest eliciting activity was recorded from xylotetraose (DP 4), followed by xylotriose (DP 3), while xylo-oligosaccharides with DP 5 and DP 2 showed reduced activity (Pring et al. 2023). In addition, Escudero et al. (2017a, b) showed that modification of the degree of xylan acetylation in plant cell walls was sufficient to induce pattern-triggered immunity (PTI). In another study, wheat hemicellulose-derived pentose arabinoxylan with DP 3–5 elicited an intense Ca^{2+} burst in treated Arabidopsis plants, whereas it was reduced in DP 2-treated groups (Mélida et al. 2020). Mélida et al. (2020) confirm that acid-hydrolyzed pentasaccharides 3^3-α-L-arabinofuranosyl-xylotetraose lost their activity in inducing Ca^{2+} bursts.

(a) **Chitosan and chitosoligomers**

Chitosan is the deacetylated derivative of chitin (CS) produced by chitin deacetylases and represents a less common natural polysaccharide found notably in zygomycete cell walls (Mohammadi et al. 2012). Deacetylases present naturally in fungi convert the cell wall chitin to chitosan in the host environment which enables the pathogen to escape from the host immune recognition and subsequent damage to its chitinous hyphae (Cord-Landwehr et al. 2016), while plant chitinases are inefficient in hydrolyzing the chitosan (Ride and Barber 1990). Thus, the polycationic nature of the deacetylated glucosamine attributes strong antimicrobial activity to chitosan and chitosligosaccharides (CSO) (Ma et al. 2017). CSO has potential implications for plant protection for various crop plants (Hadwiger 2013). CS acts against many phytopathogenic bacteria such as Pseudomonas, *Xanthomonas*, *Erwinia,* and *Agrobacterium* (Mansilla et al. 2013; Li et al. 2008; Badawy et al. 2014), and an efficient antimicrobial component against mycelial growth of many pathogenic fungi by interfering at all developmental stages (Badawy and Rabea 2011; Feliziani et al. 2015; Meng et al. 2010; Reglinski et al. 2010). Besides both effects, chitosan shows antiviral activity and limits the spread of viruses by inhibiting the replication (Davis et al. 2015; Kulikov et al. 2006) as well as induces resistance against nematode infestation in plants (Escudero et al. 2017a, b; Mwaheb et al. 2017). In addition, the addition of chitosan or chitin to the soil increases the number of chitinolytic microorganisms and in turn their ability to produce hydrolytic enzymes which is damaging to the chitinous fungal pathogens (Kongala and Kondreddy 2023), that may be due to the polycationic nature of CS which allows it to interact with the negatively charged phospholipids of the plasma membrane (Kauss et al. 1989).

(b) **Alginates**

Alginates, the main extracellular matrix and anionic heteropolysaccharide of brown algae (Phaeophyceae), are constituted of α-D-mannuronic acid (M blocks) and β-L-guluronic acid (G blocks), and alternated residues of D-mannuronic acid and L-guluronic acid (GM blocks) (Trouvelot et al. 2014; Kongala and Kondreddy 2023).

Alginate oligosaccharides (AOS) which are enzymatic depolymerization products of alginate act as defense elicitors and induce resistance against numerous plant diseases. AOS improves abiotic stress tolerance represented in the upregulation of genes involved in the abscisic acid signaling pathway, decreased lipid peroxidation, and expression of antioxidant enzyme activities (Li et al. 2018; Liu et al. 2013; Tang et al. 2011), promotes the growth of many crop plants by activating genes involved in auxin and indole-3-acetic acid (IAA) and nitric oxide (NO) biosynthetic pathways (Yang et al. 2021a, b; Iwasaki et Matsubara 2000; He et al. 2018), and enhances seed germination capabilities (Hu et al. 2004). Also, application of AOS to crop plants induces immune responses against viral, fungal, and bacterial diseases (An et al. 2009; Zhang et al. 2015, 2019; Dey et al. 2019) mainly due to enhancement of the antimicrobial products production and activation of defense-related enzymes such as ascorbate peroxidase, peroxidase, phenyl alanyme ammonia-lyase, alterna deoxidase, catalase (Chandía et al. 2004; An et al. 2009; Laporte et al. 2007; Zhang et al. 2015), as well as to salicylic acid (SA)-mediated signaling pathways (Zhang et al. 2019). However, all the biological activities are confirmed with the application of alginate oligosaccharides, notably AOS prepared by enzymatic hydrolysis is highly effective compared to chemically prepared AOS (Yamasaki et al. 2012). Alginate naïf polymer showed a very limited or no effect (Iwasaki and Matsubara 2000; Xu et al. 2003). AOS-induced disease resistance, growth promotion, and other effects are dependent on the concentration and DP between 2 and 10 being very effective in increasing the yield of different crops (Xu et al. 2003).

(c) **Xyloglucan oligomers**

Xyloglucan is a polysaccharide composed of a β-1,4-glucan backbone, elongated with side chains of xylose, fucose, and galactose residues (Park and Cosgrove 2015). It is a major component of the hemicellulose in dicotyledonous plants abundant in primary cell walls and is known to contribute to stress tolerance, stimulate plant growth, and act as a defense elicitor (Kaida et al. 2010; Páez-Watson et al. 2020). They are recognized by the plant as damage-associated molecular patterns (DAMPs) and induce defense responses such as callose deposition, and pathogenesis-related proteins (PR) gene expression leading to protection against necrotrophic and biotrophic pathogens (Claverie et al. 2018). Treating crop plants with this molecule stimulates the expression of cell walls, cellulases, and peroxidases and also induces ethylene synthesis (Larskaya and Gorshkova 2015).

(d) **Oligogalacturonides**

Homogalacturonan (HG) is an abundant polysaccharide in primary cell walls consisting of α-1,4-linked D-galacturonosyl residues, generated by the action of enzymes such as pectate lyases (PL) and polygalacturonases (PG) or by chemical hydrolysis of pectin (Kongala and Kondreddy 2023; Nothnagel et al. 1983). The resulting oligogalacturonides (OGs), which are produced during pathogenesis, act as DAMP agents to activate plant defense responses (Meresa et al. 2024) and activate a cascade of defense response events in plants such as callose deposition, accumulation of phytoalexins, ROS production, mitogen-activated protein kinases (MAPK) activation, and upregulation of PR genes (Hou et al. 2019). Several plant species have successfully developed stress tolerance with the application of pure OGs (Aziz et al. 2004; Davidsson et al. 2017; Selim et al. 2017). On the other hand, treating plants with polygalacturonoids (PGs) increases the resistance against different bacterial pathogens including *Pseudomonas syringae*, *Pectobacterium carotovorum*, and necrotrophic fungi *Botrytis cinerea* (Davidsson et al. 2017; Hou et al. 2019). Pretreatment of Arabidopsis with PGs elicited strong salicylic acid (SA) signaling pathway-mediated defense responses leading to the resistance against *Pseudomonas syringae* Pst DC3000 (Howlader et al. 2020). Also, the most effective defense and elicitor activity of PGs is dependent on DP between 10 and 15 although the lasting elicitor activity is observed with smaller oligosaccharides with DP of 3 (Davidsson et al. 2017; Hou et al. 2019). The majority of research has focused on characterizing defense responses generated by PG; however, until recently, not much progress was reported on their perception by the host plants. According to recent research, wall-associated kinases (WAKs) are possible candidates as PG receptors (Kongala and Kondreddy 2023). To confirm their candidacy, it has not been possible to create mutants of different WAKs, owing to their important role in controlling growth and development (Wagner and Kohorn 2001). Several supporting studies suggest that WAKs play an important role in PG recognition. The PGs bind to both WAK1 and WAK2 in vitro (Cabrera et al. 2008; Kohorn et al. 2009). Although the activation of WAK1 and subsequent responses are dependent on the higher DP of the PGs (Brutus et al. 2010), the general defense responses are observed with lower DP PGs also. Oligosaccharides that are generated during plant-pathogen interaction when recognized by PRR elicit an immune response but not the alone pectin neither can induce nor enhance host resistance against the pathogen (Kohorn 2016).

(e) **ß-glucans**

Beta-glucans are ubiquitous polysaccharides in plant and fungal cell walls (Trouvelot et al. 2014). Among the glucans, β-1,3-glucans, originated from the brown algal *Laminaria digitata* cell walls (Klarzynski et al. 2000), and from fungal pathogens have been implicated in triggering plant immunity (Shinya et al. 2006; Chaliha et al. 2018; Wanke et al. 2020). Plants react differently to structurally varied β-1,3-glucans. It was reported that β-1,3-, 1,6-oligoglucan was more effective than that of longer laminarinpentose in tobacco, which is a known elicitor of defense responses for this plant (Shinya et al. 2006). According to Klarzynski et al. (2000), linear laminaripentose was a strong elicitor of Phe ammonia-lyase (PAL) activity in tobacco cells; however, branched β-1,3-, 1,6-heptaglucan was inactive. Similarly, *C. rubella* responds with ROS burst to linear β-1,3-glucan but not to branched β-1,3-glucan (Wanke et al. 2020). On the contrary, both branched and non-branched β-1,3-glucans have induced immune-associated responses in *A. thaliana* (Melida et al. 2018) and tobacco (Wanke et al. 2020).

β-1,3/1,4-glucans are found in the cell walls of most members of the monocot order *Poales* as well as in evolutionary older plant lineages (Burton and Fincher 2009). Experimentally, Barghahn et al. (2021) found that those oligosaccharides have the dual capacity to act as immune-active DAMPs and/or MAMPs in monocot and dicot plant species.

(f) **Ulvans**

Ulvans are heteropolysaccharides found in green algae *Ulva* spp. (Stadnik and De Freitas 2014), composed of sulfated rhamnose residues linked to glucuronic acids, resulting in a repeated disaccharide unit β-D-glucuronosyl-(1,4)-α-L-rhamnose 3-sulfate, named aldobiouronic acid (Lahaye and Robic 2007). The bio-activity of ulvans depends on their sugar composition, molecular size, and sulfate content. It was reported that sulfate residue is required for oligoulvan to induce defense-related gene expression (Caillot et al. 2012; El Modafar et al. 2012; Abouraïcha et al. 2015). In contrast, non-sulfated ulvan dimer showed elicitation activity in suspension-cultured cells of dicot plants like sulfated polymers (Paulert et al. 2021). Besides, it was proposed that the effective elicitation activity of ulvans might be associated to their high rhamnose and/or uronic acid content (Ben Salah et al. 2018; Jaulneau et al. 2010). Recently, De Borba et al. (2021) confirmed that ulvan oligosaccharides with high rhamnose and/or uronic acid residues effectively elicited complex defense responses in wheat seedlings.

(g) **Carrageenans**

Carrageenans are sulfated polysaccharides present in some red algae, and have a potent plant defense elicitor (Mousavi et al. 2017; Le Mire et al. 2019). Among the sulfated κ, λ, and ι-carrageenan types, λ-carrageenan with the highest sulfate content induces increased disease resistance in tobacco, tomato, and Arabidopsis plants (Sangha et al. 2010; Vera et al. 2012; Sangha et al. 2015). Likewise, Le Mire et al. (2019) reported that λ-carrageenan showed greater elicitor capability in wheat to resist *Zymoseptoria tritici* compared with other bioelicitors tested. In contrast, treatment of Arabidopsis plants with ι- and κ-carrageenans demonstrated reduced leaf damage caused by *Trichoplusia ni* as compared with those λ-carrageenan-treated groups (Sangha et al. 2011). However, a sulfated λ-carrageenan exhibited the highest elicitation and resistance to *Phytophthora palmivora* infection in rubber tree leaves, while the desulphated polysaccharides did not (Pettongkhao et al. 2019)

(h) **Xanthan**

Xanthan is an extracellular polysaccharide has similar structure as cellulose, secreted by the bacterial phytopathogen *Xanthomonas campestris* (XC), composed of a linear (1,4) linked β-D-glucose polymer, and every alternate β-D-glucose is attached with a trisaccharide made of two mannose units and one glucose unit (BeMiller 2018; Kongala and Kondreddy 2023). In barley and wheat, xanthan treatment resulted in the accumulation of β-1,3-glucanase (Castro and Bach 2004; Bach et al. 2003) and triggered both systemic and local defense responses against fungus infections. Treatment of tomato plants with a crude polysaccharide extract containing purified xanthan significantly reduced the growth of leaf spot pathogen *Xanthomonas gardneri* and resulted in the accumulation of H_2O_2 and increased peroxidase levels (Kongala and Kondreddy 2023).

6.5.2 Sugars as Signaling Molecules in Plant Defense

Sugars are also involved in plant immunity as signaling molecules (Rolland et al. 2006). It concerns "small sugars," that is, mono-, di-, and small oligosaccharides such as trehalose, sucrose, galactinol, or raffinose able to activate plant defense responses and increase plant resistance to pathogens. Several papers have reported changes in plants challenged by pathogens. For example, galactinol stimulates the accumulation of defense-related gene transcripts in tobacco plants, enhances resistance against *Botrytis cinerea* and *Erwinia carotovora,* and is a signaling component of the induced systemic resistance caused by *Pseudomonas chlororaphis* (Kim et al. 2008). Sucrose induces isoflavonoid synthesis as a defense response against *Fusarium oxysporum* in lupine (Morkunas et al. 2005). Trehalose induces peroxidase activities and PAL associated with partial resistance of wheat against powdery mildew (Reignault et al. 2001). Other sugars such as D-allose or psicose can upregulate defense gene expression and stimulate plant immunity (Bolouri-Moghaddam and Van Den Ende 2012). In Arabidopsis cell suspensions, sucrose or glucose induces the expression of several PR genes and the accumulation of the corresponding proteins PR-5 and PR-2 through an SA-dependent pathway (Thibaud et al.

2004). On the other hand, glucose, sucrose, and fructose induce the PR-protein transcripts PR-Q and PAR-1 in tobacco in an SA-independent pathway (Herbers et al. 1996).

Several papers have reported changes in the sucrose/hexose ratio in plants challenged by pathogens (Bolouri-Moghaddam and Van Den Ende 2012). The modulation of apoplastic sugar content is controlled by invertases, which catalyze the conversion of sucrose to glucose and fructose, sugars play a key role in the regulation of source/sink relations of plants and response to pathogens. Such regulation could contribute to providing energy for defenses and improve the defense response against pathogens (Jeandet et al. 2022; Kanwar and Jha 2019; Su et al. 2018; Liu et al. 2015).

6.5.3 PAMPs, MAMPs, and DAMPs Type Carbohydrates

The presence of pathogens at the cell surface of plants can be perceived via the recognition of highly conserved molecular patterns as elicitors (Trouvelot et al. 2014; Souza et al. 2017) named microbe- or pathogen-associated molecular patterns (MAMPs and PAMPs) secreted by the pathogen or released from their cell wall by hydrolytic enzymes during interaction with the plant, as well as to DAMPs (damage-associated molecular patterns) that can also derive from the plant cell walls during plant-microbe interactions after hydrolysis by pathogen cell wall degrading enzymes (Wiesel et al. 2014). Pectin-derived oligogalacturonides are well-characterized DAMPs capable of activating plant immunity (Souza et al. 2017), and the oligosaccharides derived from cell wall polymers of fungi and oomycetes also act as MAMPs (Newman et al. 2013). These alarm signatures molecules are recognized by specific pattern-recognition receptors (PRR) situated on the host cell membrane (Zipfel 2014) which activate networked intracellular signaling components (Zheng et al. 2020) and are known as PAMP-triggered immunity (PTI) (Kongala and Kondreddy 2023) (Fig. 6.1). Some molecules known as MAMPs or DAMPs are used to stimulate plant immunity against some pathogens. Several works proved the efficiency of these molecules in stimulating plant immunity. Table 6.1 summarizes some of the patterns with plant host and mechanism of action.

The activation of PRR kinase domain in this first response initiates phosphorylation and the subsequent complex cascade of signaling events that leads to defense gene activation. Defense gene expression allows the synthesis of PR proteins such as the hydrolytic enzymes β-1,3-glucanases and chitinases, peroxidases, cationic defensin, lipidtransfer proteins or proteinase inhibitors, cell wall strengthening, and the accumulation of phytoalexins (Trouvelot et al. 2014). In turn, pathogens have developed a variety of virulence factors that can be delivered into host cells to interfere with PTI and dampen basal defenses (Wang et al. 2019). In response, plants have developed an additional layer of defense that enables them to intercept pathogen effectors, leading to effector triggered immunity (Yang et al. 2022). In particular, some plants possess the specific intracellular disease resistance (R) proteins to

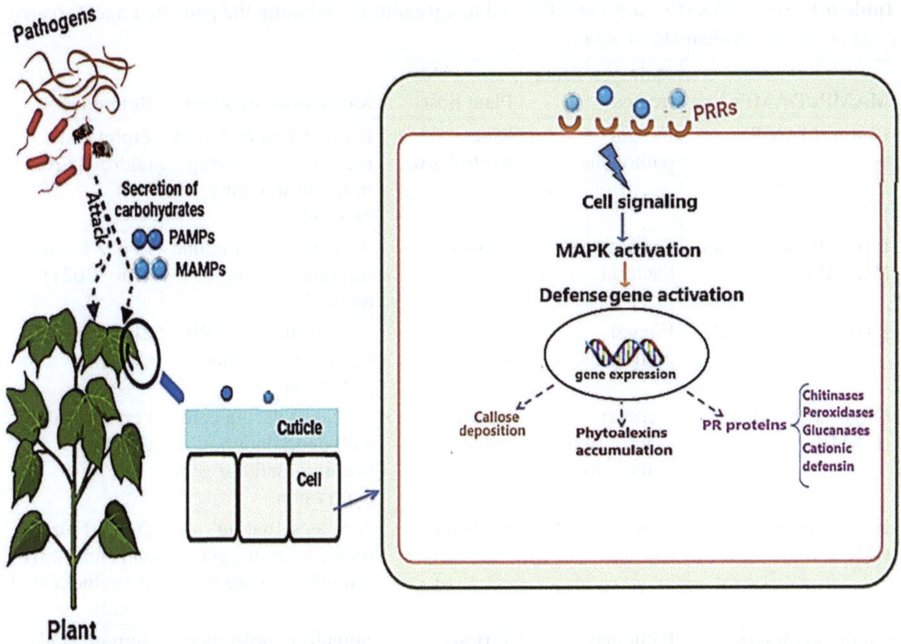

Fig. 6.1 A simplified scheme outlining the mechanisms of PAMPs/DAMPs recognition and induction of defense responses in plant

effectively counteract virulence effectors of pathogens for effector-triggered immunity (Jacob et al. 2023).

6.6 Oligosaccharides Definition and Types

Oligosaccharides are defined as short chains composed of monosaccharide units. Their length is not clearly defined, but they are characterized as typically consisting of di-, tri-, tetra-, penta-, and occasionally between 1 and 20 carbohydrate units. They are recognized for their significant roles in biological processes. Oligosaccharides are frequently present in the glycan chains of glycoproteins and glycolipids, serving roles in extracellular biological functions. Oligosaccharides serve three primary functions: Communication enables cells, tissues, organs, organisms, and ecosystems to interact with one another. Recognition plays a critical role in establishing tissue and cell identity in plants. They assist in identifying the physical characteristics of those cells. Initiation plays a role in activating a plant's defense responses, including pathogen-induced alterations in secondary cell wall structure (Cai et al. 2023; Kongala and Kondreddy 2023; Ochoa-Meza et al. 2021; Wan et al. 2021; Yang et al. 2021a, b).

Table 6.1 Some MAMPs and DAMPs used in agriculture, including the pathogen agent treated, plant host, and mechanism of action

MAMPs/DAMPs	Pathogen agent treated	Plant host	Mechanism of action	Reference
Chitin (MAMP)	Fungal pathogens	Rice, *Arabidopsis*	Recognition by LysM receptors, triggering immune signaling cascades	Zipfel and Oldroyd (2020)
Lipopolysaccharides (MAMP)	Gram-negative bacteria	Tomato, tobacco	Activation of immune responses via lipid A recognition	López-Gómez et al. (2021)
β-Glucans (MAMP)	Fungal pathogens	Various plants	Recognition by PRR leading to immune activation	Zhang et al. (2021)
Oligosaccharides (DAMP)	General (necrotrophic pathogens)	*Arabidopsis*, tomato	Released during cell well degradation, boosting defense gene expression	Ferrari et al. (2019)
Extracellular DNA (DAMP)	General stress signals	*Arabidopsis*	Acts as signaling molecule to trigger immune responses	Duran-Flores and Heil (2018); Wierzbicki et al. (2022)
Cutin monomers (DAMP)	Pathogen-induced damage	Various plants	Signaling molecules that activate plant defense pathways	Serrano et al. (2022)
Flagellin (MAMP)	Bacterial pathogens	Arabidopsis, tomato	Reecognition by FLS2 receptor, activitating MAPK pathways leading to defense responses	Boller and Flix (2009); Gómez- Gómez and Boller (2002)
Peptidoglucan (MAMP)	Baterial pathogens	Arabidopsis	Detection by LysM-containing proteins, leading to defense activation	Gust et al. (2007)
Oligogalacturonides (DAMP)	General (necrotrophic pathogens)	Tomato, Arabidopsis	Release during cell well degradation, boosting defense	Ridley et al. (2001)
ATP (DAMP)	General stress signals	Arabidopsis	Extracellular ATP acts as signaling molecule to amplify defense responses	Choi et al. (2014)
Cellobiose (DAMP)	Fungal pathogens	Various plants	Signals through its receptor to induce immunity	Souza et al. (2017)

Oligosaccharides can be categorized into various types, typically divided into two groups based on their content. Homooligosaccharides (HOS) consist of oligosaccharide units that are uniform in their sugar composition. HOS exhibits variation in its linkage structures. N-glycans commonly utilize them as the core or foundational structure to which additional moieties are appended. In plant cells, these structures serve as the location for glucose and galactose residues, which are essential for the synthesis of arabinogalactan proteins and hemicelluloses. Heterooligosaccharides (HEOS) consist of sugar residues that are distinct from each other. Heterooligosaccharides are generally associated with secreted or extracellular proteins. For instance, xyloglucan is a xylosylated glucan thought to anchor cellulose microfibrils within the load-bearing structure of the primary cell wall. Oligosaccharides exhibit diverse linkages and branching patterns, which are believed to facilitate their unique biological functions. Oligosaccharides influence cell expansion and wall porosity, facilitate the association of cellulose microfibrils, and contribute to tensile and compressive stresses. Recent studies have identified additional oligosaccharides within the plant cell wall, indicating their potential biological functions (Baek et al. 2024; Riseh et al. 2024; Perumal et al. 2023; Du et al. 2022; Morán 2021).

6.6.1 Structural Attributes of Oligosaccharides

Oligosaccharides are composed of monosaccharides connected by glycosidic bonds between an anomeric carbon atom (or a hemiacetal residue) at C-1 or C-2, and a hydroxyl group at C-1, C-2, C-3, C-4, or C-6. The configuration (L- or D-) of the monosaccharide residues in the glycosidic bonds, the position of the glycosidic linkage, and whether it is in α or β configuration determine the stereochemical features and interaction properties of glycoconjugates. The degree of polymerization (DP), defined by the number of individual sugars linked together and their sequence, determines the three-dimensional structure of the oligosaccharide, which subsequently governs its physicochemical properties, architecture, and interactions with various biological entities. Consequently, they serve as conventional transporters of chemical messages between cells, as they possess a simplistic structure yet are molecularly organized in an exceptionally diverse and specialized manner. A great diversity of particular oligosaccharides has been found in plants so far, in connection to the different receptors they should interact with. The presence of these oligosaccharides in the extracellular sections of the root, stem, leaf, flower, and sexual organs and their significance in the determination of plant development have been proven over the last 30 years. Likewise, the surfaces of microorganisms such as bacteria exhibit considerable variability, leading to a diverse array of distinct cell wall glycan structures. These participate in the recognition and communication processes at the interface between the host and the microorganism (Li et al. 2024; Ma et al. 2024; Elango et al. 2022; Franková and Fry 2021).

6.6.2 Biological Roles of Oligosaccharides in Plants

Oligosaccharides not only participate in stable physiological processes like carbon and energy metabolism but also function as signaling molecules in reaction to fluctuating external variables and internal habitats. Oligosaccharides can influence plant physiological processes, including growth, development, and stress responses, through specific or nonspecific interactions with target molecules at the system level. Numerous studies have indicated that oligosaccharides function as elicitors, stimulating plant-microbe interactions related to defense responses. Oligosaccharides specifically operate in many defense pathways to enhance plant immunity in response to biotic stress. These defense activation events, rooted in unique biological elicitation, contribute to systemic resistance, including systemic acquired resistance and induced systemic resistance. Consequently, oligosaccharides have been categorized as « bioactive compounds" (Meresa et al. 2024; Kongala and Kondreddy 2023; Meena et al. 2022; Chamkhi et al. 2021).

Conversely, oligosaccharides are crucial for preserving the structural integrity and stability of the plant cell wall, hence affecting the adaptive responses of plants to biotic and abiotic stimuli. Given their biological roles, oligosaccharides seem to participate in numerous biological processes. Oligosaccharides regulate critical cellular processes in the immune system; thus, a comprehensive understanding of this influence is vital for elucidating these interconnected systems. Additional research will be required to assess the potential application of oligosaccharides in agricultural settings to stimulate plant defenses (Xu et al. 2022).

6.6.3 Mechanisms of Plant Immunity Mediated by Oligosaccharides

This part elucidates how oligosaccharides mediate immunity by detailing the mechanisms of plant identification of pathogen-associated molecular patterns and oligosaccharide elicitors, along with the subsequent processes that result in the priming or defense of a plant. Plants possess immune surveillance mechanisms to identify patterns that are largely conserved among microorganisms. The receptors may be situated on the extracellular plasma membrane or within the cytoplasmic matrix. In plant immunity, following the identification of patterns and oligosaccharide-induced resistance, signal transduction initiates a comprehensive array of defensive responses against pathogen invasion. Oligosaccharides stimulate plant defense mechanisms via the mitogen-activated protein kinases (MAPK) signaling pathway, the generation of reactive oxygen species, the activation of genes associated with signaling and defense, salicylic acid, precursor proteins for pathogen production, jasmonic acid, ethylene, phytoalexins, and enzyme inhibitory protein systems (Kongala and Kondreddy 2023; Chaube et al. 2022; Morán 2021).

In the signal transduction route, oligosaccharides identify receptors on plant plasma membranes, initiating responses that activate signaling molecules, which subsequently activate a cascade of associated enzymes, kinases, and cytoplasmic components in plant cells. Membrane proteins, similar to oligosaccharides, can initiate G-protein-dependent processes that facilitate the release of ions within the plant cytoplasm. Oligosaccharide perception may potentially entail the initiation of passive ion outflow or the activation of membrane protein depolarisation. Following a sequence of induced protein activations, plant cells generate significant oxidative bursts as an initial defense mechanism, which are critically crucial in the defense responses of rice cells against pathogens. The production of numerous tiny signal molecules leads to the rapid accumulation of various reactive oxygen species. MAPK participates in the activation of kinase phosphorylation after oxidative bursts. This reaction can also stimulate gene expression, including pathogenesis-related proteins, via activation of receptor-like protein kinases, as well as proteins associated with cell wall reinforcement in disease-resistant systems and several candidate gene expressions (Giovannoni et al. 2025; Fuertes-Rabanal et al. 2024; Kongala and Kondreddy 2023).

6.6.3.1 Recognition and Signaling Pathways

Biotic stressors provide significant limitations on both the amount and quality of crop yields. Microorganisms associated with plants, including bacteria, fungi, and oomycetes, together with herbivores, are the primary agents of biotic stress. In reaction to these stress conditions, plants have evolved various defense mechanisms, encompassing structural and biochemical responses. Plants identify dangers posed by biotic stress agents and subsequently activate relevant molecules in signaling pathways that elicit immune defense responses, known as the plant immune system. The molecular processes of plant immunity consist of two primary steps: recognition and signaling pathways, which result in jasmonic acid (JA)-regulated systemic acquired resistance (SAR), salicylic acid (SA)-modulated systemic acquired resistance (SAR), and inducible systemic resistance (ISR) (Mishra et al. 2024; Meena et al. 2022; Wang et al. 2021).

The identification of oligosaccharide elicitors is essential for understanding the subsequent signaling events. Pattern recognition receptors (PRRs) are defined as protein molecules that facilitate the detection and binding of ligands for elicitors. Upon the binding of elicitors to the membrane-associated pattern recognition receptors (PRRs), downstream signaling processes are activated to trigger the immune response in the affected tissues. A cascade of signaling events progressively activates various signaling molecules. Subsequently, an immediate oxidative burst occurs shortly after the elicitors bind to pattern recognition receptors (PRRs), mediated by NADPH oxidase. The activation of NADPH oxidase is intricately linked to alterations in plasma membrane microdomains. A temporary elevation in Ca^{2+} concentration occurs due to the influx of this ion from the extracellular environment. The correlation between the relocalized NADPH oxidase and elevated Ca^{2+} levels at the plasma membrane facilitates the activation of NADPH oxidase, which is crucial for generating H_2O_2 necessary for the oxidative burst. Thereafter, the mitogen-activated protein kinase (MAPK) cascade is initiated. The MAPK cascade is a tri-adic protein phosphorylation pathway comprising a three-enzyme module, MAP kinase kinase kinase, MAP kinase kinase, and MAP kinase. Upon activation, MAPKs go to the nucleus and enhance the production of defense genes. Transmembrane active transport elevates cytoplasmic Ca^{2+} concentration, serving as a mediator for various physiological processes and a triggering mechanism in the signal transduction cascade. H_2O_2 serves as an intermediary that penetrates the plant cell membrane to activate the SAR defense genes and induce hypersensitive reaction cell death. The attachment of elicitors to the pattern recognition receptors (PRRs) results in the enhancement of the defense signal throughout the entire plant system, encompassing the adjacent infected tissues. The higher resistance displayed by the plants against the second pathogen challenge signals the presence of the immunological signal amplified in the whole plant system known as SAR. The elevated concentration of SA in the SAR-treated leaves triggers the synthesis of SAR (Nabi et al. 2024; Malook et al. 2022; Dalio et al. 2021; Lu and Tsuda 2021).

6.6.3.2 Initiation of Defensive Mechanisms

Upon contact with an oligosaccharide, the plant cell initiates a chain of actions resulting in a temporary elevation of extracellular H^+ and alterations in membrane polarization. This ion flux correlates with alterations in intracellular second messengers, such as Ca^{2+} and reactive oxygen species generation, alongside the inhibition of proton pumps. This leads to the generation of various intracellular signals and the upregulation of genes associated with plant defense, culminating in the accumulation of antimicrobial compounds, including phytoalexins, phenolics, and other signaling molecules pertinent to plant-microbe interactions. The production and secretion of extracellular reactive oxygen species represent another hallmark of a plant cell's interaction with an oligosaccharide. Oligosaccharides from many sources now possess the ability to function as elicitors and are being investigated for their potential to activate defense mechanisms in plants. These minor glycol-conjugates or derivatives exhibit two overlapping sites: (i) a nonspecific site that engages in the activation of general defenses by the plant cell and (ii) a specific site that is implicated in the tailored response to the distinct origin of the oligosaccharides. As previously stated, they facilitate the integration of oligosaccharides in plant hormone signaling. Furthermore, they are all interacting or potentially overlapping genes involved in abiotic stress signaling pathways within plant cells. The exogenous application of oligosaccharides to plants via the aforementioned pathways is anticipated to primarily stimulate the logarithmic growth phase of wild plants, resulting in a reduction or cessation of growth and an increase in the generation of resistant cells. The activation of more resilient cells is crucial for plant defense, thus, oligosaccharides may facilitate enduring and wide-ranging resistance in crop species (Jia et al. 2024; Bigini 2023; Liu et al. 2023).

6.6.4 Function of Oligosaccharides in Plant-Microbe Interactions

The interactions between plants and microorganisms are profoundly influenced by signaling, often mediated by carbohydrates. While monosaccharides and polysaccharides may participate in this communication, other kinds of oligosaccharides have been identified as signaling molecules. Interactions between plants and their microbiota may be advantageous or detrimental. Utilizing rhizobia or pathogens as models has yielded substantial insights into how plants equilibrate mutualistic and pathogenic responses to microorganisms. Numerous kinds of oligosaccharides have been identified as crucial in promoting plant-microbe interactions. Rhizobia synthesizes Nod factors, which are lipochitooligosaccharides (LCOs) resembling the structure of chitin oligosaccharides. Likewise, chitin oligosaccharides are generated by pathogenic fungi and oomycetes (Song et al. 2023; Xu et al. 2022).

Nondigestible oligosaccharides present in the plant cell wall are typically released passively by the plant during growth and have minimal to no involvement in plant-microbe interactions unless the cell wall is compromised. The active synthesis of oligosaccharides that function as signals in facilitating microbial interactions is also noted. In legumes, plants directly respond to symbiotic nodule-forming bacteria by commencing the synthesis of α-fucosylated LCOs. In response to pathogenic interactions with fungi, plants synthesize apoplastic cell wall-derived oligosaccharides, which effectively regulate the establishment of non-adapted pathogens, the exudation of pathogenic effector proteins, the activation of defense mechanisms, and the production of callose-based papillae surrounding initially infected cells. In addition to this pivotal research, numerous others primarily involving Arabidopsis, and to a lesser degree other plant species, are emerging to elucidate the production, perception, and functions of these oligosaccharides during pathogenesis. It is obvious that a complex communication network has evolved between plants and microorganisms, and that oligosaccharides function not only as signals for both beneficial and harmful interactions but also act as barriers. The clinical and biotechnological ramifications of oligosaccharide signaling for disease management in crops is a subject of significant attention (Yang et al. 2021a, b; Molina et al. 2024; Wanke et al. 2021).

Root-associated microbes can exert harmful, neutral, or advantageous effects on plant health, necessitating that plants utilize molecular signatures to differentiate among these symbionts, commensals, and pathogens. This is particularly crucial for the recruitment of mutualists that directly enhance nutrient absorption or indirectly shield host plants from disease threats. Beneficial microorganisms frequently commence a symbiotic connection by exuding oligosaccharides, which are recognized by other microbes as cues that activate or train symbiotic trait programs. Extensive reviews have been conducted on the molecular mechanisms governing the mutualistic relationships between plants and beneficial bacteria or mycorrhizal fungi. Root-colonizing bacteria, known as plant growth-promoting rhizobacteria, have demonstrated the ability to boost plant growth and improve resilience to biotic and abiotic stressors. The colonization by PGPR can elicit systemic responses and prepare plant cells, equipping them for swifter and more vigorous reactions to following challenges. The signaling mechanisms for the recognition of microbe-associated chemical patterns involving amino acids or peptides are well characterized by PGPR-induced defense responses (Bhupenchandra et al. 2024; Khokhani et al. 2021; Wanke et al. 2021; Thoms et al. 2021).

Given that beneficial microorganisms provide agricultural advantages and resilience against a fluctuating disease landscape, numerous strategies have been employed to augment and utilize mutualistic plant interactions for sustained crop protection. Microbial consortia as bioinoculants have been formulated to enhance plant growth and induce systemic resistance. Growing evidence indicates that advantageous connections between plants and bacteria or fungi are established or augmented through the manufacture, secretion, or sensing of oligosaccharides. Oligosaccharide signaling may serve as a way to augment the advantageous interaction between plants and beneficial microorganisms. Decoy oligosaccharides may divert pests or pathogens from plant hosts, and numerous advantageous functions of oligosaccharides have been recorded in diverse agricultural and industrial contexts. Oligosaccharides serve as biostimulants in plant nutrition, enhancing germination, root development, abiotic stress tolerance, and nutrient absorption. This review examines the impact of β-glucans and chitooligosaccharides on diminishing the vulnerability of plant hosts to biotic stress, while also briefly addressing the beneficial applications of mannooligosaccharides (Thomas-Barry et al. 2024; Alghamdi et al. 2023; Acosta-Jurado et al. 2021; Fetsiukh et al. 2021).

6.6.5 Oligosaccharides as Inducers of Plant Immune Responses

Natural oligosaccharides functioning as elicitors demonstrate significant potential in enhancing plant immune responses against diseases and pests. Certain oligosaccharides are recognized for signaling essential metabolic pathways in plants that result in the establishment of induced systemic acquired resistance. This condition of systemic acquired resistance endures for an extended duration, equipping plant cells to initiate a swift and robust defense against both biotrophic and necrotrophic diseases and pests. Systemic acquired resistance can safeguard plant tissues for several weeks posttreatment, while also inhibiting the development of endogenous viral resistance in more remote tissues (Kongala and Kondreddy 2023; He et al. 2021; Moenne and González 2021; Ochoa-Meza et al. 2021).

ESCAs are among the limited commercially available pure oligosaccharide-based elicitors that can induce an immunological response in many tissues, including plant leaves when administered to the roots. In addition to chitosan elicitors, bacterial elicitors such as oligogalacturonides and cyclic-β-1,2-D-glucan have been derived and employed as natural biopesticides against infections and pests. The application of oligosaccharides in agriculture extends beyond their role as biopesticides; they also function as resistance inducers or plant activators that initiate systemic plant defense mechanisms. The drawback of oligosaccharides is that their immunological responses are rapid and can enhance stress resistance, hence accelerating and optimizing the manufacture of endogenous resistance proteins during an actual attack. This indicates that the plant's intrinsic defense mechanism not only prevents external invasions but also enables it to mount a more robust immune response in anticipation of potential threats (Duret et al. 2024; Qi et al. 2024; Anzano et al. 2022; Cataldo and Fucile 2022; Landi et al. 2021) (Table 6.2).

6.6.5.1 Induction of Systemic Acquired Resistance (SAR)

The effective activation of the plant immune system at a specific location can lead to increased pathogen resistance in remote tissues, a phenomenon referred to as systemic acquired resistance (SAR). Recent reviews have thoroughly examined the

Table 6.2 Some oligosaccharides are used commercially as elicitors in agriculture and crop protection

Oligosaccharide	Pathogen agent treated	Plant host	Mechanism of action	References
Chitooligosaccharides	Fungal pathogens	Tomato, cucumber	Induce defense-related gene expression and phenolic compound accumulation	Katiyar et al. (2019) and Tsushima et al. (2020)
Oligogalacturonides (OGs)	Bacterial and fungal pathogens	Arabidopsis, tomato	Trigger hypersensitive response (HR), increase PR protein synthesis	Ferrari et al. (2019) and Guo et al. (2020)
Mannooligosaccharides	*Fusarium* spp.	Wheat, barley	Activate immune responses by enhancing chitinase and β-1,3-glucanase activities	Hasni et al. (2021)
Raffinose family oligosaccharides	Various pathogens	Arabidopsis, pea	Improve pathogen resistance and abiotic stress tolerance	Egert et al. (2019)
Xylooligosacchrides	Fungal pathogens	Rice, barley	Enhance defensive enzyme activities and seedling growth under biotic stress	Zhang et al. (2021)
Cellobiose	Bacterial pathogens	Lettuce	Induces accumulation of defense-related proteins and secondary metabolites	Sousa et al. (2020)
Oligo-β-glucans	*Botrytis cinerea*	Strawberry	Activate defense pathways via β-glucan receptor recognition, inducing resistance	Liu et al. (2022)

complex mechanisms generating SAR. Generally, several components infiltrate the vasculature or other communication pathways to elicit an alert state in the plant, hence augmenting defenses in other regions of the organism. Upon the arrival of mobile signals at remote cellular regions, extensive transcriptional reprogramming can enhance defenses across the plant, leading to the buildup of pathogenesis-related proteins, and antimicrobial substances, and the fortification of the cell wall before an assault. A prominent characteristic of SAR is its extensive resistance to multiple species of biotrophic and necrotrophic diseases (Malinowski et al. 2024; Jacob et al. 2023; Yu et al. 2022; Vlot et al. 2021).

Oligosaccharides are recognized mediators of systemic acquired resistance, serving as powerful inducers that enhance plants' defensive capabilities. The molecular basis of oligosaccharide-induced systemic acquired resistance (SAR) entails a delayed process that depends on jasmonic acid signaling or the interaction between salicylic acid and jasmonic acid pathways. Furthermore, both endo- and exo-oligosaccharides can elicit short-term systemic acquired resistance (SAR) responses, providing prolonged protection to plants against diverse infections. Applying exogenous oligosaccharides to foliage enhances plant resilience against several stressors, including drought and metal toxicity. Moreover, stimulating plant defensive mechanisms using oligosaccharides has garnered attention for its potential to diminish pesticide application and, consequently, the environmental effect of agricultural production. Various oligosaccharides have been assessed for their efficiency and persistence across different organisms and abiotic conditions, with encouraging results anticipated to enhance the practical uses of oligosaccharide-induced systemic acquired resistance soon. Nonetheless, despite the extensive documentation of oligosaccharides that effectively generate systemic acquired resistance, resulting in primed plants against many infections. (Meresa et al. 2024; Gao et al. 2022; Nowak et al. 2022; Yu et al. 2022) (Table 6.2).

6.6.5.2 Priming of Defensive Mechanisms

Plants can enhance their defense mechanisms to respond more rapidly and effectively to subsequent infections caused by diverse abiotic stressors. This idea, typically categorized into ecological and systematic priming, pertains to the preparation of the biological system for conditional and appropriate responses. In response to a pathogen, the mechanism of action largely diverges from the conventional plant immune response, following a primed state, the ensuing reaction is swifter and more vigorous. Oligosaccharides are recognized for their priming effects, enhancing the plant's ability to respond to pathogen exposure. Plants depend on this preparedness due to the absence of a central immune system, and immunological responses entail a considerable metabolic expense (Swaminathan et al. 2022; Versluys et al. 2022).

Oligosaccharides can prime the plant for a more rapid and robust defense response during subsequent pathogen assaults. The priming mechanism remains poorly comprehended; nevertheless, in the elite genotype of monocot crops, three effective oligosaccharides, identified as MAMPs and commonly found on pathogen cell surfaces, have been proposed to dissociate biotic and abiotic stress. The specific defense responses that are activated differ among oligosaccharides and, crucially, also among plant genotypes. The molecular and physiological mechanisms that induce the primed state are varied; oligosaccharides can modify gene expression or enzyme activity and can initiate physiological or metabolic processes via distinct routes. Priming may transpire in diverse metabolic pathways, such as those involving genes associated with detoxification and cellular membrane protection, or in genes that encode signaling molecules that initiate defense responses, including phytohormones. Priming may result in a more resilient plant suitable for sustainable agriculture. This effect can yield improved outcomes when integrated with host resistance, enhancing the efficiency of host resistance breeding. Nonetheless, priming presents restrictions and issues that must be addressed (Kongala and Kondreddy 2023; Ou et al. 2022; He et al. 2021) (Table 6.2).

6.7 Interaction between Carbohydrates and Plant Pathogens

Pathogens are microbial invaders that cause harm to a host and initiate a defense response in the host. During the coevolution of plants and pathogens, plant selective agents shape the evolution of the plant immune system, and as a consequence, pathogens develop various mechanisms to escape plant immunity and become plant pathogens. Several pathogenic microorganisms have receptors for carbohydrates on the plant cell wall surface. Infection of strains expressing certain effectors enhanced colonization of plant tissues, likely because of the deactivation of the effector target hemicellulose, thus affecting plant immunity. As a PAMP molecule, the recognition of chitin and the corresponding signaling pathway was the first pattern-triggered immunity discovered in plants (Molina et al. 2024; Kongala and Kondreddy 2023; Lorrai and Ferrari 2021).

The structure of bacterial and fungal carbohydrates binding to the receptors of PRRs is shown. Currently, much research has focused on the study of bacterial MAMP, PAMP, fungal MAMP, and EF-Tu, to improve plant resistance to pathogens through carbohydrate modification, promote the control of disease, and reduce the use of pesticides. Because of the interactions of the diverse microbial MAMP/PAMP, altered gene expression levels of susceptible and nonhost plants, and the resulting gene function loss of the resistance genes, both from the gene replacement of susceptible genes with those of resistant genes required for the synthesis of a MAMP/PAMP, can also access the potential function of the identified microbial plant combat factor. The synthetic MAMP/PAMP is consistent with the host BASL–RLCK complex and the known MAMP/PAMP-triggered immune response. Such evidence thus confirms the important role of identified microbial PAMP in plant resistance to related pathogens. The synthesis can indirectly reduce the complexity of the response triggered by effectors in addition to the role played by MAMP/PAMP in plant resistance to pathogens (Xu et al. 2022; Ghassemi et al. 2021; Wanke et al. 2021; Yu et al. 2021).

Plants recognize potentially harmful microbes through mechanisms that involve essential life processes such as membrane signaling, the biosynthesis of protective compounds, and the generation of antimicrobial peptides. The recognition of pathogen-associated molecular patterns by pattern-recognition receptors constitutes the initial event in plant-pathogen interactions. Pathogen-associated molecular patterns, which are highly conserved, are critical for the survival of pathogens, including bacterial and fungal DNAs. These patterns are detected by pattern-recognition receptors in the host. Pattern-recognition receptors can be categorized into nonself-recognition and self-recognition types, depending on their specific functions. Nonself-recognition pattern-recognition receptors identify pathogen-associated molecular patterns from external organisms, whereas self-recognition pattern-recognition receptors engage with pathogen-associated molecular patterns present on the host cell surface. A gene associated with self-pathogen recognition in rice encodes a receptor-like kinase. Nonself-recognition pattern-recognition receptors are receptor kinases of the leucine-rich repeat type, situated in the plasma membrane. One receptor is known to recognize bacterial flagellin, while another receptor recognizes bacterial EF-Tu. Both contain leucine-rich repeat motifs that facilitate protein interactions and function as binding sites for pathogen-associated molecular patterns (Hornung and Gaidt 2024; Yang et al. 2023; Zhao et al. 2023; Rumpret et al. 2021).

Pattern-recognition receptors identify pathogen-associated molecular patterns due to the glycans present on these molecules. Amorphic pathogens, including bacteria, fungi, and oomycetes, present pathogen-associated molecular patterns to the host. Fungal chitin serves as a pathogen-associated molecular pattern. Chitin oligomers with a degree of polymerisation ranging from 3 to 14 have been shown to elicit defense responses alongside the production of reactive oxygen species. Chitin, sourced from the yeast cell wall, predominantly interacts with plant LysM receptors. A chitin elicitor receptor kinase was identified as the receptor that recognizes fungal chitin. Treatment with a synthetic chitin octamer decreased powdery mildew incidence on barley and enhanced grain yield after gray mold infection due to the induction of plant defense mechanisms. A scale-down mutant exhibited deficiencies in the defense mechanisms triggered by chitin pentamers and octamers. The LysM receptor can bind up to 10 micrograms of chitin derived from fungal cells. A small protein charge patch of the LysM receptor or chitin oligosaccharides resulting from binding may activate the host immune system to counteract the evasion of pathogen-associated molecular patterns by necrotrophic fungi. Organisms that engulf chitin oligosaccharides trigger a defensive response. Fungal chitin serves as a biologically significant pathogen-associated molecular pattern at an edibiologic concentration. The fungal pathogen-associated molecular pattern β-(1,6) is a significant component in the study of host-pathogen interactions. Glucans are components of lichens and are identified by the LysM receptor, other glucans are detected by various receptors and components. Consequently, host pattern-recognition receptors identify pathogen-associated molecular pattern carbohydrates, including chitin and glucans, indicating that their carbohydrate components are crucial for the elicitor activity of these macromolecules. These plant pattern-recognition receptors typically detect glycan-containing conserved microbial proteins. To date, there have been no reported studies on self-recognition pattern-recognition receptors and plant sugars concerning pattern-recognition receptors. Knowledge of the plant pattern-recognition receptor glycan binding remains essential, despite the report of a lectin receptor kinase that binds oligosaccharides in fungi and contributes to defense against gray mold (Sey et al. 2024; Mu et al. 2023; Ngaki et al. 2023; Fang et al. 2022).

6.8 Carbohydrates Dedicated for Application in Organic Agriculture

Many carbohydrate biopolymers have been reported as potent plant immunity inducers (Meresa et al. 2024) and active ingredient protectors used in antifungal formulations to develop plant protection products against phytopathogens which will form an alternative to conventional fungicides. As reviewed, recent and suitable applications of three common polysaccharides chitosan, alginate and cellulose in organic agriculture regarding their biodegradability and non-toxicity properties (Korbecka-Glinka et al. 2022). Moreover, they can interact with many hydrophilic and hydrophobic compounds in more complex formulations (Campos et al. 2015).

Chitosan is widely used in agriculture as a trigger in plant defense against pathogenic microorganisms more effectively than fungi than bacteria (Korbecka-Glinka et al. 2022), and in pre-and postharvest treatments of crops to control microbial infections (Xing et al. 2015). CS has also been applied as a nanocomposite formulation with silver nanoparticules and tested as an anti-fungal agent against *Aspergillus flavus*, *Alternaria alternata,* and *Rhizoctonia solani* isolated from chickpea seeds (Korbecka-Glinka et al. 2022). Other antifungal activity (in vitro) of CS nanoparticles (NPs) combined with an extract obtained from leaves of *Byrsonima crassifolia* (nanche) was used, aiming at protecting bell peppers from *Alternaria alternata.* Edible coatings with these NPs sprayed on bell peppers before harvest reduced infections and improved the storability of the crop after storage reduced weight loss and better physicochemical features of peppers were observed (Gonzalez-Saucedo et al. 2019). Madanipour et al. (2019) assessed the effect of postharvest chitosan application in combination with licorice ethanol extract on the shelf-life of apple fruits. Chitosan-licorice edible coating reduced the postharvest decay rate and inhibited *Penicillium expansum* growth. In another study, in a process of searching for alternatives to sulfur dioxide fumigation of grapes, Shen et Yang (2017) developed edible coatings for grapes using chitosan in combination with salicylic acid. The coatings made of both ingredients induced the activities of chitinase, phenylalanine ammonia-lyase, β-1, and 3-glucanase, and reduced the decay of table grapes by inhibiting the growth of Botrytis cinerea. The strongest effectiveness activity against *Penicillium digitatum* was obtained with a complex edible coating of CS integrated with phenolics-rich pomegranate peel extract for controlling green mold in harvested oranges (Kharchoufi et al. 2018).

Alginates are used in agriculture as delivery systems, seed treatments, organic fertilizers, and edible coating films for vegetables and fruits (Korbecka-Glinka et al. 2022; Jarecki 2021; Jarecki and Wietecha 2021). In a new method to control oil palm tree disease damaged by the fungal infection caused by *Ganoderma boninense*. Formulation of *Streptomyces* spp. bacteria immobilized in alginate beads resulted in the lowest disease severity and the highest degree of plant vigor (Sujarit et al. 2020). Moreover, alginate oligosaccharides demonstrate excellent potential for agricultural applications to promote plant growth, alleviate growth inhibition under abiotic stress, induce defense responses in plants, and extend the shelf-life of harvested crops (Hien et al. 2000; Iwasaki and Mastsubara 2000). Zhuo et al. (2022) demonstrated that AOS treatment improved resistance to post-harvest decay and quality in kiwifruit. The results showed that in vitro AOS did not inhibit the growth of *Botrytis cinerea*, which is the causal agent of gray mold in kiwifruit, but they reduced the incidence of gray mold and diameter of lesions of kiwifruit during storage. Sodium alginate was used to develop an innovative encapsulation system for the delivery of biocontrol and protection of *Bacillus subtilis* from harmful environmental conditions. This encapsulation has enhanced the antifungal effects of *B. subtilis* and significantly increased the growth of bean plants (Saberi-Rise and Moradi-Pour 2020). Other, coating films on fruits and vegetables based on alginates and may be combined with diverse active ingredients serve as a primary packaging which is directly in contact with the fruit, wrapping it to form a gas and moisture barrier, reducing the microbial load, improving the mechanical property keeping the sensory properties intact while prolonging the shelf-life (Galus and Kadzinska 2015). Indeed, Xu et al. (2020) described the inclusion of cyclolipopeptides (CL) from *Bacillus subtilis* in the production of an easily removable alginate coating for preserving blueberries. The obtained CL-alginate coatings provided strong antifungal properties and kept blueberries fresh during 20 days of cold storage, resulting in higher firmness, reduced respiratory rate, and weight loss in the stored blueberries. As well, the addition of nanomaterials can also enhance the antifungal activity of the alginate coatings. Jiang et al. (2013) reported the efficacy of a composite alginate/nano-Ag coating in reducing the counts of different groups of microorganisms on coated shiitake mushrooms after cold storage for 16 days compared with the uncoated control.

Cellulose derivatives obtained with different functionalities through etherification or esterification (Heinze et al. 2018) are commonly used in agriculture and include ethyl cellulose (EC), hydroxypropyl cellulose (HPC), methylcellulose (ME), carboxymethyl cellulose (CMC), hydroxyethyl cellulose (HEC), and hydroxypropyl methylcellulose (HPMC). These derivatives are characterized by higher solubility in water and/or organic solvents compared to cellulose. Their properties allow for using them in formulations applied in plant protection as binders and carriers of active ingredients and biocontrol agents (Korbecka-Glinka et al. 2022). Hence, CMC can be used as the base for formulations containing microorganisms. Viji et al. (2003) used CMC as the base in formulations containing biocontrol agent-bacterial strains of *Pseudomonas aeruginosa*. Foliar sprays using these formulations effectively reduced the gray leaf spot disease in perennial ryegrass. In another study, the same polymer was combined with biocontrol microorganisms in seed treatment formulation protecting wheat against *Fusarium graminearum* (Dal Bello et al. 2002). Edible coating CMC containing moringa (*Moringa oleifera* Lam.) seed and leaf extracts developed by Tesfay et al. (2017) as a novel postharvest treatment for maintaining storage quality and inhibition of three fungi from avocado fruits: *Alternaria alternata*, *Lasiodiplodia theobromae* and *Colletotrichum gloeosporioides*. The results showed that CMC blended with moringa extracts improved storability of avocado fruits and their antifungal properties in vivo. Hydrogel microspheres made of CMC and chitosan loaded with citral showed antibacterial activity in vitro and antifungal properties (in vivo) in reducing disease incidence caused by the plant pathogenic fungus *Botrytis cinerea* in tomato plants (Ma et al. 2020). However, the recent publication of Liu et al. (2021) showed that EC microcapsules with fungicide can be effectively used as plant protection against *Botrytis cinerea*-causing gray mold that assures prolonged release and slower degradation in the environment. Valencia-Chamorro et al. (2009) developed an edible coating based on HPMC in combination with one or two preservatives and hydrophobic components for cold stored oranges and tested its antifungal properties in fruit which was first inoculated with *P. italicum* or *P. digitatum*. After cold storage, the coating had no negative effect on the fruit quality, and its antifungal properties were confirmed. The most effective coating, containing potassium sorbate and sodium propionate, controlled the development of green and blue mold on the inoculated fruit.

6.8.1 Oligosaccharides as Prospective Biocontrol Agents

Numerous researches have suggested the use of oligosaccharides in sustainable approaches for managing plant diseases. In contrast to conventional biocontrol agents, oligosaccharides are not living entities; rather, they are created by and extracted from diverse living organisms or synthesized by green chemical methods. This enhances the impression of consumers and regulators regarding these compounds as a viable low-risk element in the future of sustainable agriculture (Mondéjar-López et al. 2024; Kumar et al. 2022; Buzón-Durán et al. 2021).

Oligosaccharides, being natural chemicals, can bolster plant defense mechanisms; they can trigger induced systemic resistance (ISR), promote necrotic zone formation at disease entry points, and demonstrate broad-spectrum antifungal and insect-repellent properties. Despite being distinct from antibiotics, the immune-enhancing properties of many oligosaccharides allow them to function as disease suppressants or potential biocontrol agents. Consequently, they can assist plants in recuperating from stress via immunostimulation or avert illnesses by deterrence. Quorum quenching, chitinase inhibition, stimulation of defense-related enzymes, potentiation of ISR and SAR, and regulation of cell wall-degrading enzymes, MAPK cascades, and fungal cell resistance by oligosaccharides are routes through which these compounds can mitigate stress. Specific oligosaccharides may function as biopesticides or fungicides to impede pathogen infection and proliferation.

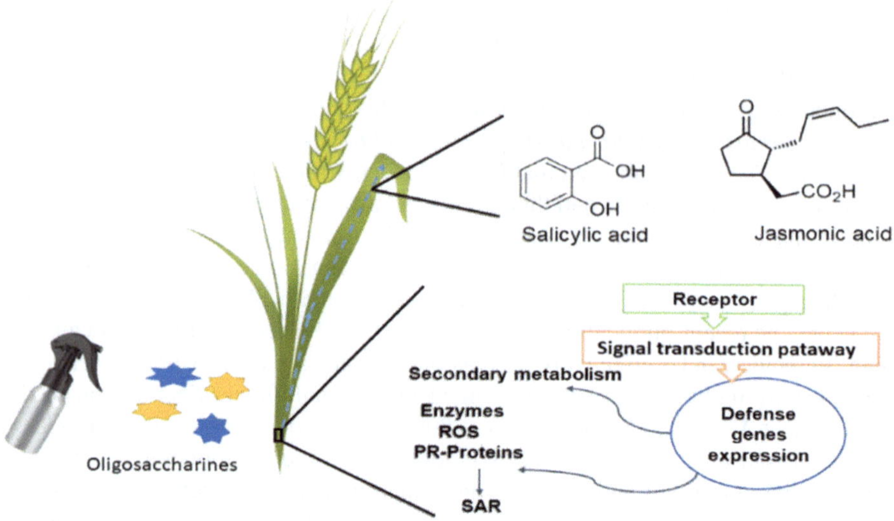

Fig. 6.2 Plant oligosaccharins are elicited by the buildup of ROS and hydroxyproline-rich glycoproteins, as well as by the stimulation of gene expression for the PR proteins PI-1, PR-1, acid PR-2, PAL, peroxidase, and proteins needed for SA production (Reprinted from Ochoa-Meza et al. 2021)

Furthermore, biostimulants composed of various oligosaccharides can enhance growth, yield, and quality. Production systems, especially organic systems that depend on natural and biological resources, are garnering significant interest from a growing number of stakeholders (Cai et al. 2023; Mukarram et al. 2023; Pring et al. 2023; Appu et al. 2021; He et al. 2021; Ochoa-Meza et al. 2021).

Numerous oligosaccharides, including chitosaccharins, chitosans, proteoglycan biocontrol agents, and laminarins, have been employed for disease management. Oligosaccharides can be categorized as immunostimulatory, deterrent, or immunostimulatory-deterrent biopesticides based on their methods of action. Oligosaccharides may serve as viable agents in several pest management tactics, including crop protection, due to their nontoxicity to animals and multiple mechanisms of action, positioning them as a potential alternative to synthetic pesticides (Fan et al. 2023; Liu et al. 2023; Elango et al. 2022). Ochoa-Meza et al. (2021) investigated the feasibility of using oligosaccharins generated from plants and fungi in wheat crops (Fig. 6.2)..

6.8.2 *Applications for Sustainable Agriculture*

Oligosaccharides, particularly glucooligosaccharides obtained from glucanases involved in plant defense mechanisms, present an opportunity to devise novel, eco-friendly strategies for pest and disease management, as well as for improving the nutritional quality of crops favored by humans. Their agricultural promise arises from their biological qualities as biostimulants and activators of plant immune systems, which enhance plants' adaptability to increased anthropogenic pressure. The study, while addressing the application potential of certain oligosaccharide families in agriculture, illustrates the significance of their use in contemporary agricultural methods to fulfill sustainable development objectives. The initiatives encompass the commercialization and advancement of plant-derived oligosaccharide products, with innovative ecological production methods. These products address the demand for items devoid of synthetic phytopharmaceuticals and fertilizers. They have a twofold purpose: firstly, as plant biostimulants and growth enhancers that maintain plant vitality, and secondly, as immune system enhancers that bolster resistance to abiotic and biotic stressors, including pests and diseases. The instances of their practical application in European and Chinese agriculture are particularly intriguing. Nonetheless, challenges exist in the commercial utilization of these compounds, including their brief duration and efficacy, alongside the prevalent use of traditional crop protection techniques. Consequently, further research is required to develop oligosaccharide-based products with an expanded range of efficacy and prolonged effectiveness, as well as to evaluate their recirculation within the ecological system to maximize their potential in the design of future agricultural systems. In the current trend toward sustainable production, this study is vital for developing an agro-ecological product portfolio and analyzing the economic and environmental impacts of incorporating oligosaccharides into agricultural practices (Du et al. 2022; Meena et al. 2022; Nowak et al. 2022).

6.9 Carbohydrate Interactions with Additional Defense Mechanisms

The interplay between carbohydrates and other defense mechanisms in plants is essential in regulating the immunological response to pathogens. Carbohydrates can engage with many defense pathways, including salicylic acid (SA), jasmonic acid (JA), and ethylene signaling pathways, to orchestrate a coordinated defense response (Fan et al. 2024).

6.9.1 Interactions Between Carbohydrate-Mediated and Alternative Defense Pathways

The interaction between carbohydrate-mediated and other defense pathways is essential for optimizing the plant's immunological response. The interplay between carbohydrate-mediated signaling and the salicylic acid pathway can trigger the activation of defense-related genes and the synthesis of pathogenesis-related proteins.

The interaction between carbohydrate-mediated signaling and the JA pathway can trigger defense responses against herbivores and necrotrophic diseases (Saleem et al. 2021).

6.9.2 Influence of Environmental Variables on Carbohydrate-Dependent Plant Defense

Environmental conditions, including temperature, light, and water availability, can profoundly influence carbohydrate-based plant defense mechanisms. Drought stress can modify the composition and structure of plant cell walls, rendering them more vulnerable to disease invasion. Certain environmental factors, such as elevated temperatures, can trigger carbohydrate-mediated defense mechanisms, resulting in increased resistance to infections (Ezquer et al. 2020).

6.9.3 Interaction Between Carbohydrates and Phytohormones

Phytohormones, including auxins, gibberellins, and cytokinins, are essential in regulating plant growth and development. Carbohydrates can engage with these phytohormones to regulate plant defense mechanisms. The interaction between carbohydrates and auxins can activate defense-related genes and stimulate the creation of defense chemicals (Khan et al. 2020).

6.9.4 Interactions Between Carbohydrates and Defense-Associated Proteins

Defense-associated proteins, including pathogenesis-related proteins and proteases, are essential for plant defense against pathogens. Carbohydrates can engage with these proteins to augment their functionality and stability. Certain carbohydrates can bind to defense-related proteins, resulting in their activation and the augmentation of their defensive capabilities. The connections between carbohydrates and other plant defense mechanisms are intricate and complicated. Understanding these connections is essential for devising innovative techniques for crop protection and enhancement. By clarifying the functions of carbohydrates in plant defense, researchers can formulate novel strategies to bolster plant immunity and advance sustainable agricultural methods (Molina et al. 2024; Naithani et al. 2021).

6.9.5 *The Influence of Environmental Variables on Carbohydrate-Dependent Plant Defense*

Environmental conditions, including temperature, light, and water availability, profoundly affect carbohydrate-based plant defense mechanisms. These variables can modify the content and structure of plant cell walls, rendering them more vulnerable to pathogen assaults. Conversely, some environmental factors can trigger carbohydrate-mediated defense mechanisms, resulting in increased resistance to infections (Novaković et al. 2018).

Temperature influences carbohydrate-based plant protection by modifying the expression of genes related to carbohydrate metabolism and defense mechanisms. For instance, elevated temperatures can stimulate the expression of defense-related genes and chemicals, whereas reduced temperatures can inhibit their expression (Devi et al. 2017).

Light is essential for carbohydrate-based plant defense. Variations in light intensity and quality can affect carbohydrate synthesis and the activation of defense mechanisms. For example, elevated light intensities can enhance the synthesis of defense-related chemicals, whereas diminished light intensities can inhibit their creation (Iqbal et al. 2021).

Water availability is a crucial environmental component influencing carbohydrate-based plant protection. Drought stress can modify the composition and architecture of plant cell walls, rendering them more vulnerable to pathogen assaults. Conversely, sufficient water availability can augment carbohydrate-mediated defense mechanisms, resulting in enhanced resistance to infections (Ganie and Ahammed 2021).

Comprehending the influence of environmental variables on carbohydrate-based plant protection is crucial for formulating successful crop protection and enhancement techniques. By clarifying the intricate relationships between environmental conditions and carbohydrate-mediated defense mechanisms, researchers can devise innovative strategies to bolster plant immunity and advance sustainable agricultural methods (Ganie and Ahammed 2021).

6.10 Utilizations in Agriculture and Biotechnology

The insights acquired from studying the function of carbohydrates in plant immunity can be utilized in multiple ways to augment crop protection and enhancement. Potential uses of carbohydrate-based techniques in agriculture and biotechnology encompass:

- ***Employment of carbohydrate-derived approaches for agricultural defense***

1. Development of carbohydrate-derived elicitors: Carbohydrate-derived elicitors can be employed to activate plant defense mechanisms, offering protec-

tion against pathogens and pests. These elicitors can be administered to crops via foliar spray or irrigation systems.

2. Engineering carbohydrate-mediated immunity in plants: Genetic engineering methods can be employed to incorporate genes that augment carbohydrate-mediated immunity in plants. This may result in the cultivation of crops with enhanced resilience to diseases and pests.
3. Carbohydrate-derived biopesticides: Carbohydrates may serve as biopesticides for the management of pests and diseases. Certain polysaccharides have been demonstrated to impede the proliferation of fungal infections (Fu et al. 2022; Guarnizo et al. 2020; Patel et al. 2020; Gowthami 2018).

- ***Biotechnological methods to augment carbohydrate-driven immunity in plants***

1. Gene editing tools, like as CRISPR/Cas9, can modify genes associated with carbohydrate-mediated immunity, facilitating the creation of crops with enhanced disease resistance.
2. RNA interference (RNAi) Technology: RNAi technology can effectively silence genes that inhibit carbohydrate-mediated immunity, hence improving disease resistance in crops (Meresa et al. 2024; Paul et al. 2021).

- ***Microbiome engineering***

This process entails the modification of the plant microbiome to augment carbohydrate-mediated immunity. This can be accomplished by employing advantageous microbes that activate plant defense mechanisms. In summary, the implementation of carbohydrate-centric approaches in agriculture and biotechnology could transform crop protection and enhancement. Researchers can utilize carbohydrates to create innovative methods for improving plant immunity and advancing sustainable agricultural techniques (Li et al. 2019).

6.11 Conclusion

In conclusion, carbohydrates are essential to the intricate framework of plant immunity and defense. Their functions in plant defense, especially through systems related to cell wall carbohydrates and carbohydrate signaling, have become increasingly apparent. Recent advancements emphasize the crucial interactions between carbohydrates and plant diseases, underscoring their importance in bolstering plant resilience. The ability to manipulate carbohydrates for enhanced plant defense signifies a promising area of research. As this evolving subject advances, future research should focus on enhancing our comprehension of the complex interactions between carbohydrates and plant defense mechanisms. This understanding is essential for formulating creative techniques that enhance plant health and resilience, therefore aiding global food security and ecological equilibrium.

References

Abouraïcha E, El Alaoui-Talibi Z, El Boutachfaiti R, Petit E, Courtois B, Courtois J, El Modafar C (2015) Induction of natural defense and protection against *Penicillium expansum* and Botrytis cinerea in apple fruit in response to bioelicitors isolated from green algae. Sci Hortic 181:121–128

Acosta-Jurado S, Fuentes-Romero F, Ruiz-Sainz JE, Janczarek M, Vinardell JM (2021) Rhizobial exopolysaccharides: genetic regulation of their synthesis and relevance in symbiosis with legumes. Inter J Mol Sci 22(12):6233

Adetunji CO, Kremer RJ, Makanjuola R, Sarin NB (2021) Application of molecular biotechnology to manage biotic stress affecting crop enhancement and sustainable agriculture. Adv Agron 168:39–81

Alghamdi AK, Parween S, Hirt H, Saad MM (2023) Genome sequence analysis of plant growth-promoting bacterium, *Isoptericola sp.* AK164 isolated from the rhizosphere of Avicennia marina growing at the Red Sea. Arch Microbiol 205:307

An QD, Zhang GL, Wu HT, Zhang ZC, Zheng GS, Luan L et al (2009) Alginate-deriving oligosaccharide production by alginase from newly isolated Flavobacterium sp. LXA and its potential application in protection against pathogens. J Appl Microbiol 106(1):161–170

Anzano A, Bonanomi G, Mazzoleni S, Lanzotti V (2022) Plant metabolomics in biotic and abiotic stress: a critical overview. Phytochem Rev 21(2):503–524

Appu M, Ramalingam P, Sathiyanarayanan A, Huang J (2021) An overview of plant defense-related enzymes responses to biotic stresses, vol 27. Plant Gene, p 100302

Aziz A, Heyraud A, Lambert B (2004) Oligogalacturonide signal transduction, induction of defense-related responses and protection of grapevine against Botrytis cinerea. Planta 218:767–774

Aziz A, Gauthier A, Bézier A, Poinssot B, Joubert JM, Pugin A et al (2007) Elicitor and resistance-inducing activities of β-1, 4 cellodextrins in grapevine, comparison with β-1, 3 glucans and α-1, 4 oligogalacturonides. J Exp Bot 58(6):1463–1472

Bacete L, Melida H, Miedes E, Molina A (2018) Plant cell wall-mediated immunity: cell wall changes trigger disease resistance responses. Plant J 93(4):614–636

Bach EE, Barros BC, Kimati H (2003) Induced resistance against Bipolaris bicolor, Bipolaris sorokiniana and Drechslera tritici-repentis in wheat leaves by xantham gum and heat-inactivated conidial suspension. J Phytopathol 151(7–8):411–418

Badawy ME, Rabea EI (2011) A biopolymer chitosan and its derivatives as promising antimicrobial agents against plant pathogens and their applications in crop protection. Int J Carbohydrate Chem 2011(1):460381

Badawy ME, Rabea EI, Taktak NE (2014) Antimicrobial and inhibitory enzyme activity of N-(benzyl) and quaternary N-(benzyl) chitosan derivatives on plant pathogens. Carbohydr Polym 111:670–682

Baek KR, Ramakrishnan SR, Kim SJ, Seo SO (2024) Yeast cell wall mannan structural features, biological activities, and production strategies. Heliyon 10:e27896

Barghahn S, Arnal G, Jain N, Petutschnig E, Brumer H, Lipka V (2021) Mixed linkage β-1, 3/1, 4-glucan oligosaccharides induce defense responses in Hordeum vulgare and Arabidopsis thaliana. Front Plant Sci 12:682439

Bauer S, Mekonnen DW, Hartmann M, Yildiz I, Janowski R, Lange B et al (2021) UGT76B1, a promiscuous hub of small molecule-based immune signaling, glucosylates N-hydroxypipecolic acid, and balances plant immunity. Plant Cell 33(3):714–734

BeMiller JN (2018) Carbohydrate chemistry for food scientists

Ben Salah I, Aghrouss S, Douira A, Aissam S, El Alaoui-Talibi Z, Filali-Maltouf A, El Modafar C (2018) Seaweed polysaccharides as bio-elicitors of natural defenses in olive trees against verticillium wilt of olive. J Plant Interact 13(1):248–255

Bhavani S, Singh RP, Hodson DP, Huerta-Espino J, Randhawa MS (2022) Wheat rusts: current status, prospects of genetic control and integrated approaches to enhance resistance durability. In: Wheat improvement: food security in a changing climate. Springer, Cham, pp 125–141

Bhupenchandra I, Chongtham SK, Devi AG, Dutta P, Sahoo MR, Mohanty S et al (2024) Unlocking the potential of arbuscular mycorrhizal fungi: exploring role in plant growth promotion, nutrient uptake mechanisms, biotic stress alleviation, and sustaining agricultural production systems. J Plant Growth Regul:1–39

Bigini V (2023) Engineering danger sensing and signaling in plant immunity: use of oligosaccharins to enhance durum wheat resistance to fusariosis

Blumstein M, Sala A, Weston DJ, Holbrook NM, Hopkins R (2022) Plant carbohydrate storage: intra- and inter-specific trade-offs reveal a major life history trait. New Phytol 235(6):2211–2222

Boller T, Felix G (2009) A renaissance of elicitors: perception of microbe-associated molecular patterns and danger signals by pattern-recognition receptors. Annu Rev Plant Biol 60:379–406

Bolouri-Moghaddam MR, Van Den Ende W (2012) Sugars and plant innate immunity. J Exp Bot 63:3989–3998

Breia R, Conde A, Badim H, Fortes AM, Gerós H, Granell A (2021) Plant SWEETs: from sugar transport to plant–pathogen interaction and more unexpected physiological roles. Plant Physiol 186(2):836–852

Brown J (2022) Applications of carbohydrates on plant protection and plant growth. Int J Chem Sci 17(1):164

Brutus A, Sicilia F, Macone A, Cervone F, De Lorenzo G (2010) A domain swap approach reveals a role of the plant wall-associated kinase 1 (WAK1) as a receptor of oligogalacturonides. Proc Natl Acad Sci USA 107(20):9452–9457

Burton RA, Fincher GB (2009) (1, 3; 1, 4)-β-d-glucans in cell walls of the Poaceae, lower plants, and fungi: a tale of two linkages. Mol Plant 2(5):873–882

Buzón-Durán L, Langa-Lomba N, González-García V, Casanova-Gascón J, Martín-Gil J, Pérez-Lebeña E, Martín-Ramos P (2021) On the applicability of chitosan oligomers-amino acid conjugate complexes as eco-friendly fungicides against grapevine trunk pathogens. Agronomy 11(2):324

Cabrera JC, Boland A, Messiaen J, Cambier P, Van Cutsem P (2008) Egg box conformation of oligogalacturonides: the time-dependent stabilization of the elicitor-active conformation increases its biological activity. Glycobiology 18(6):473–482

Cai M, Zhang S, Su L, Li B, Li F, Borrás-Hidalgo O et al (2023) Application of oligosaccharides to induce innate immunity in plants. In: BIO web of conferences, vol 72. EDP Sci, p 01004

Caillot S, Rat S, Tavernier ML, Michaud P, Kovensky J, Wadouachi A et al (2012) Native and sulfated oligoglucuronans as elicitors of defence-related responses inducing protection against Botrytis cinerea of Vitis vinifera. Carbohydr Polym 87(2):1728–1736

Campos EVR, de Oliveira JL, Fraceto LF, Singh B (2015) Polysaccharides as safer release systems for agrochemicals. Agron Sustain Dev 35:47–66

Carpita NC (2000) The cell wall. Biochemistry and molecular biology of plants

Castro OL, Bach EE (2004) Increased production of β-1,3 glucanase and proteins in Bipolaris sorokiniana pathosystem treated using commercial xanthan gum. Plant Physiol Biochem 42(2):165–169

Cataldo E, Fucile M (2022) And Mattii GB, vol 11. Biostimulants in viticulture, A sustainable approach against biotic and abiotic stresses. Plants

Chaliha C, Rugen MD, Field RA, Kalita E (2018) Glycans as modulators of plant defense against filamentous pathogens. Front Plant Sci 9:928

Chandía NP, Matsuhiro B, Mejías E, Moenne A (2004) Alginic acids in Lessonia vadosa: partial hydrolysis and elicitor properties of the polymannuronic acid fraction. J Appl Phycol 16(2):127–133

Chaube MA, Trattnig N, Lee DH, Belkhadir Y, Pfrengle F (2022) Synthesis of fungal cell wall oligosaccharides and their ability to trigger plant immune responses. Eur J Org Chem 2022(27):e202200313

Chen X, Zou K, Li X, Chen F, Cheng Y, Li S et al (2023) Transcriptomic analysis of the response of susceptible and resistant bitter melon (Momordica charantia L.) to powdery mildew infection revealing complex resistance via multiple signaling pathways. Int J Mol Sci 24(18):14262

Chiquito-Contreras CJ, Meza-Menchaca T, Guzmán-López O, Vásquez EC, Ricaño-Rodríguez J (2024) Molecular insights into plant–microbe interactions: a comprehensive review of key mechanisms. Frontiers Bioscience-Elite 16(1):9

Choi J et al (2014) Extracellular ATP is a central signaling molecule in plant stress responses. Curr Opin Plant Biol 20:82–87

Claverie J, Balacey S, Lemaître-Guillier C, Brulé D, Chiltz A, Granet L et al (2018) The cell wall-derived xyloglucan is a new DAMP triggering plant immunity in Vitis vinifera and Arabidopsis thaliana. Front Plant Sci 9:1725

Cord-Landwehr S, Melcher RL, Kolkenbrock S, Moerschbacher BM (2016) A chitin deacetylase from the endophytic fungus Pestalotiopsis sp. efficiently inactivates the elicitor activity of chitin oligomers in rice cells. Sci Rep 6(1):38018

Costa JS, de Mello T, Fagundes DP, Simões IM, de Araujo CP, Ferreira A et al (2024) Dynamics of biochemical and hormonal reserves degradation in immature seeds of Euterpe edulis Martius (Arecaceae) during in vitro germination. Plant Cell, Tissue Organ Culture (PCTOC) 156(1):10

Dal Bello GM, Mónaco CI, Simón MR (2002) Biological control of seedling blight of wheat caused by Fusarium graminearum with beneficial rhizosphere microorganisms. World J Microbiol Biotechnol 18:627–636

Dalio RJ, Paschoal D, Arena GD, Magalhães DM, Oliveira TS, Merfa MV et al (2021) Hypersensitive response: from NLR pathogen recognition to cell death response. Ann Appl Biol 178(2):268–280

Davidsson P, Broberg M, Kariola T, Sipari N, Pirhonen M, Palva ET (2017) Short oligogalacturonides induce pathogen resistance-associated gene expression in Arabidopsis thaliana. BMC Plant Biol 17:1–17

Davis R, Zivanovic S, Michael Davidson P, D'Souza DH (2015) Enteric viral surrogate reduction by chitosan. Food Environ Virol 7(4):359–365

De Borba MC, Velho AC, Maia-Grondard A, Baltenweck R, Magnin-Robert M, Randoux B et al (2021) The algal polysaccharide ulvan induces resistance in wheat against Zymoseptoria tritici without major alteration of leaf metabolome. Front Plant Sci 12:703712

Delplace F, Huard-Chauveau C, Berthomé R, Roby D (2022) Network organization of the plant immune system: from pathogen perception to robust defense induction. Plant J 109(2):447–470

Devi LE et al (2017) Adaptation Strategies and Defence Mechanisms of Plants During Environmental Stress. In: Ghorbanpour M, Varma A (eds) Medicinal Plants and Environmental Challenges. Springer, Cham https://doi.org/10.1007/978-3-319-68717-9_20

Dey P, Ramanujam R, Venkatesan G, Nagarathnam R (2019) Sodium alginate potentiates antioxidant defense and PR proteins against early blight disease caused by Alternaria solani in Solanum lycopersicum Linn. PLoS One 14(9):e0223216

Du J, Anderson CT, Xiao C (2022) Dynamics of pectic homogalacturonan in cellular morphogenesis and adhesion, wall integrity sensing and plant development, vol 8. Nature Plants, p 332

Duran-Flores D, Heil M (2018) Extracellular self-DNA as a damage-associated molecular pattern (DAMP) signal that triggers plant immunity. J Plant Physiol 214:153–157

Duret M, Wallner A, Buée M, Aziz A (2024) Rhizosphere microbiome assembly, drivers and functions in perennial ligneous plant health. Microbiol Res 287:127860

Egert A et al (2015) Carbohydrates in plant immunity-the sweet side of biotic interactions. Plant Physiol Biochem 82:42–50

Egert A et al (2019) The role of raffinose family oligosaccharides in plant stress responses: a review. Plan Theory 8(6):173

El Modafar C, Elgadda M, El Boutachfaiti R, Abouraicha E, Zehhar N, Petit E et al (2012) Induction of natural defence accompanied by salicylic acid-dependant systemic acquired resistance in tomato seedlings in response to bioelicitors isolated from green algae. Sci Hortic 138:55–63

El Omari B (2022) Accumulation versus storage of total non-structural carbohydrates in woody plants. Trees 36:869

Elango D, Rajendran K, Van der Laan L, Sebastiar S, Raigne J, Thaiparambil NA et al (2022) Raffinose family oligosaccharides: friend or foe for human and plant health? Front Plant Sci 13:829118

Ellouzi H, Ben Slimene Debez I, Amraoui S, Rabhi M, Hanana M, Alyami NM et al (2024) Effect of seed priming with auxin on ROS detoxification and carbohydrate metabolism and their relationship with germination and early seedling establishment in salt stressed maize. BMC Plant Biol 24(1):704

Escudero N, Lopez-Moya F, Ghahremani Z, Zavala-Gonzalez EA, AlagueroCordovilla A, Ros-Ibanez C, Lacasa A, Sorribas FJ, Lopez-Llorca LV (2017b) Chitosan increases tomato root colonization by pochonia chlamydosporia and their combination reduces root-knot nematode damage. Front Plant Sci 8

Escudero V, Jordá L, Sopeña-Torres S, Melida H, Miedes E, Muñoz-Barrios A et al (2017a) Alteration of cell wall xylan acetylation triggers defense responses that counterbalance the immune deficiencies of plants impaired in the β-subunit of the heterotrimeric G-protein. Plant J 92(3):386–399

Ezquer I, Salameh I, Colombo L, Kalaitzis P (2020) Plant cell walls tackling climate change: insights into plant cell wall remodeling, its regulation, and biotechnological strategies to improve crop adaptations. Plants. https://air.unimi.it/handle/2434/711540

Fan S, Zhao B, Dhondt-Cordelier S, Vaillant-Gaveau N (2024) Plant-growth-promoting Rhizobacteria modulate carbohydrate metabolism in connection with host plant defense mechanism. Int J Mol Sci 25(3):1465

Fan Z, Wang L, Qin Y, Li P (2023) Activity of chitin/chitosan/chitosan oligosaccharide against plant pathogenic nematodes and potential modes of application in agriculture: a review, vol 306. Carbohydrate Polymers, p 120592

Fang X, Lian H, Bi S, Liu S, Yuan X, Liao C (2022) Roles of pattern recognition receptors in response to fungal keratitis. Life Sci 24:320522

Feliziani E, Landi L, Romanazzi G (2015) Preharvest treatments with chitosan and other alternatives to conventional fungicides to control postharvest decay of strawberry. Carbohydr Polym 132:111–117

Ferrari S et al (2019) Oligogalacturonides in plant immunity. J Exp Bot 70(5):1293–1301

Fetsiukh A, Conrad J, Bergquist J, Timmusk S (2021) Silica particles trigger the exopolysaccharide production of harsh environment isolates of growth-promoting rhizobacteria and increase their ability to enhance wheat biomass in drought-stressed soils. Int J Mol Sci 22(12):6201

Franková L, Fry SC (2021) Hemicellulose-remodelling transglycanase activities from charophytes: towards the evolution of the land-plant cell wall. Plant J 108:7

Fu R, Meng D, Song B, Wang H, Zhang J, Li J (2022) The carbohydrate elicitor Riclinoctaose facilitates defense and growth of potato roots by inducing changes in transcriptional and metabolic profiles. Plant Sci 322:111349

Fuertes-Rabanal M, Largo-Gosens A, Fischer A, Munzert KS, Carrasco-López C, Sánchez-Vallet A et al (2024) Linear β-1, 2-glucans trigger immune hallmarks and enhance disease resistance in plants. J Exp Bot 75(22):7337–7350

Gai Y, Wang H (2024) Plant disease: a growing threat to global food security. Agronomy 14

Galindo-Trigo S, Khandare V, Roosjen M, Adams J, Wangler A-M, Bayer M, Borst JW, Smakowska-Luzan E, and Butenko MA (2024) A multifaceted kinase axis regulates plant organ abscission through conserved signaling mechanisms. Curr Biol 34:3020–3030. https://doi.org/10.1016/j.cub.2024.05.057

Gallusci P, Agius DR, Moschou PN, Dobránszki J, Kaiserli E, Martinelli F (2023) Deep inside the epigenetic memories of stressed plants. Trends Plant Sci 28(2):142–153

Galus S, Kadzińska J (2015) Food applications of emulsion-based edible films and coatings. Trends Food Sci Technol 45(2):273–283

Ganie SA, Ahammed GJ (2021) Dynamics of cell wall structure and related genomic resources for drought tolerance in rice. Plant Cell Rep 40:437–459

Gao H, Ma K, Ji G, Pan L, Wang Z, Cui M, Zhou Q (2022) Protein glycosylation changes during systemic acquired resistance in Arabidopsis thaliana. Int J Biol Macromol 212:381–392

Gao H, Zhou Q, Yang L, Zhang K, Ma Y, Xu ZQ (2020) Metabolomics analysis identifies metabolites associated with systemic acquired resistance in *Arabidopsis*. PeerJ 8:e10047

Garon D (2014) Biodiversité et évolution du monde végétal. EDP Sciences

Ghassemi N, Poulhazan A, Deligey F, Mentink-Vigier F, Marcotte I, Wang T (2021) Solid-state NMR investigations of extracellular matrixes and cell walls of algae, bacteria, fungi, and plants. Chem Rev 122(10):10036–10086

Giovannoni M, Scortica A, Scafati V, Piccirilli E, Sorio D, Benedetti M, Mattei B (2025) The reducing end of cell wall oligosaccharides is critical for DAMP activity in Arabidopsis thaliana and can be exploited by oligosaccharide oxidases in the reduction of oxidized phenolics. Plant Physiol Biochem 109466:109466

Gómez-Gómez L, Boller T (2002) Flagellin perception: a paradigm for innate immunity. Trends Plant Sci 7(6):251–256

González-Saucedo A, Barrera-Necha LL, Ventura-Aguilar RI, Correa-Pacheco ZN, Bautista-Baños S, Hernández-López M (2019) Extension of the postharvest quality of bell pepper by applying nanostructured coatings of chitosan with Byrsonima crassifolia extract (L.) Kunth. Postharv Biol & Technol 149:74–82. https://doi.org/10.1016/j.postharvbio.2018.11.019

Gowthami L (2018) Role of elicitors in plant defense mechanism. J Pharmacognosy Phytochem 7(6):2806–2812

Guarnizo N, Oliveros D, Murillo-Arango W, Bermúdez-Cardona MB (2020) Oligosaccharides: defense inducers, their recognition in plants, commercial uses and perspectives. Molecules 25(24):5972

Guo F et al (2020) Oligogalacturonides activate plant immune responses and induce resistance against pathogens in Arabidopsis. Plant Physiol 182(2):1529–1545

Gust AA et al (2007) Bacteria-derived peptidoglycans constitute pathogen-associated molecular patterns triggering innate immunity in Arabidopsis. J Biol Chem 282(44):32338–32348

Ha CM, Rao X, Saxena G, Dixon RA (2021) Growth–defense trade-offs and yield loss in plants with engineered cell walls. New Phytol 231:60

Hadwiger LA (2013) Multiple effects of chitosan on plant systems: solid science or hype. Plant Sci 208:42–49

Hasni I et al (2021) Mannooligosaccharides activate defense responses and enhance resistance against Fusarium in wheat. Front Plant Sci 12:673611

Hassan SU, Donia A, Sial U, Zhang X, Bokhari H (2020) Glycoprotein-and lectin-based approaches for detection of pathogens. Pathogens 9(9):694

He J, Han W, Wang J, Qian Y, Saito M, Bai W et al (2021) Functions of oligosaccharides in improving tomato seeding growth and chilling resistance. J Plant Growth Regul:1–11

He J, Li R, Sun X, Wang W, Hu J, Xie H, Yin H (2018) Effects of calcium alginate submicroparticles on seed germination and seedling growth of wheat (Triticum aestivum L.). Polymers 10(10):1154

Heinze T, El Seoud OA, Koschella A (2018) Cellulose derivatives: synthesis, structure, and properties. Springer, Cham, pp 1–531

Herbers K, Meuwly P, Metraux JP, Sonnewald U (1996) Salicylic acid-independent induction of pathogenesis-related protein transcripts by sugars is dependent on leaf developmental stage. FEBS Lett 397:239–244

Hien NQ, Nagasawa N, Tham LX, Yoshii F, Dang VH, Mitomo H et al (2000) Growth-promotion of plants with depolymerized alginates by irradiation. Radiat Phys Chem 59(1):97–101

Hiltbrunner E, Arnaiz J, Körner C (2021) Biomass allocation and seasonal non-structural carbohydrate dynamics do not explain the success of tall forbs in short alpine grassland. Oecologia 197:1063

Hornung V, Gaidt MM (2024) Friendly fire: recognition of self by the innate immune system. Curr Opin Immunol 90:102457
Hou S, Liu Z, Shen H, Wu D (2019) Damage-associated molecular pattern-triggered immunity in plants. Frontiers. Plant Sci 10
Howlader P, Bose SK, Zhang C, Jia X, Wang W, Yin H (2020) Pectin oligosaccharide (POS) induce resistance in Arabidopsis thaliana through salicylic acid-mediated signaling pathway against *pseudomonas syringae* pv. Tomato DC3000. Physiol Mol Plant Pathol 110:101483
Hu X, Jiang X, Hwang H, Liu S, Guan H (2004) Promotive effects of alginate-derived oligosaccharide on maize seed germination. J Appl Phycol 16:73–76
Iqbal A, Qiang D, Xiangru W, Huiping G, Hengheng Z, Xiling Z, Meizhen S (2023) Phosphorus and carbohydrate metabolism contributes to low phosphorus tolerance in cotton. BMC Plant Biol 23(1):97
Iqbal Z, Iqbal MS, Hashem A, Abd-Allah EF, Ansari MI (2021) Plant defense responses to biotic stress and its interplay with fluctuating dark/light conditions. Front Plant Sci 12:631810
Iwasaki KI, Matsubara Y (2000) Purification of alginate oligosaccharides with root growth-promoting activity toward lettuce. Biosci Biotechnol Biochem 64(5):1067–1070
Jacob P, Hige J, Dangl JL (2023) Is localized acquired resistance the mechanism for effector-triggered disease resistance in plants? Nature Plants 9(8):1184–1190
Jarecki W (2021) Soybean response to seed coating with chitosan+ alginate/PEG and/or inoculation. Agronomy 11(9):1737
Jarecki W, and Wietecha J (2021) Effect of seed coating on the yield of soybean Glycine max (L.) Merr
Jaulneau V, Lafitte C, Jacquet C, Fournier S, Salamagne S, Briand X et al (2010) Ulvan, a sulfated polysaccharide from green algae, activates plant immunity through the jasmonic acid signaling pathway. Biomed Res Int 2010(1):525291
Jeandet P, Formela-Luboińska M, Labudda M, Morkunas I (2022) The role of sugars in plant responses to stress and their regulatory function during development. Int J Mol Sci 23(9):5161
Ji HM, Mao HY, Li SJ, Feng T, Zhang ZY, Cheng L et al (2021) Fol-milR1, a pathogenicity factor of Fusarium oxysporum, confers tomato wilt disease resistance by impairing host immune responses. New Phytol 232(2):705–718
Jia Y, Tang J, Dang Y, Yang Y, He S, Jiang Y et al (2024) Lansiumamide B induced the degradation of polysaccharides and oligosaccharides showing antifungal activity against Rhizoctonia solani. Ind Crop Prod 214:118529
Jiang T, Feng L, Wang Y (2013) Effect of alginate/nano-ag coating on microbial and physicochemical characteristics of shiitake mushroom (Lentinus edodes) during cold storage. Food Chem 141(2):954–960
Kaida R, Sugawara S, Negoro K, Maki H, Hayashi T, Kaneko TS (2010) Acceleration of cell growth by xyloglucan oligosaccharides in suspension-cultured tobacco cells. Mol Plant 3(3):549–554
Kanwar P, Jha G (2019) Alterations in plant sugar metabolism: signatory of pathogen attack. Planta 249:305–318
Kashyap B, Kumar R (2021) Sensing methodologies in agriculture for monitoring biotic stress in plants due to pathogens and pests. Inventions
Katiyar A et al (2019) Chito-oligosaccharides as elicitors of plant defense mechanisms: a review. Plant Signaling and Behavior 14(7):e1615818
Kaur S, Samota MK, Choudhary M, Choudhary M, Pandey AK, Sharma A, Thakur J (2022) How do plants defend themselves against pathogens-biochemical mechanisms and genetic interventions. Physiol Mol Biol Plants 28(2):485–504
Kauss H, Jeblick W, Domard A (1989) The degrees of polymerization and N-acetylation of chitosan determine its ability to elicit callose formation in suspension cells and protoplasts of Catharanthus roseus. Planta 178:385–392
Khan N, Bano A, Ali S, Babar MA (2020) Crosstalk amongst phytohormones from planta and PGPR under biotic and abiotic stresses. Plant Growth Regul 90:189–203

Kharchoufi S, Parafati L, Licciardello F, Muratore G, Hamdi M, Cirvilleri G, Restuccia C (2018) Edible coatings incorporating pomegranate peel extract and biocontrol yeast to reduce Penicillium digitatum postharvest decay of oranges. Food Microbiol 74:107–112

Khokhani D, Carrera Carriel C, Vayla S, Irving TB, Stonoha-Arther C, Keller NP, Ané JM (2021) Deciphering the chitin code in plant symbiosis, defense, and microbial networks. Ann Rev Microbiol 75(1):583–607

Kim MS, Cho SM, Kang EY, Im YJ, Hwangbo H, Kim YC et al (2008) Galactinol is a signaling component of the induced systemic resistance caused by pseudomonas chlororaphis O6 root colonization. Mol Plant-Microbe Interact 21:1643–1653

Klarzynski O, Plesse B, Joubert JM, Yvin JC, Kopp M, Kloareg B, Fritig B (2000) Linear β-1, 3 glucans are elicitors of defense responses in tobacco. Plant Physiol 124(3):1027–1038

Kohorn BD (2016) Cell wall-associated kinases and pectin perception. J Exp Bot 67(2):489–494

Kohorn BD, Johansen S, Shishido A, Todorova T, Martinez R, Defeo E, Obregon P (2009) Pectin activation of MAP kinase and gene expression is WAK2 dependent. Plant J 60(6):974–982

Kongala SI, Kondreddy A (2023) A review on plant and pathogen derived carbohydrates, oligosaccharides and their role in plant's immunity. Carbohydrate Polymer Technol Appl 6:100330

Korbecka-Glinka G, Piekarska K, Wiśniewska-Wrona M (2022) The use of carbohydrate biopolymers in plant protection against pathogenic fungi. Polymers 14(14):2854

Kozieł E, Otulak-Kozieł K, Bujarski JJ (2021) Plant cell wall as a key player during resistant and susceptible plant-virus interactions. Frontiers. in Microbiology 12

Kulikov SN, Chirkov SN, Il'ina AV, Lopatin SA, Varlamov VP (2006) Effect of the molecular weight of chitosan on its antiviral activity in plants. Appl Biochem Microbiol 42(2):200–203

Kumar R, Kumar N, Rajput VD, Mandzhieva S, Minkina T, Saharan BS et al (2022) Advances in biopolymeric nanopesticides: a new eco-friendly/eco-protective perspective in precision agriculture. Nano 12(22):3964

Lahaye M, Robic A (2007) Structure and functional properties of ulvan, a polysaccharide from green seaweeds. Biomacromolecules 8(6):1765–1774

Landi L, Foglia R, Murolo S, and Romanazzi G (2021) The mycorrizal status in vineyards affected by esca. Journal of Fungi,

Laporte D, Vera J, Chandía NP, Zúniga EA, Matsuhiro B, Moenne A (2007) Structurally unrelated algal oligosaccharides differentially stimulate growth and defense against tobacco mosaic virus in tobacco plants. J Appl Phycol 19(1):79–88

Larskaya IA, Gorshkova TA (2015) Plant oligosaccharides—outsiders among elicitors? Biochem Mosc 80:881–900

Le Mire G, Siah A, Marolleau B, Gaucher M, Maumené C, Brostaux Y et al (2019) Evaluation of λ-carrageenan, CpG-ODN, glycine betaine, *Spirulina platensis*, and ergosterol as elicitors for control of *Zymoseptoria tritici* in wheat. Phytopathology 109(3):409–417

Li B, Wang X, Chen R, Huangfu W, Xie G (2008) Antibacterial activity of chitosan solution against Xanthomonas pathogenic bacteria isolated from Euphorbia pulcherrima. Carbohydr Polym 72(2):287–292

Li D, Wu M (2021) Pattern recognition receptors in health and diseases. Signal transduction and targeted therapy

Li J, Wang X, Lin X, Yan G, Liu L, Zheng H et al (2018) Alginate-derived oligosaccharides promote water stress tolerance in cucumber (Cucumis sativus L.). Plant Physiol Biochem 130:80–88

Li L, Shi X, Qi S, Zhang X, Fung HY, Li Q, Han Q (2024) Strategies, techniques and applications for food authentication based on carbohydrates: a review, vol 344. Carbohydrate Polymers, p 122533

Li R, Liu C, Zhao R, Wang L, Chen L, Yu W et al (2019) CRISPR/Cas9-mediated SlNPR1 mutagenesis reduces tomato plant drought tolerance. BMC Plant Biol 19:1–13

Lin B, Qing X, Liao J, Zhuo K (2020) Role of protein glycosylation in host-pathogen interaction. Cells 9(4):1022

Liu H, Zhang YH, Yin H, Wang WX, Zhao XM, Du YG (2013) Alginate oligosaccharides enhanced Triticum aestivum L. tolerance to drought stress. Plant Physiol Biochem 62:33–40

Liu J, Han L, Huai B, Zheng P, Chang Q, Guan T et al (2015) Down-regulation of a wheat alkaline/neutral invertase correlates with reduced host susceptibility to wheat stripe rust caused by Puccinia striiformis. J Exp Bot 66(22):7325–7338

Liu M, Zhang F, Liu B, Meng X (2023) Mechanisms and technology of marine oligosaccharides to control postharvest disease of fruits. Food Chem 404:134664

Liu Q, Liu P, Xu Y, Wang B, Liu P, Hao J, Liu X (2021) Encapsulation of fluazinam to extend efficacy duration in controlling Botrytis cinerea on cucumber. Pest Manag Sci 77(6):2836–2842

Liu X et al (2022) Oligo-β-glucans activate plant immunity against *Botrytis cinerea* in strawberries. Mol Plant 15(1):119–134

Locci F, Benedetti M, Pontiggia D, Citterico M, Caprari C, Mattei B et al (2019) An Arabidopsis berberine bridge enzyme-like protein specifically oxidizes cellulose oligomers and plays a role in immunity. Plant J 98(3):540–554

López-Gómez B et al (2021) The complexity of lipopolysaccharide recognition in plants. Front Plant Sci 11:1279

Lorrai R, Ferrari S (2021) Host cell wall damage during pathogen infection: mechanisms of perception and role in plant-pathogen interactions, vol 10. Plants

Lu Y, Tsuda K (2021) Intimate association of PRR-and NLR-mediated signaling in plant immunity. Mol Plant-Microbe Interact 34:3

Lubbe FC, Klimeš A, Doležal J, Jandová V, Mudrák O, Janeček Š et al (2021) Carbohydrate storage in herbs: the forgotten functional dimension of the plant economic spectrum. Ann Bot 127(6):813–825

Ma H, Zhao Y, Lu Z, Xing R, Yao X, Jin Z et al (2020) Citral-loaded chitosan/carboxymethyl cellulose copolymer hydrogel microspheres with improved antimicrobial effects for plant protection. Int J Biol Macromol 164:986–993

Ma Y, Zhang L, Ma X, Bai K, Tian Z, Wang Z, et al. (2024) Saccharide mapping as an extraordinary method on characterization and identification of plant and fungi polysaccharides: a review. Int J Biol Macromol 133350

Ma Z, Garrido-Maestu A, Jeong KC (2017) Application, mode of action, and in vivo activity of chitosan and its micro-and nanoparticles as antimicrobial agents: a review. Carbohydr Polym 176:257–265

Madanipour S, Alimohammadi M, Rezaie S, Nabizadeh R, Jahed Khaniki G, Hadi M et al (2019) Influence of postharvest application of chitosan combined with ethanolic extract of liquorice on shelflife of apple fruit. J Environ Health Sci Eng 17:331–336

Malinowski R, Singh D, Kasprzewska A, Blicharz S, Basińska-Barczak A (2024) Vascular tissue–boon or bane? How pathogens usurp long-distance transport in plants and the defence mechanisms deployed to counteract them. New Phytol 243(6):2075–2092

Malook SU, Maqbool S, Hafeez M, Karunarathna SC, Suwannarach N (2022) Molecular and biochemical mechanisms of elicitors in pest resistance. Life 12(6):844

Mansilla AY, Albertengo L, Rodríguez MS, Debbaudt A, Zúñiga A, Casalongué CA (2013) Evidence on antimicrobial properties and mode of action of a chitosan obtained from crustacean exoskeletons on pseudomonas syringae pv. Tomato DC3000. Appl Microbiol Biotechnol 97:6957–6966

Meena M, Yadav G, Sonigra P, Nagda A, Mehta T, Swapnil P, Marwal A (2022) Role of elicitors to initiate the induction of systemic resistance in plants to biotic stress. Plant Stress 5:100103

Mélida H, Bacete L, Ruprecht C, Rebaque D, del Hierro I, López G, Brunner F, Pfrengle F and Molina A (2020) Arabinoxylan-Oligosaccharides Act as Damage Associated Molecular Patterns in Plants Regulating Disease Resistance. Front Plant Sci 11:1210. https://doi.org/10.3389/fpls.2020.01210

Melida H, Sopeña-Torres S, Bacete L, Garrido-Arandia M, Jordá L, Lopez G et al (2018) Non-branched β-1, 3-glucan oligosaccharides trigger immune responses in Arabidopsis. Plant J 93(1):34–49

Meng X, Yang L, Kennedy JF, Tian S (2010) Effects of chitosan and oligochitosan on growth of two fungal pathogens and physiological properties in pear fruit. Carbohydr Polym 81(1):70–75

Meresa BK, Aymut KM, Weldemichael MY, Geberemedhin KH, Kassegn HH, Geberemikael BA, Egigu EM (2024) Carbohydrate Elicitor-induced plant immunity: advances and prospects. Heliyon

Mijailovic N, Richet N, Villaume S, Nesler A, Perazzolli M, Aït Barka E, Aziz A (2022) D-tagatose-based product triggers sweet immunity and resistance of grapevine to downy mildew, but not to gray mold disease. Plan Theory 11(3):296

Mishra S, Roychowdhury R, Ray S, Hada A, Kumar A, Sarker U, et al (2024) Salicylic acid (SA)-mediated plant immunity against biotic stresses: an insight on molecular components and signaling mechanism. Plant Stress, 100427

Mittler R, Zandalinas SI, Fichman Y, Van Breusegem F (2022) Reactive oxygen species signalling in plant stress responses. Nat Rev Mol Cell Biol 23(10):663–679

Moenne A, González A (2021) Chitosan-, alginate-carrageenan-derived oligosaccharides stimulate defense against biotic and abiotic stresses, and growth in plants: a historical perspective, vol 503. Carbohydrate Res, p 108298

Mohammadi M, Zamani A, Karimi K (2012) Determination of glucosamine in fungal cell walls by high-performance liquid chromatography (HPLC). J Agric Food Chem 60(42):10511–10515

Molina A, Sánchez-Vallet A, Jordá L, Carrasco-López C, Rodríguez-Herva JJ, López-Solanilla E (2024) Plant cell walls: source of carbohydrate-based signals in plant-pathogen interactions. Curr Opin Plant Biol 82:102630

Mondéjar-López M, García-Simarro MP, Navarro-Simarro P, Gómez-Gómez L, Ahrazem O, and Niza E (2024) A review on the encapsulation of "eco-friendly" compounds in natural polymer-based nanoparticles as next generation nano-agrochemicals for sustainable agriculture and crop management. Int J Biol Macromol 136030

Morán D R (2021) Novel cell wall-derived oligosaccharides trigger immune responses and disease resistance in plants

Morkunas I, Marczak L, Stachowiak J, Stobiecki M (2005) Sucrose-induced lupine defense against Fusarium oxysporum. Sucrose-stimulated accumulation of isoflavonoids as a defense response of lupine to Fusarium oxysporum. Plant Physiol Biochem 43:363–373

Mousavi EA, Nasibi F, Manouchehri Kalantari K, Oloumi H (2017) Stimulation effect of carrageenan on enzymatic defense system of sweet basil against Cuscuta campestris infection. J Plant Interact 12(1):286–294

Mu R, Dong L, Wang C (2023) Carbohydrates as putative pattern recognition receptor agonists in vaccine development. Trends Immunol 44:845

Mukarram M, Ali J, Dadkhah-Aghdash H, Kurjak D, Kačík F, Ďurkovič J (2023) Chitosan-induced biotic stress tolerance and crosstalk with phytohormones, antioxidants, and other signalling molecules. Front Plant Sci 14:1217822

Mwaheb MAMA, Hussain M, Tian J, Zhang X, Hamid MI, El-Kassim NA, Hassan GM, Xiang M, Liu X (2017) Synergetic suppression of soybean cyst nematodes by chitosan and Hirsutella minnesotensis via the assembly of the soybean rhizosphere microbial communities. Biol Control 115:85–94

Nabi Z, Manzoor S, Nabi SU, Wani TA, Gulzar H, Farooq M et al (2024) Pattern-triggered immunity and effector-triggered immunity: crosstalk and cooperation of PRR and NLR-mediated plant defense pathways during host–pathogen interactions. Physiol Mol Biol Plants 30:1–18

Naithani S, Komath SS, Nonomura A, Govindjee G (2021) Plant lectins and their many roles: carbohydrate-binding and beyond. J Plant Physiol 266:153531

Newman MA, Sundelin T, Nielsen JT, Erbs G (2013) MAMP (microbe-associated molecular pattern) triggered immunity in plants. Front Plant Sci 4:139

Ngaki MN, Srivastava SK, Feifei W, Bhattacharyya MK (2023) A soybean pattern recognition receptor conferring broad-spectrum pathogen and pest resistance regulates expression of several NLR receptor proteins. bioRxiv

Nothnagel EA, McNeil M, Albersheim P, Dell A (1983) Host-pathogen interactions: XXII. A galacturonic acid oligosaccharide from plant cell walls elicits phytoalexins. Plant Physiol 71(4):916–926

Novaković L, Guo T, Bacic A, Sampathkumar A, Johnson KL (2018) Hitting the wall-sensing and signaling pathways involved in plant cell wall remodeling in response to abiotic stress. Plan Theory 7(4):89
Nowak A, Tyśkiewicz R, Wiater A, Jaroszuk-Ściseł J (2022) (1→3)-α-D-glucooligosaccharides As elicitors influencing the activity of plant resistance pathways in wheat tissues. Agronomy
Ochoa-Meza LC, Quintana-Obregón EA, Vargas-Arispuro I, Falcón-Rodríguez AB, Aispuro-Hernández E, Virgen-Ortiz JJ, Martínez-Téllez MÁ (2021) Oligosaccharins as elicitors of defense responses in wheat. Polymers 13(18):3105
Ou L, Zhang Q, Ji D, Li Y, Zhou X, Jin L (2022) Physiological, transcriptomic investigation on the tea plant growth and yield motivation by chitosan oligosaccharides. Horticulturae 8
Páez-Watson T, Álvarez-Suárez JM, Rivas-Romero F, Estrada L, López D, Pérez Pelea L et al (2020) Increased salinity stress tolerance of Nicotiana tabacum L. in vitro plants with the addition of xyloglucan oligosaccharides to the culture medium. In Vitro Cellular Developmental Biology-Plant 56:325–334
Park YB, Cosgrove DJ (2015) Xyloglucan and its interactions with other components of the growing cell wall. Plant Cell Physiol 56(2):180–194
Patel ZM, Mahapatra R, Jampala SSM (2020) Role of fungal elicitors in plant defense mechanism. In: Molecular aspects of plant beneficial microbes in agriculture. Academic Press, pp 143–158
Paul NC, Park SW, Liu H, Choi S, Ma J, MacCready JS et al (2021) Plant and fungal genome editing to enhance plant disease resistance using the CRISPR/Cas9 system. Front Plant Sci 12:700925
Paulert R, Brunel F, Melcher RL, Cord-Landwehr S, Niehues A, Mormann M, Moerschbacher BM (2021) The non-sulfated ulvanobiuronic acid of ulvans is the smallest active unit able to induce an oxidative burst in dicot cells. Carbohydr Polym 270:118338
Perumal PK, Dong CD, Chauhan AS, Anisha GS, Kadri MS, Chen CW et al (2023) Advances in oligosaccharides production from algal sources and potential applications. Biotechnol Adv 67:108195
Pettongkhao S, Bilanglod A, Khompatara K, Churngchow N (2019) Sulphated polysaccharide from Acanthophora spicifera induced Hevea brasiliensis defense responses against Phytophthora palmivora infection. Plan Theory 8(3):73
Pring S, Kato H, Imano S, Camagna M, Tanaka A, Kimoto H et al (2023) Induction of plant disease resistance by mixed oligosaccharide elicitors prepared from plant cell wall and crustacean shells. Physiol Plant 175(5):e14052
Qi P, Qi B, Gu C, Huo S, Dang X, Liu Y, Zhao B (2024) Construction of an immune-related prognostic model and potential drugs screening for esophageal cancer based on bioinformatics analyses and network pharmacology. Immunity Inflammation Disease 12(5):e1266
Reglinski T, Elmer PAG, Taylor JT, Wood PN, Hoyte SM (2010) Inhibition of Botrytis cinerea growth and suppression of botrytis bunch rot in grapes using chitosan. Plant Pathol 59(5):882–890
Reignault PH, Cogan A, Muchembled J, Lounes-Hadj Sahraoui A, Durand R, Sancholle M (2001) Trehalose induces resistance to powdery mildew in wheat. New Phytol 149:519–529
Ride JP, Barber MS (1990) Purification and characterization of multiple forms of endochitinase from wheat leaves. Plant Sci 71(2):185–197
Ridley BL, O'Neill MA, Mohnen D (2001) Pectins: structure, biosynthesis, and oligogalacturonide-related signaling. Phytochemistry 57(6):929–967
Riseh RS, Vazvani MG, Taheri A, Kennedy JF (2024) Pectin-associated immune responses in plant-microbe interactions: a review. Int J Biol Macromol 132790
Rolland F, Baena-Gonzalez E, Sheen J (2006) Sugar sensing and signaling in plants: conserved and novel mechanisms. Annu Rev Plant Biol 57:675–709
Rumpret M, von Richthofen HJ, Peperzak V, Meyaard L (2021) Inhibitory pattern recognition receptors. J Exp Med 219(1):e20211463

Saberi-Rise R, Moradi-Pour M (2020) The effect of Bacillus subtilis Vru1 encapsulated in alginate–bentonite coating enriched with titanium nanoparticles against Rhizoctonia solani on bean. Int J Biol Macromol 152:1089–1097

Saleem M, Fariduddin Q, Castroverde CDM (2021) Salicylic acid: a key regulator of redox signalling and plant immunity. Plant Physiol Biochem 168:381–397

Sangha JS, Kandasamy S, Khan W, Bahia NS, Singh RP, Critchley AT, Prithiviraj B (2015) Λ-Carrageenan suppresses tomato chlorotic dwarf viroid (TCDVd) replication and symptom expression in tomatoes. Mar Drugs 13(5):2875–2889

Sangha JS, Khan W, Ji X, Zhang J, Mills AA, Critchley AT, Prithiviraj B (2011) Carrageenans, sulphated polysaccharides of red seaweeds, differentially affect Arabidopsis thaliana resistance to Trichoplusia ni (cabbage Looper). PLoS One 6(10):e26834

Sangha JS, Ravichandran S, Prithiviraj K, Critchley AT, Prithiviraj B (2010) Sulfated macroalgal polysaccharides λ-carrageenan and ι-carrageenan differentially alter Arabidopsis thaliana resistance to Sclerotinia sclerotiorum. Physiol Mol Plant Pathol 75(1–2):38–45

Selim S, Sanssené J, Rossard S, Courtois J (2017) Systemic induction of the defensin and phytoalexin pisatin pathways in pea (Pisum sativum) against Aphanomyces euteiches by acetylated and nonacetylated oligogalacturonides. Molecules 22(6):1017

Serrano M et al (2022) Cutin monomers: a new player in plant immunity. Plant Cell 34(3):955–964

Sey E, Willment JA, Brown GD (2024) Mammalian pattern recognition receptors (PRRs) involved in recognition of fungi. Human and Animal Relationships

Shah S, Li X, Jiang Z, Fahad S, Hassan S (2022) Exploration of the phytohormone regulation of energy storage compound accumulation in microalgae. Food Energy Sec 11(4):e418

Shen Y, Yang H (2017) Effect of preharvest chitosan-g-salicylic acid treatment on postharvest table grape quality, shelf life, and resistance to Botrytis cinerea-induced spoilage. Sci Hortic 224:367–373

Shinya T, Ménard R, Kozone I, Matsuoka H, Shibuya N, Kauffmann S et al (2006) Novel β-1, 3-, 1, 6-oligoglucan elicitor from Alternaria alternata 102 for defense responses in tobacco. FEBS J 273(11):2421–2431

Singh A, Lim GH, Kachroo P (2017) Transport of chemical signals in systemic acquired resistance. J Integr Plant Biol 59(5):336–344

Song Q, Li X, Zhao Z, Li D, Li Y, Hou N (2023) Effects of PAHs and ecosystem responses of remediation strategies under cold stress: from PAHs degradation to ecological restoration regulated by signal molecular. Chem Eng J 475:146042

Sousa FC et al (2020) Cellobiose induces defense responses in lettuce plants against bacterial pathogens. Physiol Mol Plant Pathol 113:101605

Souza CDA et al (2017) Cellobiose-dehydrogenase-derived cellobiose protects plants by dampening antagonistic interaction with pathogens. Plant Cell Physiol 58(2):284–298

Stadnik MJ, Freitas MBD (2014) Algal polysaccharides as source of plant resistance inducers. Tropical Plant Pathol 39:111–118

Su F, Zhao B, Dhondt-Cordelier S, Vaillant-Gaveau N (2024) Plant-growth-promoting Rhizobacteria modulate carbohydrate metabolism in connection with host plant defense mechanism. Int J Mol Sci 25(3):1465

Su T, Han M, Min J, Chen P, Mao Y, Huang Q et al (2018) Genome-wide survey of invertase encoding genes and functional characterization of an extracellular fungal pathogen-responsive invertase in Glycine max. Int J 19(8):2395. https://www.mdpi.com/1422-0067/19/8/2395

Sujarit K, Pathom-aree W, Mori M, Dobashi K, Shiomi K, Lumyong S (2020) Streptomyces palmae CMU-AB204T, an antifungal producing-actinomycete, as a potential biocontrol agent to protect palm oil producing trees from basal stem rot disease fungus. Ganoderma boninense Biol Contr 148:104307

Swaminathan S, Lionetti V, Zabotina OA (2022) Plant cell wall integrity perturbations and priming for defense, vol 11. Plants

Tang J, Zhou Q, Chu H, Nagata S (2011) Characterization of alginase and elicitor-active oligosaccharides from Gracilibacillus A7 in alleviating salt stress for Brassica campestris L. J Agric Food Chem 59(14):7896–7901

Tesfay SZ, Magwaza LS, Mbili N, Mditshwa A (2017) Carboxyl methylcellulose (CMC) containing moringa plant extracts as new postharvest organic edible coating for avocado (Persea americana mill.) fruit. Sci Hortic 226:201–207

Thibaud MC, Gineste S, Nussaume L, Robaglia C (2004) Sucrose increases pathogenesis-related PR-2 gene expression in Arabidopsis thaliana through an SA-dependent but NPR1-independent signaling pathway. Plant Physiol Biochem 42:81–88

Thomas-Barry G, Martin CS, Ramsubhag A, Eudoxie G, Miller JR (2024) Multi-trait efficiency and interactivity of bacterial consortia used to enhance plant performance under water stress conditions. Microbiol Res 281:127610

Thoms D, Liang Y, Haney CH (2021) Maintaining symbiotic homeostasis: how do plants engage with beneficial microorganisms while at the same time restricting pathogens? Mol Plant-Microbe Interact 34(5):462–469

Tiwari M, Pati D, Mohapatra R, Sahu BB, Singh P (2022) The impact of microbes in plant immunity and priming induced inheritance: a sustainable approach for crop protection, vol 4. Plant Stress, p 100072

Trouvelot S, Héloir MC, Poinssot B, Gauthier A, Paris F, Guillier C et al (2014) Carbohydrates in plant immunity and plant protection: roles and potential application as foliar sprays. Front Plant Sci 5:592

Tsushima S et al (2020) Induction of systemic resistance in cucumber plants by chito-oligosaccharides against bacterial pathogens. J Gen Plant Pathol 86(4):309–319

Valencia-Chamorro SA, Pérez-Gago MB, del Río MÁ, Palou L (2009) Effect of antifungal hydroxypropyl methylcellulose (HPMC)-lipid edible composite coatings on postharvest decay development and quality attributes of cold-stored 'Valencia'oranges. Postharvest Biol Technol 54(2):72–79

Vera J, Castro J, Contreras RA, González A, Moenne A (2012) Oligo-carrageenans induce a long-term and broad-range protection against pathogens in tobacco plants (var. Xanthi). Physiol Mol Plant Pathol 79:31–39

Versluys M, Toksoy Öner E, Van den Ende W (2022) Fructan oligosaccharide priming alters apoplastic sugar dynamics and improves resistance against Botrytis cinerea in chicory. J Exp Bot 73(12):4214–4235

Viji G, Uddin W, Romaine CP (2003) Suppression of gray leaf spot (blast) of perennial ryegrass turf by Pseudomonas aeruginosa from spent mushroom substrate. Biol Control 26(3):233–243

Vlot AC, Sales JH, Lenk M, Bauer K, Brambilla A, Sommer A et al (2021) Systemic propagation of immunity in plants. New Phytol 229(3):1234–1250

Wagner TA, Kohorn BD (2001) Wall-associated kinases are expressed throughout plant development and are required for cell expansion. Plant Cell 13(2):303–318

Wan J, He M, Hou Q, Zou L, Yang Y, Wei Y, Chen X (2021) Cell wall associated immunity in plants. Stress Biology 1:3

Wang Y, Mostafa S, Zeng W, Jin B (2021) Function and mechanism of jasmonic acid in plant responses to abiotic and biotic stresses. Int J Mol Sci 22(16):8568

Wang Y, Tyler BM, Wang Y (2019) Defense and counterdefense during plant-pathogenic oomycete infection. Ann Rev Microbiol 73(1):667–696

Wanke A, Malisic M, Wawra S, Zuccaro A (2021) Unraveling the sugar code: the role of microbial extracellular glycans in plant–microbe interactions. J Exp Bot 72(1):15–35

Wanke A, Rovenich H, Schwanke F, Velte S, Becker S, Hehemann JH et al (2020) Plant species-specific recognition of long and short β-1, 3-linked glucans is mediated by different receptor systems. Plant J 102(6):1142–1156

Wierzbicki AT et al (2022) Emerging roles of extracellular DNA in plant defense. Trends Plant Sci 27(6):515–526

Wiesel L, Newton AC, Elliott I, Booty D, Gilroy EM, Birch PRJ, Hein I (2014) Molecular effects of resistance elicitors from biological origin and their potential for crop protection. Frontiers. Plant Sci 5. NOV

Xing K, Zhu X, Peng X, Qin S (2015) Chitosan antimicrobial and eliciting properties for pest control in agriculture: a review. Agron Sustain Dev 35:569–588

Xu L, Zhang B, Qin Y, Li F, Yang S, Lu P et al (2020) Preparation and characterization of antifungal coating films composed of sodium alginate and cyclolipopeptides produced by Bacillus subtilis. Int J Biol Macromol 143:602–609

Xu N, Zhao Q, Zhang Z, Zhang Q, Wang Y, Qin G et al (2022) Phyllosphere microorganisms: sources, drivers, and their interactions with plant hosts. J Agric Food Chem 70(16):4860–4870

Xu X, Iwamoto Y, Kitamura Y, Oda T, Muramatsu T (2003) Root growth-promoting activity of unsaturated oligomeric uronates from alginate on carrot and rice plants. Biosci Biotechnol Biochem 67(9):2022–2025

Yamasaki Y, Yokose T, Nishikawa T, Kim D, Jiang Z, Yamaguchi K, Oda T (2012) Effects of alginate oligosaccharide mixtures on the growth and fatty acid composition of the green alga Chlamydomonas reinhardtii. J Biosci Bioeng 113(1):112–116

Yang B, Yang S, Zheng W, Wang Y (2022) Plant immunity inducers: from discovery to agricultural application. Stress Biology 2(1):5

Yang C, Liu R, Pang J, Ren B, Zhou H, Wang G et al (2021a) Poaceae-specific cell wall-derived oligosaccharides activate plant immunity via OsCERK1 during Magnaporthe oryzae infection in rice. Nat Commun 12(1):2178

Yang J, Shen Z, Sun Z, Wang P, Jiang X (2021b) Growth stimulation activity of alginate-derived oligosaccharides with different molecular weights and mannuronate/guluronate ratio on Hordeum vulgare L. J Plant Growth Regul 40:91–100

Yang Y, Steidele CE, Rössner C, Löffelhardt B, Kolb D, Leisen T et al (2023) Convergent evolution of plant pattern recognition receptors sensing cysteine-rich patterns from three microbial kingdoms. Nat Commun 14(1):3621

Yu J, Wang D, Geetha N, Khawar KM, Jogaiah S, Mujtaba M (2021) Current trends and challenges in the synthesis and applications of chitosan-based nanocomposites for plants: a review. Carbohydr Polym 261:117904

Yu Y, Gui Y, Li Z, Jiang C, Guo J, Niu D (2022) Induced systemic resistance for improving plant immunity by beneficial microbes, vol 11. Plants

Zarattini M, Corso M, Kadowaki MA, Monclaro A, Magri S, Milanese I et al (2021) LPMO-oxidized cellulose oligosaccharides evoke immunity in Arabidopsis conferring resistance towards necrotrophic fungus B. Cinerea. Communications Biol 4(1):727

Zaynab M, Fatima M, Sharif Y, Zafar MH, Ali H, Khan KA (2019) Role of primary metabolites in plant defense against pathogens. Microb Pathog 137:103728

Zhang C, Howlader P, Liu T, Sun X, Jia X, Zhao X et al (2019) Alginate oligosaccharide (AOS) induced resistance to Pst DC3000 via salicylic acid-mediated signaling pathway in Arabidopsis thaliana. Carbohydr Polym 225:115221

Zhang S, Tang W, Jiang L, Hou Y, Yang F, Chen W, Li X (2015) Elicitor activity of algino-oligosaccharide and its potential application in protection of rice plant (*Oryza saliva* L.) against magnaporthe grisea. Biotechnol Biotechnol Equip 29(4):646–652

Zhang W et al (2021) Activation of plant immune responses through β-glucan recognition. Plant J 106(1):73–88

Zhao L, Niu J, Feng D, Wang X, Zhang R (2023) Immune functions of pattern recognition receptors in Lepidoptera. Frontiers Immunol 14

Zheng F, Chen L, Zhang P, Zhou J, Lu X, Tian W (2020) Carbohydrate polymers exhibit great potential as effective elicitors in organic agriculture: a review. Carbohydr Polym 230(15):115637. https://doi.org/10.1016/j.carbpol.2019.115637

Zhuo R, Li B, Tian S (2022) Alginate oligosaccharide improves resistance to postharvest decay and quality in kiwifruit (Actinidia deliciosa cv. Bruno). Horticultural. Plant J 8(1):44–52

Zipfel C, Oldroyd GED (2020) Understanding the plant immune system. Curr Biol 30(12):R500–R512

Zipfel C (2014) Plant pattern-recognition receptors. Trends Immunol 35(7):345–351

Chapter 7
The Protective Effects of Melatonin in Biotic Stress Mitigation

Syed Tanveer Shah, Abdul Basit, Ayesha Khan, Heba I. Mohamed, and Muhammad Adil Younas

Abstract Plants frequently encounter biotic stresses due to pathogens (including viruses, fungi, oomycetes, bacteria, and insects), whereas the interaction between plants and pathogens represents an ongoing evolutionary battle, as both strive to outcompete each other in natural ecosystems. To counteract these threats, to minimize pest and pathogen attacks, plants have developed intricate resistance mechanisms. Plant hormones such as melatonin, salicylic acid, jasmonates, abscisic acid, gibberellins, auxins, brassinosteroids, ethylene, cytokinins, and melatonin operate within a well-coordinated network that underpins plant resistance to pests and diseases shows its role as the major driving factor of numerous plant physiological processes. Melatonin (N-acetyl-5-methoxytryptamine; derived from tryptophan) is synthesized by various organisms, including bacteria, plants, animals, and fungi. Since its identification in plants, significant progress has been made in understanding phytomelatonin functions. This multifunctional molecule, often referred to as a master regulator, influences certain mechanisms of plants (such as seeds and fruits ripening, plant growth, parthenocarpy, rooting, flowering, leaf senescence, biomass production, and photosynthetic efficiency). One of its most important roles is to protect against biotic stress. This chapter reviews the metabolism and biosynthesis of melatonin in plants, the defensive mechanism at the cellular level against dis-

S. T. Shah (✉) · M. A. Younas
Department of Agriculture, Hazara University Mansehra, Khyber Pakhtunkhwa, Dhodial, Pakistan
e-mail: dr.syedtanveershah@hu.edu.pk

A. Basit
Department of Horticulture, Graduate School, Kyungpook National University, Daegu, South Korea

A. Khan
Laboratory of Plant Physiology, Lithuanian Research Centre for Agriculture and Forestry, Kėdainiai, Lithuania

H. I. Mohamed
Biological and Geological Sciences Department, Faculty of Education, Ain Shams University, Cairo, Egypt

K. A. Abd-Elsalam, H. I. Mohamed (eds.), *Elicitors for Sustainable Crop Production*, Sustainability Sciences in Asia and Africa,
https://doi.org/10.1007/978-981-96-6569-3_7

eases, and biotic stresses. Moreover, the interaction of melatonin with other plant substances integrated the melatonin-plant biocontrol model against various plant defense processes. In the end, knowledge regarding melatonin as an antiviral, antibacterial, antifungal, and pesticide agent against various plant pathogens.

Keywords Antiviral · Antibacterial · Antifungal · Biosynthesis · Biotic stress · Biocontrol · Pesticide

7.1 Introduction

Plants frequently encounter biotic stresses due to pathogens (including viruses, fungi, oomycetes, bacteria, and insects), whereas the interaction between plants and pathogens represents an ongoing evolutionary battle, as both strive to outcompete each other in natural ecosystems. To counteract these threats, in order to minimize pest and pathogen attacks, plants have developed intricate resistance mechanisms (Panstruga et al. 2009; Denancé et al. 2013). Their molecular resistance system enables them to detect attacks through specialized proteins called plant recognition receptors (PRRs), which helps them to identify preserved pathogen constituents named as pathogen-associated molecular patterns (PAMPs). In response, to suppress PAMP-triggered immunity (PTI), pathogens release effector proteins, while plants, in response, secrete proteins capable of recognizing these effectors through resistance proteins, activating effector-triggered immunity (Dodds and Rathjen 2010).

This immune mechanism, while protective, may disrupt other physiological processes which plants mitigate through controlling their resistance responses using phytohormones, which in turn balance defense mechanisms with environmental alteration and growth (Figueroa-Macías et al. 2021). Plant hormones [such as melatonin, salicylic acid (SA), jasmonates (JA), abscisic acid (ABA), gibberellins (GA), auxins, brassinosteroids (BR), ethylene (ET), and cytokinins (CK)] operate within a well-coordinated network that underpins plant resistance to pests and diseases, and show their role as a major driving factor of numerous plant physiological processes (including defense signaling, photomorphogenesis, and stress responses) (Munné-Bosch and Müller 2013).

Melatonin (N-acetyl-5-methoxytryptamine; derived from tryptophan) is synthesized by various organisms, including bacteria (Tilden et al. 1997; Rodriguez-Naranjo et al. 2012; Hardeland 2019; Que et al. 2020), plants (Arnao 2014; Hardeland 2019; Cheng et al. 2021) animals (Zeman and Herichová 2011; Claustrat and Leston 2015; Singh et al. 2020; Cheng et al. 2021), and fungi (Cheng et al. 2021; Rehman et al. 2022). Initially identified in the pineal gland of mammals in 1958 (Lerner et al. 1958), its chemical structure was elucidated in 1959 (Lerner et al. 1959). It was termed as phytomelatonin by 1995 after its discovery in plants (Kolar et al. 1995; Dubbels et al. 1995; Hattori et al. 1995).

Since its identification in plants, significant progress has been made in understanding phytomelatonin's functions. This multifunctional molecule, often referred to as a master regulator, influences certain mechanism of plants (such as seeds and fruits ripening, plant growth, parthenocarpy, rooting, flowering, leaf senescence, biomass production, and photosynthetic efficiency) (Arnao and Hernández-Ruiz 2018a, b, 2020a, b, 2021; Arnao et al. 2019a, b, 2021, 2022). Among its most notable roles is its protective function under biotic and abiotic stresses (Buttar et al. 2020; Altaf et al. 2021; Zeng et al. 2022; Menhas et al. 2022). Given the economic losses caused by numerous pathogens, there has been growing interest in exploring melatonin's role in plant defense systems (Tiwari et al. 2020; Tiwari et al. 2021; Himanshu et al. 2024). To minimize economic impacts and control pathogen contagions, several approaches have been developed including improved crop cultivars with enhanced immune using chemical priming (Zhao et al. 2019a, b), virus-free plants usage (Hadidi et al. 2023), and genetic alterations (Onaga et al. 2016). This chapter aimed to discuss the metabolism and biosynthesis of melatonin in plants. Moreover, it explores the role and function of melatonin against biotic and abiotic stresses and other plant processes at the cellular level. It also highlights the role of melatonin as an antiviral, antibacterial, antifungal, and pesticide agent.

7.2 The Plant's Defensive Mechanism Against Diseases

In natural environment, plant defense mechanisms have a major role in nutrients translocation and resistance against diseases (Miller et al. 2010). Aside from plants protecting themselves, there are two major pathways [systemic acquired resistance (SRS) and induced systematic resistance (ISR)] through which beneficial microorganisms and bacterial species promote plant development. The variation in gene expression is driven by SAR through PR proteins, while ISR is supported by beneficial bacteria. However, the same system regulates and activates their genetic expression in different ways. Moreover, SAR is induced by beneficial microorganisms under biotic stress (accumulating salicylic acid and PR proteins), while ethylene- and jasmonate-regulated pathways depend on ISR (Nawrocka and Maolepsza 2013).

7.2.1 Generation of ROS and its Significance in the Protective Cycle

ROS generation is critical for executing a sensitized host defense reaction. The stability between clearance and ROS directly disrupts due to abiotic or biotic stress (Mittler 2017). As a result of biotic stress, the cell produces ROS that may cause severe oxidative damage in plant tissues, such as peroxide of hydrogen (H_2O_2), anion superoxide radical ($O_2^{\bullet-}$), oxygen ($^1O_2$), and hydroxyl ($^\bullet OH$) radicals. Equilibrium of redox, systemic responses, differentiation control, and stress signals are controlled by minute amount of ROS in higher plants, whereas ROS in greater

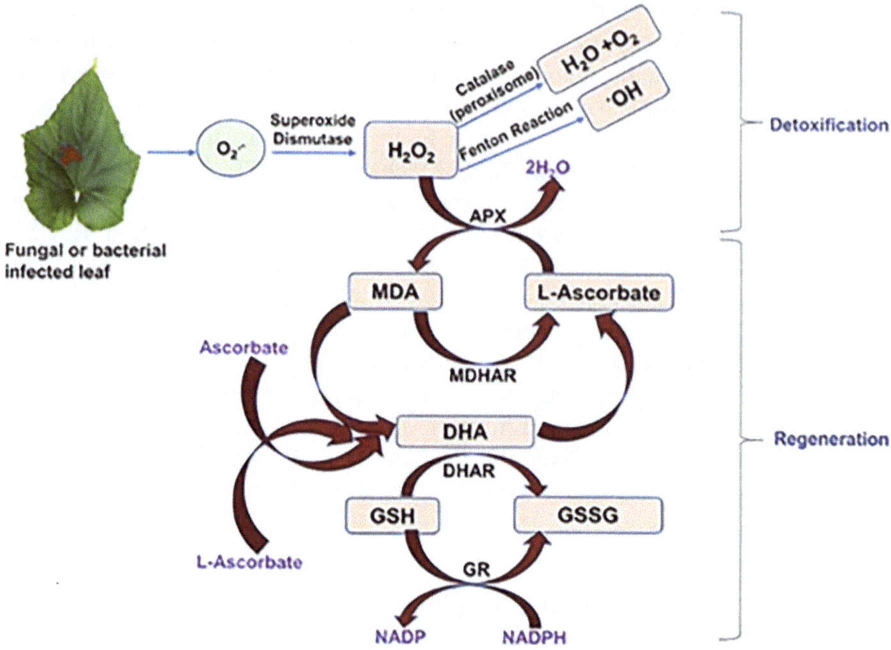

Fig. 7.1 Detoxification of reactive oxygen radicals using antioxidant defense system by plants (Kaur et al. 2022a, b)

amounts obstruct cell components with membrane damage, lipid oxidation, and protein breakdown (Das and Roychoudhury 2014). These results in the hampering of regular cell function under high rates of ROS (Asthir et al. 2010) while plants change their metabolic processes by sending signals which help to activate defense genes to heal the affected plant parts under stress conditions (Gill et al. 2019) (Fig. 7.1).

7.2.2 *Structural Modifications in Host Plants*

For infiltration purposes, the primary target of pathogens is the plant cell membrane because the cell wall acts as a sophisticated defensive tool against microorganisms due to millions of years of coevolution. Microbes establish toxic connections with the plants they invade by avoiding the plant cell walls, which demands accurate host recognition, followed by the creation of suitable infectious components and/or chemical liquids (Turra et al. 2015). Non-pathogens or non-adapted pathogen are microbes that have not developed adequate means for breaching a host plant's cell wall. The membrane of a cell is used by the host plant's cells as a defense barrier against hazardous bacteria that may breach developed barriers.

7.2.3 *Antimicrobial Compound Production and Metabolic Change*

Plants produce numerous secondary metabolites that are harmful to pathogens enhance plants resistance against pathogens. Secondary metabolites have several roles in plant defense, including deterrent and toxic effects, and functioning as a prelude to the physiological protection response. Numerous antibacterial proteins, enzymes, and metabolites are produced during the biochemical shift in hosts. All these kinds of host responses are strong and stiff enough to decrease pathogen-caused lesions (Kaur et al. 2022a, b).

7.3 Melatonin

Melatonin is a mammalian neurohormone, antioxidant, and signaling molecule (Dubbels et al. 1995), which makes this biological substance widely used in various research nowadays (Reiter et al. 2010; Reiter 1995; Jan et al. 2009). It usually affects multiple processes, that is, immune system, circadian rhythms, sleep, and body temperature. Apart from this, it is also used as an antioxidant in different cell functions (Brainard et al. 2001; Rodriguez et al. 2004; Jan et al. 2009). In the 1990s, plant scientists were greatly surprised by the discovery of melatonin in the plant kingdom, in addition to its major role in the animal kingdom (Hattori et al. 1995; Dubbels et al. 1995). However, two major research groups started working on it with different perceptions. It is thought to be found in about 20 plant families.

The swiftly evolving recognition technology and the addition of materials separation, genome sequencing, and molecular biology techniques in plants have made it possible to identify features of melatonin such as physiological properties, metabolic processes traits, biosynthesis, dissemination, quantity, and biochemical characteristics of melatonin (Erland et al. 2016; Back et al. 2016; Byeon and Back 2014a, b; Arnao and Hernández-Ruiz 2014; Tan et al. 2012). Melatonin helps in different plant processes, that is, plant growth, and overcoming various stresses (Arnao and Hernández-Ruiz 2014, 2015, 2019a; Byeon and Back 2014b; Wang et al. 2018; Wei et al. 2015; Sun et al. 2015). Melatonin was usually known to be involved in different regulatory pathways and functions of plants and is hence termed "phytomelatonin" (Blask et al. 2004; Khan et al. 2024). Melatonin is a plant hormone proved by its discovery in *Arabidopsis thaliana* (very initial plant melatonin sensor (Arnao and Hernández-Ruiz 2019a, b). Aside from that important fact, melatonin regulates the development of seeds, root development and growth, senescence of foliage, circadian cycles, afterward ripening of fruit, and plant reactions to detrimental surroundings (Arnao and Hernández-Ruiz 2015; Zhang et al. 2014).

7.4 Metabolism and Biosynthesis of Melatonin in Plants

The melatonin to be phytohormone was validated through isotope tracer investigation in *Hypericum perforatum* L. cv. Anthos plantlets as well as in higher plants (which was disagreed at that time) and the hypothesis was coined by St. John's wort (Arnao and Hernández-Ruiz 2006; Zhao et al. 2019a, b; Tan et al. 2013). Four biochemical phases of melatonin catalyzed by six enzymes, including N-acetylserotonin methyltransferase (ASMT), tryptophan hydroxylase (TPH), caffeic acid O-methyltransferase (COMT), tryptophan decarboxylase (TDC), tryptamine 5-hydroxylase (T5H), and serotonin N-acetyltransferase (SNAT) (Back et al. 2016).

Similar to herbivores, two enzymes have been recognized by plants that help in tryptophan decarboxylation and hydroxylation (essential to produce melatonin) (Park and Back 2012). However, as compared to vertebrates, the first two enzymatic stages to produce melatonin in plants from tryptophan are inverted. TDC catalyzed decarboxylation of tryptophan is the first stage in melatonin production in rice (*Oryza sativa*) after which T5H catalyzed hydroxylation (Park et al. 2013). Surprisingly, even though T5H is required for serotonin production, melatonin levels were increased in transgenic rice by T5H silencing, suggesting that serotonin levels are not to melatonin levels. Serotonin acetylation, SNAT in plants performs the final time in melatonin production (Back et al. 2016). The introduction of *C. reinhardtii* AANAT homolog (crAANAT) into MicroTom tomato (Okazaki et al. 2009) and human AANAT (hAANAT) into rice enhanced the melatonin and genes appearance (Kang et al. 2010).

It is currently believed that AANAT paralogs facilitate the N-acetylation of serotonin in plants. This process appears distinct from homologs acquired through horizontal gene transfer and derived from chlorophyceans. However, there is no clue about gene orders and particular isoenzymes responsible for this activity. Melatonin production is thought to end with the omethylation of N-acetylserotonin. Cloning and expressing ASMT gene in rice is associated with other genes in the plant (O-methyltransferase) (Kang et al. 2011). ASMT kinetics revealed significant differences in animal and rice (lacks substrate inhibition capabilities by ASMT) (Kang et al. 2011). Contradictory evidence has been observed regarding ASMT subforms in rice, as recent studies suggest that caffeic acid O-methyltransferase (COMT) participates in dinmelatonin production in *Arabidopsis thaliana* (Byeon and Back et al. 2014a; Byeon et al. 2015). Cpying these melatonin biosynthesis genes could clear the way for deeper exploration into melatonin-mediated functions. ASMT and COMT, which first methylate serotonin to produce 5-methoxytryptamine, demonstrate substrate affinity for both N-acetylserotonin and serotonin while interactions with SNAT led to melatonin synthesis (Back et al. 2016; Arnao and Hernández-Ruiz 2015). Based on these results, it is clarified that plants have complex way to synthesize melatonin rather than other organism (such as mammals, bacteria, and yeast). To fully understand the molecular mechanisms underlying melatonin production, diverse analytical systems should be employed concurrently. Figure 7.2 shows the melatonin biosynthesis.

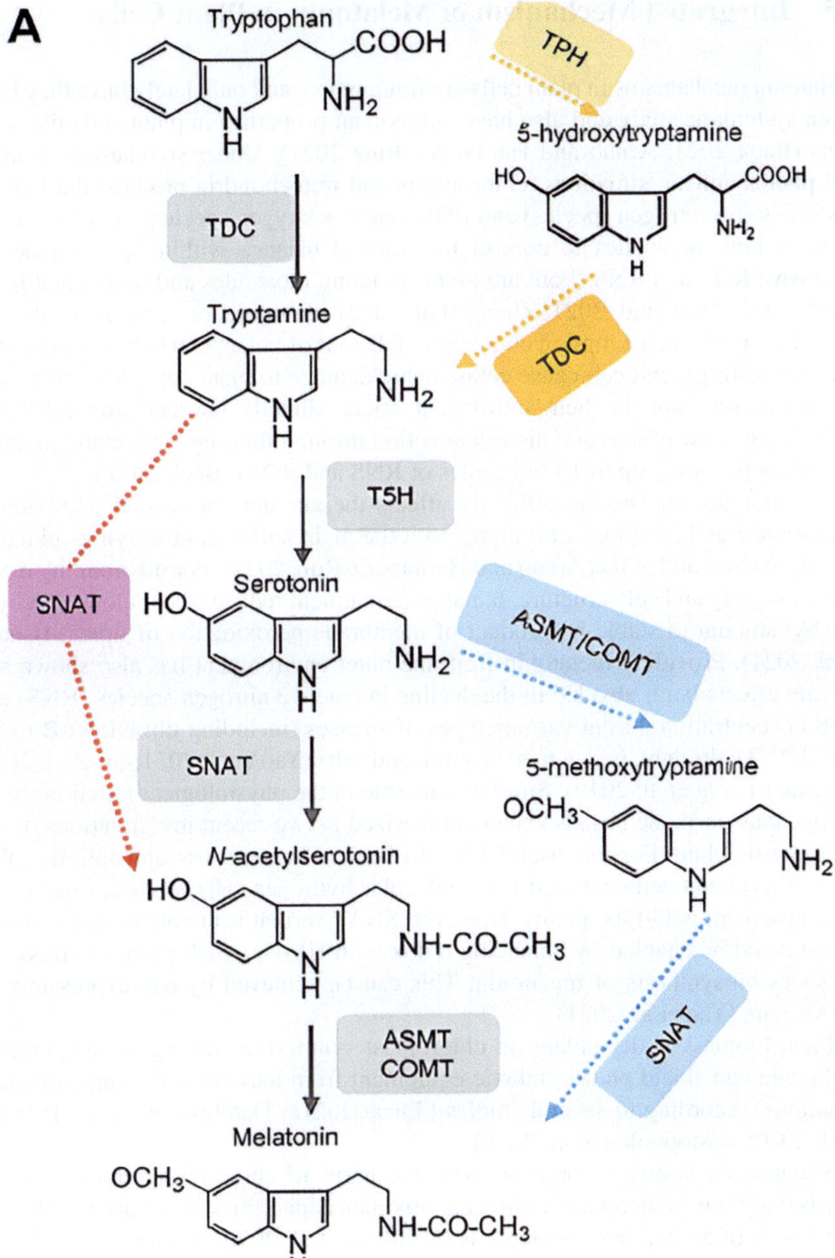

Fig. 7.2 Melatonin biosynthesis (Back et al. 2016)

7.5 Integrated Mechanism of Melatonin in Plant Cells

Melatonin mechanisms in plant cells are fragmented, and only lately have they been given systematic study and also have antioxidant properties in plant cells like vitamins (Back 2021; Arnao and Hernandez-Ruiz 2021). Under standard respiration and photosynthetic situations, chloroplasts and mitochondria produce the bulk of RNS (reactive nitrogen species) and ROS (reactive oxygen species). It is biosynthesized in both organelles to control free radical balance within both organelles. Excessive ROS and RNS (both are plant signaling molecules and toxic chemicals) (Back 2021; Altaf et al. 2021; Zhang et al. 2022) can injure the structure of the cell when the supply-consumption connection falls out of balance, which is frequent in stressful settings, and can cause catastrophic damage to plant cells. Melatonin and its metabolites, due to their antioxidant effect, directly interact with RNS and ROS. They show powerful antioxidant action through their nonenzymatic method, melatonin (removes up to 10 molecules of RNS and ROS) (Back 2021).

Additionally, melatonin indirectly affects the amounts of several antioxidants (glutathione and ascorbic acid) along with the main antioxidant enzymes glutathione peroxidase and SOD (Arnao and Hernandez-Ruiz 2021). For maintaining membrane integrity and cell structure, melatonin treatment reduces the malondialdehyde (MDA) amount (a stable by-product of membrane peroxidation of lipids) (Chang et al. 2021). Providing melatonin from the outer environment has also shown significant effects but it also led to the decline in reactive nitrogen species (RNS) and ROS concentration during various types of stresses (including ultraviolet B radiation (UV-B), drought, heavy metals, cold, and salt) (Yao et al. 2021; Su et al. 2021; Moustafa-Farag et al. 2019). Similarly, its role in the physiological functions as an endogenous hormone has also been emphasized in two recent investigations on the Arabidopsis plant. For the establishment of redox homeostasis and salinity tolerance, it has been demonstrated that molecular hydrogen (H_2) promotes melatonin biosynthesis in wild-type plants. However, SNAT variant is unable to synthesize it, which could be enabled by balancing redox with UV-B, which prompts stress and enhances biosynthesis of melatonin. This can be achieved by overexpressing the SNAT gene (Yao et al. 2021).

Plant biomass is dependent on chloroplast synthesized during photosynthesis. Melatonin can shield photosynthetic equipment from harm under a range of stress situations, according to several studies (Talaat 2021a; Danilova et al. 2021; Wang et al. 2020; Kostopoulou et al. 2015).

Melatonin's ability to prevent oxidative harm to chloroplasts can be largely ascribed to both its direct and indirect antioxidant capabilities as in barley cells with high levels of antioxidant carotenoids (Danilova et al. 2020; Danilova et al. 2021). However, antioxidant capabilities and reduced photosynthesis in tomato plants with reduced levels of melatonin are due to virus-induced suppression of the TDC gene (Wang et al. 2020). Under salt and drought stress conditions, melatonin enhanced photosynthetic efficiency and chlorophyll content while suppressing the expression of chlorophyll-degrading enzymes, such as pheophorbide A oxygenase, in soybean

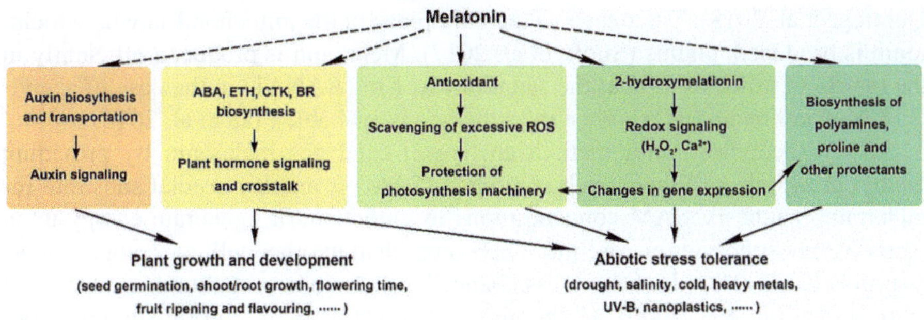

Fig. 7.3 Melatonin's function and mechanism in the regulation of plant growth, development, and abiotic stress tolerance (Yang et al. 2022)

while upregulating the key genes associated with photosystem I (PsaG and PsaK) and photosystem II (PsbP and PsbO) (Wei et al. 2015). Additionally, polyamines play a crucial role in plant stress responses by stabilizing nucleic acids, and proteins and regulating the activity of various enzymes (Aghdam et al. 2019; Zhang et al. 2019).

Polyamines are potent antioxidants that go through free radicals and protect membrane integrity same as melatonin (Aghdam et al. 2019; Zhang et al. 2019). Melatonin can enhance polyamine levels in wheat, which primarily consists of spermine, putrescine, and spermidine, by promoting their synthesis and inhibiting their degradation. This effect is achieved through the upregulation of genes encoding key polyamine biosynthetic enzymes, such as ornithine decarboxylase (ODC) and arginine decarboxylase (ADC) while downregulating genes linked with their breakdown [including diamine oxidase (DAO) and polyamine oxidase (PAO)]. The breakdown-causing enzymes for polyamines were inhibited (Talaat 2021b). Alfalfa (Zhang et al. 2019), carrots (Lei et al. 2004), and tomatoes (Aghdam et al. 2019) have all had comparable results. Additionally, under stress conditions (such as high salinity, heavy metals, and low temperatures), water-soluble sugars are enhanced by melatonin, free proline, as well as appropriate protectants (Talaat 2021a; Chang et al. 2021). Figure 7.3 provides an overview of how melatonin regulates plant tolerance to biotic stress.

7.6 Mitochondria and Melatonin

As previously stated, mitochondria originated from proteobacteria via endosymbiosis. Interestingly, it has already been established that these bacteria can produce melatonin (Manchester et al. 1995; Tilden et al. 1997). Based on these details, it was hypothesized that the ability of these bacteria to synthesize melatonin (in advanced mitochondria) and its synthesizing enzymes has been observed in plants and mammals too (Tan et al. 2013; He et al. 2016; Wang et al. 2017; Suofu et al. 2017;

Quintela et al. 2018). The matrix is the only place in the mitochondria where melatonin is produced, claims (Suofu et al. 2017). Melatonin is produced efficiently in the matrix of mitochondria at the same rate as Km (AANAT) in the case of acetyl-CoA (melatonin co-factor) when the substrate is available (Tan et al. 2016).

ASMT (acetylserotonin methyltransferase) catalyzes melatonin by providing methyl group where S-adenosylmethionine (SAM) is another crucial substrate for melatonin synthesis. SAM concentrations in mitochondria (generate energy upon oxidative phosphorylation) are more persistent than in other cellular compartments (Agrimi et al. 2004), which leads to enhanced melatonin in mitochondria rather than cytosol (Martín et al. 2000). Melatonin levels must be high to optimize mitochondrial activity led to superoxide like ROS (electron escape from transport chain and pair up with oxygen) and other reactive substances formation. Under normal conditions, the mitochondrial antioxidant system gently controls ROS levels and keeps them within a particular range suited for signaling molecules, while excessive ROS levels have detrimental effects (such as tissue and cell injury).

Melatonin (an indirect mitochondrial antioxidant and direct free radical scavenger) enhances certain mitochondrial enzymes' (catalase, glutathione reductase, and SODs) workability and also declines ROS generation by enhancing specific proteins (uncoupling protein; UCP1 and UCP3 in adipose tissue and cardiomyocytes) action (Reiter et al. 2017; Banerjee et al. 2020; Pal et al. 2019; Tan et al. 2011; Fernandez Vazquez et al. 2018; Aslan et al. 2020; Tan and Reiter 2019). Melatonin's capacity to reduce ROS generation (known as ROS avoidance action) is dependent on numerous processes, where all such activities help in shielding from injuries produced by ROS (Hardeland 2009). Certainly, several experiments have shown melatonin's preventive effects on tissue damage induced by oxidative stress. These include brain, lung, heart, gut, testes, and skeletal muscle injuries produced by ischemia/reperfusion, as well as tissue and organ damage caused by hypoxia (Berger et al. 2019; Dong et al. 2020).

Melatonin's protective effects are mostly linked with mitochondria such as stabilizing its membrane capacity, enhancing mitochondrial complexes (1, 3, and 4), consistency of ATP production, and less time of membrane pore opening (López et al. 2017; Jauhari et al. 2020). Several diseases (cardiovascular, systemic metabolic disorders, neurodegenerative, breast and prostate cancer, diabetes, and multiple ovarian cysts) are also caused due to lower melatonin production (Su et al. 2017; Tamtaji et al. 2020), but melatonin therapy has also favorable effects on the majority of these disorders (Cardinali 2019; Sanchez-Barcelo et al. 2017). Mitochondrial dynamics such as fission and fusion (enlarged action and magnitude of mitochondria) events with the help of the host cell occur in mitochondria (self-renewing organelles). Fission boosts mitochondrial numbers and is required for cell reproduction. Fission serves to differentiate malfunctioning mitochondria from normal mitochondria (in the case of postmitotic cells).

Mitophagy is an autophagic mechanism that occurs when defective segments are separated (Ma et al. 2020). On the injured point, fission has been linked to the lower capacity of the mitochondrial membrane (Cho and Sun 2019). Drp 1 receptors, dynamin-related protein 1 (such as mitochondrial fission protein, factor, elongation factors 1 and 2), implement fission activity by converting cytosolic Drp 1 (GTPase) upon sensing mitochondrial membrane changes (Yu et al. 2020). Thus, melatonin always keeps mitochondria working properly in many ways (reducing fission and stimulating fusion) (Tan et al. 2019). Melatonin decreases mitochondrial fission primarily by inhibiting mtPTP, which preserves the mitochondrial membrane potential. mtPTP appears to feature numerous heterogeneous structures, including holes or channels in mitochondria (either outer or inner). Membrane folding occurs due to leakages (inside the cytosol, matrix, and intermembranous space); therefore, mtPTP assembly is not clearly defined yet.

Certain things have been found in mitochondria's inner membrane (ATP/ADP carrier and mtPTP substances), in the matrix (cyclophilin-D, a major controlling element), and in the outer membrane (translocator protein and voltage-dependent anion channel 1) (Fayaz et al. 2015). Upon ROS increment, cardiolipin pairs with AAC to transfer ACC, which opens pores, while ACC closes the pore otherwise. mtPTP opening (ACC and VDAC1-associated) by activating AMPK and control of cardiolipin oxidation occurs under melatonin (Zhou et al. 2017). Rinpk3-PGAM5-CypD signaling path and other cellular actions inhibit mtPTP opening (cyclophilin-D) under melatonin and excrete cytochrome C into cytoplasm that activates caspase 3 leading to apoptosis (Zhou et al. 2018).

Severe illnesses and disorders are linked with excessive apoptosis, which also activate several elements to inhibit apoptosis in organs, primarily by inhibiting cytochrome C excretion through controlling mtPTP opening (Asghari et al. 2017). Melatonin's effects on mitophagy increment have been documented but it is highly dependent on research settings and cell type (Yoon et al. 2019; Chen et al. 2020). When the data is thoroughly examined, it (mitophagy increment, either valuable or harmful) shows a great role in melatonin, such as wasted cell cleaning and peripheral mitochondria decline, which leads to neural linkage damage (Palikaras et al. 2017; Hardeland 2018). Melatonin's prevention of mitophagy may not be a direct action, but rather an indirect outcome of melatonin's protective effects on mitochondria. Melatonin's preservation of mitochondrial structure and function should result in less mitochondrial fission and consequently less mitophagy. To prove this claim, well-designed research is required. Melatonin appears to be required for the maintenance of mitochondria workability. Figure 7.4 depicts the link between melatonin and mitochondria, which have been thoroughly researched (Tan and Reiter 2019).

7.7 Role of Melatonin Against Biotic Stress

Melatonin appears to protect plants against biotic stress, significantly affecting the production and reducing financial loss in crops globally (Moustafa-Farag et al. 2019; Arnao and Hernández-Ruiz 2014, 2015). Considering its renowned significance as a biocide that suppresses the growth of microbes, Yin et al. (2013) first identified the value of melatonin in agricultural microbe pathogenicity by treating apple orchards with melatonin and found less damage to apple leaves with slowed

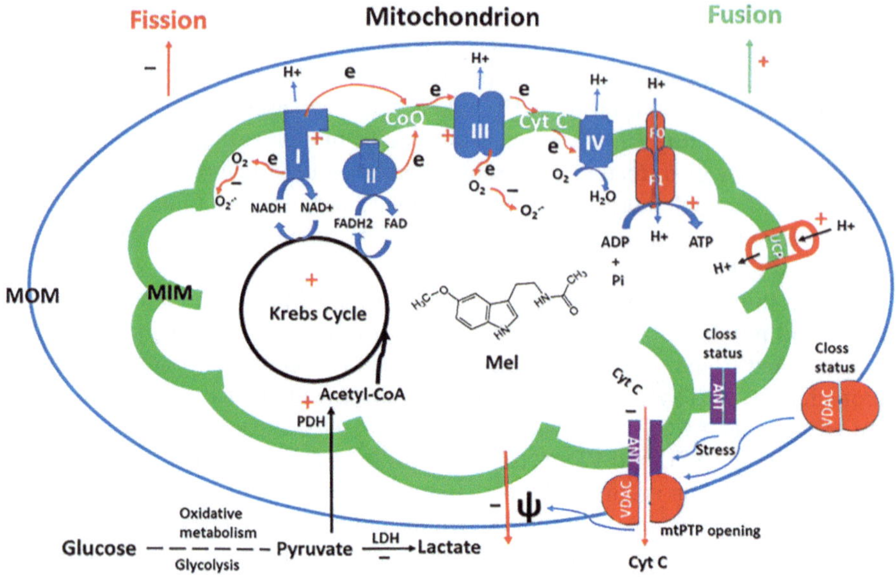

Fig. 7.4 Potential effects of melatonin on mitochondrial function (Tan and Hardeland 2020). Abbreviations: *LDH* lactate dehydrogenase, *ANT* adenine nucleotide translocase or ADP/ATP carrier, *PDH* pyruvate dehydrogenase *UCP* uncoupling protein, *MMP* mitochondrial membrane potential, *Cyt C* cytochrome C, *VDAC* voltage-dependent anion channel, *mPTP* mitochondrial permeability transition pore, *MOM* mitochondrial outer membrane, and *MIM* mitochondrial inner membrane. Melatonin is represented as Mel. Red "+" indicates the stimulatory effects of melatonin, while black "−" signifies suppressive effects

spread of infection compared to control. The scientists revealed that chitinase, phenylalanine ammonia lyase (PAL), and β1,3 glucanase help with resistance to disease in this particular crop by examining pathogenesis-related proteins. Melatonin protects plants from pathogens by modulating plant innate immunity, according to growing data. Immunity can be used to activate defense responses (Zhao et al. 2015; Shi et al. 2015). In Arabidopsis variant (SNAT knockout), the lower melatonin quantity was recorded where it showed to be more susceptible to *P. syringae* rather than wild type along lowered defense genes (ICS1, PR1, and PDF1.2) (Lee et al. 2015). Furthermore, lignin and gossypol accumulation as a result of enhanced metabolic flux into gossypol, phenylpropanoid, and mevalonate pathways due to melatonin administration enhanced the cotton resistance to *Verticillium dahlia* (Shi et al. 2015).

Moreover, it improves crop protection efficacy while decreasing pesticide usage (Dayoub et al. 2011; Casado-Zapico et al. 2010), while combining melatonin with other fungicides constitutes a potential biocontrol method for agricultural application. Melatonin's importance in defending plants from stress caused by bacteria, particularly its signaling transmission and method of action in building innate resistance toward bacterial invasion by pathogens, has been widely studied. SA and JA/

ethylene-related signaling cascades enhanced pathogen-associated molecular patterns which in turn improved the plant resistance against microbes (Glazebrook 2005; Hernández-Ruiz and Arnao 2018). Those plant variants with a lack of ethylene and SA signaling led to inborn immunity inhibition due to melatonin while enhanced pathogen-related, SA, and ethylene-dependent genes in the Arabidopsis *P.* syringae prototype plant–bacterial disease interactions system (Lee and Back 2017; Lee et al. 2014).

Based on findings accumulated during the past 10 years, melatonin tends to help prevent infections in combination with all of these established systems. In accordance with a detailed examination of genome-wide gene regulation in melatonin-treated *Arabidopsis* (Lee et al. 2014), melatonin shows a close connection with other phytohormones (salicylic acid and jasmonic acid) due to genes engaged in signal transduction pathways. Phytomelatonin concentration gets higher in host cells due to oxidative bursts during the initial phases of microbes' contamination (Arnao and Hernández-Ruiz 2015).

The combined influence of H_2O_2 and NO activation makes melatonin a major factor in the defense signaling component in Arabidopsis *P. syringae* framework. 1 (OXI1) kinases get activated by oxidative signal and MAPK networks to run melatonin-induced inherent immunological signal transmission (Lee and Back 2017; Arnao and Hernández-Ruiz 2015). Figure 7.5 shows SA/JA/ethylene and NO dependent pathways to demonstrate role of phytomelatonin in plant inherent defense toward pathogens.

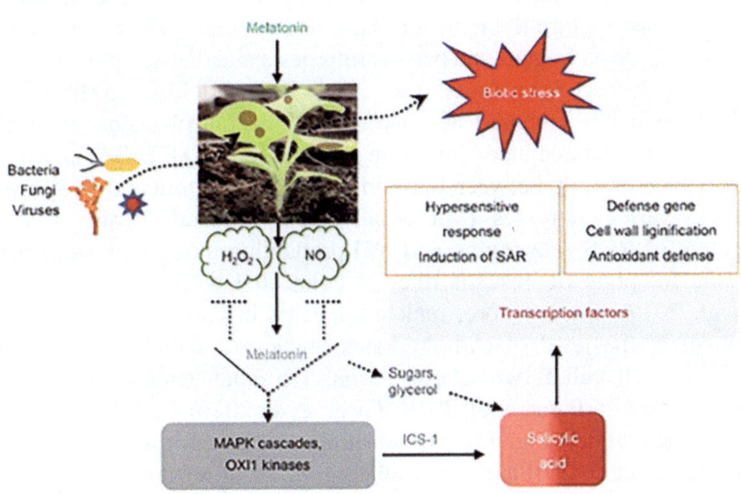

Fig. 7.5 Protection of plants using melatonin following signal transduction pathways against biotic stress (Sun et al. 2021)

7.7.1 Melatonin-Integrated Plant Biocontrol Model

The plant possesses both active and passive defense mechanisms to defend itself from the attack of pathogens. The occurrence of cuticle (waxy) on the cell wall and leaves, and secondary metabolites synthesis, make up a passive defense for plants. The elements of the primary defense line against infections are chemical and physical hindrances. Still, some pathogens possess the ability to cross the obstruction caused by the first line of defense. In reaction to these pathogens, an empirical defense mechanism is triggered known as activated immunity to reduce the development of pathogens to cause disease. Plants prosper two levels of pathogen identification: The initial defense mechanism in plants relies on recognizing PAMPs through PRRs on plant cells which further activates a resistance mechanism known as PTI. Certain pathogens, however, can suppress this initial response by deploying effector proteins which then leads to the activation of a secondary defense mechanism, effector-triggered immunity (ETI), through intracellular receptors with nucleotide-binding domains and leucine-rich repeats (NLRs). ETI is often escorted by programmed cell death (PCD) and hypersensitive responses (HR) which restrict the spread of viruses and biotrophic pathogens (Dodds and Rathjen 2010).

ETI and PTI are responsible pathways to develop resistance against microbes (get regulated by plant hormones such as JA, SA, and ET) as well as activation of signaling molecules such as nitric oxide (NO), SA, and reactive oxygen species (ROS), thereby protecting plants from severe pathogen attacks (Pieterse et al. 2012; Shi et al. 2014; Yang et al. 2015). SA is primarily involved in defense against biotrophic pathogens (require living host tissue), while JA and ET are crucial for combating necrotrophic pathogens and herbivorous pests that damage plant tissue during contamination (Robert-Seilaniantz et al. 2011; Bari and Jones 2009; Glazebrook 2005). Melatonin has emerged as a promising biocontrol agent, enhancing the appearance of genes associated with these phytohormones (ET, GA, auxin, and JA) signaling. This cross-talk between hormonal pathways significantly improves plant tolerance to various stresses (Ma et al. 2018; Moustafa-Farag et al. 2019). Enhancement of ROS scavenging and NO production along with gene activation (responsible for plant defense) is directly connected with melatonin (Moustafa-Farag et al. 2019). Furthermore, melatonin helps in callose removal as well as enhancement in the level of carbohydrates (galactose, cellulose, and xylose) in attacked leaves cell wall of *Arabidopsis thaliana* to enhance resistance ability (Zeng et al. 2022; Moustafa-Farag et al. 2019; Tiwari et al. 2020).

The proposed model (Fig. 7.6) explains the involvement of the melatonin-ROS-NO triad (Arnao et al. 2018a, b) in regulating the plant hormonal network and the interactions between pathogens (such as fungi, bacteria, and viruses) and melatonin-mediated defense responses. This model highlights the role of PRRs in detecting PAMPs to activate PTI (Zhao et al. 2019a, b). It also underscores melatonin's role in inducing the expression of pathogenesis-related (PR) genes (PR1 and PR5), thereby enhancing resistance against pathogens. However, no studies have conclusively explained melatonin's direct effect on regulating the genetic material of plant-infecting viruses.

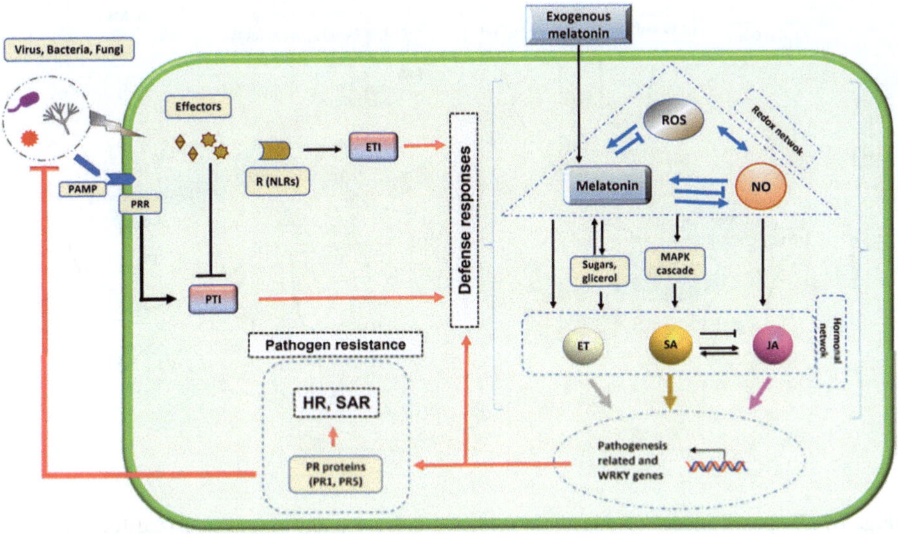

Fig. 7.6 The model outlines the stepwise interplay that occurs between plant pathogens and melatonin-mediated defense responses (Hernandez-Ruiz et al. 2023)

7.7.2 *Melatonin as an Antiviral Agent*

In plants, there are limited scientific findings about melatonin's antiviral activity thus far, such as pathogenic RNA reduction in plant samples (*Solanum lycopersicum* and *Nicotiana glutinosa*) following the application of 2x melatonin from outside. The role of melatonin as a healing agent is connected with SA level increment in NO-dependent channels (Zhao et al. 2019a, b). ASGV (apple stem grooving virus) inducing infection in GALA cultivars has been successfully removed by melatonin (Chen et al. 2019a, b). Figure 7.7 and Table 7.1 provide examples of melatonin's advantageous antiviral effects on plants. Melatonin's impact on plant–virus interactions requires further investigation.

Although their hosts are typically biotrophic illnesses, plant viruses cause local lesions as well as systemic injury, leading to yellowing of plant tissues, misshapen, and dwarfing (Tiwari et al. 2021; Moustafa-Farag et al. 2019). Plant viral infections have long prevented global sustainable agricultural expansion, leading to lower production and economic losses. Control of diseases caused by microorganisms (fungi and bacteria) is easier than viruses (Sofy et al. 2019). Several approaches for managing viral infections have been developed recently, such as eco-friendly chemicals usage (having general shielding influence), and conventional virus-resistant/tolerant cultivar breeding (Sofy et al. 2021).

There have been a few publications (since the first research were completed in 2019) on virus infection in plants to recognize plant–virus linkage with melatonin to identify its shielding role in crops. Chen and colleagues conducted one of these investigations on the possible impact on apple (propagate through asexual means)

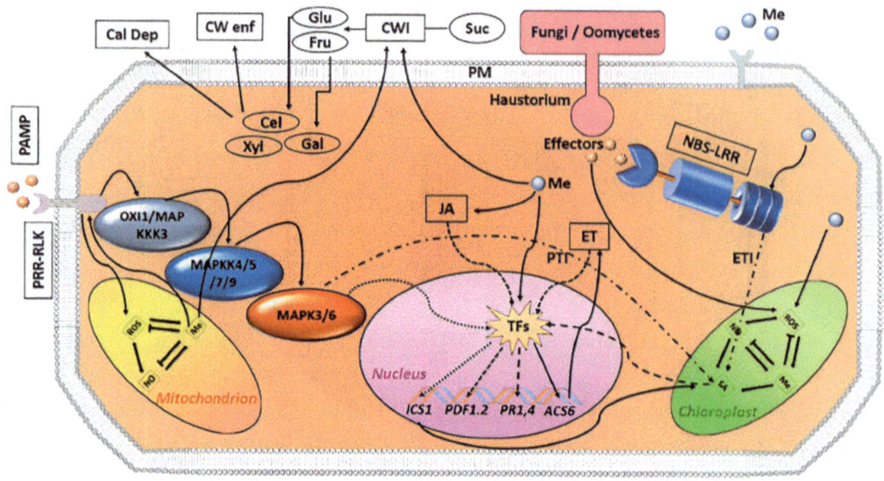

Fig. 7.7 Proposed model of melatonin-mediated defense pathways, synthesized from findings across multiple studies (Moustafa-Farag et al. 2019)

Table 7.1 Melatonin affects plant–virus pathosystem

Plant name	Pathogen name	Melatonin dosage (μM)	Involved mechanism	Effect	Reference
Nicotiana glutinosa and *Solanum lycopersicum*	Tobacco mosaic virus (TMV)	100 (twice)	Salicylic acid (SA) concentrations lead to increased expression of PR1 and PR5 genes	Reduction of virus concentration in infected plants	Zhao et al. (2019a, b)
Malus domestica	Apple stem grooving virus (ASGV)	15	Not reconnoitered	Eradication of virus from previously infected shoot tips	Chen et al. (2019a, b)
Solanum melongena	Alfalfa mosaic virus (AMV)	100	Increase antioxidant enzymes, and gene expression of some enzymes	Promotion resistance infection	Sofy et al. (2021)

SA Salicylic acid, *PR1* and *PR5* plant resistance genes 1 and 5

mentioned above for its prevention for further breeding processes (Chen et al. 2019a, b). Where contagious stem part was put in different concentrations of melatonin (16 h at 22 °C), and each treatment had three replicates with 10 samples. The 15-M melatonin sample performed best under controlled conditions (95% virus free, enhanced, and lengthy branches). Similarly, melatonin application from outside showed that it has great influence in viral infection reduction to remain 4 weeks in media. In shoot tip culture, this approach resulted in 100% survival, 85% shoot growth, and 95% virus-free plants. Melatonin treatments may give an alternate biotechnological tool in order to lower viral infection in fauna. in case of susceptible rice to RSV (rice strip virus), the pot soil (14 days old) were kept dry to apply different melatonin concentrations (10, 50, 100, 500, and 1000-M) and rice samplings were kept in dark (for 12 h and injected with RSV for 3 days) at greenhouse where 10 and 100 M melatonin and sodium nitroprusside (SNP) were recorded to excellently prevent its spread by OsWRYK45 and OsPR1b action causing PR genes initiation (Lu et al. 2019).

Experimental study on two rice cultivars (Zhendao-88 and Nipponbare) infected with RSV, where melatonin was applied from outside which has greatly reduced infection by enhancing NO synthesis (NO-based pathway) (Lu et al. 2019). TMV (tobacco mosaic virus; melatonin mediated) were studied in two crops (*Solanum lycopersicum* and *Nicotiana glutinosa*) by germinating seeds having greenhouse net (insect proof) for 16 h light by applying different melatonin concentrations (5–6 leaf stage) to samplings roots. The leaves were manually infected with TMV after one or two melatonin treatments. In three separate studies, ten seedlings were utilized for each treatment, where 100-M melatonin reduced viral spread in Nicotiana (37.4%) while enhancing gene levels (PR1 and 5) in *lycopersicum* (declined viral RNAs) (Zhao et al. 2019a, b). Furthermore, H_2O_2 production and RNA reduction of TMV in viral *Nicotiana* enhanced SA and NO buildup, while cPTIO declined in the case of melatonin-induced viral spread (Zhao et al. 2020).

Additionally, melatonin influence was also observed in egg plants seeds [alfalfa mosaic virus (AMV)] by providing 12-h light (27 °C and 23 °C), and samples (14 days old) were foliar applied with 100-M melatonin or SA. To evaluate the response to infection, foliar-sprayed plants were infected with AMV after three days. Each treatment was replicated five times. Foliar spray with 100 M melatonin or SA notably lower virus levels along with notable improvement in certain morphological features including leave surface area, gene expressions such as dehydroascorbate reductase and glutathione reductase, leaves quantity as well as chlorophyll and biomass, antioxidant, root to shoot length, and carotenoid, while AMV severity was observed in noticeable leaves. Melatonin and SA, on the other hand, mitigated AMV's damage from oxidation is mitigated by reducing the amount of other substances (superoxide anion, peroxide of hydrogen, malondialdehyde, and hydroxyl radical). As a result, SA and melatonin are eco-friendly having a great role in viral spread prevention (Sofy et al. 2021).

7.7.3 Melatonin as an Antifungal Agent

In animal studies, the presence of immune-regulatory properties in Candida sepsis has demonstrated the therapeutic benefits of melatonin. It has been found to shorten the recovery time and interleukin-6 levels in rats undergoing typical antimycotic therapy (Yavuz et al. 2007). Additionally, evidence from various studies suggests that during candidiasis, melatonin mitigates oxidative stress and enhances phagocytic action (Cengiz et al. 2012). By controlling the jasmonic acid signaling pathway and hydrogen peroxide (H_2O_2) production by melatonin, plant–fungi interaction model has been shown to enhance tomato fruit resistance against *Botrytis cinerea* (Liu et al. 2019). Moreover, melatonin has been reported to lower fungicide dosage requirements, boost the efficacy of treatments for potato late blight, increase the susceptibility of phytophthora infestans to fungicides, and improve tolerance to environmental stresses (Zhang et al. 2017). In cucurbits and especially watermelon, improved resistance against foliar pathogens (soilborne oomycetes and powdery mildew) has been recorded under elevated melatonin buildup in plants. This resistance led to an alteration in the appearance of genes involved in PAMP and ETI-facilitated defense pathways (Mandal et al. 2018). In *Arabidopsis thaliana*, upregulation of JA-responsive genes (PR4 and PR3) under melatonin application led to a reduction in pathogen sporangia quantity and incidence of *Plasmodiophora brassicae* infections (Chang et al. 2018). Similarly, enhanced expression of related genes and antioxidant enzyme action were observed under 100 μM melatonin treatment in cucumber plants to lower the severity of powdery mildew (Sun et al. 2019).

The disruption of the fungal amino acid metabolism inhibited *Phytophthora nicotianae*, both in vivo and in vitro, when treated with a combination of ethylicin and melatonin, demonstrating a synergistic antifungal effect (Zhang et al. 2018a, b). Enhanced photosynthesis, apple seedling growth, potassium concentration, and alleviated replant disease symptoms have been recorded by exogenous treatment of melatonin to replant soil (Li et al. 2018). Other fungal pathogens (*Alternaria* spp., *Botrytis* spp., *Penicillium* spp., *P. nicotianae*, and *Fusarium* spp.) have also shown similar protective effects (Zhang et al. 2018a, b; Li et al. 2018; Arnao et al. 2015). Additionally, studies suggest that endophytic rhizobacteria can enhance plant capacity for melatonin production (Jiao et al. 2016; Ma et al. 2018). During storage periods of 4–12 h, melatonin has also been found to inhibit microbial growth (including bacteria, molds, and yeasts) in stored apple juice (Zhang et al. 2018a, b). However, contrasting findings showed that melatonin scavenges ROS (linked to defense responses) in disease-ridden fruits, thus making citrus fruit susceptible to green mold (caused by *Penicillium digitatum*) (Lin et al. 2019). Several theories explain melatonin's protective role against plant fungal pathogens, with researchers suggesting its capacity to enhance antioxidant enzyme activity as a key mechanism as well as regulating cellular H_2O_2 levels (Aghdam and Fard 2017; Yin et al. 2013).

Current transcriptomic analyses have revealed that PTI and ETI-related genes in *Arabidopsis thaliana* and watermelon get activated under melatonin exogenous application (Mandal et al. 2018; Weeda et al. 2014). Moreover, besides melatonin's role as a signaling agent in various cellular and physiological responses to various stresses (Arnao et al. 2019), additional defense mechanisms associated with melatonin are illustrated in Table 7.2 and Fig. 7.7.

Table 7.2 Exogenous melatonin treatment affects plant–fungal interaction

Plant name	Pathogen name	Melatonin dosage (μM)	Involved mechanism	Effect	Reference
Litchi chinensis	*Peronophythora litchii*	250	Melatonin enhanced the activities of phenylalanine ammonia-lyase, cinnamate-4-hydroxylase and 4-hydroxycinnamate CoA ligase while promoting the accumulations of phenolics and flavonoids	Improved disease resistance and restricted lesion expansion	Zhang et al. (2021)
Oryza sativa and *Hordeum vulgare*	*Magnaporthe oryzae*	5000, 10,000	Melatonin entered fungal cells efficiently and colocalized with the critical mitogen-activated protein kinase named Mps1, suppressing phosphorylation of Mps1	Reduced disease severity and inhibited pathogen growth	Li et al. (2022a, b)
Lilium cultivar	*Botrytis elliptica*	20, 200, 2000, 20,000	Improved plant resistance by MAPK signaling cascade	Melatonin may play a role in the disease resistance	Xie et al. (2022)
Solanum tuberosum	*Phytophthora infestans*	1000, 3000, 6000, 8000, 10,000	Gene expression	Suppression of the virulence and control disease	Zhang et al. (2017)
Gossypium hirsutum	*Verticillium dahliae*	10, 25, 50, 100	Melatonin affects lignin and gossypol synthesis genes in phenylpropanoid, MVA, and gossypol pathways	Improved disease resistance	Li et al. (2019)
Fusarium oxysporum	*Cucumis sativus*	100	Increased the activity and transcripts of peroxidase and catalase, the content of phenols and flavonoids	Melatonin enhances resistance	Ahammed et al. (2020)

(continued)

Table 7.2 (continued)

Plant name	Pathogen name	Melatonin dosage (μM)	Involved mechanism	Effect	Reference
Malus prunifolia	*Diplocarpon mali*	50–500	Maintain intracellular H_2O_2 concentrations Activities of plant defense-related enzymes	Alleviating disease damage Fungal infection resistance Lesion reduction	Yin et al. (2013)
Musaacuminata	*Fusarium oxysporum*	100	Resistance induced via regulating the expression of MaHSP90s gene	Improvement of disease resistance	Wei et al. (2017)
Fragaria ananassa	*Botrytis cinerea* and *Rhizopus stolonifer*	100	H_2O_2 levels and antioxidant enzyme activities	Reduction of postharvest decay in stored strawberry fruits	Aghdam and Fard (2017)
Citrullus lanatus	*Podosphaera xanthii* and *Phythophthora capsici*	100	Upregulation of PTI- and ETI-associated genes	Disease resistance	Mandal et al. (2018)
A. Thaliana	–	1×10^{-4} or 100	Upregulation of genes involved in abscisic acid (ABA), ET, SA, and JA pathways	Increase plant resilience to biotic and abiotic stress	Weeda et al. (2014)

H_2O_2 hydrogen peroxide, *SA* salicylic acid, *ET* ethylene, *PTI* pattern-triggered immunity, *ETI* effector-triggered immunity, *JA* jasmonic acid, *ABA* abscisic acid

7.7.4 Melatonin as an Antibacterial Agent

Both in vivo and in vitro studies demonstrate melatonin's antibacterial properties, which have been evaluated in animals. Under in vitro environments, it has demonstrated bactericidal activity against multidrug-resistant Gram-positive and Gram-negative bacteria (including methicillin-resistant *Staphylococcus aureus*, carbapenem-resistant *Pseudomonas aeruginosa*, and *Acinetobacter baumannii*). Furthermore, isoniazid and melatonin combined application have greatly (three- to fourfold) suppressed bacteria, while a concentration of 10 μM melatonin was also found to strongly prevent *Mycobacterium tuberculosis*; H37Rv strain (Wiid et al. 1999). Proposed mechanisms for melatonin's antibacterial activity in animals include the destabilization of bacterial cell walls through cytoderm binding, the regulation of intracellular Ca^{2+} and cAMP, the production of free radicals, and the depletion of intracellular substrates (Nabavi et al. 2019; Huang et al. 2019; Boga et al. 2019). Melatonin has a significant antibacterial role during plant–bacteria interactions, such as a 17% reduction of rice bacterial leaf streak (BLS caused by *Xanthomonas oryzae* pv. Oryzicola) under exogenous melatonin application (Chen et al. 2019a, b).

Similarly, against *X. oryzae* pv. oryzicola and *X. oryzae* pv. oryzae, the direct inhibitory influence of melatonin has also been reported (Chen et al. 2018, 2019a, b). Additionally, melatonin supplementation in the insect vector of Candidatus *Liberibacter asiaticus* (CLas, the causal agent of citrus greening disease) was found to counteract its adverse effects. This treatment reduced the CLas bacterial populace in psyllids and prolonged the lifespan of infected and healthy vectors (Nehela and Killiny 2018). In cassava crops, melatonin biosynthesis genes (MeRAV2 and MeRAV1) are critical for driving resistance against bacterial blight (Wei et al. 2018). Through signal transduction pathways, the antibacterial defense mechanisms involve elevated NO levels that enhance the expression of SA-associated genes (such as plant defensin 1.2; PDF1.2, EDS1, plant resistance 1; PR1, PR2, PR5, and PAD4) (Shi et al. 2015).

In Arabidopsis infected with *Pseudomonas syringae* pv. tomato (Pst) DC3000, melatonin was found to arouse MAPK cascades, which upregulate the SA biosynthesis gene ICS1 (isochorismate synthase 1) (Lee et al. 2014; Lee and Back 2016; Lee and Back 2017). Zhao et al. (2015) reported that melatonin-treated plants showed increased cell wall invertase (CWI) activity, enhancing cell wall strengthening through the accumulation of galactose, cellulose, and xylose accumulation and callose removal. Additionally, against Pst DC3000 via NO and SA-dependent pathways, glycerol and sugar were implicated in melatonin-related defenses (Qian et al. 2015). Melatonin treatment also induces ACS6 (a key enzyme in ethylene biosynthesis), which further promotes PDF1.2 expressions, indicating the potential involvement of both JA and ethylene in pathogen resistance (Lee et al. 2014; Zhao et al. 2015; Arnao and Hernández-Ruiz 2018b). Table 7.3 and Fig. 7.7 represent a comprehensive summary of melatonin's bactericidal activities, underlying mechanisms, and positive impacts on plants.

7.7.5 *Melatonin as a Pesticide Agent*

Melatonin has demonstrated significant protective capabilities against oxidative stress induced by heavy metals, drugs, and toxins (such as sulfur mustard, a chemical warfare compound) in animals (Reiter et al. 2008; Ucar et al. 2007; Korkmaz et al. 2008). For example, melatonin coadministration reduced lipid peroxidation markers like malondialdehyde and 4-hydroxyalkenals in paraquat-treated rats and reduced acute mortality in diquat-exposed mice (Xu et al. 2007). In honeybees, improved survival rates were noted under overexpression of T5H, a key enzyme in melatonin biosynthesis led to enhanced tolerance to abiotic stressors, including pesticides, UV and exposure (Fan et al. 2020).

Table 7.3 Antibacterial properties of melatonin in plants

Plant name	Pathogen name	Melatonin dosage (μM)	Involved mechanism	Effect	Reference
Arabidopsis thaliana, Nicotiana benthamiana	*Pseudomonas syringae*	1 or 10	Expression of defense genes PR1, PR5, and PDF1.2	Inhibition of pathogen propagation	Lee et al. (2014)
A. thaliana	*P. syringae*	10	Pathogenesis-related (PR) genes by the harmonizing signaling between SA and ET	Increase resistance against pathogen	Lee et al. (2015)
A. thaliana, N. benthamiana	*P. syringae*	1	Induction of PR genes through MAPK signaling cascades	Disease resistance	Lee and Back (2016)
A. thaliana	*P. syringae*	20	Involvement of sugars and glycerol in melatonin-mediated innate immunity in SA- and NO-dependent Pathways	Disease resistance	Qian et al. (2015)
A. thaliana	*P. syringae*	20	NO and melatonin levels in leaves Defense-related genes	Improvement of disease resistance	Shi et al. (2015)
A. thaliana	*P. syringae*	50	Activities of CWI and vacuolar invertase (VI)	Cell-wall reinforcement and callose deposition during infection	Zhao et al. (2015)
A. thaliana	*P. syringae*	1	Induction of PR1 and ICS1 expression genes through MAPK cascades in the coexistence H_2O_2 and NO	Disease resistance	Lee and Back (2017)
Oryza sativa and *N. benthamiana*	*Xanthomonas oryzae pv. oryzae*	200	The biofilm formation, swimming motility and extracellular protease activities of Xoc also decreased under melatonin treatment	Reduces pathogenicity and incidence disease	Chen et al. (2019a, b)

(continued)

Table 7.3 (continued)

Plant name	Pathogen name	Melatonin dosage (μM)	Involved mechanism	Effect	Reference
Oryza sativa	*Xanthomonas oryzae pv. oryzae*	20–100 μg/ml	Increased NO and activation of defense-related genes	Improved plant resistance	Chen et al. (2020)
Manihot esculenta Crantz	*Xanthomonas axonopodis pv. Manihotis*	50, 100, 150	Induced gene expression	Increased disease resistance by ethylene	Wei et al. (2022)
Cucumis sativus	*Pseudomonas syringae pv. Lachrymans*	10,100, 1000, 10,000	Induced expressed genes were involved in some distinct signal transduction pathways, such as calmodulin (CaM), polyamines (PAs), nitric oxide (NO), and salicylic acid (SA)	Reduces pathogenicity and incidence disease	Li et al. (2022a, b)

SA salicylic acid, *PR1* and *PR5* plant resistance genes 1 and 5, *ET* ethylene, *MAPK* mitogen-activated protein kinase, *NO* nitric oxide, *PDF1.2* plant defensin 1.2 gene, *ICS1* isochorismate synthase 1 gene, H_2O_2 hydrogen peroxide, *NO* nitric oxide

Melatonin acts as a potential safener against pesticide-induced stress in plants and has been highlighted through various studies, as explained in Table 7.4. Among the evaluated herbicides, butafenacil (uracil-based herbicide) was the first to be tested. This compound was developed by Syngenta Co. in Basel, Switzerland in 2003, chemically identified as 1-(allyloxycarbonyl)- 1-methylethyl-2 -chloro-5-[1,2,3,6- tetrahydro- 3- methyl- 2,6-dioxo-4-(trifluoromethyl) pyrimidin-1-yl] benzoate. Butafenacil is designed to manage broad-leaved weeds (both annual and perennial) by preventing the enzyme protoporphyrinogen IX oxidase (a key component in chlorophyll biosynthesis). This inhibition leads to cellular death by membrane lipid peroxidation, accumulating chlorophyll precursors, resulting in the production of harmful singlet oxygen molecules (Jung et al. 2004). Studies have shown that melatonin-rich transgenic rice plants (which overproduce endogenous melatonin), as compared to wild-type rice plants, exhibit resistance to butafenacil. These plants maintain higher chlorophyll levels, catalase and superoxide dismutase (SOD) activities, reduced hydrogen peroxide content and malondialdehyde (MDA). This groundbreaking research paved the way for considering melatonin as a potential safener against herbicide-induced stress (Park et al. 2013).

Severe cellular damage due to enhanced ROS, as a result of disruption in photosynthesis at the photosystem I level, occurs under Paraquat (also known as dimethyl viologen; 1,1′-dimethyl-4,4′-bipyridinium dichloride; broad-spectrum herbicide) application (Moustaka and Moustakas 2014). Similarly, melatonin-primed pea

Table 7.4 Verification for safener melatonin activity in plants using different pesticides

Pesticide	Common name	Plant	Effects	Reference
Herbicide	Butafenacil	Rice	High tolerance to herbicide	Park et al. (2013)
Herbicide	Paraquat	Pea	High photosynthesis	Szafranska et al. (2016)
		Poplar	High tolerance to stress and low damage	Ding et al. (2018)
Herbicide	Bentazone	Batata	High growth and tolerance to herbicide Low damage	Caputo et al. (2020)
Fungicide	Carbendazim	Tomato	Low damage, high stress tolerance, and fungicide metabolizing	Yan et al. (2019)
Fungicide	Ethylicin	Tobacco	Synergistic action, suppression of virulence, low fungicide doses, and eco-friendly alternative	Zhang et al. (2018a, b)
Fungicide	Infinito	Potato	Synergistic action, low dosage, and high efficacy	Jung et al. (2004)

seeds exhibited improved oxidative stress and photosynthetic stability under paraquat stress (Szafranska et al. 2016, 2017; Yan et al. 2019; Caputo et al. 2020). Furthermore, paraquat-induced oxidative stress was significantly lowered in poplar leaves pretreated with melatonin. Melatonin application boosts antioxidant enzyme activities (ascorbate peroxidase, catalase, superoxide dismutase, SOD, and peroxidase) while declining lipid peroxidation and membrane damage in poplar leaf discs. Additionally, in paraquat-exposed leaves, it elevated antioxidant metabolites (such as proline, ascorbate, ASC, and glutathione; GSH). These findings highlight the protective role of melatonin against paraquat-induced damage in plants (Ding et al. 2018).

Bentazone (3-isopropyl-1H-2,1,3-benzothiadiazin-4(3H)-one 2,2-dioxide), a postemergence diazinone herbicide that inhibits photosystem II at the D1 protein level, is a potential herbicide for controlling yellow nutsedge (*Cyperus esculentus*) in sweet potato (*Ipomoea batatas*) crops. While effective against annual weeds across various crops, bentazone caused significant injuries (even at low doses @ 0.1 mM) to sweet potato seedlings with losses ranging from 41% to 75% depending on varietal tolerance (Motsenbocker and Monaco 1991). However, compared to individual bentazone treatments, the application of melatonin as a safener doubled biomass yield and reduced damage by 30%, showing its potential to control weeds (Caputo et al. 2020). The "safener," originally referred to compounds protecting plants from herbicide-induced damage, was first introduced in 1971. This term can be extended to include other pesticides such as fungicides, insecticides, and acaricides. For instance, melatonin has been explored as a protective agent against fungicides like carbendazim (methyl benzimidazole-2-ylcarbamate), which inhibits fungal mitosis but also causes phytotoxicity and oxidative stress in tomato plants. Reduced lipid peroxidation and ROS production and modulation of the ASC-GSH cycle have been noted under co-treatment with exogenous melatonin. This resulted in increased levels of antioxidative enzymes, glutathione, and ascorbate, leading to enhanced detoxification capacity and enabling plants to metabolize the fungicide with minimal adverse effects. Researchers suggest that melatonin enables plants to better withstand pesticide-induced stress by enhancing their adaptive responses (Yan et al. 2019).

Oomycete-related diseases could be controlled by ethylicin (S-ethylethanethiosulfonate; biofungicide with broad-spectrum activity). Pathogenic oomycetes (such as phytophthora species) are among the most destructive agricultural threats. Co-treatments of ethylicin and melatonin effectively control the virulence of *P. nicotianae* and hyphal growth and reduce cell viability in tobacco black shank (caused by *P. nicotianae*). Both compounds targeted similar metabolic pathways, dysregulating virulence-related genes, overexpressing apoptosis-inducing factors, and disrupting amino acid metabolism. The findings suggest that combining ethylicin and melatonin provides an eco-friendly and effective substitute for managing oomycete diseases, including tobacco black shank (Zhang et al. 2018a, b). A synergistic antifungal effect of melatonin was also noted in controlling potato late blight (caused by *Phytophthora infestans*) when combined with fungicides (Infinito®), suggesting its potential to reduce chemical dosages (Zhang et al. 2017). Figure 7.1 incorporates the established role of melatonin in mitigating abiotic and biotic stress, highlighting its potential as a natural safener. For synthetic safeners, their function is suggested to extend beyond their traditional role in subsequent vacuolar storage, glutathione transferases (GST), and inactivation and detoxification within the ASC-GSH cycle. They are also proposed to act as modulators of the redox network, further enhancing plant stress tolerance mechanisms (Dixon et al. 2002; Acosta et al. 2022).

7.8 Conclusion

Pathogen infections are a major limiting factor in global crop production, leading to significant yield losses. Melatonin (N-acetyl-5-methoxytryptamine) has attracted considerable interest in plant science research due to its diverse roles in plant systems. Two primary mechanisms by which melatonin helps plants cope with pathogen infections are the efficient scavenging of reactive oxygen species and the activation of antioxidant defense responses. Melatonin plays a crucial role in enhancing the expression of stress-specific genes, antioxidant enzyme genes, and pathogenesis-related genes, making it an important regulator of plant defense responses during pathogen infections. Its complementary effect when used with fungicides offers new hope for reducing the reliance on harmful chemicals in vegetable and fruit crops. This chapter highlights the functions of melatonin in mitigating biotic stresses—such as those caused by fungi, viruses, bacteria, and insects—in crops, while also explaining the associated physiological and molecular mechanisms. The versatile role of melatonin in regulating stress demonstrates successful instances of growth promotion and increased resistance, positioning it as a suitable option for environmentally friendly crop production and food safety.

References

Aghdam MS, Fard JR (2017) Melatonin treatment attenuates postharvest decay and maintains nutritional quality of strawberry fruits (*Fragaria anannasa* cv. Selva) by enhancing GABA shunt activity. Food Chem 221:1650–1657

Aghdam MS, Luo Z, Jannatizadeh A, Sheikh-Assadi M, Sharafi Y, Farmani B, Fard JR, Razavi F (2019) Employing exogenous melatonin applying confers chilling tolerance in tomato fruits by upregulating ZAT2/6/12 giving rise to promoting endogenous polyamines, proline, and nitric oxide accumulation by triggering arginine pathway activity. Food Chem 275:549–556

Agrimi G, Di Noia MA, Marobbio CMT, Fiermonte G, Lasorsa FM, Palmieri F (2004) Identification of the human mitochondrial S-adenosylmethionine transporter: bacterial expression, reconstitution, functional characterization and tissue distribution. Biochem J 379:183–190

Ahammed GJ, Mao Q, Yan Y, Wu M, Wang Y, Ren J, Guo P, Liu A, Chen S (2020) Role of melatonin in arbuscular mycorrhizal fungi-induced resistance to fusarium wilt in cucumber. Phytopathology 110:999–1009

Altaf MA, Shahid R, Ren MX, Mora-Poblete F, Arnao MB, Naz S, Anwar M, Altaf MM, Shahid S, Shakoor A (2021) Phytomelatonin: an overview of the importance and mediating functions of melatonin against environmental stresses. Physiol Plant 172:820–846

Arnao MB (2014) Phytomelatonin: discovery, content, and role in plants. Adv Bot 2014:e815769

Arnao MB, Hernández-Ruiz J (2006) The physiological function of melatonin in plants. Plant Signal Behav 1:89–95

Arnao MB, Hernández-Ruiz J (2014) Melatonin: plant growth regulator and/or biostimulator during stress? Trends Plant Sci 19:789–797

Arnao MB, Hernández-Ruiz J (2015) Functions of melatonin in plants: a review. J Pineal Res 59(2):133–50

Arnao MB, Hernández-Ruiz J (2018a) Melatonin and its relationship to plant hormones. Ann Bot 121:195–207

Arnao MB, Hernández-Ruiz J (2018b) The multi-regulatory properties of melatonin in plants. In: Ramakrishna A, Roshchina VV (eds) Neurotransmitters in plants. CRC Press, Boca Raton, p 448. ISBN 978-0-203-71148-4

Arnao MB, Hernández-Ruiz J (2019a) Melatonin as a chemical substance or as Phytomelatonin rich-extracts for use as plant protector and/or biostimulant in accordance with EC legislation. Agron 9:570

Arnao MB, Hernández-Ruiz J (2019b) Melatonin: a new plant hormone and/or a plant master regulator? Trends Plant Sci 24:38–48

Arnao MB, Hernández-Ruiz J (2020a) Is Phytomelatonin a new plant hormone? Agron 10:95

Arnao MB, Hernández-Ruiz J (2020b) Melatonin in Flowering, Fruit Set and Fruit Ripening. Plant Reprod 33:77–87

Arnao MB, Hernandez-Ruiz J (2021) Melatonin as a plant biostimulant in crops and during post-harvest: a new approach is needed. J Sci Food Agri 101:5297–5304

Arnao MB, Hernández-Ruiz J (2021) Melatonin as a regulatory hub of plant hormone levels and action in stress situations. Plant Biol 23:7–19

Arnao MB, Hernández-Ruiz J, Cano A, Reiter RJ (2021) Melatonin and carbohydrate metabolism in plant cells. Plan Theory 10:1917

Arnao MB, Cano A, Hernández-Ruiz J (2022) Phytomelatonin: an unexpected molecule with amazing performances in plants. J Exp Bot 73:5779–5800

Asghari MH, Abdollahi M, de Oliveira MR, Nabavi SM (2017) A review of the protective role of melatonin during phosphine-induced cardiotoxicity: focus on mitochondrial dysfunction, oxidative stress and apoptosis. J Pharm Pharmacol 69:236–243

Aslan G, Gül HF, Tektemur A, Sahna E (2020) Ischemic postconditioning reduced myocardial ischemia-reperfusion injury: the roles of melatonin and uncoupling protein 3. Anatol J Cardiol 23:19–27

Asthir B, Koundal A, Bains NS, Mann SK (2010) Stimulation of antioxidative enzymes and polyamines during stripe rust disease of wheat. Biol Plant 54:329–333

Back K (2021) Melatonin metabolism, signaling and possible roles in plants. Plant J 105:376–391

Back K, Tan DX, Reiter RJ (2016) Melatonin biosynthesis in plants: multiple pathways catalyze tryptophan to melatonin in the cytoplasm or chloroplasts. J Pineal Res 61:426–437

Banerjee A, Chattopadhyay A, Pal PK, Bandyopadhyay D (2020) Melatonin is a potential therapeutic molecule for oxidative stress induced red blood cell (RBC) injury: a review. Melatonin Res 3:1–31

Bari R, Jones JDG (2009) Role of plant hormones in plant Defence responses. Plant Mol Biol 69:473–488

Berger HR, Nyman AKG, Morken TS, Widerøe M (2019) Transient effect of melatonin treatment after neonatal hypoxic-ischemic brain injury in rats. PLoS One 14

Boga JA, Caballero B, Potes Y, Perez-Martinez Z, Reiter RJ, Vega-Naredo I, Coto-Montes A (2019) Therapeutic potential of melatonin related to its role as an autophagy regulator: a review. J Pineal Res 66:12534

Brainard GC, Hanifin JP, Greeson JM, Byrne B, Glickman G, Gerner E, Rollag MD (2001) Action spectrum for melatonin regulation in humans: evidence for a novel circadian photoreceptor. J Neurosci 21:6405–6412

Buttar ZA, Wu SN, Arnao MB, Wang C, Ullah I, Wang C (2020) Melatonin suppressed the heat stress-induced damage in wheat seedlings by modulating the antioxidant machinery. Plants 9:809

Byeon Y, Back K (2014a) An increase in melatonin in transgenic rice causes pleiotropic phenotypes, including enhanced seedling growth, delayed flowering, and low grain yield. J Pineal Res 56:408–414

Byeon Y, Back K (2014b) Melatonin synthesis in rice seedlings in vivo is enhanced at high temperatures and under dark conditions due to increased serotonin N acetyltransferase and N-acetylserotonin methyltransferase activities. J Pineal Res 56:189–195

Byeon Y, Choi GH, Lee HY, Back K (2015) Melatonin biosynthesis requires N-acetylserotonin methyltransferase activity of caffeic acid Omethyltransferase in rice. J Exp Bot 66:6917–6925

Caputo GA, Wadl PA, McCarty L, Adelberg J, Jennings KM, Cutulle M (2020) In vitro Safening of Bentazon by melatonin in Sweetpotato (*Ipomoea batatas*). HortScience 55:1406–1410

Cardinali DP (2019) Melatonin: clinical perspectives in neurodegeneration. Front Endocrinol 2019:10

Casado-Zapico S, Rodriguez-Blanco J, García-Santos G, Martín V, Sánchez-Sánchez AM, Antolín I, Rodriguez C (2010) Synergistic antitumor effect of melatonin with several chemotherapeutic drugs on human Ewing sarcoma cancer cells: potentiation of the extrinsic apoptotic pathway. J Pineal Res 48:72–80

Cengiz MI, Cengiz S, Wang HL (2012) Melatonin and oral cavity. Inter J Dent 2012(1):491872

Chang F, Cao X, Peng J, Tan X, Chen T, Zeng J, Cao Y, Qiao D (2018) Induced resistance to *Plasmodiophora brassicae* in Arabidopsis by melatonin. Chin J Appl Environ Bio 24:75–80

Chang T, Zhao Y, He H, Xi Q, Fu J, Zhao Y (2021) Exogenous melatonin improves growth in hulless barley seedlings under cold stress by influencing the expression rhythms of circadian clock genes. Peer J 9:e10740

Chen X, Sun C, Laborda P, Zhao Y, Palmer I, Fu ZQ, Qiu J, Liu F (2018) Melatonin treatment inhibits the growth of *Xanthomonas oryzae* pv. Oryzae. Front Microbiol 9:2280

Chen X, Sun C, Laborda P, He Y, Zhao Y, Li C, Liu F (2019a) Melatonin treatments reduce the pathogenicity and inhibit the growth of *Xanthomonas oryzae* pv. oryzicola. Plant Patho 68:288–296

Chen L, Wang M, Li J, Feng C, Cui Z, Zhao L, Wang Q (2019b) Exogenous application of melatonin improves eradication of apple stem grooving virus from the infected in vitro shoots by shoot tip culture. Plant Pathol 68:997–1006

Chen WR, Zhou YJ, Yang JQ, Liu F, Wu XP, Sha Y (2020) Melatonin attenuates calcium deposition from vascular smooth muscle cells by activating mitochondrial fusion and mitophagy via an AMPK/OPA1 signaling pathway. Oxidative Med Cell Longev 2020:5298483

Cheng G, Ma T, Deng Z, Gutiérrez-Gamboa G, Ge Q, Xu P, Zhang Q, Zhang J, Meng J, Reiter RJ (2021) Plant-derived melatonin from food: a gift of nature. Food Funct 12:2829–2849

Cho HM, Sun W (2019) The coordinated regulation of mitochondrial structure and function by Drp1 for mitochondrial quality surveillance. BMB Rep 52

Claustrat B, Leston J (2015) Melatonin: physiological effects in humans. Neurochirurgie 61:77–84

Danilova ED, Efimova MV, Kolomeichuk LV, Kuznetsov VV (2020) Melatonin supports photochemical activity of assimilation apparatus and delays senescence of leaves of monocotyledonous plants. Dokl Biochem Biophys 495:271–275

Danilova ED, Zlobin IE, Kuznetsov VV, Efimova MV (2021) Exogenic melatonin reduces the toxic effect of polymetallic stress on barley plants. Dokl Biochem Biophys 499:228–232

Das K, Roychoudhury A (2014) Reactive oxygen species (ROS) and response of antioxidants as ROS-scavengers during environmental stress in plants. Front Environ Sci 2(53):1–13

Dayoub JC, Ortiz F, LópezL C, Venegas C, del Pino-Zumaquero A, Roda O, Sánchez-Montesinos I, Acuña-Castroviejo D, Escames G (2011) Synergism between melatonin and atorvastatin against endothelial cell damage induced by lipopolysaccharide. J Pineal Res 51:324–330

Denancé N, Sánchez-Vallet A, Goffner D, Molina A (2013) Disease resistance or growth: the role of plant hormones in balancing immune responses and fitness costs. Front Plant Sci 24:155

Ding F, Wang G, Zhang S (2018) Exogenous melatonin mitigates methyl Viologen-triggered oxidative stress in poplar leaf. Molecules 23:2852

Dixon DP, Lapthorn A, Edwards R (2002) Plant glutathione transferases. Genome Biol 3:1–10

Dodds PN, Rathjen JP (2010) Plant immunity: towards an integrated view of plant pathogen interactions. Nat Rev Genet 11:539–548

Dong SA, Gong LR, Yu JB, Kan YX (2020) The role of melatonin in electroacupuncture alleviating lung injury induced by limb ischemia-reperfusion in rabbits. Med Sci Monit 2020:26

Dubbels R, Reiter RJ, Klenke E, Goebel A, Schnakenberg E, Ehlers C, Schiwara HW, Schloot W (1995) Melatonin in edible plants identified by radioimmunoassay and by high performance liquid chromatography-mass spectrometry. J Pineal Res 18:28–31

Erland LA, Chattopadhyay A, Jones AMP, Saxena PK (2016) Melatonin in plants and plant culture systems: variability, stability and efficient quantification. Front Plant Sci 7:1721

Fan W, Li G, Zhang X, Wang Y, Wang C, Xu B, Guo X, Li H (2020) The role of melatonin and Tryptophan-5-Hydroxylase-1 in different abiotic stressors in Apis cerana cerana. J Insect Physiol 128:104180

Fayaz SM, Raj YV, Krishnamurthy RG (2015) CypD: the key to the death door. CNS Neurol Disord Drug Targets 14:654–663

Fernandez Vazquez G, Reiter RJ, Agil A (2018) Melatonin increases brown adipose tissue mass and function in Zücker diabetic fatty rats: implications for obesity control. J Pineal Res 64:e12472

Figueroa-Macías JP, García YC, Núñez M, Díaz K, Olea AF, Espinoza L (2021) Plant growth-defense trade-offs: molecular processes leading to physiological changes. Inter J Mol Sci 22(2):693

Gill MB, Zeng F, Shabala L, Zhang G, Yu M, Demidchik V, Shabala S, Zhou M (2019) Identification of QTL related to ROS formation under hypoxia and their association with waterlogging and salt tolerance in barley. Int J Mol Sci 20:1–15

Giraldo Acosta M, Cano A, Hernández-Ruiz J, Arnao MB (2022) Melatonin as a possible natural safener in crops. Plants 11(7):890

Glazebrook J (2005) Contrasting mechanisms of defense against biotrophic and necrotrophic pathogens. Annu Rev Phytopathol 43:205–227

Hadidi A, Barba M, Rubino L (2023) The fine legacy of giovanni P. martelli (1935–2020), a pre-eminent plant virologist and the founder of modern grapevine virology. Viruses 15(1):210

Hardeland R (2009) Neuroprotection by radical avoidance: search for suitable agents. Molecules 14:5054–5102

Hardeland R (2018) Recent findings in melatonin research and their relevance to the CNS. Cent Nerv Syst Agents Med Chem 2018:18

Hardeland R (2019) Melatonin in the evolution of plants and other phototrophs. Melatonin Res 2:10–36

Hattori A, Migitaka H, Iigo M, Itoh M, Yamamoto K, Ohtani-Kaneko R, Hara M, Suzuki T, Reiter RJ (1995) Identification of melatonin in plants and its effects on plasma melatonin levels and binding to melatonin receptors in vertebrates. Biochem Mol Biol Int 35:627–634

He C, Wang J, Zhang Z, Yang M, Li Y, Tian X, Ma T, Tao J, Zhu K, Song Y (2016) Mitochondria synthesize melatonin to ameliorate its function and improve mice oocyte's quality under in vitro conditions. Int J Mol Sci 17:939

Hernández-Ruiz J, Arnao MB (2018) Relationship of melatonin and salicylic acid in biotic/abiotic plant stress responses. Agronomy 8:33

Hernandez-Ruiz J, Giraldo-Acosta M, El-Mihyaoui A, Cano A, Arnao MB (2023) Melatonin as a possible natural anti-viral compound in plant biocontrol. Plan Theory 12:781

Himanshu SS, Rana VS et al (2024) Unlocking the sustainable role of melatonin in fruit production and stress tolerance: a review. CABI Agric Biosci 5:103

Huang SH, Liao CL, Chen SJ, Shi LG, Lin L, Chen YW, Cheng CP, Sytwu HK, Shang ST, Lin GJ (2019) Melatonin possesses an anti-influenza potential through its immune modulatory effect. J Funct Foods 58:189–198

Jan JE, Reiter RJ, Wasdell MB, Bax M (2009) The role of the thalamus in sleep, pineal melatonin production, and circadian rhythm sleep disorders. J Pineal Res 46:1–7

Jauhari A, Baranov SV, Suofu Y, Kim J, Singh T, Yablonska S, Li F, Wang X, Oberly P, Minnigh MB (2020) Melatonin inhibits cytosolic mitochondrial DNA-induced neuroinflammatory signaling in accelerated aging and neurodegeneration. J Clin Invest 130:3124–3136

Jiao J, Ma Y, Chen S, Liu C, Song Y, Qin Y, Yuan C, Liu Y (2016) Melatonin-producing endophytic bacteria from grapevine roots promote the abiotic stress-induced production of endogenous melatonin in their hosts. Front Plant Sci 7:542

Jung S, Lee Y, Yang K, Lee SB, Jang SM, Ha SB, Back K (2004) Dual targeting of Myxococcus xanthus Protoporphyrinogen oxidase into chloroplasts and mitochondria and high level Oxyfluorfen resistance. Plant Cell Environ 27:1436–1446

Kang K, Lee K, Park S, Kim YS, Back K (2010) Enhanced production of melatonin by ectopic overexpression of human serotonin Nacetyltransferase plays a role in cold resistance in transgenic rice seedlings. J Pineal Res 49:176–182

Kang K, Kong K, Park S, Natsagdorj U, Kim YS, Back K (2011) Molecular cloning of a plant N-acetylserotonin methyltransferase and its expression characteristics in rice. J Pineal Res 50:304–309

Kaur S, Samota MK, Choudhary M, Choudhary M, Pandey AK, Sharma A, Thakur J (2022a) How do plants defend themselves against pathogens-biochemical mechanisms and genetic interventions. Physiol Mol Biol Plants28(2):485–504

Kaur S, Samota MK, Choudhary M, Choudhary M, Pandey AK, Sharma A, Thakur J (2022b) How do plants defend themselves against pathogens-biochemical mechanisms and genetic interventions. Physiol Mol Biol Plants 28(2):485–504

Khan M, Hussain A, Yun BW, Mun BG (2024) Melatonin: the multifaceted molecule in plant growth and defense. Int J Mol Sci 25:6799

Kolar J, Machackova I, Illnerova H, Prinsen E, van Dongen W, van Onckelen H (1995) Melatonin in higher plant determined by radioimmunoassay and liquid chromatography-mass spectrometry. Biol Rhythm Res 26:406–409

Korkmaz A, Kunak ZI, Paredes SD, Yaren H, Tan DX, Reiter RJ (2008) The use of melatonin to combat mustard toxicity. Neuroendocrinol Lett 29:614–619

Kostopoulou Z, Therios I, Roumeliotis E, Kanellis AK, Molassiotis A (2015) Melatonin combined with ascorbic acid provides salt adaptation in Citrus aurantium L. seedlings. Plant Physiol Biochem 86:155–165

Lee HY, Back K (2016) Mitogen-activated protein kinase pathways are required for melatonin-mediated defense responses in plants. J Pineal Res 60:327–335

Lee HY, Back K (2017) Melatonin is required for H_2O_2-and NO-mediated defense signaling through MAPKKK3 and OXI1 in *Arabidopsis thaliana*. J Pineal Res 62:e12379

Lee HY, Byeon Y, Back K (2014) Melatonin as a signal molecule triggering defense responses against pathogen attack in Arabidopsis and tobacco. J Pineal Res 57:262–268

Lee HY, Byeon Y, Tan DX, Reiter RJ, Back K (2015) Arabidopsis serotonin N-acetyltransferase knockout mutant plants exhibit decreased melatonin and salicylic acid levels resulting in susceptibility to an avirulent pathogen. J Pineal Res 58:291–299

Lei X, Zhu R, Zhang G, Dai Y (2004) Attenuation of cold-induced apoptosis by exogenous melatonin in carrot suspension cells: the possible involvement of polyamines. J Pineal Res 36:126–131

Lerner AB, Case JD, Takahashi Y, Lee TH, Mori W (1958) Isolation of melatonin, the pineal gland factor that lightens melanocyteS1. J Am Chem Soc 80:2587–2587

Lerner AB, Case JD, Heinzelman RV (1959) Structure of melatonin. J Am Chem Soc 81:6084–6085

Li C, Zhao Q, Gao T, Wang H, Zhang Z, Liang B, Wei Z, Liu C, Ma F (2018) The mitigation effects of exogenous melatonin on replant disease in apple. J Pineal Res 65:12523

Li R, Bi R, Cai H, Zhao J, Sun P, Xu W, Zhou Y, Yang W, Zheng L, Chen XL et al (2022a) Melatonin functions as a broad-Spectrum antifungal by targeting a conserved pathogen protein kinase. J Pineal Res 74:e12839

Li L, Du C, Wang L, Lai M, Fan H (2022b) Exogenous melatonin improves the resistance to cucumber bacterial angular leaf spot caused by *pseudomonas syringae* pv. Lachrymans Physiol Plant 174:e13724

Lin Y, Fan L, Xia X, Wang Z, Yin Y, Cheng Y, Li Z (2019) Melatonin decreases resistance to postharvest green mold on citrus fruit by scavenging defense-related reactive oxygen species. Postharvest Boil Technol 153:21–30

Liu C, Chen L, Zhao R, Li R, Zhang S, Yu W, Sheng J, Shen L (2019) Melatonin induces disease resistance to Botrytis cinerea in tomato fruit by activating Jasmonic acid signaling pathway. J Agric Food Chem 67:6116–6124

López A, Ortiz F, Doerrie C, Venegas C, Fernández-Ortiz M, Aranda P, Díaz-Casado ME, Fernández-Gil B, Barriocanal-Casado E, Escames G (2017) Mitochondrial impairment and melatonin protection in parkinsonian mice do not depend of inducible or neuronal nitric oxide synthases. PLoS One 12:e0183090

Lu R, Liu Z, Shao Y, Sun F, Zhang Y, Cui J, Zhou Y, Shen W, Zhou T (2019) Melatonin is responsible for Rice resistance to Rice stripe virus infection through a nitric oxide-dependent pathway. Virol J 16:141

Ma X, Zhang J, Burgess P, Rossi S, Huang B (2018) Interactive effects of melatonin and Cytokinin on alleviating drought-induced leaf senescence in creeping Bentgrass (*Agrostis stolonifera*). Environ Exp Bot 145:1–11

Ma K, Chen G, Li W, Kepp O, Zhu Y, Chen Q (2020) Mitophagy, mitochondrial homeostasis, and cell fate. Front Cell Dev Biol 8:467

Manchester LC, Poeggeler B, Alvares FL Ogden GB, Reiter RJ (1995) Melatonin immunoreactivity in the photosynthetic prokaryote *Rhodospirillum rubrum*: implications for an ancient antioxidant system. Cell Mol Biol Res 41: 391–395

Mandal MK, Suren H, Ward B, Boroujerdi A, Kousik C (2018) Differential roles of melatonin in plant-host resistance and pathogen suppression in cucurbits. J Pineal Res 65:12505

Martín M, Macías M, Escames G, Reiter RJ, Agapito MT, Ortiz GG, Acuña-Castroviejo D (2000) Melatonin-induced increased activity of the respiratory chain complexes I and IV can prevent mitochondrial damage induced by ruthenium red in vivo. J Pineal Res 28:242–248. (PubMed)

Menhas S, Yang X, Hayat K, Aftab T, Bundschuh J, Arnao MB, Zhou Y, Zhou P (2022) Exogenous melatonin enhances cd tolerance and phytoremediation efficiency by ameliorating cd-induced stress in oilseed crops: a review. J Plant Growth Regul 41:922–935

Miller GAD et al (2010) Reactive oxygen species homeostasis and signaling during drought and salinity stresses. Plant Cell Environ 33:453–467

Mittler R (2017) ROS are good. Trends Plant Sci 22:11–19

Motsenbocker CE, Monaco TJ (1991) Sweet potatoes (*Ipomoea batatas*) differ in response to Bentazon. Weed Technol 5:345–350

Moustafa-Farag M, Almoneafy A, Mahmoud A, Elkelish A, Arnao MB, Li L, Ai S (2019) Melatonin and its protective role against biotic stress impacts on plants. Biomol Ther 10:54

Moustaka J, Moustakas M (2014) Photoprotective mechanism of the non-target organism Arabidopsis thaliana to Paraquat exposure. Pestic Biochem Physiol 111: 1–6

Munné-Bosch S, Müller M (2013) Hormonal cross-talk in plant development and stress responses. Front Plant Sci 4:529

Nabavi SM, Nabavi SF, Sureda A, Xiao J, Dehpour AR, Shirooie S, Silva AS, Baldi A, Khan H, Daglia M (2019) Anti-inflammatory effects of melatonin: a mechanistic review. Crit Rev Food Sci Nutr 59:4–16

Nawrocka J, Małolepsza U (2013) Diversity in plant systemic resistance induced by Trichoderma. Biol Control 67:149–156

Nehela Y, Killiny N (2018) Infection with phytopathogenic bacterium inhibits melatonin biosynthesis, decreases longevity of its vector, and suppresses the free radical-defense. J Pineal Res 65:e12511. (PubMed)

Onaga G, Wydra KD, Koopmann B, Séré Y, von Tiedemann A (2016) Elevated temperature increases in planta expression levels of virulence related genes in Magnaporthe oryzae and compromises resistance in Oryza sativa cv. Nipponbare. Functional Plant Biology 44(3):358–371

Okazaki M, Higuchi K, Hanawa Y, Shiraiwa Y, Ezura H (2009) Cloning and characterization of a *Chlamydomonas reinhardtii* cDNA arylalkylamine N-acetyltransferase and its use in the genetic engineering of melatonin content in the micro-tom tomato. J Pineal Res 46:373–382

Pal PK, Bhattacharjee B, Chattopadhyay A, Bandyopadhyay D (2019) Pleiotropic roles of melatonin against oxidative stress mediated tissue injury in the gastrointestinal tract: an overview. Melatonin Res 2:158–184

Palikaras K, Daskalaki I, Markaki M, Tavernarakis N (2017) Mitophagy and age-related pathologies: development of new therapeutics by targeting mitochondrial turnover. Pharmacol Therap 178:157–174

Panstruga R, Parker JE, Schulze-Lefert P (2009) SnapShot: plant immune response pathways. Cell 136:978.e1–978.e3

Park S, Back K (2012) Melatonin promotes seminal root elongation and root growth in transgenic rice after germination. J Pineal Res 53:385–389

Park S, Lee DE, Jang H, Byeon Y, Kim YS, Back K (2013) Melatonin-rich transgenic Rice plants exhibit resistance to herbicide-induced oxidative stress. J Pineal Res 54:258–263

Pieterse CMJ, Van der Does D, Zamioudis C, Leon-Reyes A, VanWees SCM (2012) Hormonal modulation of plant immunity. Annu Rev Cell Dev Biol 28:489–521

Qian Y, Tan DX, Reiter RJ, Shi H (2015) Comparative metabolomic analysis highlights the involvement of sugars and glycerol in melatonin-mediated innate immunity against bacterial pathogen in Arabidopsis. Sci Rep 5:15815

Que Z, Ma T, Shang Y, Ge Q, Zhang Q, Xu P, Zhang J, Francoise U, Liu X, Sun X (2020) Microorganisms: producers of melatonin in fermented foods and beverages. J Agric Food Chem 68:4799–4811

Quintela T, Gonçalves I, Silva M, Duarte AC, Guedes P, Andrade K, Freitas F, Talhada D, Albuquerque T, Tavares S, Passarinha LA (2018) Choroid plexus is an additional source of melatonin in the brain. J Pineal Res 65(4):e12528

Rehman RS, Hussain M, Ali M, Zafar SA, Pasha AN, Bashir H, Ashraf NA, Javed A, Shah WAA (2022) Comprehensive review on melatonin compound and its functions in different fungi and plants. Int J Pathog Res 10:9–21

Reiter RJ (1995) Oxygen radical detoxification processes during aging: the functional importance of melatonin. Aging Clin Exp Res 7:340–351

Reiter RJ, Korkmaz A, Paredes SD, Manchester LC, Tan DX (2008) Melatonin reduces oxidative/Nitrosative stress due to drugs, toxins, metals, and herbicides. Neuro Endocrinol Lett 29:609–613

Reiter RJ, Tan DX, Fuentes-Broto L (2010) Melatonin: a multitasking molecule. Pro Brain Res 181:127–151

Reiter RJ, Rosales-Corral S, Tan DX, Jou MJ, Galano A, Xu B (2017) Melatonin as a mitochondria-targeted antioxidant: one of evolution's best ideas. Cell Mol Life Sci 74:3863–3881

Robert-Seilaniantz A, Grant M, Jones JDG (2011) Hormone crosstalk in plant disease and defense: more than just Jasmonate-salicylate antagonism. Annu Rev Phytopathol 49:317–343

Rodriguez C, Mayo JC, Sainz RM, Antolín I, Herrera F, Martín V, Reiter RJ (2004) Regulation of antioxidant enzymes: a significant role for melatonin. J Pineal Res 36:1–9

Rodriguez-Naranjo MI, Torija MJ, Mas A, Cantos-Villar E, Garcia-Parrilla MC (2012) Production of melatonin by saccharomyces strains under growth and fermentation conditions. J Pineal Res 53:219–224

Sanchez-Barcelo EJ, Rueda N, Mediavilla MD, Martinez-Cue C, Reiter RJ (2017) Clinical uses of melatonin in neurological diseases and mental and behavioural disorders. Curr Med Chem 24(35):3851–3878

Shi CS, Qi HY, Boularan C, Huang NN, Abu-Asab M, Shelhamer JH, Kehrl JH (2014) SARS-coronavirus open reading frame-9b suppresses innate immunity by targeting mitochondria and theMAVS/TRAF3/TRAF6 signalosome. J Immunol 193:3080–3089

Shi H, Chen Y, Tan DX, Reiter RJ, Chan Z, He C (2015) Melatonin induces nitric oxide and the potential mechanisms relate to innate immunity against bacterial pathogen infection in Arabidopsis. J Pineal Res 59:102–108

Singh A, Singh R, Tripathi MK (2020) Photoperiodic manipulation modulates the innate and cell mediated immune functions in the fresh water Snake. Natrix Piscator Sci Rep 10:14722

Sofy AR, Sofy MR, Hmed AA, El-Dougdoug NK (2019) Potential effect of plant growth-promoting Rhizobacteria (PGPR) on enhancing protection against viral diseases. In: Maheshwari DK, Dheeman (eds) Field crops: sustainable management by PGPR; sustainable development and biodiversity. Springer, Cham, pp 411–445

Sofy AR, Sofy MR, Hmed AA, Dawoud RA, Refaey EE, Mohamed HI, El-Dougdoug NK (2021) Molecular characterization of the alfalfa mosaic virus infecting *Solanum melongena* in Egypt and the control of its deleterious effects with melatonin and salicylic acid. Plan Theory 10:459

Su H, Chen T, Li J, Xiao J, Wang S, Guo X, Bu P (2017) Correlations of serum cyclophilin a and melatonin concentrations with hypertension-induced left ventricular hypertrophy. Arch Med Res 2017:48

Su J, Yang X, Shao Y, Chen Z, Shen W (2021) Molecular hydrogen–induced salinity tolerance requires melatonin signalling in Arabidopsis thaliana. Plant Cell Environ 44:476–490. (PubMed)

Sun Q, Zhang N, Wang J, Zhang H, Li D, Shi J, Li R, Weeda S, Zhao B, Ren S, Guo YD (2015) Melatonin promotes ripening and improves quality of tomato fruit during po stharvest life. J Exp Bot 66:657–668

Sun Y, Liu Z, Lan G, Jiao C, Sun Y (2019) Effect of exogenous melatonin on resistance of cucumber to downy mildew. Sci Hortic 255:231–241

Sun C, Liu L, Wang L, Li B, Jin C, Lin X (2021) Melatonin: a master regulator of plant development and stress responses. J Integr Plant Biol 63:126–145

Suofu Y, Li W, Jean-Alphonse FG, Jia J, Khattar NK, Li J, Baranov SV, Leronni D, Mihalik AC, He Y, Cecon E (2017) Dual role of mitochondria in producing melatonin and driving GPCR signaling to block cytochrome c release. Proc Natl Acad Sci USA 114:E7997–E8006

Szafranska K, Reiter RJ, Posmyk MM (2016) Melatonin application to Pisum sativum L. seeds positively influences the function of the photosynthetic apparatus in growing seedlings during Paraquat-induced oxidative stress. Front Plant Sci 7:1663

Szafrańska K, Reiter RJ, Posmyk MM (2017) Melatonin improves the photosynthetic apparatus in pea leaves stressed by Paraquatvia chlorophyll breakdown regulation and its accelerated de novo synthesis. Front Plant Sci 8:878

Talaat NB (2021a) Co-application of melatonin and salicylic acid counteracts salt stress-induced damage in wheat (*Triticum aestivum* L.) photosynthetic machinery. J Soil Sci Plant Nutr 21:2893–2906

Talaat NB (2021b) Polyamine and nitrogen metabolism regulation by melatonin and salicylic acid combined treatment as a repressor for salt toxicity in wheat (*Triticum aestivum* L.) plants. Plant Growth Regul 95:315–329

Tamtaji OR, Reiter RJ, Alipoor R, Dadgostar E, Kouchaki E, Asemi Z (2020) Melatonin and Parkinson disease: current status and future perspectives for molecular mechanisms. Cell Mol Neurobiol 40:15–23

Tan XL, Fan ZQ, Kuang JF, Lu WJ, Reiter RJ, Lakshmanan P, Su XG, Zhou J, Chen JY, Shan W (2019) Melatonin delays leaf senescence of Chinese flowering cabbage by suppressing ABFs-mediated abscisic acid biosynthesis and chlorophyll degradation. J Pineal Res 67(1):e12570

Tan D, Hardeland R (2020) Targeting host defense system and rescuing compromised mitochondria to increase tolerance against pathogens by melatonin may impact outcome of deadly virus infection pertinent to COVID-19. Molecules 25:4410

Tan DX, Reiter RJ (2019) Mitochondria: the birth place, battle ground and the site of melatonin metabolism in cells. Melatonin Res 2:44–66

Tan DX, Manchester LC, Fuentes-Broto L, Paredes SD, Reiter RJ (2011) Significance and application of melatonin in the regulation of brown adipose tissue metabolism: relation to human obesity. Obes Rev 12:167–188

Tan DX, Hardeland R, Manchester LC, Korkmaz A, Ma S, Rosales-Corral S, Reiter RJ (2012) Functional roles of melatonin in plants, and perspectives in nutritional and agricultural science. J Exp Bot 63:577–597

Tan DX, Manchester LC, Liu X, Rosales-Corral SA, Acuna-Castroviejo D, Reiter RJ (2013) Mitochondria and chloroplasts as the original sites of melatonin synthesis: a hypothesis related to melatonin's primary function and evolution in eukaryotes. J Pineal Res 54:127–138

Tan DX, Manchester LC, Qin L, Reiter RJ (2016) Melatonin: a mitochondrial targeting molecule involving mitochondrial protection and dynamics. Int J Mol Sci 17:2124

Tilden AR, Becker MA, Amma LL, Arciniega J, McGaw AK (1997) Melatonin production in an aerobic photosynthetic bacterium: an evolutionarily early association with darkness. J Pineal Res 22:102–106

Tiwari RK, Lal MK, Naga KC, Kumar R, Chourasia KN, Subhash S, Kumar D, Sharma S (2020) Emerging roles of melatonin in mitigating abiotic and biotic stresses of horticultural crops. Sci Hortic 272:109592

Tiwari RK, Lal MK, Kumar R, Mangal V, Altaf MA, Sharma S, Singh B, Kumar M (2021) Insight into melatonin-mediated response and signaling in the regulation of plant defense under biotic stress. Plant Mol Biol 109:385–399

Turra D, ElGhalid M, Rossi F, DiPietro A (2015) Fungal pathogen uses sex pheromone receptor for chemotropic sensing of host plant signals. Nat 527:521–524

Ucar M, Korkmaz A, Reiter RJ, Yaren H, Öter S, Kurt B, Topal T (2007) Melatonin alleviates lung damage induced by the chemical warfare agent nitrogen mustard. Toxicol Lett 173:124–131

Wang L, Feng C, Zheng X, Guo Y, Zhou F, Shan D, Liu X, Kong J (2017) Plant mitochondria synthesize melatonin and enhance the tolerance of plants to drought stress. J Pineal Res 63:e12429

Wang Y, Reiter RJ, Chan Z (2018) Phytomelatonin: a universal abiotic stress regulator. J Exp Bot 69:963–974

Wang M, Zhang S, Ding F (2020) Melatonin mitigates chilling-induced oxidative stress and photosynthesis inhibition in tomato plants. Antioxidants 9:218

Weeda S, Zhang N, Zhao X, Ndip G, Guo Y, Buck GA, Fu C, Ren S (2014) Arabidopsis transcriptome analysis reveals key roles of melatonin in plant defense systems. PLoS One 9:e93462

Wei W, Li QT, Chu YN, Reiter RJ, Yu XM, Zhu DH, Zhang WK, Ma B, Lin Q, Zhang JS (2015) Melatonin enhances plant growth and abiotic stress tolerance in soybean plants. J Exp Bot 66:695–707

Wei Y, Hu W, Wang Q, Zeng H, Li X, Yan Y, Reiter RJ, He C, Shi H (2017) Identification, transcriptional and functional analysis of heat-shock protein 90s in banana (Musa acuminata L.) highlight their novel role in melatonin-mediated plant response to fusarium wilt. J Pineal Res 62:12367

Wei Y, Chang Y, Zeng H, Liu G, He C, Shi H (2018) RAV transcription factors are essential for disease resistance against cassava bacterial blight via activation of melatonin biosynthesis genes. J Pineal Res 64:12454

Wei Y, Zhu B, Ma G, Shao X, Xie H, Cheng X, Zeng H, Shi H (2022) The coordination of melatonin and anti-bacterial activity by EIL5 underlies ethylene-induced disease resistance in cassava. Plant J 111:683–697

Wiid I, Hoal-van Helden E, Hon D, Lombard C, van Helden P (1999) Potentiation of isoniazid activity against Mycobacterium tuberculosis by melatonin. Antimicrobial agents and chemotherapy 43(4):975–977

Xie X, Han Y, Yuan X, Zhang M, Li P, Ding A, Wang J, Cheng T, Zhang Q (2022) Transcriptome analysis reveals that exogenous melatonin confers Lilium disease resistance to *botrytis Elliptica*. Front Genet 13:892674

Xu J, Sun S, Wei W, Fu J, Qi W, Manchester LC, Tan DX, Reiter RJ (2007) Melatonin reduces mortality and Oxidatively mediated hepatic and renal damage due to Diquat treatment. J Pineal Res 42:166–171

Yan Y, Sun S, Zhao N, Yang W, Shi Q, Gong B (2019) COMT1 overexpression resulting in increased melatonin biosynthesis contributes to the alleviation of Carbendazim Phytotoxicity and residues in tomato plants. Environ Pollut 252:51–61

Yang YX, Ahammed GJ, Wu C, Fan SY, Zhou YH (2015) Crosstalk among Jasmonate, salicylate and ethylene signaling pathways in plant disease and immune responses. Curr Protein Pept Sci 16:450–461

Yang X, Chen J, Ma Y, Huang M, Qiu T, Bian H, Han N, Wang J (2022) Function, mechanism, and application of plant melatonin: an update with a focus on the cereal crop, barley (*Hordeum vulgare* L.). Antioxidants 11:634

Yao JW, Ma Z, Ma YQ, Zhu Y, Lei MQ, Hao CY, Chen LY, Xu ZQ, Huang X (2021) Role of melatonin in UV-B signaling pathway and UV-B stress resistance in *Arabidopsis thaliana*. Plant Cell Environ 44:114–129

Yavuz T, Kaya D, Behçet M, Ozturk E, Yavuz Ö (2007) Effects of melatonin on Candida sepsis in an experimental rat model. Adv Ther 24:91–100

Yin L, Wang P, Li M, Ke X, Li C, Liang D, Wu S, Ma X, Li C, Zou Y, Ma F (2013) Exogenous melatonin improves malus resistance to Marssonina apple blotch. J Pineal Res 54:426–434

Yoon YM, Kim HJ, Lee JH, Lee SH (2019) Melatonin enhances mitophagy by upregulating expression of heat shock 70 kDa protein 1L in human mesenchymal stem cells under oxidative stress. Int J Mol Sci 20:4545

Yu R, Lendahl U, Nistér M, Zhao J (2020) Regulation of mammalian mitochondrial dynamics: opportunities and challenges. Front Endocrinol 11:374

Zeman M, Herichová I (2011) Circadian melatonin production develops faster in birds than in mammals. Gen Comp Endocrinol 172:23–30

Zeng H, Bai Y, Wei Y, Reiter RJ, Shi H (2022) Phytomelatonin as a central molecule in plant disease resistance. J Exp Bot 73:5874–5885

Zhang HJ, Zhang NA, Yang RC, Wang L, Sun QQ, Li DB, Cao YY, Weeda S, Zhao B, Ren S, Guo YD (2014) Melatonin promotes seed germination under high salinity by regulating antioxidant systems, ABA and GA 4 interaction in cucumber (*Cucumis sativus* L.). J Pineal Res 57:269–279

Zhang S, Zheng X, Reiter RJ, Feng S, Wang Y, Liu S, Jin L, Li Z, Datla R, Ren M (2017) Melatonin attenuates potato late blight by disrupting cell growth, stress tolerance, fungicide susceptibility and homeostasis of gene expression in *Phytophthora infestans*. Front Plant Sci 8:1993

Zhang H, Liu X, Chen T, Ji Y, Shi K, Wang L, Zheng X, Kong J (2018a) Melatonin in apples and juice: inhibition of Browning and Microorganism growth in apple juice. Molecules 23:521

Zhang S, Liu S, Zhang J, Reiter RJ, Wang Y, Qiu D, Luo X, Khalid AR, Wang H, Feng L, Lin Z (2018b) Synergistic anti-oomycete effect of melatonin with a biofungicide against Oomycetic black shank disease. J Pineal Res 65:e12492

Zhang Q, Liu X, Zhang Z, Liu N, Li D, Hu L (2019) Melatonin improved waterlogging tolerance in alfalfa (*Medicago sativa*) by reprogramming polyamine and ethylene metabolism. Front Plant Sci 10:44

Zhang Z, Wang T, Liu G, Hu M, Yun Z, Duan X, Cai K, Jiang G (2021) Inhibition of downy blight and enhancement of resistance in litchi fruit by postharvest application of melatonin. Food Chem 347:129009

Zhang T, Wang J, Sun Y, Zhang L, Zheng S (2022) Versatile roles of melatonin in growth and stress tolerance in plants. J Plant Growth Regul 41:507–523

Zhao H, Xu L, Su T, Jiang Y, Hu L, Ma F (2015) Melatonin regulates carbohydrate metabolism and defenses against *pseudomonas syringae* pv. Tomato DC3000 infection in Arabidopsis thaliana. J Pineal Res 59:109–119

Zhao L, Chen L, Gu P, Zhan X, Zhang Y, Hou C, Wu Z, Wu YF, Wang QC (2019a) Exogenous application of melatonin improves plant resistance to virus infection. Plant Pathol 68:1287–1295

Zhao D, Yu Y, Zhao Z, Sharma R, Reiter RJ (2019b) Melatonin synthesis and function: evolutionary history in animals and plants. Front Endocrinol 10:249

Zhou H, Zhang Y, Hu S, Shi C, Zhu P, Ma Q, Jin Q, Cao F, Tian F, Chen Y (2017) Melatonin protects cardiac microvasculature against ischemia/reperfusion injury via suppression of mitochondrial fission-VDAC1-HK2-mPTP-mitophagy axis. J Pineal Res 63(1):e12413

Zhou H, Li D, Zhu P, Ma Q, Toan S, Wang J, Hu S, Chen Y, Zhang Y (2018) Inhibitory effect of melatonin on necroptosis via repressing the Ripk3-PGAM5-CypD-mPTP pathway attenuates cardiac microvascular ischemia-reperfusion injury. J Pineal Res 65(3):e12503

Chapter 8
Jasmonic Acid for Induced Plant Defense Against Biotic Stress

Bader Un Nisa, Jannat Akram, Sadia Majeed, and Muhammad Shafiq

Abstract Jasmonic acid (JA) is a significant plant hormone that plays a defensive role in plants under stress conditions. The complex production, signaling mechanisms, and regulatory functions of JA are highlighted in this chapter, highlighting its function to mitigate biotic stress, such as disease, pest, and herbivore attacks. A complex network of MYC transcription factors, COI1 receptors, and JAZ proteins indicates the functioning of JA. This network coordinates under conditions of systemic and localized resistance, including induced systemic resistance (ISR). The chapter further emphasizes the ecological and agricultural applications of JA and its derivatives, highlighting their significant role in sustainable crop protection and pest management. Despite its promising prospects, various challenges also arise, such as environmental variability and scalability, which require further investigation. Further research on JA-mediated defense mechanisms illustrates a pathway for the enhancement of global food security and promoting resilient agricultural practices during rising challenges. Additionally, JA interacts with other plant hormones, such as salicylic acid (SA), auxins (IAA), gibberellins (GA), and abscisic acid (ABA), which all work together to fine-tune the balance between plant growth and defense.

Keywords Jasmonic acid · Plant defense mechanisms · Phytohormone signaling · Hormonal cross talk · Crop resilience

B. U. Nisa · S. Majeed
Department of Botany, Lahore College for Women University, Lahore, Pakistan

J. Akram
Department of Plant Breeding and Genetics, University of the Punjab, Lahore, Pakistan

M. Shafiq (✉)
Department of Horticulture, Faculty of Agricultural Sciences, University of the Punjab, Lahore, Pakistan
e-mail: shafiq.iags@pu.edu.pk

K. A. Abd-Elsalam, H. I. Mohamed (eds.), *Elicitors for Sustainable Crop Production*, Sustainability Sciences in Asia and Africa,
https://doi.org/10.1007/978-981-96-6569-3_8

8.1 Introduction

The lives of plants are endangered by various abiotic and biotic stresses. Since plants are sessile organisms, evolution has helped them evolve a variety of defense mechanisms and behaviors against predators and adverse environmental conditions. Within seconds, cellular sensors perceive a threat and transmit messages. Ca^{2+}, reactive oxygen species (ROS), hydrogen sulfide (H_2S), nitrogen oxide (NO), and inositol triphosphate (IP3) are examples of these signals referred to as second messengers, which interact with various proteins to improve the response of the plant against stress (Iqbal et al. 2021). Numerous mechanisms, including osmoregulation, osmotic adjustment, growth and elongation of cells, mineral nutrition, apoptosis, and hormonal signal transduction (HST), are regulated by the ROS-Ca^{2+} hub in plants, which has been the subject of extensive research in response to various environmental constraints (Guo et al. 2022).

It has been proposed that the plant stress signaling cascades are simultaneously activated by ROS-Ca^{2+} hub/phytohormone signaling (Devireddy et al. 2021). Additionally, ROS functions as the primary signaling pathway in plants in stressful situations by interacting with the other signaling components and secondary messengers like transcription factors, plant hormones, and mitogen-activated protein kinases (MAPKs) (Devireddy et al. 2021). On the other hand, environmental factors such as soil salinity and condition, high temperatures, and pathogen attacks disrupt the intricate biological network that is responsible for plant growth, development, and survival (Muhammad et al. 2024). ROS and its derivatives are among the many intermediate molecules that may build up in the cytoplasm and interfere with the pathways involved in signaling, particularly when they are changed into the dangerous substances that compromise cell viability (Juan et al. 2021).

Phytohormones that take part in ROS-Ca^{2+} mediated signaling include strigolactones, jasmonic acid (JA), salicylic acid (SA), gibberellins (GA), brassinosteroids (BRs), ethylene (ET), cytokinins (CKs), auxins (IAA), and abscisic acid (ABA) (EL Sabagh et al. 2022). In reaction to a stressor, each of those hormones can have a distinct effect and work either antagonistically event synergistically. JA and its derivatives, including 12-O-glucosyl-JA, 12-carboxy-JA-Ile, 12-hydroxyjasmonic acid sulfate (12-HSO_4-JA), JA-glucosyl ester, cis-jasmone, methyl jasmonate (MeJA), jasmonoyl-amino acid, 12-O-glucosyl-JAIle, JA-Ile glucosyl ester, and JA-Ile methyl ester, are examples of phytohormones. Numerous developmental processes are regulated by JAs, which are signaling molecules generated from lipids. They are found in practically all plants and are linked to how plants react to biotic and abiotic stress (Ali and Baek 2020).

Fatty acids known as JAs are members of the family of oxidized lipids. These compounds, called oxylipins, are produced from cyclopentanones (Ghorbel et al. 2021). The autoxidation of the polyunsaturated fatty acids or the activities of the alpha-dioxygenases or lipoxygenases produce the biologically active compounds known as oxylipin signaling molecules (Ghorbel et al. 2021). JAs control a variety of important biological activities in plants, including stomatal opening, leaf expansion, stamen and trichome development, cell cycle regulation, glucose transport, and hook formation (Ali and Baek 2020).

The aims of this chapter are to explore the regulatory mechanisms of jasmonic acid in mitigating abiotic stress-induced damage in plants, to analyze the role of jasmonic acid in modulating plant physiological and biochemical responses under drought, salinity, and temperature extremes, to examine the cross talk between jasmonic acid and other phytohormones in orchestrating stress tolerance, and to highlight the potential applications of jasmonic acid in improving crop resilience through genetic and biotechnological approaches.

8.2 Jasmonic Acid Metabolism

Higher plant species have been shown to have JAs, which are ubiquitous (Wang et al. 2021). In plants, the JA endogenous level is very low in the matured leaves and roots but higher in reproductive organs and flowers (Hewedy et al. 2023). The importance of JA as a phytohormone involved in both normal and stressed plant growth has come to light in recent years (Jang et al. 2020). Various research has been undertaken on plants from several families, including monocotyledons, Brassicaceae (the model plant *Arabidopsis thaliana*), Solanaceae, tomato (*Lycopersicum esculentum*), and tobacco: (*Nicotiana* spp.). To activate the defensive response machinery, several studies have shown that systemin and wounding raise the internal level of the JAs, which involve metabolic precursor of the (12-OPDA) or 12-oxo-phytodienoic acid and its products methyl jasmonate and jasmonic acid (Wu et al. 2024).

8.3 Biosynthesis of Jasmonic Acid

The biosynthesis pathway of jasmonic acid has already been reviewed in several publications. These reviews include important details on this pathway's genes, enzymes, and biosynthesis activities. They also explain the mechanisms controlling substrate specificity and enzymatic crystal structure, and they conclude by describing how JA biosynthesis is regulated (Jang et al. 2020). It recently demonstrated that the process of the biosynthesis of jasmonic acid involves several membrane transporters, such as jasmonate transporter 1 (JAT1), comatose (CTS), and JASSY. A modernized explanation of the synthesis of JA and their procedure, incorporating the reaction phases and the names of the substrates and enzymes, is provided in Fig. 8.1. Phospholipase A1 (PLA1) releases (18:3) (α-LeA) or α-linolenic acid, which is necessary for the synthesis of the JA, from the galactolipids' sn1 location

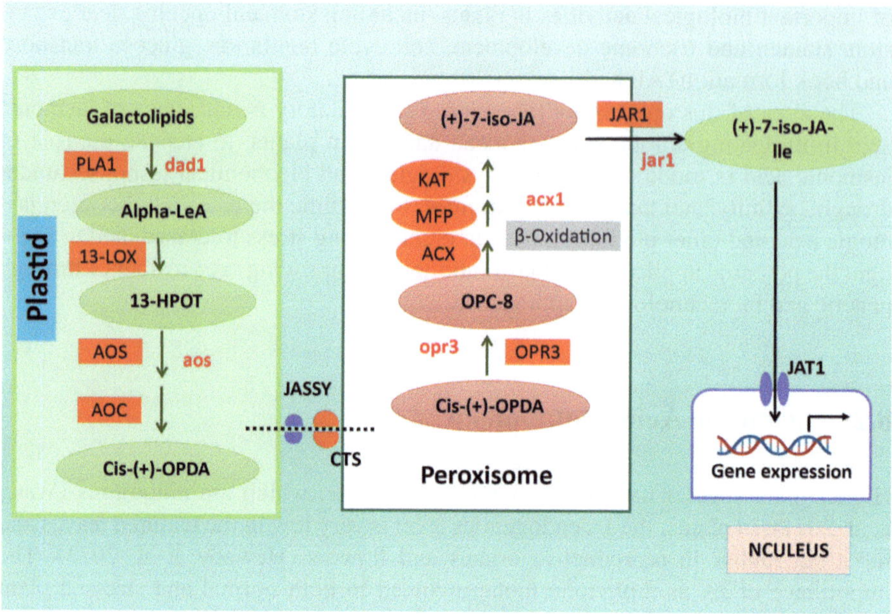

Fig. 8.1 JA/JA-Ile biosynthesis begins when galactolipid releases α-linolenic acid. Plasmids, peroxisomes, and the cytoplasm all produce JA from scratch. Then JA-Ile is transfer to the nucleus for the gene expression. Enzyme names are emphasized in orange, while major genes are noted in red

on chloroplast membranes (Wang et al. 2021). Then, allene oxide synthase (AOS), 13-lipoxygenase (LOX), and allene oxide cyclase (AOC) convert the substrate to (OPDA) or (9S, 13S)-12-oxo-phytodienoic acid. Chloroplasts contain every enzyme and product involved in this process. OPDA reductase 3 (OPR3) subsequently transports OPDA to the peroxisome from the chloroplast, where it is converted into the (OPC-8:0) or 3-oxo-2-(cis-20-pentenyl)-cyclopentane-1-octanoic acid (Khan et al. 2023a, b).

It is still unknown how OPDA is transported to the peroxisomes from chloroplasts. Guan et al. discovered and named JASSY, a protein found on the external membranes of chloroplasts that is in charge of the OPDA export (Wang et al. 2021). The recipient's primary mediator of OPDA imports into the peroxisomes is CTS, a type of ABC transporter in the peroxisomal membrane (Wang et al. 2021). To create (+)-7-iso-JA, OPC-8 goes through three β-oxidation processes using multifunctional protein (MEP), L-3-ketoacyl CoA thiolase (KAT), and acyl CoA oxidase (ACX) (Jang et al. 2020). After (+)-7-iso-JA is transferred to cytoplasm and combined with isoleucine (Ile), JAR1 (a JA-Ile synthesizing enzyme) produces the most JA bioactive molecule, (+) 7-iso-JA-Ile. Finally, ABC transporter JAT1, transports JA-Ile into the nucleus where it takes part in the JA signaling pathways in latter stages (Liu and Timko 2021) (Fig. 8.1).

8.4 Signaling of Jasmonic Acid

The JA signaling system has two modalities: repression under normal circumstances and activation under stressful circumstances (Fig. 8.2). Genes involved in JA production remain dormant under suppression (normal) circumstances because the cytoplasm contains relatively little JA-Ile (Ali and Baek 2020). The jasmonate zinc-finger inflorescence meristem (ZIM) domain identifies a class of transcriptional repressors known as JAZ proteins, which suppress the gene promoters under this environment by attaching to several transcription factors (Xu et al. 2021). JAZ proteins modeled here cooperate with their adaptor proteins (NINJA) and populate in the assembly of an efficient closed-complex of JAZ transcriptional repressors (the JAZ–NINJA–TPL complex). This tight complex of JAZ repressors requires the additional recruitment of the histone deacetylases 6 (HDA6) and 19 (HDA19) to maintain it. This compound may inhibit jasmonate-dependent gene activation (Ali and Baek 2020; Zhou et al. 2023).

Under stressful situations, JA-Ile builds up into the cytosol and is transported to nucleus by jasmonic acid transfer proteins (AtJAT1/AtABCG16) through the nuclear and plasma membrane (Ali and Baek 2020). The JA signaling pathway is

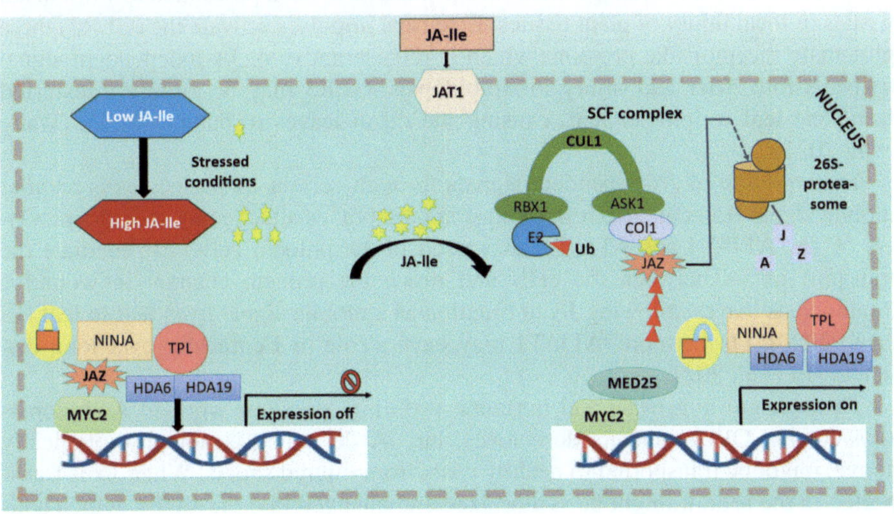

Fig. 8.2 A functional signal transduction model of jasmonic acid. The JAZ proteins block transcription factors, a small amount of JA-Ile inhibits the stimulation of the promoters of genes that respond to jasmonate under normal circumstances (without abiotic stress stimuli). An active transcriptional repression complex is formed by the combination of the adaptor protein NINJA and TPL. Additionally, HDA6 and HDA19 are enlisted to inhibit JA reactions by transforming the open complex into a closed one. JA-Ile, an epimer of JA, is transported into the nucleus by JAT1 and becomes more abundant under stress. Proteases then destroy JAZ as a result of JA-Ile usage of the JAZ connection with the F-box protein COI1 inside the SCF complex. The G-box region attracts MED25 once TFs are encouraged to bind to it, raising the number of genes that respond to jasmonate

thought to have begun at this point. The E3 ubiquitin ligase known as the SCF complex [kinetochore protein 1 (SKP1)-Cullin 1 (CUL1)-F-box] controls JA reactions inside the core. To begin with, JA-Ile has been conveyed to the core and recognizable through the (COI1) F-box protein interior of the SCF complex. JA-Ile is recognized through the COI1-JAZ co-receptor complex, which encourages JAZ to be associated with COI1 (Wang et al. 2021). The 26S proteasome degrades (ubiquitinates) JAZs as a result of this interaction. As a result, the activation of TFs and gene expression responsive to jasmonic acid occur from the absence of JAZ protein (repressor degradation) (Li et al. 2022). Furthermore, the Arabidopsis mediator complex includes mediator 25 (MED25), often referred to as the phytochrome and flowering time 1 (PFT1), which binds the TF to RNA polymerase II and affects gene transcription (Prusty et al. 2023) (Fig. 8.2).

The process by which JA signaling is sent to different plant parts across long distances is another fascinating topic of JA research. JA and JA-Ile levels rise from the site of the local damage to the distant healthy and unharmed tissues when plants are subjected to insect herbivory, necrotrophic diseases, or mechanical trauma. As a result, the entire plant develops broad-spectrum immunity against subsequent threats (Li et al. 2020). In general, JA can spread systematically through the air or vascular bundles (Sun and Zhang 2021). To date, the two types of mobile signals (JA-dependent and JA-independent) in the plant have been postulated to warn distal tissues of local injury of plant tissues. Electrical impulses activate the (GLR3) three glutamate receptor-like proteins, which is representative of JA-independent signal transmission. After leaf injury in Arabidopsis, GLR3 may be rapidly transferred from one leaf to another leaf, causing the distal leaves to burst with JA (Wang et al. 2021).

Transmission of JA-dependent signals through certain JA molecules is another mobile signal (Li et al. 2020). Two plasma membranes attached by specific importers of JA (ATJAT4 and ATJAT3, expressed in the phloem cells) can mediate the transport of JA between the cells and down the phloem channel in wounded Arabidopsis leaves, allowing JA to function as a mobile signal from leaf to leaf (Li et al. 2020). Furthermore, ATJAT1 may have a role in JA transfer over long distances (Li et al. 2021).

Through grafting trials and hormone profiling, previous work in Arabidopsis showed that OPDA and its derivatives, but JA-Ile which is not physiologically active, might be transported to healthy roots from injured shoots. When JA is transported to the root, it might be synthesized and changed into the JA-Ile, which activates the signal transduction pathway of JA (Wang et al. 2021). According to a thorough investigation, airborne transfer can allow a threat signal to be transmitted across vast distances to activate plant defenses. Without carrier assistance, jasmonic acid cannot easily enter the cell membrane, but MeJA may because of its high volatility. MeJA can travel in the air from damaged to undamaged leaves or from the injured plants to nearby plants, as has been shown in recent years (Wang et al. 2021).

8.5 Role of Jasmonic Acid in Induced Systemic Resistance

Jasmonic acid (JA) is an important phytohormone in induced systemic resistance (ISR), a defense mechanism that increases plant resistance to various pathogens and herbivores (Table 8.1). ISR has been suggested to be caused by beneficial bacteria such as plant growth promoters (PGPR) in the rhizosphere and some fungi that interact with roots and activate JA-mediated signaling pathways, enabling plants to respond more strongly to future attacks (Ghorbel et al. 2021).

8.5.1 *JA and Activation of ISR Paths*

Jasmonic acid together with its derivative methyl jasmonate (MeJA) serves as a signaling molecule in ISR pathways. Following microbial interaction or insect feeding, plant JA levels increase, which in turn activates defense responses in tissues away from the site of initial interaction.

The JA-activated ISR involves signal transduction pathways that trigger downstream defense gene expression, leading to systemic resistance throughout the plant. This systemic activation ensures that even unexposed parts of the plant are strengthened against potential pathogens or herbivores (Wang et al. 2020).

Table 8.1 Role of jasmonic acid against biotic stresses

Biotic stress	Plant system	Role of jasmonic acid	References
Insect herbivory	*Arabidopsis thaliana*, tomato	Activates defense genes encoding protease inhibitors and secondary metabolites	Ali and Baek (2020)
Fungal pathogens	Rice, barley	Regulates defense pathways and promotes the production of phytoalexins	Bhatla and Lal (2023)
Bacterial infections	Tomato, tobacco	Induces hypersensitive response (HR) and defense-related gene expression	Bian et al. (2020)
Nematode infestation	Soybean, potato	Modulates callose deposition and defense signaling to restrict nematode movement	Bian et al. (2020)
Viral infections	Arabidopsis, cucumber	Suppresses viral replication by inducing RNA silencing and defense gene expression	Ding et al. (2022a, b)
Mycorrhizal interaction	Maize, soybean	Regulates mutualistic interactions, balancing growth-defense trade-offs	Ghorbel et al. (2021)
Parasitic plants	Cuscuta species (dodder)	Triggers localized defense responses to limit parasitic plant attachment	Gomi (2021)
Insect-vectored diseases	Grapevine, citrus	Promotes production of volatiles and secondary metabolites deterring insect vectors	Hu et al. (2020)
Induced systemic resistance (ISR)	Arabidopsis, cucumber	Enhances plant systemic immunity in conjunction with beneficial rhizobacteria	Jang et al. (2020)

8.5.2 *JA and Primary Effect in ISR*

Priming is a phenomenon where plants prepare for an increased defense response after recognizing beneficial microbes or minor stress signals. When activated, basal levels of defense-related proteins and metabolites remain low until an actual attack occurs, allowing the plant to conserve resources.

JA plays a key role in the primary integument of plants. When attacked by a pathogen or insect, JA-mediated signaling pathways rapidly mobilize stored resources, leading to a rapid and robust defense response that includes callose deposition, production of reactive oxygen species (ROS), and antimicrobial compounds (Bhatla and Lal 2023).

8.5.3 *JA-Precipitated Defense Responses*

JA stimulates the manufacturing of secondary metabolites (which includes alkaloids and terpenoids) and shielding proteins (like proteinase inhibitors and polyphenol oxidases) that deter herbivores and restriction pathogen increase.

JA additionally induces the buildup of pathogenesis-related (PR) proteins, antimicrobial compounds, and volatile natural compounds (VOCs), which serve to push back herbivores and entice natural predators of pests, creating an oblique protection mechanism (Bhatla and Lal 2023).

8.5.4 *JA Interaction with Other Hormones in ISR*

8.5.4.1 A–Salicylic Acid (SA) Cross Talk

JA and salicylic acid (SA) pathways regularly have adverse interactions. While SA is in general associated with systemic received resistance (SAR) in opposition to biotrophic pathogens, JA-mediated ISR is typically powerful toward necrotrophic pathogens and herbivores. This opposed cross talk lets in the plant to fine-music its defense primarily based on the kind of chance.

8.5.4.2 JA–Ethylene (ET) Synergy

JA regularly works synergistically with ethylene (ET) to decorate protection responses, especially below herbivore attacks or necrotrophic pathogen infections. JA–ET co-regulation turns on a particular set of genes worried in generating shielding compounds and proteins (Ali and Baek 2020).

8.5.5 JA Receptor and Signaling Additives in ISR

8.5.5.1 COI1-JAZ Receptor Complicated

The coronatine-insensitive protein 1 (COI1) is a key JA receptor that interacts with JAZ (jasmonate ZIM-domain) proteins. Below non-burdened conditions, JAZ proteins repress JA-responsive genes. While JA degrees increase upon pressure, JA binds to COI1, leading to JAZ protein degradation, freeing the repression and permitting the expression of JA-responsive genes crucial for ISR (Wang et al. 2020).

8.5.5.2 MYC Transcription Elements

As soon as JAZ proteins are degraded, MYC transcription factors (including MYC2) are activated. MYC transcription elements bind to promoters of JA-responsive genes, which then spark off the plant's immune responses, such as the expression of protection enzymes and antimicrobial proteins (Scalschi et al. 2020).

8.5.5.3 JA-Structured Defense Gene Activation

JA signaling results in the activation of a range of protection genes, lots of which encode enzymes concerned with the synthesis of shielding secondary metabolites.

A few key genes in JA-mediated ISR consist of:

- **LOX (lipoxygenase):** Worried inside the biosynthesis of JA from fatty acid precursors.
- **AOS (allene oxide synthase) and AOC (allene oxide cyclase):** Catalyze steps inside the JA biosynthesis pathway. PR genes: Produce pathogenesis-related proteins with antimicrobial homes (Scalschi et al. 2020).

8.6 Ecological Function of JA in ISR

JA also regulates the release of unstable natural compounds (VOCs) that could entice natural predators or parasitoids of herbivores attacking the plant. This ecological aspect of ISR blessings the plant with the aid of recruiting allies toward herbivores.

JA's position in ISR enhances the plant's potential to discourage or manage herbivory and pathogen pressure, permitting it to survive and reproduce greater efficiently beneath strain situations (Gomi 2020).

8.7 Detailed Mechanisms of Jasmonic Acid (JA) in Plant Defense

Jasmonic acid (JA) operates via intricate molecular mechanisms that form a core of plant protection responses to biotic stresses like necrotrophic pathogens and herbivores. JA not only activates defense-associated gene expression but additionally coordinates with different hormone signaling pathways, allowing plants to optimize their protection structures based totally on the nature (Wang et al. 2021).

Here, we provide an updated overview of the latest findings on the expression patterns of genes encoding enzymes in the jasmonic acid (JA) biosynthesis pathway, along with insights into the subcellular localization of their corresponding gene products (Wang et al. 2021).

8.7.1 ω-3 Fatty Acid Desaturases

The layer of natural film is rich in unsaturated greasy acids get from polyunsaturation of layer chemicals. There are two major pathways in charge of the biosynthetic of polyunsaturated greasy acids in higher plants: the prokaryotic pathway box (related to the chloroplast) and the eukaryotic pathway box (contained in the endoplasmic reticulum, ER). The major galactolipids generated in chloroplasts, oleic acid (18:1), are converted to 18:3 and in the endoplasmic reticulum, phosphatidylcholine is the primary lipid derived from trienoic acids (Ma et al. 2020).

Three different genes, FAD3, FAD7, and FAD8, encode ω-3 fatty acid desaturases, which are essential for fatty acid metabolism in Arabidopsis thaliana. While FAD7 and FAD8 are found in chloroplasts, the FAD3 enzyme is found in the endoplasmic reticulum (ER). Significant amounts of linolenic acid (18:3) are preserved in cellular membranes despite defects in any one of these genes because of lingering desaturase activity and the movement of freshly produced 18:3 between cellular compartments. However, since 18:3 is a precursor for the synthesis of jasmonic acid (JA), the triple mutant that lacks all three desaturases (fad3-fad7-fad8) shows a significant decrease in trienoic fatty acids and produces little to no JA. External supplementation with either 18:3 or JA can reverse male sterility, which is caused by this deficit. Male sterility is similarly exhibited by the JA-insensitive coi1 mutant in *A. thaliana*, underscoring the crucial function of JA in pollen formation (Ma et al. 2020).

18:3's function as a precursor for the creation of jasmonic acid (JA) implies that ω-3 fatty acid desaturases play a crucial role in controlling how plants react to wounds. Research has indicated that following injury, ω-3 desaturase mRNA levels, particularly those of tobacco plastidial and mung bean microsomal variants, rise. Transgenic research using an Arabidopsis FAD7 promoter in tobacco also showed that wounding changes the plastidial ω-3 desaturase gene (FAD7)'s spatial expression. While leaf expression happens without JA, the octadecanoid pathway controls the gene's activity in roots. Increased 18:3 levels in the main lipids of injured tomato and tobacco leaves, as well as raised 18:2 levels, which show activation of the whole fatty acid desaturation pathway, suggest that this induction likely improves 18:3 availability for JA production (Genva et al. 2019).

Furthermore, in tomato and potato, exogenous 18:3 causes wound-responsive pin gene expression, indicating that JA production enzymes exist even in plants that are not under stress. Pin gene activation during wounding is decreased by antisense inhibition of potato ω-3 desaturase, highlighting the likelihood that newly produced 18:3 is the source of wound-induced JA. They are crucial for JA-dependent wound signaling pathways, although the precise lipid sources and subcellular sites of 18:3 needed for JA production are still unknown (Howlader et al. 2020).

8.7.2 Lipoxygenases (LOX)

The dioxygenation of unsaturated greasy acids is catalyzed by lipoxygenases (LOX), which are nonheme iron-containing chemicals that specifically target free greasy acids with a 1, 4-pentadiene theme. Although the main LOX substrates in plants are 18:2 and 18:3, some research indicates that LOX can also operate on fatty acids that are lipid-bound, raising doubts about the need for lipase activity to start the octadecanoid pathway. Large gene families encode LOX enzymes, which vary in their kinetic behavior, substrate preferences, and stereospecificity. Some of these enzymes are found in the cytoplasm, mitochondria, vacuoles, or chloroplasts (Kim et al. 2015).

In many plants, including Arabidopsis, soybean, barley, and potatoes, LOX gene expression is induced by wounding and jasmonic acid (JA) treatment. For instance, in Arabidopsis, wounding robustly induces the chloroplast-targeted isoform AtLOX2, whereas the cytoplasmic isoform AtLOX1 reacts to JA but not to wounding. Likewise, in tomato and potato, JA-inducible 13-LOX isoforms localize to chloroplasts, indicating a function in systemic and wound signaling. According to transgenic research, certain LOX isoforms—such as AtLOX2 in Arabidopsis and 13-LOX in potatoes—are essential for wound-induced JA production and pest resistance. Nonetheless, it seems that some LOX isoforms' defense-related functions are unrelated to the synthesis of JA (Ghorbel et al. 2021).

Products generated from LOX are crucial for plant-pathogen interactions in addition to wound signaling. Tobacco, for example, contains a 9-LOX isoform that helps with resistance to *Phytophthora parasitica.* Furthermore, in response to nitrogen availability, tissue-specific and developmental variables govern the involvement of soybean vegetative LOX in nitrogen and assimilate partitioning. These various tasks demonstrate the LOX enzymes' complex significance in plant defense and physiology (Ghorbel et al. 2021).

8.7.3 Allene Oxide Synthase and Cyclase

Fatty acid hydroperoxides need to be neutralized right away because they are hazardous, hydroperoxide lyase (HPL) breaks down 9- and 13-hydroperoxides in non-stressed plants into aldehydes and oxoacids, some of which have antibacterial qualities. In contrast, a crucial stage in the production of jasmonic acid (JA) is the conversion of 13-hydroperoxy-linolenic acid (13-HPLNA) into 12,13-epoxylinolenic acid by the cytochrome P450 enzyme allene oxide synthase (AOS), which belongs to the CYP74A family. AOS is found in plastids and has been found in a number of plants, including flax and Arabidopsis. It is identified as a regulatory point in the octadecanoid system because its expression rises in response to injury or jasmonate therapy (Bian et al. 2020).

With the help of allene oxide cyclase (AOC), the unstable 12,13-epoxylinolenic acid quickly cyclizes into 12-oxophytodienoic acid (12-oxo-PDA). Given the short half-life of the intermediate, it is thought that this enzyme associates with AOS for efficiency. It stereospecifically generates the 9(S),13(S)-12-oxo-PDA isomer. 12-oxo-PDA itself serves as a signaling molecule in processes such as tendril coiling and alkaloid synthesis, in addition to being a precursor for JA. Its dual role in plant defense signaling is highlighted by the fact that 12-oxo-PDA conversion to JA is necessary for activating particular genes, such as the tomato pin genes (Bian et al. 2020).

8.7.4 12-Oxo-Phytodienoic Acid Reductase and β-Oxidation

The cytosolic enzyme 12-oxo-PDA reductase (OPDR) converts 12-oxo-phytodienoic acid (12-oxo-PDA) to OPC-8:0. Transporting 12-oxo-PDA from chloroplasts to the cytoplasm is important for this phase (Table 8.2). With the cloning of the Arabidopsis OPDR gene, its characteristic within the manufacture of jasmonic acid (JA) can also now be investigated further (Bian et al. 2020).

To create JA, OPC-8:0 goes through three rounds of β-oxidation, most likely in peroxisomes. OPC-8:0 must be transported into peroxisomes and activated by fatty acyl-CoA synthetase to an acyl-CoA derivative. This is followed by a series of enzymatic reactions, such as oxidation, hydration, and thiolysis. There are known genes that encode these enzymes, such as Arabidopsis' wound-inducible acyl-CoA oxidase. Peroxisomal β-oxidation-deficient mutants offer instruments to investigate its function in JA production. After forming, JA is released into the cytoplasm to engage in signaling interactions with as-yet-unidentified receptors (Hu et al. 2020).

Table 8.2 Key components in the JA defense pathway

Step	Molecules/proteins involved	Function	Reference
JA biosynthesis	LOX, AOS, AOC, OPR3	Converts linolenic acid to jasmonic acid	Gomi (2021)
JA conjugation	JA-Ile	Bioactive form of JA recognized by receptors	Jang et al. (2020)
Receptor activation	COI1, JAZ proteins	JA-Ile binds COI1, leading to JAZ degradation	Li et al. (2020)
Signal transduction	MYC2, MYC-related transcription factors	Activates expression of JA-responsive genes	Scalschi et al. (2020)
Defense genes	PR proteins, proteinase inhibitors, VOC enzymes	Direct defense against pathogens and herbivores	Wang et al. (2020)
Crosstalk	SA (antagonistic), ET (synergistic)	Fine-tunes defense based on threat type	Zhang et al. (2022)

8.8 Cross Talk Between Jasmonic Acid and Other Phytohormones During the Biotic Stress Response

Plant hormones utilize their limited resources to respond to the most severe stresses under varying environmental conditions. As a result, they create distinct signaling pathways that assist in balancing the defense response and plant growth (Khan et al. 2023a, b). This perplexing hormonal crosstalk is frequently inducer-, measurements-, species-, and tissue-specific. As a fundamental stretch hormone pathway, the JA signaling pathway interatomic with other plant hormones to create a complex signaling arrangement that's connected to other signaling pathway (Hewedy et al. 2023). Numerous likenesses between JA and other plant hormone signaling pathways have been found in later a long time. The viability of JA is affected by its intuition with other phytohormones, counting auxins, gibberellins (GAs), and abscisic corrosive (ABA) (Lv et al. 2021). Plants can prioritize their responses and hence control distinct environmental reactions because of the hormonal cross talk that is essential for several signaling pathways (Wang et al. 2021).

8.8.1 *JA–Auxins*

JA and auxin undergo adverse interactions to manipulate several plant developmental approaches, inclusive of senescence, flower development, and root increase. Auxin is crucial for controlling mobile department and starting up seed improvement (Mazzoni-Putman et al. 2021).

8.8.1.1 JA and Auxin Synergy

Plant defense mechanisms depend on JA and auxin signaling, which work in coordination to protect plants from necrotrophic diseases. Both pathways make use of repressors (AUX/IAA for auxin, JAZ for JA) and F-box proteins (COI1 for JA, TIR1 for auxin) that are ubiquitinated to activate the pathway (Wang et al. 2024). Resistance to necrotrophic diseases and vulnerability to *Pseudomonas syringae* depend on JA and auxin signaling (Ding et al. 2022a, b).

8.8.1.2 JA and Auxin Regulate Root Growth

Auxin and JA inhibit root growth by activating the JAZ–MYC2 signaling pathway. Auxin–TIR–AUX/IAA–ARF signaling and JA biosynthesis are induced by exogenous auxin. In turn, JA promotes JAZ1 activity, influences root growth, and increases auxin biosynthesis through auxin synthase gene (ASA1) expression. Furthermore, myelocytomatosis-related transcription factor 2 (MYC2) inhibits microtubule-associated protein (PTL1 and PTL2) expression, which inhibits auxin signaling and prevents root development (Luo et al. 2023).

8.8.1.3 Auxin and JA in the Development of Flowers

The growth of stamens and petals depends on the interactions between auxin signaling elements (ARF6 and ARF8) and JA signaling components (JAZ, MYB21, and MYB24). These developmental pathways are also impacted when JA is administered endogenously (Kumar et al. 2024).

8.8.2 JA-ABA

ABA serves as a powerful regulator of plant responses to stressful conditions. Through JAZs-MYC2, ABA and JA interact to control plant development, protect against herbivorous insect feeding, and respond to water, drought, cold, and salt stress (Hewedy et al. 2023). In general, ABA and JA work antagonistically in different developmental processes and synergistically in others, such as herbivory defense and seed germination inhibition. The discovery of the JA-induced ABA receptor PYL4 in early molecular research demonstrated a connection between core ABA signaling and JA signaling. In response to JA, two ABA receptor mutants, pyl4, and pyl5, show decreased anthocyanin accumulation and JA-hypersensitive shoot growth (Zeng et al. 2021).

8.8.2.1 Antagonistic Interaction Between ABA and JA–Ethylene Pathway

Plant defensin (PDF1.2), a marker gene for the JA–ethylene defensive signaling pathway, is significantly reduced in production by ABA. When compared to mock-treated plants, exogenous ABA treatment resulted in a tenfold reduction in baseline PDF1.2 transcript levels, even when co-treated with methyl jasmonate (MJ) or ethylene (Wei et al. 2020). ABA stopped the activation of PDF1.2 and different JA-ethylene sensitive genes, including CHI (basic chitinase), HEL (hevein-like protein), and LEC (lectin-like protein), after MJ or ethylene treatments ABA stopped the activation of PDF1.2 and different JA-ethylene sensitive genes, which include CHI (primary chitinase), HEL (hevein-like protein), and LEC (lectin-like protein), after MJ or ethylene remedies (Loo et al. 2021). The antagonistic impact of ABA was confirmed by the considerable reduction in MJ-induced expression of these genes with co-application of ABA (Rai et al. 2024).

8.8.2.2 Physiological Behavior of ABA

As evidenced by increased defensive gene expression in ABA-deficient mutants, ABA plays a crucial role in regulating JA-ethylene-responsive defense systems. The results imply that by antagonizing JA-ethylene signaling pathways, ABA can reduce plant disease resistance under stressful circumstances like drought or water stress (Jiang et al. 2024).

8.8.3 JA-GA

The relationship between gibberellins (GAs) and jasmonic acid (JA) is essential for maintaining a balance between plant defense and development under stress. GAs control the growth of plants by encouraging developmental programs. In contrast, JA triggers defensive mechanisms in response to environmental stresses. The accumulation of aspartic acid, glutamic acid, leucine, and alanine (DELLAs) results from a decrease in GA levels during salinity, which inhibits growth. In contrast, GA biosynthesis mutants have greater stress tolerance, and DELLA mutants lessen the growth inhibition brought on by stress. It implied that GAs are essential for coordinating defense and growth responses. The idea that DELLAs are essential to this coordination process is supported by additional research on DELLA mutants (Castro-Camba et al. 2022).

8.8.3.1 GA Biosynthesis Is Downregulated by JA

It has been demonstrated that JA signaling suppresses GA biosynthesis in response to both abiotic and biotic stresses, including pathogen infection and herbivore attack. GA 20-oxidases have a role in GA precursor activation, and JA signaling

Table 8.3 Presenting examples of JA's interactions with other hormones and its impact on plant immunity

Hormonal interaction	Plant species	Outcome on immunity	Reference
JA–auxin	Arabidopsis	Auxin signaling suppression improves tolerance to herbivores	Zhang et al. (2022)
JA–GA	Rice	Pathogen resistance is decreased by antagonistic interaction	Zhu et al. (2024)
JA–ABA	Tomato	Synergistic reaction increases resistance to pathogens and drought and cold stress	Ding et al. (2022a, b)

inhibits their expression. One important mechanism by which JA redirects resources from growth-related processes toward defense is the downregulation of GA biosynthesis (Figueroa-Macías et al. 2021).

8.8.3.2 Antagonistic Relation Between JA and GA During Stress Responses

The relationship between gibberellins (GAs) and jasmonic acid (JA) is essential for maintaining a balance between plant defense and development during stress conditions (Table 8.3). GAs control the growth of plants by encouraging developmental programs. In contrast, JA triggers defensive mechanisms in response to environmental stresses. The accumulation of aspartic acid, glutamic acid, leucine, and alanine (DELLAs) results from a decrease in GA levels during salinity, which inhibits growth. In contrast, GA biosynthesis mutants have greater stress tolerance, and DELLA mutants lessen the growth inhibition brought on by stress. It implied that GAs are essential for coordinating defense and growth responses. The idea that DELLAs are essential to this coordination process is supported by additional research on DELLA mutants (Xue et al. 2022). Therefore, under stress, the balance between JA and GA is crucial to maximize plant development and defense (Zhang et al. 2020a) (Fig. 8.3).

8.9 Application of JA and Its Derivatives to Boost Resistance in Different Crops

The lipid-derived phytohormones known as jasmonates (JAs), which include molecules of jasmononic acid and its derivatives including methyl jasmonate (MeJA) and jasmonoylisoleucine (JA-Ile), are crucial for controlling plant development and different environmental responses. They have a well-established function in plant's resilience to herbivores and other biotic stresses. By triggering both direct and indirect plant defense mechanisms, they can be used in sustainable pest management.

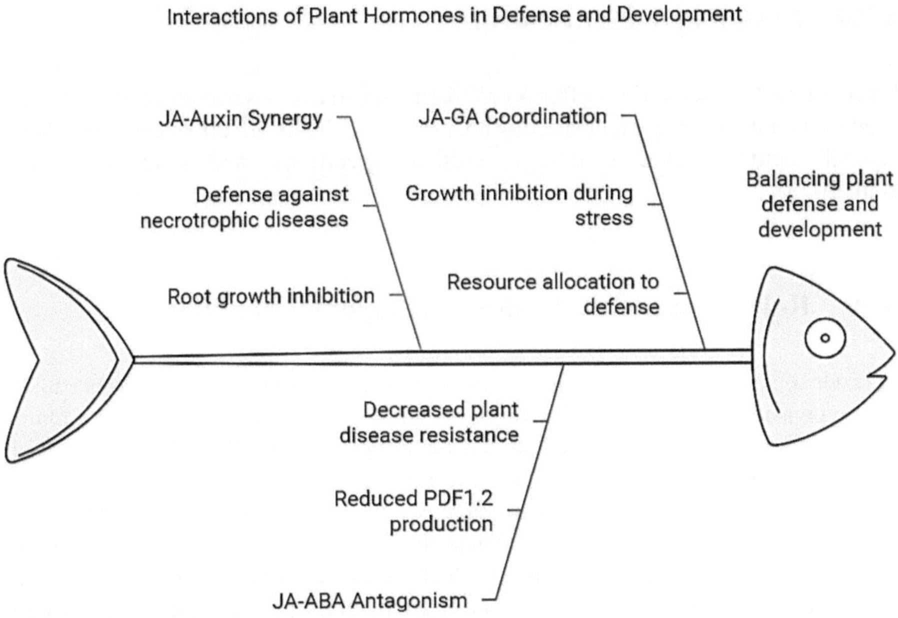

Fig. 8.3 Interactions of JA with other plant hormones in defense and development

8.9.1 *Activation of Defense Mechanisms*

Several jasmonates, such as methyl jasmonate (MeJA), cis-jasmone (CJ), and prohydrojasmon (PDJ), are useful in triggering defense reactions against insect attacks (Ali et al. 2024). These jasmonates increase defense against herbivores and pathogens by inducing defense-related genes and secondary metabolites (Li et al. 2022).

8.9.2 *Emission Control of Volatile Substances*

JA-treated plants produce several volatile compounds that resemble those of plants injured by herbivores. These volatile chemicals aid in the management and control of pests by attracting herbivores' natural enemies (Razo-Belman and Ozuna 2023).

8.9.3 *Production of Extrafloral Nectar*

By utilizing jasmonate, the production of extrafloral nectar increases, which feeds natural enemies and improves their fitness. This indicates employing the "attract and reward" tactic to bolster the defense mechanism of plants (Sobhy et al. 2024).

8.9.4 Sustainable Pest Control

Since JA and its derivatives offer a promising approach in agriculture, they can be used in the pest management strategy (Ali et al. 2023). Its effectiveness depends on several factors, including dosage, cultivar specificity, and time (Doostkam et al. 2023).

8.10 Role of Jasmonic Against Pathogenic Infections

In order to survive the changing natural environment and possible diseases caused by the invasion of pathogens like fungi, bacteria, viruses, viroids, and so on, plants must undergo various physiological and metabolic changes (Roychowdhury et al. 2020). The life cycle of plants is being negatively impacted by the global climate change. As a result, there is a greater chance of disease pressure, especially in major crops, which could lead to crop loss. These diseases pose a serious risk to plants and ultimately impact the development, productivity, and general well-being of entire crops. These plant diseases frequently infiltrate plant tissues because they impair the cells' ability to function and transport nutrients. Additionally, these infections impair the plant's ability to flourish by interfering with vital biological processes including photosynthesis and water intake (Roychowdhury et al. 2024). Plant hormones trigger defensive mechanisms because they are crucial in establishing a link between host-pathogen identification, their interaction, and subsequent cellular responses (Nambara and Van Wees 2021). JA generally stimulates plant defense mechanisms against herbivores and necrotrophic diseases, and it is advantageous to the development of systemic resistance (Ghozlan et al. 2020). Numerous studies have shown that JA is crucial for plants' responses to biotic stress, primarily from insects or pathogens. This section covers recent findings on the role of JA in plants' ability to withstand different kinds of biotic stress.

8.10.1 JA Role During Bacterial Infection

Induced systemic resistance (ISR) prepares plants to defend against pathogens more quickly and effectively, while JA controls the expression of ISR. The JA signaling pathways are activated by the helpful rhizobacteria. Unlike the hypersensitive response (HR) linked to salicylic acid (SA), ISR does not result in cell death. Because necrotrophic infections, such as certain bacteria, feed on dead cells, it works well against them. As a result, JA has the beneficial capacity to create resistance without causing cell death (Yu et al. 2022). JA triggers the activation of several defense genes. These defensive genes encode pathogenesis-related (PR) proteins, plant defensins, and proteinase inhibitors. Because they target bacterial infections, these genes are crucial for defense. By rupturing cell membranes or blocking vital bacterial enzymes, these compounds damage the bacterial cell (Khan et al. 2024).

8.10.2 JA Role During Fungal Infection

Plant defense mechanisms against fungal diseases, especially necrotrophic fungi, depend extensively on jasmonic acid (JA). According to the findings, plant defense genes can be expressed when JA or MeJA is applied exogenously. So also, treatment with JAs moved forward resistance to necrotrophic diseases by essentially expanding the expression of a few qualities within the JA signaling framework. MeJA treatment altogether expanded the expression of PR4, PR5, and peroxidase (PEROX) to ensure wheat from *Fusarium graminearum* (Fernandes and Ghag 2022). When JA was administered exogenously to citrus fruits, peroxidase and PPO activity increased. As a result, it prevents green and blue molds from growing on citrus. When wheat was infected with *Fusarium culmorum*, the administration of MeJA also led to a decrease in lipid peroxidation and H_2O_2 concentration. Numerous genes, including those linked to terpenoid biosynthesis, phenylalanine metabolism, plant-pathogen interaction, JA biosynthesis, and the JA signaling pathway, were markedly upregulated following MeJA treatment. This improved *Panax notoginseng*'s resistance to *Fusarium solani* during growth (Ning et al. 2021).

8.10.3 JA Role in Insect Resistance

JA is the most significant signaling molecule among those that control the plant defense responses to the insects. Insects, which are mostly herbivores, commonly cause harm to plants, which causes endogenous JA levels to rise quickly (Al-Zahrani et al. 2020). It's interesting to note that, despite JA's non-toxicity to insects, when plants are sprayed with it, systemic resistance and secondary metabolites are produced to prevent insect damage (Zhang et al. 2020a, b). Plants always accumulate defensive proteins in response to herbivore attacks, which alter pests' digestive physiologies and provide protection to the plants from herbivory. JAs may stimulate these protective proteins. For instance, exogenous JA or MeJA can stimulate the polyphenol oxidase (PPO) enzyme important for defense, which successfully defends plants against herbivores. MeJA boosted plant defense mechanisms against *Ostrinia furnacalis* and increased the synthesis of toxic proteins in maize (Wang et al. 2021). Furthermore, a variety of secondary metabolites that are produced via intricate branching metabolic pathways and are stimulated by jasmonates actively contribute to plants' resistance to insects. In the last ten years, secondary metabolites primarily tannins, total flavonoids, total phenols, and lignin that create resistance to herbivorous insect attacks have been discovered (Yan et al. 2021).

8.10.4 JA Role During Viral Infection

In the plant antiviral defense, mechanisms JA is essential (Wu and Ye 2020). Methyl jasmonate (MeJA), an unstable methyl ester of JA, successfully produces resistance in *Nicotiana benthamiana* to tobacco mosaic infection (TMV, sort Tobamovirus) and tomato spotted shrivel infection (TSWV, sort Tospovirus), to rice worn-out stunt infection (RRSV, sort Oryzavirus), rice stripe infection (RSV, sort Tenuivirus), and rice dark-streaked overshadow infection (RBSDV, sort Fijivirus), and to beet wavy beat infection (BCTV, class Begomovirus) in *Arabidopsis thaliana*. On the other hand, JA signaling concealment makes a difference in anticipating viral contamination: rice with JA-insensitive mutant coi1 and myc3 is more powerless to infections that taint rice, such as RBSDV and RSV, than the normal sort (Han et al. 2020).

There is growing evidence that the viruses interfere with the JA pathway, particularly JA signaling, to increase viral infection to undermine the plant JA-mediated defense (Zhao and Li 2021). Various ponders have illustrated that the plants contaminated with the tomato yellow leaf twist Sardinia infection (TYLCSV, sort Begomovirus), RRSV, and RBSDV show smothered expression of qualities included within the biosynthesis of the JA. This proposes that infections may also target JA biosynthesis to extend viral contamination. The precise nature of this regulation's operation is still vague.

8.11 Conclusion

Jasmonic acid (JA) has a crucial role in plant defense mechanisms against many biotic stresses, including pathogens, pests, and herbivores. The regulation of transcriptional reprogramming and the synthesis of particular protective chemicals turn on this defense mechanism. Its production and signaling pathways involve numerous essential factors, including MYC transcription factors, COI1 receptors, and JAZ proteins. These elements work together to create an effective network that activates both localized and systemic resistance, including induced systemic resistance (ISR). Auxins, gibberellins, abscisic acid, salicylic acid, and other plant hormones interact with JA to allow plants to cooperate and balance immunity and growth. For plants to adapt to challenging biotic stress situations, this hormone signaling is essential. Additionally, numerous case studies and field applications show that JA and its derivatives deliver promising outcomes for sustainable agriculture and function in concert with crop protection.

However, additional research is necessary to address several issues, including environmental variability, potential trade-offs with growth, and constraints in large-scale use. To improve crop resilience sustainably, more research should concentrate on optimizing JA-based tactics through genetic engineering, precision agriçulture, and integrated pest management techniques. Utilizing the knowledge of JA-mediated defense, scientists along with industrial professionals might utilize this phytohormone's potential to create novel approaches to sustainable crop protection. Thus, it will promote global food security amidst growing agricultural difficulties.

References

Al-Zahrani W, Bafeel SO, El-Zohri M (2020) Jasmonates mediate plant defense responses to *Spodoptera exigua* herbivory in tomato and maize foliage. Plant Signal Behavior 15:1746898

Ali J, Mukarram M, Ojo J, Dawam N, Riyazuddin R, Ghramh HA, Khan KA, Chen R, Kurjak D, Bayram A (2024) Harnessing phytohormones: advancing plant growth and defence strategies for sustainable agriculture. Physiol Plant 176:e14307

Ali J, Wei D, Mahamood M, Zhou F, King PJH, Zhou W, Shamsi IH (2023) Exogenous application of methyl salicylate induces Defence in *brassica* against peach potato aphid *Myzus persicae*. Plan Theory 12:1770–1778

Ali MS, Baek KH (2020) Jasmonic acid signaling pathway in response to abiotic stresses in plants. Int J Mol Sci 21:621

Bhatla SC, Lal MA (2023) Jasmonic acid. In: Plant physiology, development and metabolism. Springer, pp 473–478

Bian C, Duan Y, Wang J, Xiu Q, Wang J, Hou Y, Song X, Zhou M (2020) Validamycin a induces broad-spectrum resistance involving salicylic acid and jasmonic acid/ethylene signaling pathways. Mol Plant-Microbe Interact 33:1424–1437

Castro-Camba R, Sánchez C, Vidal N, Vielba JM (2022) Interactions of gibberellins with phytohormones and their role in stress responses. Horticulturae 8:241

Devireddy AR, Zandalinas SI, Fichman Y, Mittler R (2021) Integration of reactive oxygen species and hormone signaling during abiotic stress. Plant J 105:459–476

Ding F, Wang X, Li Z, Wang M (2022b) Jasmonate positively regulates cold tolerance by promoting ABA biosynthesis in tomato. Plan Theory 12:60–67

Ding L, Li YT, Wu YZ, Li T, Geng R, Cao J, Zhang W, Tan XL (2022a) Plant disease resistance-related signaling pathways: recent progress and future prospects. Int J Mol Sci 23:16200–16210

Doostkam M, Sohrabi F, Modarresi M, Kohanmoo MA, Bayram A (2023) Genetic variation of cucumber (*Cucumis sativus* L.) cultivars to exogenously applied jasmonic acid to induce resistance to *Liriomyza sativae*. Arthropod Plant Interact 17:289–299

El Sabagh A, Islam MS, Hossain A, Iqbal MA, Mubeen M, Waleed M, Reginato M, Battaglia M, Ahmed S, Rehman A (2022) Phytohormones as growth regulators during abiotic stress tolerance in plants. Frontiers Agronomy 4:765068

Fernandes LB, Ghag SB (2022) Molecular insights into the jasmonate signaling and associated defense responses against wilt caused by *Fusarium oxysporum*. Plant Physiol Biochem 174:22–34

Figueroa-Macías JP, García YC, Núñez M, Díaz K, Olea AF, Espinoza L (2021) Plant growth-defense trade-offs: molecular processes leading to physiological changes. Int J Mol Sci 22:693

Genva M, Obounou AF, Andersson MX, Deleu M, Lins L, Fauconnier ML (2019) New insights into the biosynthesis of esterified oxylipins and their involvement in plant defense and developmental mechanisms. Phytochem Rev 18:343–358

Ghorbel M, Brini F, Sharma A, Landi M (2021) Role of jasmonic acid in plants: the molecular point of view. Plant Cell Rep 40:1471–1494

Ghozlan MH, Eman EA, Tokgöz S, Lakshman DK, Mitra A (2020) Plant defense against necrotrophic pathogens. Am J Plant Sci 11:2122–2138

Gomi K (2020) Jasmonic acid: an essential plant hormone, vol 21. MDPI, p 1261

Gomi K (2021) Jasmonic acid pathway in plants 2.0, vol 22. MDPI, p 3506

Guo Q, Han J, Li C, Hou X, Zhao C, Wang Q, Wu J, Mur LA (2022) Defining key metabolic roles in osmotic adjustment and ROS homeostasis in the recretohalophyte Karelinia caspia under salt stress. Physiol Plant 174:13663

Han K, Huang H, Zheng H, Ji M, Yuan Q, Cui W, Zhang H, Peng J, Lu Y, Rao S (2020) Rice stripe virus coat protein induces the accumulation of jasmonic acid, activating plant defence against the virus while also attracting its vector to feed. Mol Plant Pathol 21:1647–1653

Hewedy OA, Elsheery NI, Karkour AM, Elhamouly N, Arafa RA, Mahmoud GAE, Dawood MFA, Hussein WE, Mansour A, Amin DH (2023) Jasmonic acid regulates plant development and orchestrates stress response during tough times. Environ Exp Bot 208:105260

Howlader P, Bose SK, Jia X, Zhang C, Wang W, Yin H (2020) Oligogalacturonides induce resistance in *Arabidopsis thaliana* by triggering salicylic acid and jasmonic acid pathways against Pst DC3000. Int J Biol Macromol 164:4054–4064

Hu J, Huang J, Xu H, Wang Y, Li C, Wen P, You X, Zhang X, Pan G, Li Q (2020) Rice stripe virus suppresses jasmonic acid-mediated resistance by hijacking brassinosteroid signaling pathway in rice. PLoS Pathog 16:1008801

Iqbal Z, Iqbal MS, Hashem A, Abdullah EF, Ansari MI (2021) Plant defense responses to biotic stress and its interplay with fluctuating dark/light conditions. Front Plant Sci 12:631810

Jang G, Yoon Y, Choi YD (2020) Crosstalk with jasmonic acid integrates multiple responses in plant development. Int J Mol Sci 21:305

Jiang Y, Yue Y, Wang Z, Lu C, Wang Z, Yin Z, Li Y, Hao GF, Ding X (2024) A novel ABA structural analogues enhanced plant resistance by inducing the plant immunity and inactivating ABA signaling pathway. Advanced Agrochem 3:64–73

Juan CA, Lastra JM, Plou FJ, Pérez-Lebeña E (2021) The chemistry of reactive oxygen species (ROS) revisited: outlining their role in biological macromolecules (DNA, lipids and proteins) and induced pathologies. Int J Mol Sci 22:4642

Khan MIR, Kumari S, Nazir F, Khanna RR, Gupta R, Chhillar H (2023a) Defensive role of plant hormones in advancing abiotic stress-resistant rice plants. Rice Sci 30:15–35

Khan T, Shah LR, Khan S, Rashid S, Rashid R, Afroza B (2023b) Key enzymes of Oxylipins pathway. In: Phyto-Oxylipins. CRC Press, pp 43–63

Khan V, Jha A, Seth T, Iqbal N, Umar S (2024) Exploring the role of jasmonic acid in boosting the production of secondary metabolites in medicinal plants: pathway for future research. Ind Crop Prod 220:119227

Kim J, Chang C, Tucker ML (2015) To grow old: regulatory role of ethylene and jasmonic acid in senescence. Front Plant Sci 6:20–27

Kumar A, Partap M, Warghat AR (2024) Jasmonic acid: a versatile phytohormone regulating growth, physiology, and biochemical responses. J Plant Growth Regul 1:1–24

Li C, Xu M, Cai X, Han Z, Si J, Chen D (2022) Jasmonate signaling pathway modulates plant defense, growth, and their trade-offs. Int J Mol Sci 23:3945

Li M, Wang F, Li S, Yu G, Wang L, Li Q, Zhu X, Li Z, Yuan L, Liu P (2020) Importers drive leaf-to-leaf jasmonic acid transmission in wound-induced systemic immunity. Mol Plant 13:1485–1498

Li M, Yu G, Cao C, Liu P (2021) Metabolism, signaling, and transport of jasmonates. Plant Commun 2:50–63

Liu H, Timko MP (2021) Jasmonic acid signaling and molecular crosstalk with other phytohormones. Int J Mol Sci 22:2914

Loo S, Tay SV, Kam A, Tang F, Fan JS, Yang D, Tam JP (2021) Anti-fungal Hevein-like peptides biosynthesized from quinoa cleavable hololectins. Molecules 26:5909

Luo L, Wang Y, Qiu L, Han X, Zhu Y, Liu L, Man M, Li F, Ren M, Xing Y (2023) MYC2: a master switch for plant physiological processes and specialized metabolite synthesis. Int J Mol Sci 24:3511

Lv ZY, Sun WJ, Jiang R, Chen JF, Ying X, Zhang L, Chen WS (2021) Phytohormones jasmonic acid, salicylic acid, gibberellins, and abscisic acid are key mediators of plant secondary metabolites. World J Trad Chinese Med 7:307–325

Ma F, Yang X, Shi Z, Miao X (2020) Novel crosstalk between ethylene-and jasmonic acid-pathway responses to a piercing–sucking insect in rice. New Phytol 225:474–487
Mazzoni-Putman SM, Brumos J, Zhao C, Alonso JM, Stepanova AN (2021) Auxin interactions with other hormones in plant development. Cold Spring Harb Perspect Biol 13:39990
Muhammad M, Waheed A, Wahab A, Majeed M, Nazim M, Liu YH, Li L, Li WJ (2024) Soil salinity and drought tolerance: An evaluation of plant growth, productivity, microbial diversity, and amelioration strategies. Plant Stress 11:100319
Nambara E, Van Wees SC (2021) Plant hormone functions and interactions in biological systems, vol 105. Wiley Online Library, pp 287–289
Ning K, Li M, Wei G, Zhou Y, Zhang G, Huai H, Wei F, Chen Z, Wang Y, Dong L (2021) Genomic and transcriptomic analysis provide insights into root rot resistance in *Panax notoginseng*. Front Plant Sci 12:775019
Prusty A, Sharma S, Malik N, Kapoor S, Tyagi AK (2023) Functional versatility of multi-protein mediator complex in plant growth and development. Crit Rev Plant Sci 42:138–176
Rai GK, Khanday DM, Choudhary SM, Kumar P, Kumari S, Martínez-Andújar C, Martínez-Melgarejo PA, Rai PK, Pérez-Alfocea F (2024) Unlocking nature's stress buster: Abscisic acid's crucial role in defending plants against abiotic stress. Plant Stress 11:100359
Razo-Belman R, Ozuna C (2023) Volatile organic compounds: a review of their current applications as pest biocontrol and disease management. Horticulturae 9:441
Roychowdhury R, Choudhury S, Hasanuzzaman M, Srivastava S (2020) Sustainable agriculture in the era of climate change. Springer, pp 1–37
Roychowdhury R, Hada A, Biswas S, Mishra S, Prusty MR, Das SP, Ray S, Kumar A, Sarker U (2024) Jasmonic acid (JA) in plant immune response: unravelling complex molecular mechanisms and networking of Defence Signalling against pathogens. J Plant Growth Regul 24:1–26
Scalschi L, Llorens E, García-Agustín P, Vicedo B (2020) Role of jasmonic acid pathway in tomato plant-*pseudomonas syringae* interaction. Plan Theory 9:136–145
Sobhy IS, Gurr GM, Jones TH (2024) Induced plant resistance and its influence on natural enemy use of plant-derived foods. Curr Opinion Insect Sci 1:101–118
Sun T, Zhang Y (2021) Short-and long-distance signaling in plant defense. Plant J 105:505–517
Wang H, Zha W, Huang A, Wu Y, Shi S, Zhou L, You A (2024) The roles of Phytohormones in plant defense mechanisms against the Brown Planthopper. Genes 15:1579–1589
Wang J, Song L, Gong X, Xu J, Li M (2020) Functions of jasmonic acid in plant regulation and response to abiotic stress. Int J Mol Sci 21:1446–1455
Wang Y, Mostafa S, Zeng W, Jin B (2021) Function and mechanism of jasmonic acid in plant responses to abiotic and biotic stresses. Int J Mol Sci 22:8568–8577
Wei H, Movahedi A, Xu C, Sun W, Wang P, Li D, Yin T, Zhuge Q (2020) Characterization, expression profiling, and functional analysis of PtDef, a defensin-encoding gene from *Populus trichocarpa*. Front Microbiol 11:106–112
Wu J, Chen Y, Xu Y, An Y, Hu Z, Xiong A, Wang G (2024) Effects of Jasmonic acid on stress response and quality formation in vegetable crops and their underlying molecular mechanisms. Plan Theory 13:1557–1568
Wu X, Ye J (2020) Manipulation of jasmonate signaling by plant viruses and their insect vectors. Viruses 12:148–160
Xu DB, Ma YN, Qin TF, Tang WL, Qi XW, Wang X, Liu RC, Fang HL, Chen ZQ, Liang CY (2021) Transcriptome-wide identification and characterization of the jaz gene family in *Mentha canadensis* L. Int J Mol Sci 22:8859–8870
Xue H, Gao X, He P, Xiao G (2022) Origin, evolution, and molecular function of DELLA proteins in plants. The Crop Journal 10:287–299
Yan J, Tan Y, Lv Y, Wang F, Zhang Y, Chi D (2021) Effect of jasmonate treatments on leaves of *Rosa rugosa* 'plena'and detoxification enzymes and feeding of adult Monolepta hieroglyphica. J For Res 32:1253–1261
Yu Y, Gui Y, Li Z, Jiang C, Guo J, Niu D (2022) Induced systemic resistance for improving plant immunity by beneficial microbes. Plan Theory 11:386

Zeng Y, Verstraeten I, Trinh HK, Heugebaert T, Stevens CV, Garcia-Maquilon I, Rodriguez PL, Vanneste S, Geelen D (2021) Arabidopsis hypocotyl adventitious root formation is suppressed by ABA signaling. Genes 12:1141–1150

Zhang H, Zhao Y, Zhu JK (2020b) Thriving under stress: how plants balance growth and the stress response. Dev Cell 55:529–543

Zhang PJ, Zhao C, Ye ZH, Yu XP (2020a) Trade-off between defense priming by herbivore-induced plant volatiles and constitutive defense in tomato. Pest Manag Sci 76:1893–1901

Zhang Q, Gong M, Xu X, Li H, Deng W (2022) Roles of auxin in the growth, development, and stress tolerance of horticultural plants. Cells 11:2761–2790

Zhao S, Li Y (2021) Current understanding of the interplays between host hormones and plant viral infections. PLoS Pathog 17:109–124

Zhou XE, Zhang Y, Yao J, Zheng J, Zhou Y, He Q, Moreno J, Lam VQ, Cao X, Sugimoto K (2023) Assembly of JAZ–JAZ and JAZ–NINJA complexes in jasmonate signaling. Plant Communications 4:1–10

Zhu X, Zhao Y, Shi CM, Xu G, Wang N, Zuo S, Ning Y, Kang H, Liu W, Wang R (2024) Antagonistic control of rice immunity against distinct pathogens by the two transcription modules via salicylic acid and jasmonic acid pathways. Dev Cell 59:1609–1622

Chapter 9
Alginate-Induced Immunity: A New Frontier in Plant Health

Mona F. A. Dawood, Murad Muhammed, Debasis Mitra, Mohammad Imran Mir, and Samar G. Thabet

Abstract Plants are constantly challenged by diverse pathogens and have evolved sophisticated defense reactions to survive these attacks. Treatment of diverse biotic and abiotic agents can strengthen pathogen infection resistance in plants. Recently, abiotic resistance inducers proved to be important components of the integrated disease management programs. Natural inducers among others include polysaccharides and their derived oligosaccharides, representing eco-friendly compounds that promote plant growth beside a concurrent stimulation of pathogen-induced resistance. Alginates (ALG) are gaining interest by virtue of their eco-friendliness and dual roles as a plant stimulant for defense mechanisms and growth promotions. Alginate and its degradation products initiate defense responses by eliciting plant immunity, as it acts as a pathogen-associated molecular pattern. Moreover, ALG regulates the biosynthesis and metabolism of abscisic acid, which helps maintain fruit quality and prolong shelf life. Alginate and its biological activity in plants, effectiveness of alginates against pathogen infestation. Identifying of the mechanisms

M. F. A. Dawood (✉)
Botany and Microbiology Department, Faculty of Science, Assiut University, Assiut, Egypt
e-mail: Mo_fa87@aun.edu.eg

M. Muhammed
State Key Laboratory of Desert and Oasis Ecology, Key Laboratory of Ecological Safety and Sustainable Development in Arid Lands, Xinjiang Institute of Ecology and Geography, Chinese Academy of Sciences, Xinjiang, China

University of Chinese Academy of Sciences, Beijing, People's Republic of China
e-mail: muradbotany1@uop.edu.pk

D. Mitra
Department of Microbiology, Graphic Era (Deemed to be University), Dehradun, Uttarakhand, India

M. I. Mir
Department of Plant Sciences, University of Hyderabad, Hyderabad, Telangana, India

S. G. Thabet
Department of Botany, Faculty of Science, Fayoum University, Fayoum, Egypt

K. A. Abd-Elsalam, H. I. Mohamed (eds.), *Elicitors for Sustainable Crop Production*, Sustainability Sciences in Asia and Africa,
https://doi.org/10.1007/978-981-96-6569-3_9

under biotic and abiotic stress will assist in developing research and additional application practices of alginate-derived oligosaccharides (AOS) providing sustainable and economical plant protection methods in the future.

Keywords Oligosaccharide · Plant immunity · Postharvest · Alginate · Natural products

9.1 Introduction

Using various pathogenic mechanisms, plant pathogens cause diseases that can lead to notable economic losses (Al-Ani and Furtado 2020). Pathogen recognition of plant surface receptors, the product of virulent proteins and the invasion of host defense barriers are often key hallmarks of a successful infection (Dawood et al. 2023a, b).

During coevolution, pathogens develop new virulent races, while plants adapt with defense mechanisms. Pathogen attacks trigger signal pathways that elicit the expression of plant defense genes, activating resistance mechanisms (Dawood et al. 2024).

These responses, when activated, interrupt pathogen infection and downregulate pathogen gene expression. Both endogenous and exogenous elicitors can trigger defense responses (Bagy et al. 2019). The use of exogenous agents is common in agriculture to reduce losses caused by pathogens. Genetic resistance that can be obtained from breeding and engineering crops is an availing long-term solution; however, at a shorter timescale to active plant immunity, biotic inducers can also be used but are quite complex due to interactions among pathogens, plants, biological inducers, and environmental factors (Dawood et al. 2025, Abdelmoneim et al. 2025, Mawad et al. 2024). These complex interactions, however, also reduce the efficacy of the field relative to controlled conditions (Bagy et al. 2019). The role of abiotic agents in improving defense responses against pathogen infection was reported such as hydrogen-peroxide mixed compounds which were an effective example for defense priming (Abdelrhim et al. 2021).

Plants possess innate immune systems that share in defending against phytopathogens. For developing this natural defense system, researchers have applied various chemical compounds and natural products, known as plant immunity inducers. These inducers include oligosaccharides, proteins, peptides, unsaturated fatty acids, plant hormones, and their functional analogs (Zhu et al. 2023; Wang et al. 2024). Unlike conventional chemical pesticides, these inducers potentially induce the plant's innate immune responses against infestation (Wang et al. 2024). The action mechanisms of plant immunity inducers involve multiple processes, such as the production of reactive oxygen species (ROS), upregulation of defense-related genes, and stimulation of disease-resistant compound synthesis (Sallam et al. 2021; Ochoa-Meza et al. 2021). These agents are employed in contemporary agricultural practices to minimize dependence on chemical pesticides, thereby promoting eco-friendly practices. Not only does this help to reduce pollution and maintain ecological balance, but it also improves the safety and quality of agricultural products (Younes et al. 2024, Abdelrhim et al. 2024).

Hydrolysis of natural polysaccharides is a strategy to produce oligosaccharides, which can also act as elicitors of resistance and change the expression patterns of plant defense genes (Wang et al. 2024). In this context, oligogalacturonide initiates a plant defense response, such as production of phytoalexins, β-1,3-glucanase, chitinase, ROS, and inducing nitric oxide (NO) production (Rasul et al. 2013).

The oligosaccharide, oligochitosan, showed its efficacy against fungi, bacteria, and viruses by activating salicylic acid (SA) and jasmonic acid–ethylene pathways besides improve abiotic stresses resistance via an abscisic acid (ABA)-dependent mechanism (Moenne et al. 2021). Also, the microbial oligosaccharide such as oligocurdlan, extracted from *Agrobacterium* spp., activate immune defense responses against *Phytophthora infestans* in potatoes (Li et al. 2014). Furthermore, oligosaccharides derived from green and brown algae activate defense signaling and are commonly used in agriculture as fertilizers and soil conditioners (Stengel et al. 2015). FytoSave® and Stemicol®, are commercial products of oligosaccharides used for plant protection, emphasizing their potential as alternatives to fungicides. van Aubel et al. (2014) reported that the commercial products of oligosaccharides such as FytoSave® (LIDA Plant Research, Valencia, Spain), a mix of oligochitosans and oligopectates have been marketed for plant protection and applied against downy mildew infection in grape and cucumber. In 2018, FytoSave® product (LIDA Plant Research, Valencia, Spain) as the first plant phytovaccine with phytosanitary registration was admitted by the European Commission for use in organic agriculture (https://www.infoagro.com, accessed on January 8, 2022). Also, FytoSave® (LIDA Plant Research, Valencia, Spain), COS–OGA, active against *Phytophthora infestans,* the causal agent of potato late blight, by enhancing pathogenesis-related (PR) proteins via SA pathway. COS–OGA combines cationic chitosan oligomers, chitooligosaccharides (COSs), with anionic pectin oligomers, oligogalacturonides (OGAs) (van Aubel et al. 2014, 2016). The product, Stemicol® (LIDA Plant Research, Valencia, Spain), is a combination of chitooligosaccharides applied against fruit rot in tomatoes, strawberries, and grapes (https://www.lidaplantresearch.com/phytovaccines/stemicol, accessed on January 8, 2022).

Alginate oligosaccharides (AOS), derived from seaweed, enhance seed germination, shoot elongation, and pathogen resistance by activating ROS production and defense enzymes (Zhang et al. 2015). With a global market value of $442.6 million in 2020, alginate and its derivatives are gaining prominence as environmentally friendly solutions in sustainable agriculture (Zhang et al. 2021). Sodium alginate, a polysaccharide produced mainly from seaweed, has emerged as a plant defense elicitor of significant interest. Oligosaccharides produced from ALG, are known as alginate oligosaccharides (AOS), and have been known for their potential as functional biomaterials in various applications such as promoting seed germination, shoot elongation, root development, and activate defense mechanisms against phytopathogens (Zhang et al. 2019; Popletaeva et al. 2018; Dey et al. 2019). ALG has been observed to enhance seed germination (Luan et al. 2009), improve plant growth and development (Bhattacharyya et al. 2015), and upregulate ion uptake and transport (Ali et al. 2014). In addition, ALG promotes photosynthetic efficiency (Aftab et al. 2011) via increasing pigment content (Idrees et al. 2011) which ultimately boosts protein assimilation (Idrees et al. 2015). Studies also showed that ALG activates the production of secondary metabolites, such as alkaloids and essential oils, ensuring its influential in sustainable agriculture (Aftab et al. 2011; Idrees et al. 2011). ALG has shown promise in plant disease protection by inducing inhibition of phytopathogens (Sano 1999). The role of AOS in activating plant defense mechanisms was reported by attenuating ROS as well as promoting the production of pathogenesis-related proteins and antioxidants like peroxidase (POD), catalase (CAT), polyphenol oxidase (PPO), and phenylalanine ammonia-lyase (PAL) (Zhang et al. 2021). Moreover, AOS activates phosphodiesterase production in plant cell suspension cultures and stimulates a natural defense inducer (Zhang et al. 2021). These studies display the significance of ALG and AOS as eco-friendly inducers for improving crop resilience and productivity against infestation, thus can be applied as valuable substitution to synthetic pesticides in sustainable agriculture. The exploration of natural inducers such as ALG and AOS provides a promising choice for resistance management strategies against plant diseases while reducing the application of chemically derived pesticides.

With a global market value projected to reach $547.2 million by 2027, alginate and its derivatives hold promise for sustainable agriculture (Zhang et al. 2021). Their ability to enhance plant defenses and growth draw attention to their potential as viable choices to synthetic pesticides. Thus, this chapter focuses on the defense mechanisms induced by ALG or their derivatives in combating pathogen attacks and enhancing plant resilience. Further investigations of natural inducers like ALG and AOS could significantly advance plant disease management practices.

9.2 Plant Immune System and Defense Mechanisms

The molecular basis of plant immunity is well-studied and relies on the ability of plant cells to recognize pathogen-associated molecular patterns, or microbe-associated molecular patterns (PAMPs/MAMPs), which are characteristic for microbes, through their cell surface pattern recognition receptors (PRRs) (Wang et al. 2024). PAMPs play an important role for the pathogen to thrive, and the recognition of the PAMPs mounts a defense response known as PAMP-triggered immunity (PTI), an innate immunity that confers plant against broad-spectrum protection against nonhost pathogens and attenuates virulent pathogen induced diseases (Naveed et al. 2020). But pathogens can develop strategies to overcome PTI. In contrast, plants mount resistance by producing resistance proteins (R proteins) that directly recognize effector proteins from the pathogen, which triggers effector-triggered immunity (ETI), a highly specific response often involving hypersensitive responses (HR) and systemic acquired resistance (SAR) (Noman et al. 2020). Additionally, damage-associated molecular patterns (DAMPs), including cell wall fragments formed during pathogen infestation, can trigger immune responses

(Noman et al. 2020). The pathways PTI and ETI trigger the defense responses such as reactive oxygen species (ROS) accumulation, cell wall fortification, and upregulation of defense hormones [SA, JA, and ethylene (ET)], with regulatory cross talk among these pathways (Singh et al. 2024). Under normal conditions, plant defense genes remain inactive but can be induced by various stimuli, leading to induced resistance. This involves systemic acquired resistance (SAR) or induced systemic resistance (ISR), both promote plant capability to withstand subsequent infection (Yue et al. 2021).

9.2.1 Systemic Acquired Resistance (SAR)

SAR provides long-term protection against multiple pathogenic microbes depending on signals like SA, ROS, and lipid-based molecules (Sallam et al. 2021). SA is a crucial regulator of SAR, activating the expression of pathogenesis-related (PR) gene and signaling pathways through the *isochorismate synthase 1 (ICS1)* (Abdelrhim et al. 2024, 2023). ROS and nitric oxide (NO) act as chemical signals that activate SAR, causing induction of defense strategies (Sofy et al 2022, Sheteiwy et al. 2022, Dawood et al. 2023a, b, 2024).

9.2.2 Induced Systemic Resistance (ISR)

Plant growth-promoting rhizobacteria induces ISR that colonize roots and activate systemic defenses against phytopathogens. In this process PGPR can utilize SA-, JA-, ET-dependent pathways (De Vleesschauwer and Höfte 2009). NPR1, a key regulator of SAR, also plays a partial role in ISR, though its specific functions remain under investigation (Pieterse et al. 2014). The encapsulation of plant probiotic agents, like PGPR, is a novel technique in agriculture, notably in the treatment of plant diseases, and it has great promise, especially when combined with useful, affordable, and accessible chemicals like ALG (Saberi Riseh et al. 2021). For instance, microorganisms that promote plant development, including Bacillus and Pseudomonas, and biocontrol agents were co-encapsulated in a biodegradable polymer covering, particularly ALG, to improve plant health (Fig. 9.1).

9.2.3 Defense Priming

Priming technique improves the tolerance of plants to stress, enabling faster and stronger defense responses upon subsequent pathogen attacks. Biological and chemical products can act as priming agents, strengthening resistance across various plant species (Mauch-Mani et al. 2017). This framework of PTI, ETI, SAR,

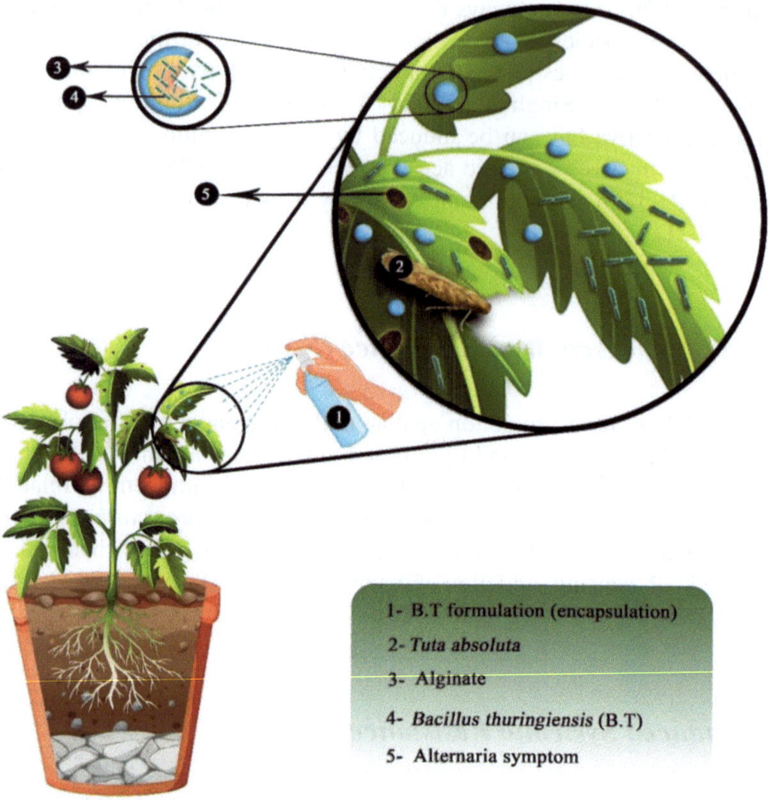

Fig. 9.1 Diagram illustrates the impact of alginate (ALG) encapsulation technology on agricultural practices. It highlights how microcapsules containing biologically relevant bacteria and active substances can effectively immobilize these agents, enhancing their functionality. The graph demonstrates the efficacy of biocontrol bacteria microcapsules in managing pest populations and disease outbreaks in crops, showcasing a significant improvement in plant health and resilience (reprinted from Saberi Riseh et al. 2021)

ISR, and priming highlights the dynamic and interconnected nature of plant immune responses to biotic stressors.

9.3 Alginate: Properties and Preparation

Algal polysaccharides are important organic molecules with paramount structural diversity and prospect for activating resistance in plants (Mani and Nagarathnam 2018). Sodium alginate (ALG), a polysaccharide produced mainly from brown macroalgae such as *Macrocystis pyrifera*, *Laminaria hyperborea*, and *Ascophyllum nodosum* and some bacteria, including *Azotobacter vinelandii* and *Pseudomonas* species, forming up to 40% of their dry weight (Abasalizadeh et al. 2020).

Structurally, ALG is a linear biopolymer formed of 1,4-linked β-D-mannuronate (M) and α-L-guluronate (G), which have homopolymeric and heteropolymeric blocks (Dey et al. 2019). Alginate is a linear anionic biopolymer composed of β-D-mannuronic acid (M) and its C5-epimer α-L-guluronic acid (G) linked by 1 → 4 glycosidic bonds. This polymer can exist as a homopolysaccharide (poly-M or poly-G) or as a heteropolysaccharide containing M/G in varying ratios and sequences. Alginate was first identified as a major component of brown algae cell walls in the late nineteenth century, during efforts to explore valuable products derived from kelp (Nussinovitch 2012). The term "alginic acid" was coined to describe the gelatinous material isolated from seaweed (Draget et al. 2005). However, the molecular structure of alginate was elucidated much later, primarily through the foundational work of Fischer and Dörfel in 1955 (Fischer and Dörfel 1955).

Due to its hydrophilic nature, ALG hydrogels can absorb large amounts of water or biological fluids while keeping their structural integrity, making them suitable for many biological applications (Abasalizadeh et al. 2020). ALG has the advantage of being both nontoxic and environmentally friendly, offering a sustainable alternative to synthetic polymers in numerous fields.

Since then, the structural variability of alginate, including its M/G content and sequential organization, has been extensively studied in various macroalgae (Haug et al. 1974). This variability, combined with alginate's capacity to form gels with diverse physicochemical properties, demonstrates its extensive application in various industries, including food, cosmetics, wastewater treatment, pharmaceuticals, and biomedicine (Szekalska et al. 2016; Giridhar 2022; Yarkent et al. 2022).

Alginate is commercially produced from seaweed, certain microorganisms such as *Pseudomonas* and *Azotobacter* have been extensively studied for their alginate production capabilities (Hay et al. 2013). The first production of ALG was produced by *Pseudomonas aeruginosa* during the mid-1960s by Linker and Jones (1964, 1966). Research on this opportunistic pathogen has demonstrated that ALG plays a critical role in the lung infections of cystic fibrosis patients, where it acts as a key virulence factor (Yorgey et al. 2008).

Also, *Azotobacter vinelandii*, a nitrogen-fixing bacterium associated with plants, was identified as ALG producer after *Pseudomonas aeruginosa* (Gorin and Spencer 1966). After that these genera, including *P. putida* and *P. fluorescens* (Conti et al. 1994), *P. mendocina* (Sengha et al. 1989), *P. syringae* (Fakhr et al. 1999), and *A. chroococcum* (Cote and Krull 1988; Pecina and Paneque 1994) have reported to produce ALG.

The biosynthetic mechanisms of bacterial alginates are considerably similar among species but some differences from seaweed have been reported in a key characteristic that microbial alginates are often O-acetylated, with esterification occurring at O-2 and/or O-3 of the D-mannuronate units (Skjåk-Bræk et al. 1986). This acetylation influences the physicochemical properties of polymers relative to macroalgal alginate (Smidsrød and Draget 2004). Furthermore, the seaweed-derived alginates, bacterial alginates exhibit variations in M-, G-, and MG-block arrangements that affect gel-forming properties and viscosity (Jiao et al. 2019; Abkakhajouei et al. 2022).

Mucoid strains of *P. aeruginosa* synthesize ALG rich with M- and MG-blocks, which are often O-acetylated. These alterations promote the interaction between polymers and water molecules, causing increased polymer extension and water retention. Thus, this facilitates the production of a well-developed biofilm matrix (Gloag et al. 2018). On the other hand, ALG produced by *Azotobacter* spp. are closely linked with cells and maintain high levels of L-guluronate residues (G-blocks). This structure allows the bond of divalent ions, such as Ca^{2+}, producing hard, brittle gels that support the development of desiccation-resistant cysts under unfavorable environmental conditions (Díaz-Barrera et al. 2021).

The production of tailor-made alginates, supposed to have particular molecular weights, M/G ratios, and O-acetyl contents, has taken prominent interest because of its significance for use-driven design in various industrial and medical applications (Ertesvåg 2015; Kirdponpattara et al. 2015). Advances in biotechnology improved the physical and chemical characteristics of ALG, to be applicable for various applications such as drug delivery, tissue engineering, and food technology (Moradali et al. 2015; Dudun et al. 2022).

The ability to improve fermentation parameters and growth conditions to induce ALG production and structural modifications has been reported, particularly in *Azotobacter vinelandii* which is nonpathogenic, making it highly applicable for biotechnological processes (Pacheco-Leyva et al. 2016; Bonartseva et al. 2017). The yield and structure of ALG including molecular weight and viscosity has been affected by the level of oxygen, nutrient availability, and pH has demonstrated significant effects on ALG yield and structure, which are crucial for application-specific tailoring (Bonartseva et al. 2017).

Recent research in genomic sequencing and molecular tools have increased the knowledge on fine structural adjustment of ALG. Genome sequencing of *Pseudomonas aeruginosa* and *A. vinelandii* has showed the genes involved in alginate biosynthesis, enabling targeted modifications through genetic engineering (Setubal et al. 2009). The isolation of alginate-modifying enzymes, such as mannuronan C-5-epimerases and alginate lyases, has further allowed on-demand chemical tailoring, enabling to enhance or suppress specific characteristics of ALG (Gaardløs et al. 2022; Gawin et al. 2020; Liang et al. 2022). For instance, alginate lyases have been applied to degrade overproduced alginate in clinical settings, elucidating concerns such as biofilm formation in cystic fibrosis patients (Mahajan et al. 2021; Tavafi et al. 2018).

Understanding the molecular mechanisms affecting ALG biosynthesis, transport, and metabolism—which are mainly linked with regulatory networks—is needed for optimizing production processes and therapeutic applications (Hay et al. 2014; Urtuvia et al. 2017). Regulatory proteins such as AlgT and c-di-GMP signaling molecules play critical roles in modulating gene expression related to alginate production (Borges et al. 2020). These insights have facilitated the development of mutant strains with enhanced biosynthetic capacities for large-scale production or therapeutic use against alginate-overproducing pathogens (García et al. 2018).

This chapter highlights the major steps of microbial ALG biosynthesis, from precursor assembly and polymerization to chemical modifications and transport across bacterial membranes. We explore the role of environmental factors, genetic regulation, and enzymatic modifications in shaping the final alginate product. Special emphasis is placed on the integration of biotechnological and molecular tools to achieve tailored alginate production for specific industrial and medical applications.

The unique physicochemical properties of ALG allow it to serve as an efficient delivery vehicle in applications such as drug delivery systems, wound dressings, and 3D bioprinting (Abasalizadeh et al. 2020). While insoluble in water and organic solvents in its acid form, ALG's monovalent alginate salts are water-soluble, forming stable solutions that enhance their applicability across different domains (Abasalizadeh et al. 2020). These characteristics make ALG a valuable component in plant disease management and sustainable agriculture. At physiological temperatures and pH, and in the presence of specific chemical initiators, the linear polymer of ALG can undergo free-radical polymerization to form a three-dimensional hydrogel network (Abasalizadeh et al. 2020). This polymerization process enables the incorporation of various compounds into the hydrogel matrix, creating a liquid–solid phase under physiological conditions (Varghese et al. 2008; Garagorri et al. 2008).

In medicine, ALG hydrogels are extensively used to encapsulate drugs, facilitating targeted delivery to specific tissues and organs (Abasalizadeh et al. 2020). Their formation as hydrogels allows ALG to serve as carriers for proteins, DNA, and live cells while preserving their biological activity (Kaur et al. 2024). Beyond medical applications, ALG has demonstrated its ability to promote plant growth and development and to enhance resistance against both biotic and abiotic stresses (Mughunth et al. 2024).

In plants, ALG can also interact with phenolic compounds, secondary metabolites that cross-link with ALG to fortify cell walls, thereby increasing resistance to pathogen infestation (Mughunth et al. 2024). These properties highlight ALG's versatility and its significant potential in agricultural and biomedical fields.

9.4 The Role of Alginate in Controlling Plant Diseases

Sodium alginate (ALG) and their derivatives like alginate oligosaccharides (AOS) have garnered considerable attention due to its environmental compatibility and nontoxic properties, particularly as an elicitor in controlling plant diseases (Dey et al. 2019). In a study investigating its role against *Alternaria solani*, the causal agent of tomato blight, ALG treatments were applied to tomato leaves two days before pathogen inoculation. The 0.4% concentration was particularly effective in suppressing fungal hyphal growth and significantly reducing pathogen colonization, as evidenced by Uvitex-2B fluorescence staining. ALG treatment also enhanced the activity of key defense enzymes, including superoxide dismutase (SOD), guaiacol peroxidase (GPX), and catalase (CAT), in infected tomato plants. Furthermore, defense-related genes associated with salicylic acid (SA), ethylene (ET), and jasmonic acid (JA) signaling pathways were regulated. Specifically, the expression of β-1,3-glucanase (PR2), nonexpresser of pathogenesis-related protein 1 (NPR1),

1-aminocyclopropane-1-carboxylate oxidase (ACO1), and lipoxygenase D (LoxD) was significantly elevated in plants treated with 0.4–0.6% ALG and infected with *A. solani* (Dey et al. 2019). Interestingly, PR4 expression increased in pathogen-infected control leaves but was lower in ALG-pretreated, pathogen-infected leaves compared to untreated ones. These findings highlight ALG's potential to induce systemic resistance in plants through enhanced enzyme activity and upregulation of multiple defense marker genes, making it a valuable tool in sustainable agriculture.

The major cell wall components of many phytopathogenic fungi, such as *Alternaria solani*, include chitin and glucans. Plant enzymes, such as β-1,3-glucanases and chitinases, play critical roles in plant defense by degrading these fungal cell wall components. These enzymes can also indirectly activate plant immune responses by generating oligosaccharides, which serve as elicitors to trigger further expression of pathogenesis-related (PR) proteins (Lawrence et al. 2000). In the case of ALG, its role in inducing defense responses in tomato seedlings against *A. solani* was shown to enhance the activity of antioxidant enzymes and PR proteins, which inhibit disease development (Dey et al. 2019). This demonstrates the potential of ALG as an elicitor in plant disease control by activating both direct and indirect defense mechanisms.

Other research has focused on combining protein elicitors with biopolymers like ALG to protect them from external stresses, improve their interaction with plant cell receptors, and induce disease resistance (Keefe and Jiang 2011). Peptidylprolyl isomerases (PPIases) are a class of protein elicitors that facilitate protein folding and regulate immune responses, transcription, and the cell cycle (Shaw 2002). In one study, PPIases from *Pseudomonas fluorescens* were encapsulated in ALG microparticles to assess their eliciting activity. The ALG-PPIase complex demonstrated enhanced resistance in tobacco plants infected with *A. longipes* and wheat plants infected with *Stagonospora nodorum* (Popletaeva et al. 2018). This complex not only induced resistance but also promoted plant growth by acting as an antipathogenic agent, with ALG enhancing the efficacy of encapsulated PPIase compared to free PPIase in some plant–pathogen models.

Furthermore, the role of reactive oxygen species (ROS) and nitric oxide (NO) in the plant immune response has been extensively studied. Zhang et al. (2015) showed that algino-oligosaccharides (AOS) act as elicitors in rice plants, enhancing disease resistance against *Magnaporthe grisea.* Their application led to increased production of phytoalexins and antioxidant enzymes, such as phenylalanine ammonia-lyase (PAL), catalase (CAT), and peroxidase (POD), which collectively bolster the plant's defenses. The induction of phytoalexin accumulation, particularly oryzalexin C, served as a sensitive marker for AOS elicitor activity (Zhang et al. 2015). Additionally, AOS treatments activated the salicylic acid (SA) signaling pathway, significantly enhancing the expression of the PR1 gene, a key marker of systemic acquired resistance (Zhang et al. 2015). This work illustrates the multifaceted roles of biopolymer-elicitor complexes like ALG in promoting plant disease resistance. By enhancing the activity of enzymes and activating key immune signaling pathways, ALG can serve as an effective tool for sustainable agricultural practices, offering protection against a range of phytopathogens.

The potential of alginate (ALG) and alginate oligosaccharides (AOS) as bio-elicitors for enhancing plant defense mechanisms has been extensively studied. For example, ALG derived from the brown seaweed *Bifurcaria bifurcata* and its oligosaccharide derivatives were shown to stimulate the natural defense responses of tomato seedlings by increasing phenylalanine ammonia-lyase (PAL) activity and polyphenol levels in treated leaves. PAL activity, a key indicator of phenylpropanoid metabolism activation, was notably enhanced 12 h posttreatment. This metabolic pathway plays a critical role in fortifying plants against phytopathogens. These findings highlight ALG and its derivatives as promising eco-sustainable resources for plant protection (Aitouguinane et al. 2020). In another study, a hydrogel combining ALG, lentinan, and aminooligosaccharides demonstrated strong resistance induction against tobacco mosaic virus (TMV) in *Nicotiana benthamiana* by increasing calcium ion release and promoting plant growth. Such results underline the multifaceted benefits of ALG-based biopolymers, including enhanced disease resistance and growth promotion (Xiang et al. 2019).

Further research corroborates the efficacy of ALG and AOS as elicitors against various plant pathogens, including fungi, bacteria, and viruses. These elicitors activate multiple defense pathways, including salicylic acid (SA) and jasmonic acid–ethylene (JA–ET) signaling. Evidence also suggests that AOS can induce abscisic acid (ABA)-dependent pathways, which are pivotal for abiotic stress responses, such as drought, salinity, and heavy metal stress (Tang et al. 2011; Liu et al. 2013; Moenne and González 2021). For instance, AOS application significantly enhanced wheat's tolerance to drought stress by modulating key physiological parameters (Liu et al. 2013). Similarly, in Brassica plants, AOS derived from alginase activity helped alleviate salt stress by activating critical stress-responsive pathways (Tang et al. 2011). These findings demonstrate the broad-spectrum utility of AOS in mitigating both biotic and abiotic stresses in plants. Moreover, studies have demonstrated that AOS treatments increase resistance to postharvest decay in fruits like kiwifruit and pears, further expanding their potential applications in agriculture (Han et al. 2021; Zhuo et al. 2021). For example, combining AOS with beneficial microbes such as *Meyerozyma guilliermondii* significantly reduced fungal infections in pears, indicating synergistic effects in pathogen control (Han et al. 2021). The potential of alginate (ALG) and alginate oligosaccharides (AOS) as bio-elicitors for enhancing plant defense mechanisms has been extensively studied. For example, ALG derived from the brown seaweed *Bifurcaria bifurcata* and its oligosaccharide derivatives were shown to stimulate the natural defense responses of tomato seedlings by increasing phenylalanine ammonia-lyase (PAL) activity and polyphenol levels in treated leaves. PAL activity, a key indicator of phenylpropanoid metabolism activation, was notably enhanced 12 h posttreatment. This metabolic pathway plays a critical role in fortifying plants against phytopathogens. These findings highlight ALG and its derivatives as promising eco-sustainable resources for plant protection (Aitouguinane et al. 2020).

In another study, a hydrogel combining ALG, lentinan, and aminooligosaccharides demonstrated strong resistance induction against tobacco mosaic virus (TMV) in *Nicotiana benthamiana* by increasing calcium ion release and promoting plant growth. Such results underline the multifaceted benefits of ALG-based biopolymers, including enhanced disease resistance and growth promotion (Xiang et al. 2019).

Further research corroborates the efficacy of ALG and AOS as elicitors against various plant pathogens, including fungi, bacteria, and viruses. These elicitors activate multiple defense pathways, including salicylic acid and jasmonic acid–ethylene (JA–ET) signaling. Evidence also suggests that AOS can induce abscisic acid-dependent pathways, which are pivotal for abiotic stress responses, such as drought, salinity, and heavy metal stress (Tang et al. 2011; Liu et al. 2013; Moenne and González 2021).

For instance, AOS application significantly enhanced wheat's tolerance to drought stress by modulating key physiological parameters (Liu et al. 2013). Similarly, in Brassica plants, AOS derived from alginase activity helped alleviate salt stress by activating critical stress-responsive pathways (Tang et al. 2011). These findings demonstrate the broad-spectrum utility of AOS in mitigating both biotic and abiotic stresses in plants. Moreover, studies have demonstrated that AOS treatments increase resistance to postharvest decay in fruits like kiwifruit and pears, further expanding their potential applications in agriculture (Zhuo et al. 2021; Han et al. 2021). For example, combining AOS with beneficial microbes such as *Meyerozyma guilliermondii* significantly reduced fungal infections in pears, indicating synergistic effects in pathogen control (Han et al. 2021).

The biological activities of alginate oligosaccharides (AOS) have been extensively studied, highlighting their roles as growth regulators and pathogen-associated molecular patterns (PAMPs). Current research trends have shifted from simply identifying biological activities to understanding the underlying mechanisms, although structure-activity relationship studies remain critical. Due to their polar nature and complex structures, including multiple isomers, connection modes, and branching, optimizing degradation, separation, and identification methods is essential for characterizing the relationship between AOS structure and physiological activity.

Another key research focus is understanding how plants recognize AOS, which remains a challenging topic. Small biomolecules like AOS typically interact with receptors in the cell membrane, nucleus, or cytoplasm. Previous studies using Eu^{3+} complexes as luminescence probes visualized AOS binding to epidermal and stomatal guard cells, suggesting the presence of binding sites on the cell surface (Guo et al. 2012). However, no specific AOS receptor has been definitively identified.

Elicitor-active oligosaccharides, including AOS, may interact with pattern recognition receptors (PRRs). To date, PRRs for chitin and oligogalacturonides (OGAs) have been identified (Trouvelot et al. 2014). Wall-associated kinase 1 (WAK1) serves as a receptor for OGAs, with interactions dependent on calcium-mediated "eggbox" complexes (Brutus et al. 2010; Kohorn and Kohorn 2012). Interestingly, alginates can also form "eggbox" structures via calcium bridges (Decreux and Messiaen 2005; Cabrera et al. 2008). In vitro studies showed that alginates, particularly those rich in G residues, bind to the extracellular domain of WAK1 (Decreux and Messiaen 2005). While WAK's role as an AOS receptor remains unconfirmed, exploring its function could provide significant insights into the AOS signaling mechanism. More recent studies on the role of ALG in inducing various mechanisms of disease resistance in various plants have been represented in Table 9.1.

Table 9.1 The role of alginate on managing disease resistance of plants

Plant	Infection	Alginate	Role	References
	Phytophthora infestans (*P. infestans*)	Alginate oligo obtained by a newly reported enzyme Aly2saccharides	AOS significantly induced reactive oxygen species (ROS) accumulation, calcium influx, stomata closure, callose deposition, upregulated the salicylic acid (SA) synthesis related gene and the defense-related genes. AOS-treated seedlings verified the SA pathway and suggested the presence of chitin elicitor receptor kinase (CERK) that AtCERK1 bind AOS tightly, suggesting AtCERK1 is responsible for AOS recognition	Peng et al. (2025)
Tomato	Verticillium wilt of tomato	Natural alginate extracted from a brown seaweed *Bifurcaria bifurcata*	The alginate reduced disease severity and mortality rate and increased the natural defenses of tomatoes like stimulating phenylalanine ammonia-lyase activity and the accumulation of phenolic compounds and lignin in leaves and roots	Ziane et al. (2024)
Rice	Southern rice black-streaked dwarf virus (SRBSDV)	Oligosaccharins	Oligosaccharins enhanced rice antiviral defense by activating multiple defense pathways like jasmonic acid (JA) signaling pathway. Oligosaccharins treatment also induced a burst of reactive oxygen species	Wang et al. 2024
Date palm	*Fusarium oxysporum* f. sp. *Albedinis*	Alginate extracted from *Bifurcaria bifurcata*	Alginate activates date palm defenses, which involve phenylalanine ammonia-lyase (PAL), and overexpressed PAL gene and the genes encoding oxidative enzymes (SOD and LOX)	Bouissil et al. (2022)
Rubber tree (*Hevea brasiliensis*)	*Phytophthora palmivora*	Sulfated polysaccharide from *Acanthophora spicifera*	Alginate induced SA and scopoletin accumulation and SA-responsive gene expression but suppressed JA-responsive gene expression	Pettongkhao et al. (2019)
Arabidopsis thaliana	*Pseudomonas syringae pv. Tomato* DC3000 (*Pst* DC3000)	Alginate oligosaccharide	(AOS) decreased disease index and bacteria colonies, burst of ROS and NO, high transcription expression of resistance genes *PR1* and increased the level of SA. Our results provide a reference for the profound research and application of AOS in agriculture	Zhang et al. (2019)
Solanum lycopersicum	*Alternaria solani*, which causes early blight disease	An algal biopolymer (sodium alginate)	Sodium alginate has effectively controlled the growth of *A. Solani* by increasing the expression levels of SOD, decreasing cell death and increasing protease inhibitors	Dey et al. (2019)

9.5 Other Biological Activities Function of Alginate

Alginate or its derivatives have been applied to plants to enhance development of plants in terms of germination as well as growth of shoots and roots, mitigating abiotic stress, and extending shelf life (Fig. 9.2).

9.5.1 Promoting Germination, Root and Shoot Growth

The positive impact of ALG and AOS on plant growth has been extensively documented. These compounds play significant roles in promoting seed germination, root and shoot growth, and the production of secondary metabolites. Yonemoto et al. (1993) used the bacterial alginate lyase from *Eisenia bicyclis*, an edible seaweed, to degrade ALG into AOS. The study concluded the crucial role of AOS in promoting proliferation and differentiation and improved germination of rice plants (Yonemoto et al. 1993). Similarly, Hu et al. (2004) demonstrated that ALG oligosaccharides promoted the activity of β-amylase, thereby enhancing maize seed germination. AOS was also shown to improve faba bean seed germination (El-Mohdy 2017), and Luan et al. (2012) reported that alginate degradation products enhanced germination activity in mustard greens (*Brassica juncea* var. rugosa) and lettuce (*Lactuca sativa* var. lollo rossa).

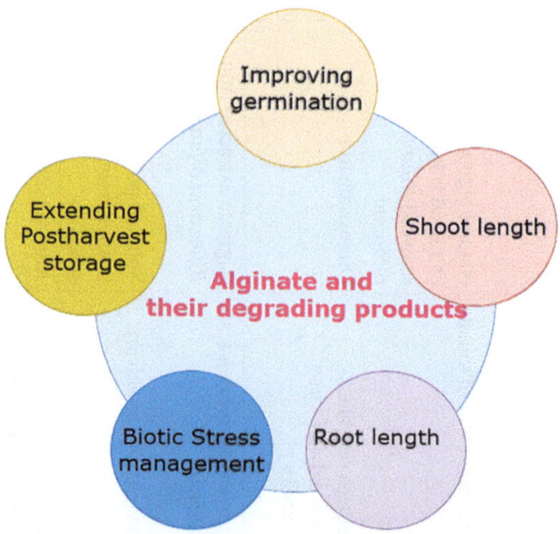

Fig. 9.2 The biological actions of alginate include biostimulation, microbial interaction, wound healing, nutrient release, disease resistance, biofilm development, and soil structure enhancement. It increases soil structure, stimulates systemic resistance, facilitates wound healing, controls nutrient release, encourages plant development, and strengthens biocontrol applications. Because of its use in sustainable farming methods, alginate is a useful tool for improving plant health and boosting the effectiveness of microorganisms

For root growth, Iwasaki et al. (2000) find out that fractionated AOS synthesized using an enzyme from *Corynebacterium* sp. increased the root elongation of lettuce plant. Additionally, Zhang et al. (2013) reported that alginate lyase from *Sphingomonas* sp. enhanced wheat root development. Xu et al. (2003) noted that oligoglucuronate (OG) and oligomannuronate (OM) derived from alginate digestion improved root elongation in carrot (*Daucus sativa*) and rice, with OG exhibiting greater benefits. The promotive effects of AOS on shoot elongation have also been recorded. Laporte et al. (2006) obtained OM and OG from alginates extracted from *Lessonia trabeculata* and *Lessonia vadosa*, observing improved shoot length in tobacco (*Nicotiana tabacum*). However, Xu et al. (2003) noted that only OM promoted tobacco shoot elongation, while OG did not. Furthermore, Luan et al. (2012) demonstrated that alginate products produced using *Flavobacterium* sp. and gamma irradiation stimulated barley and soybean seedling development. AOS has shown promotive effects on medicinal plants, such as *Mentha arvensis* L. (Naeem et al. 2012), *Artemisia annua* L. (Aftab et al. 2011), *Catharanthus roseus* L. (Idrees et al. 2011), *Papaver somniferum* L. (Khan et al. 2011), and *Eucalyptus citriodora* (Ali et al. 2014). Optimal AOS concentrations (80–120 μg/mL) varied slightly among plants but consistently promoted growth, enhanced physiological activities, and improved secondary metabolite production, including essential oils and alkaloids.

The mechanisms behind these effects have been attributed to nitric oxide (NO) generation, which enhances TaMPK6 gene expression and nitrate reductase activity, thereby stimulating wheat root growth (Zhang et al. 2013). Additionally, AOS treatments have been shown to accelerate auxin biosynthesis and transport by upregulating auxin-related genes (OsYUCCA1, OsYUCCA5, OsIAA11, and OsPIN1) and reducing indole-3-acetic acid (IAA) oxidase activity, leading to increased shoot fresh weight in *Brassica campestris* L. (Zhang et al. 2015). Furthermore, calcium ion-activated enzymes such as nitrate reductase (NR), glutamine synthetase (GS), and glutamate synthase (GOGAT) contribute to regulating root growth through nitrogen metabolism (Zhang et al. 2015).

9.5.2 Role of Alginate and Derived Oligosaccharides in Postharvest Storage

Alginate and its derived oligosaccharides are gaining recognition as effective agents for extending the shelf life of postharvest products. As a biomacromolecule, ALG exhibits a strong water-binding capacity, making it an ideal component for edible coatings. These coatings act as a barrier to moisture loss and oxygen exchange, effectively delaying the ripening process in various fruits. For instance, ALG coatings have been reported to extend the shelf life of peaches (Maftoonazad et al. 2008), tomatoes (*Solanum lycopersicum*; Zapata et al. 2008), blueberries (*Vaccinium* spp.; Chiabrando and Giacalone 2017), and apples (*Malus domestica*; Olivas et al. 2007; Rojas-Graü et al. 2007). In addition to its macromolecular applications, AOS allows unique characteristics as a postharvest treatment. AOS, being smaller molecules, exhibit multiple biological activities that contribute to maintaining the quality of fresh produce during storage. For example, Hussein et al. demonstrated that spraying harvested potatoes (*Solanum tuberosum*) with irradiated ALG resulted in higher total soluble solids, total phenols, and starch content compared to untreated potatoes after storage. Similarly, AOS application has been shown to prolong the shelf life of strawberries (*Fragaria* × *ananassa*) by mediating ABA metabolism. Bose et al. (2019) reported that AOS delayed ABA accumulation and repressed the expression of ABA signaling pathway genes, which are central to fruit ripening. Moreover, AOS enhances the biochemical properties of fruits, including hardness, titratable acidity (TA), pH, TSS, vitamin C, anthocyanins, total phenols, and flavonoids. These enhancements delay the degradation of cell wall components and suppress the expression of genes associated with cell wall breakdown. By influencing the inherent ripening pathways, both ALG and AOS elucidate significant prospects as environmentally friendly agents for improving fruit quality during postharvest storage.

9.6 Conclusion

Alginate and their derivatives represent a promising solution for improving plant resilience to biotic stress in terms of disease management and plant development. ALG or AOS work on improving antioxidant systems, regulating stress-responsive genes, and improving physiological parameters, which are the main mechanisms that highlight their potential role in sustainable agriculture. Further studies on ALG or degrading product applications could lead to innovative strategies to mitigate the impacts of biotic stress or maybe emerging stresses challenges plants and ensure food security.

References

Abasalizadeh F, Moghaddam SV, Alizadeh E, Akbari E, Kashani E, Fazljou SMB et al (2020) Alginate-based hydrogels as drug delivery vehicles in cancer treatment and their applications in wound dressing and 3D bioprinting. J Biol Eng 14(1):8

Abdelrhim AS, Dawood MFA, Galal AA (2021) Hydrogen peroxide-mixed compounds and/or microwave radiation as alternative control means against onion seed associated pathogens, *aspergillus Niger* and *fusarium oxysporum*. J Plant Pathol:1–15

Abdelrhim AS, Abdellatif YMR, Hossain MA, Alamri S, Pessarakli M, Lessy AMN, Dawood MFA (2023) Comparative study of three biological control agents and two conventional fungicides against coriander damping-off and root rot caused by *Rhizoctonia solani*. Plan Theory 12(8):1694

Abdelrhim AS, Hemeda NF, Mwaheb MA, Omar MO, Dawood MF (2024) The role of *Trichoderma koningii* and *Trichoderma harzianum* in mitigating the combined stresses motivated by *Sclerotinia sclerotiorum* and salinity in common bean (*Phaseolus vulgaris*). Plant Stress. 11:100370

Abdelmoneim MS, Dawood MF, Hafez EE, Hammad SF, Ghazy MA (2025) Unveiling the role of cadaverine in mitigating salinity and/or Bisphenol A toxicity in tomato plants and reduced Bisphenol A accumulation in tomato roots. Plant Physio Biochem 224:109799.

Abka-Khajouei R, Tounsi L, Shahabi N, Patel AK, Abdelkafi S, Michaud P (2022) Structures, properties, and applications of alginates. Mar Drugs 20(6):364

Abramovitch R, Janjusevic R, Stebbins E, Martin GB (2006) Type III effector AvrPtoB requires intrinsic E3 ubiquitin ligase activity to suppress plant cell death and immunity. Proc Natl Acad Sci USA 103(8):2851–2856

Aftab T, Khan MMA, Idrees M, Naeem M, Moinuddin HN (2011) Enhancing the growth, photosynthetic capacity, and artemisinin content in *Artemisia annua* L. by irradiated sodium alginate. Radiat Phys Chem 80(7):833–836

Aitouguinane M, Bouissil S, Mouhoub A, Rchid H, Fendri I, Abdelkafi S et al (2020) Induction of natural defenses in tomato seedlings by using alginate and oligoalginates derivatives extracted from Moroccan brown algae. Mar Drugs 18(10):521

Akimoto C, Aoyagi H, Dicosmo F, Tanaka H (2000) Synergistic effect of active oxygen species and alginate on chitinase production by *Wasabia japonica* cells and its application. J Biosci Bioeng 89(2):131–137

Al-Ani LKT, Furtado EL (2020) The effect of incompatible plant pathogens on the host plant. In: Molecular aspects of plant beneficial microbes in agriculture. Academic Press, pp 47–57

Ali A, Khan MMA, Uddin M, Naeem M, Idrees M, Hashmi N et al (2014) Radiolytically depolymerized sodium alginate improves physiological activities, yield attributes, and composition of essential oil of *Eucalyptus citriodora* hook. Carbohydr Polym 112:134–144

An QD, Zhang GL, Wu HT, Zhang ZC, Li X (2009) Alginate-deriving oligosaccharide production by alginase from newly isolated *Flavobacterium* sp. LXA and its potential application in protection against pathogens. J Appl Microbiol 106(1):161–170

Aranega-Bou P, de la O Leyva M, Finiti I, García-Agustín P, González-Bosch C (2014) Priming of plant resistance by natural compounds: Hexanoic acid as a model. Front Plant Sci 5:488

Ausubel FM (2005) Are innate immune signaling pathways in plants and animals conserved? Nat Immunol 6(10):973–979

Bagy HMK, Hassan EA, Nafady NA, Dawood MF (2019) Efficacy of arbuscular mycorrhizal fungi and endophytic strain Epicoccum nigrum ASU11 as biocontrol agents against blackleg disease of potato caused by bacterial strain Pectobacterium carotovora subsp. atrosepticum PHY7. Biological Control 134:103–113

Bakker PAHM, Pieterse CMJ, van Loon LC (2007) Induced systemic resistance by fluorescent *pseudomonas* spp. Phytopathology 97(2):239–243

Baysal Ö, Gürsoy YZ, Örnek H, Duru A (2005) Enhanced tomato resistance to bacterial canker by application of turtle oil. J Gen Plant Pathol 71(3):204–210. https://doi.org/10.1007/s10327-004-0187-y

Bhattacharyya D, Babgohari MZ, Rathor P, Prithiviraj B (2015) Seaweed extracts as biostimulants in horticulture. Sci Hortic (Amsterdam) 196:39–48

Bittel P, Robatzek S (2007) Microbe-associated molecular patterns (MAMPs) probe plant immunity. Curr Opin Plant Biol 10(4):335–341

Boller T, Felix G (2009) A renaissance of elicitors: perception of microbe-associated molecular patterns and danger signals by pattern-recognition receptors. Annu Rev Plant Biol 60:379–406

Bonartseva GA, Akulina EA, Myshkina VL, Voinova VV, Makhina TK, Bonartsev AP (2017) Alginate biosynthesis by *Azotobacter* bacteria. Appl Biochem Microbiol 53:52–59

Borges AL, Castro B, Govindarajan S, Solvik T, Escalante V, Bondy-Denomy J (2020) Bacterial alginate regulators and phage homologs repress CRISPR-Cas immunity. Nat Microbiol 5:679–687

Bose SK, Howlader P, Jia X, Wang W, Yin H (2019) Alginate oligosaccharide postharvest treatment preserves fruit quality and increases storage life via abscisic acid signaling in strawberry. Food Chem 283:665–674

Bouissil S, Guérin C, Roche J, Dubessay P, El Alaoui-Talibi Z, Pierre G, Michaud P, Mouzeyar S, Delattre C, El Modafar C (2022) Induction of defense gene expression and the resistance of date palm to *fusarium oxysporum* f. sp. *albedinis* in response to alginate extracted from *Bifurcaria bifurcata*. Mar Drugs 20(2):88

Brutus A, Sicilia F, Macone A, Cervone F, De Lorenzo G (2010) A domain swap approach reveals a role of the plant wall-associated kinase 1 (WAK1) as a receptor of oligogalacturonides. Proc Natl Acad Sci U S A.107:9452–9457

Cabrera JC, Boland A, Messiaen J, Cambier P, Van Cutsem P (2008) Egg box conformation of oligogalacturonides: the time-dependent stabilization of the elicitor-active conformation increases its biological activity. Glycobiology 18:473–482

Chandia NP, Matsuhiro B, Mejías E, Moenne A (2004) Alginic acids in *Lessonia vadosa*: partial hydrolysis and elicitor properties of the polymannuronic acid fraction. J Appl Phycol 16(2):127–133. https://doi.org/10.1023/B:JAPH.0000044900.28775.91

Chiabrando V, Giacalone G (2017) Quality evaluation of blueberries coated with chitosan and sodium alginate during postharvest storage. Int Food Res J 2017(24):1553–1561

Cohen YR (2002) B-Aminobutyric acid-induced resistance against plant pathogens. Plant Dis 86(5):448–457

Conrath U (2009) Priming of induced plant defense responses. In: Advances in botanical research, vol 51. Academic, pp 361–395

Conti E, Flaibani A, O'Regan M, Sutherland IW (1994) Alginate from *Pseudomonas fluorescens* and *P. Putida*: production and properties. Microbiology 140:1125–1132

Cote GL, Krull LH (1988) Characterization of the exocellular polysaccharides from *Azotobacter chroococcum*. Carbohydr Res 181:143–152

Cui H, Wang Y, Xue L, Chu J, Yan C, Fu J et al (2010) *Pseudomonas syringae* effector protein AvrB perturbs *Arabidopsis* hormone signaling by activating MAP kinase 4. Cell Host Microbe 7(2):164–175

Dawood MFA, Moursi YS, Abdelrhim AS, Hassan AA (2023a) An overview of the biology of rice bacterial blight pathogens and prospects of conventional methods for their management. In: Bacterial diseases of rice and their management. Apple Academic Press, In

Dawood MF, Sofy MR, Mohamed HI, Sofy AR, Abdel-Kader HA (2023b) N-or/and P-deprived *Coccomyxa chodatii* SAG 216–2 extracts instigated mercury tolerance of germinated wheat seedlings. Plant Soil 483(1):225–253

Dawood MFA, Moursi YS, Abdelrhim AS, Hassan AA (2024) Investigation of ecology, molecular, and host–pathogen interaction of rice blast pathogen and management approaches. In: Fungal diseases of rice and their management. Apple Academic Press

Dawood MF, Tahjib-Ul-Arif M, Shirazy BJ, Latef AAHA (2025) Unraveling the role of κ-carrageenan on the combined effect of drought and chromium stress in wheat (Triticum aestivium L.). Plant Physiology and Biochemistry 223:109781

Decreux A, Messiaen J (2005) Wall-associated kinase WAK1 interacts with cell wall pectins in a calcium-induced conformation. Plant Cell Physiol 46:268–278

De Vleesschauwer D, Höfte M (2009) Rhizobacteria-induced systemic resistance. In Advances in Botanical Research; Academic Press: Cambridge, MA, USA, 51:223–281

Dey P, Ramanujam R, Venkatesan G, Nagarathnam R (2019) Sodium alginate potentiates antioxidant defense and PR proteins against early blight disease caused by *Alternaria solani* in *Solanum lycopersicum* Linn. PLoS One 14(9):e0223216

Díaz-Barrera A, Sanchez-Rosales F, Padilla-Córdova C, Andler R, Peña C (2021) Molecular weight and guluronic/mannuronic ratio of alginate produced by *Azotobacter vinelandii* at two bioreactor scales under diazotrophic conditions. Bioprocess Biosyst Eng 44:1275–1287

Draget KI, Smidsrød O, Skjåk-Bræk G (2005) Alginates from algae. In: Polysaccharides and polyamides in the food industry: properties, production and patents. Wiley-Blackwell, Hoboken, pp 1–30

Dudun AA, Akoulina EA, Zhuikov VA, Makhina TK, Voinova VV, Belishev NV et al (2022) Competitive biosynthesis of bacterial alginate using *Azotobacter vinelandii* 12 for tissue engineering applications. Polymers 14:131

Durrant WE, Dong X (2004) Systemic acquired resistance. Annu Rev Phytopathol 42:185–209

El Hadrami A, Adam LR, El Hadrami I, Daayf F (2010) Chitosan in plant protection. Mar Drugs 8:968–987

El-Mohdy HLA (2017) Radiation-induced degradation of sodium alginate and its plant growth promotion effect, Arb J Chem 10:S431–S438

Ertesvåg H (2015) Alginate-modifying enzymes: biological roles and biotechnological uses. Front Microbiol 2015:6

Fakhr MK, Alejandro PV, Chakrabarty AM, Bender CL (1999) Regulation of alginate biosynthesis in *pseudomonas syringae* pv. *syringae*. J Bacteriol 181:3478–3485

Fischer FG, Dörfel H (1955) Die Polyuronsäuren der Braunalgen (Kohlenhydrate der Algen I). Hoppe Seylers Z Physiol Chem 302:186–203

Gaardløs M, Heggeset TM, Tøndervik A, Tezé D, Svensson B, Ertesvåg H et al (2022) Mechanistic basis for understanding the dual activities of the bifunctional *Azotobacter vinelandii* mannuronan C-5-epimerase and alginate lyase AlgE7. Appl Environ Microbiol 88(3)

Gacesa P (1998) Bacterial alginate biosynthesis—recent progress and future prospects. Microbiology 144:1133–1143. https://doi.org/10.1099/00221287-144-5-1133

Garagorri N, Fermanian S, Thibault R, Ambrose WM, Schein OD, Chakravarti S et al (2008) Keratocyte behavior in three-dimensional photopolymerizable poly (ethylene glycol) hydrogels. Acta Biomater 4(5):1139–1147

García A, Ferrer P, Albiol J, Castillo T, Segura D, Peña C (2018) Metabolic flux analysis and the NAD(P)H/NAD(P)+ ratios in chemostat cultures of Azotobacter vinelandii. Microb Cell Factories 17(10)

García A, Castillo T, Ramos D, Ahumada-Manuel CL, Núñez C, Galindo E et al (2020) Molecular weight and viscosifying power of alginates produced by mutant strains of *Azotobacter vinelandii* under microaerophilic conditions. Biotechnol Rep 26:e00436

Gawin A, Tietze L, Aarstad OA, Aachmann FL, Brautaset T, Ertesvåg H (2020) Functional characterization of three *Azotobacter chroococcum* alginate-modifying enzymes related to the Azotobacter vinelandii AlgE mannuronan C-5-epimerase family. Sci Rep 2020:10

Gill JF, Deretic V, Chakrabarty AM (1987) Alginate production by the mucoid *Pseudomonas aeruginosa* associated with cystic fibrosis. Microbiol Sci 4(10):296–299

Giridhar RS (2022) Alginates—a seaweed product: its properties and applications. In: Properties and applications of alginates. IntechOpen, London

Gloag ES, German GK, Stoodley P, Wozniak DJ (2018) Viscoelastic properties of Pseudomonas aeruginosa variant biofilms. Sci Rep 8:9691

Goh CH, Heng PWS, Chan LW (2012) Alginates as a useful natural polymer for microencapsulation and therapeutic applications. Carbohydr Polym 88(1):1–12

González A, Castro J, Vera J, Moenne A (2012) Seaweed oligosaccharides stimulate plant growth by enhancing carbon and nitrogen assimilation, basal metabolism, and cell division. J Plant Growth Regul 32(2):443–448

Gorin PAJ, Spencer JFT (1966) Exocellular alginic acid from Azotobacter vinelandii. Can J Chem 44:993–998

Gozzo F, Faoro F (2013) Systemic acquired resistance (50 years after discovery): moving from the lab to the field. J Agric Food Chem 61(50):12473–12491

Guarnizo N, Oliveros D, Murillo-Arango W, Bermúdez-Cardona MB (2020) Oligosaccharides: defense inducers, their recognition in plants, commercial uses, and perspectives. Molecules 25:5972

Guo W, Yin H, Ye Z, Zhao X, Yuan J, Du Y (2012) A comparison study on the interactions of two oligosaccharides with tobacco cells by time-resolved fluorometric method. Carbohydr Polym 90:491–495

Han J, Zhao L, Zhu H, Dhanasekaran S, Zhang X, Zhang H (2021) Study on the effect of alginate oligosaccharide combined with *Meyerozyma guilliermondii* against *Penicillium expansum* in pears and the possible mechanisms involved. Physiol Mol Plant Pathol 115:101654

Haug A, Larsen B, Smidsrød O (1974) Uronic acid sequence in alginate from different sources. Carbohydr Res 32:217–225

Hay ID, Rehman ZU, Moradali MF, Wang Y, Rehm BHA (2013) Microbial alginate production, modification and its applications. Microb Biotechnol 6:637–650

Hay ID, Wang Y, Moradali MF, Rehman ZU, Rehm BH (2014) Genetics and regulation of bacterial alginate production. Environ Microbiol 16:2997–3011

Hernández-Herrera RM, Virgen-Calleros G, Ruiz-López M, Zañudo-Hernández J, Délano-Frier JP, Sánchez-Hernández C (2013) Extracts from green and brown seaweeds protect tomato (*Solanum lycopersicum*) against the necrotrophic fungus Alternaria solani. J Appl Phycol 26:1607–1616

Hu X, Gao C (2008) Photoinitiating polymerization to prepare biocompatible chitosan hydrogels. J Appl Polym Sci 110(2):1059–1067

Hu X, Jiang X, Hwang H, Liu S, Guan H (2004) Promotive effects of alginate-derived oligosaccharide on maize seed germination. J Appl Phycol 16(2):73–76

Idrees M, Naeem M, Alam M, Aftab T, Hashmi N, Khan MMA et al (2011) Utilizing the γ-irradiated sodium alginate as a plant growth promoter for enhancing growth, physiological activities, and alkaloids production in Catharanthus roseus L. Agric Sci China 10(8):1213–1221

Idrees M, Dar TA, Naeem M, Aftab T, Khan MMA, Ali A et al (2015) Effects of gamma-irradiated sodium alginate on lemongrass: field trials monitoring production of essential oil. Ind Crop Prod 63:269–275

Ifkovits JL, Burdick JA (2007) Photopolymerizable and degradable biomaterials for tissue engineering applications. Tissue Eng 13(10):2369–2383. https://doi.org/10.1089/ten.2007.0093

Iwasaki K, Matsubara Y (2000) Purification of alginate oligosaccharides with root growth-promoting activity toward lettuce, Biosci. Biotechnol. Biochem 64:1067–1070.

Iriti M, Castorina G, Vitalini S, Mignani I, Soave C, Fico G et al (2010) Chitosan-induced ethylene-independent resistance does not reduce crop yield in bean. Biol Control 54(3):241–247

Jiao W, Chen W, Mei Y, Yun Y, Wang B, Zhong Q et al (2019) Effects of molecular weight and guluronic acid/mannuronic acid ratio on the rheological behavior and stabilizing property of sodium alginate. Molecules 24:4374

Jones JDG, Dangl JL (2006) The plant immune system. Nature 444(7117):323–329

Kaur H, Gogoi B, Sharma I, Das DK, Azad MA, Pramanik DD, Pramanik A (2024) Hydrogels as a potential biomaterial for multimodal therapeutic applications. Mol Pharm 21(10):4827–4848

Keefe AJ, Jiang S (2011) Poly(zwitterionic) protein conjugates offer increased stability without sacrificing binding affinity or bioactivity. Nat Chem 4(1):59–63

Keith RC, Keith LMW, Hernandez-Guzman G, Uppalapati SR, Bender CL (2003) Alginate gene expression by pseudomonas syringae pv. Tomato DC3000 in host and non-host plants. Microbiology 149(5):1127–1133

Khalil HMM, Hassan EA, Nafady NA, Dawood MFA (2019) Efficacy of arbuscular mycorrhizal fungi and endophytic strain *Epicoccum nigrum* ASU11 as biocontrol agents against blackleg disease of potato caused by bacterial strain *Pectobacterium carotovora* subsp. *atrosepticum* PHY7. Biol Control 134:103–113

Khan ZH, Khan MMA, Aftab T, Idrees M, Naeem M (2011) Influence of alginate oligosaccharides on growth, yield and alkaloid production of opium poppy (*Papaver somniferum* L.), Front Agric China 5:122–127

Kirdponpattara S, Khamkeaw A, Sanchavanakit N, Pavasant P, Phisalaphong M (2015) Structural modification and characterization of bacterial cellulose–alginate composite scaffolds for tissue engineering. Carbohydr Polym 132:146–155

Kohorn BD, Kohorn SL (2012) The cell wall-associated kinases, WAKs, as pectin receptors. Front Plant Sci 2012:3

Laporte D, Vera J, Chandía NP, Zúñiga EA, Matsuhiro B, Moenne A (2006) Structurally unrelated algal oligosaccharides differentially stimulate growth and defense against tobacco mosaic virus in tobacco plants. J Appl Phycol 19(1):79–88

Lawrence CB, Singh NP, Qiu J, Gardner RG, Tuzun S (2000) Constitutive hydrolytic enzymes are associated with polygenic resistance of tomato to Alternaria solani and may function as an elicitor release mechanism. Physiol Mol Plant Pathol 57:211–220

Li J, Zhu L, Lu G, Zhan X-B, Lin C-C, Zheng Z-Y (2014) Curdlan β-1,3-Glucooligosaccharides Induce the Defense Responses against Phytophthora infestans Infection of Potato (Solanum tuberosum L. cv. McCain G1) Leaf Cells. PLoS ONE 9:e97197.

Linker A, Jones RS (1964) A polysaccharide resembling alginic acid from a Pseudomonas microorganism. Nature 204(4954):187–188

Linker A, Jones RS (1966) A new polysaccharide resembling alginic acid isolated from pseudomonads. J Bio Chem 241(16):3845–3851

Luan le Q, Ha VT, Uyen NH, Trang le TT, Hien NQ (2012) Preparation of oligoalginate plant growth promoter by gamma irradiation of alginate solution containing hydrogen peroxide, J Agric Food Chem 60:1737–1741

Ma LJ, Zhang Y, Bu N, Li N, Wang SH (2010) An alginate-derived oligosaccharide enhanced wheat tolerance to cadmium stress. Plant Growth Regul 62:71–76

Macho AP, Zipfel C (2014) Plant PRRs and the activation of innate immune signaling. Mol Cell 54(2):263–272. https://doi.org/10.1016/j.molcel.2014.03.028

Maftoonazad N, Ramaswamy HS, Marcotte M (2008) Shelf-life extension of peaches through sodium alginate and methyl cellulose edible coatings. Int J Food Sci Technol 43:951–957

Mahajan S, Sunsunwal S, Gautam V, Singh M, Ramya TNC (2021) Biofilm inhibitory effect of alginate lyases on mucoid P. Aeruginosa from a cystic fibrosis patient. Biochem Biophys Rep 26:101028

Mani SD, Nagarathnam R (2018) Sulfated polysaccharide from Kappaphycus alvarezii primes defense responses against anthracnose disease of *Capsicum annuum*. Algal Res 32:121–130

Mauch-Mani B, Baccelli I, Luna E, Flors V (2017) Defense priming: An adaptive part of induced resistance. Annu Rev Plant Biol 68:485–512

Mawad AM, Aldaby ES, Madany MM, Dawood MF (2024) The application of PAHs-degrading Pseudomonas aeruginosa to mitigate the phytotoxic impact of pyrene on barley (*Hordeum vulgare* L.) and broad bean (*Vicia faba* L.) plants. Plant Physiol Biochem 215:108959

May TB, Chakrabarty AM (1994) Isolation and assay of Pseudomonas aeruginosa alginate. In: Methods Enzymol. Elsevier, Cambridge, MA, USA, pp 295–304

Moenne A, González A (2021) Chitosan-, alginate-, and carrageenan-derived oligosaccharides stimulate defense against biotic and abiotic stresses, and growth in plants: a historical perspective. Carbohydr Res 503:108298

Moradali MF, Donati I, Sims I, Ghods S, Rehm BHA (2015) Alginate polymerization and modification are linked in *Pseudomonas aeruginosa*. MBio 6(3):e00453–e00415

Mughunth RJ, Velmurugan S, Mohanalakshmi M, Vanitha K (2024) A review of seaweed extract's potential as a biostimulant to enhance growth and mitigate stress in horticulture crops. Sci Horti 334:113312

Naeem M, Idrees M, Aftab T, Khan MMA (2012) L. Varshney Moinuddin, Irradiated sodium alginate improves plant growth, physiological activities and active constituents in Mentha arvensis L, J Appl Pharmaceut Sci 28–35

Naveed ZA, Wei X, Chen J, Mubeen H, Ali GS (2020) The PTI to ETI continuum in *Phytophthora*-plant interactions. Front Plant Sci 11:593905

Noman A, Aqeel M, Qari SH, Al Surhanee AA, Yasin G, Alamri S, Hashem M, Al-Saadi AM (2020) Plant hypersensitive response vs pathogen ingression: death of few gives life to others. Microbial Pathog 145:104224

Nussinovitch A (2012) Alginates. In: Hydrocolloid applications: gum technology in the food and other industries. Springer, Berlin, Germany, pp 19–39

Ochoa-Meza LC, Quintana-Obregón EA, Vargas-Arispuro I, Falcón-Rodríguez AB, Aispuro-Hernández E, Virgen-Ortiz JJ et al (2021) Oligosaccharins as elicitors of defense responses in wheat. Polymers 13:3105

Olivas GI, Mattinson DS, Barbosa-Cánovas GV (2007) Alginate coatings for preservation of minimally processed 'gala' apples. Postharvest Biol Technol 45:89–96

Pacheco-Leyva I, Guevara Pezoa F, Díaz-Barrera A (2016) Alginate biosynthesis in *Azotobacter vinelandii*: overview of molecular mechanisms in connection with oxygen availability. Int J Polym Sci 2016:1–12

Pecina A, Paneque A (1994) Studies on some enzymes of alginic acid biosynthesis in mucoid and non-mucoid Azotobacter chroococcum strains. Appl Biochem Biotechnol 49:51–58

Pedersen SS, Hoiby N, Espersen F, Koch C (1992) Role of alginate in infection with mucoid Pseudomonas aeruginosa in cystic fibrosis. Thorax 47:6–13

Peña C, Miranda L, Segura D, Núñez C, Espín G, Galindo E (2002) Alginate production by Azotobacter vinelandii mutants altered in poly-β-hydroxybutyrate and alginate biosynthesis. J Ind Microbiol Biotechnol 29:209–213

Peng C, Xu W, Wang X, Meng F, Zhao Y, Wang Q, Peng L (2025) Alginate oligosaccharides trigger multiple defense responses in tobacco and induce resistance to Phytophthora infestans. Front Plant Sci 16:1506873.

Pettongkhao S, Bilanglod A, Khompatara K, Churngchow N (2019) Sulphated Polysaccharide from Acanthophora spicifera Induced Hevea brasiliensis Defense Responses Against Phytophthora palmivora Infection. Plants 8:73

Pieterse CM, Zamioudis C, Berendsen RL, Weller DM, van Wees SC, Bakker PA (2014) Induced systemic resistance by beneficial microbes. Annu Rev Phytopathol 52:347–375

Popletaeva SB, Statsyuk NV, Voinova TM, Arslanova LR, Zernov AL, Bonartsev AP et al (2018) Evaluation of eliciting activity of peptidyl-prolyl cis/trans isomerase from *Pseudomonas fluorescens* encapsulated in sodium alginate regarding plant resistance to viral and fungal pathogens. AIMS Microbiol 4(2):192–208

Qiao M, Ouyang LM (2013) Recombinant production of alginate lyase for improved stress resistance in plants. J Pure Appl Microbiol 7(4):1615–1622

Rasul S, Dubreuil-Maurizi C, Lamotte O, Koen E, Poinssot B, Alcaraz G et al (2012) Nitric oxide production mediates oligogalacturonide-triggered immunity and resistance to Botrytis cinerea in Arabidopsis thaliana. Plant Cell Environ 35:1483–1499

Rehm BHA, Valla S (1997) Bacterial alginates: biosynthesis and applications. Appl Microbiol Biotechnol 48:281–288

Rojas-Graü MA, Tapia MS, Rodríguez FJ, Carmona AJ, Martin-Belloso O (2007) Alginate and gellan-based edible coatings as carriers of antibrowning agents applied on fresh-cut Fuji apples. Food Hydrocoll 21:118–127

Saberi Riseh R, Skorik YA, Thakur VK, Moradi Pour M, Tamanadar E, Noghabi SS (2021) Encapsulation of plant biocontrol bacteria with alginate as a Main polymer material. Int J Mol Sci 22(20):11165. https://doi.org/10.3390/ijms222011165

Sadoff HL (1975) Encystment and germination in *Azotobacter vinelandii*. Bacteriol Rev 39:516–539. https://doi.org/10.1128/br.39.4.516-539.1975

Salachna P, Grzeszczuk M, Meller E, Sobol M (2018) Oligo-alginate with low molecular mass improves growth and physiological activity of *Eucomis autumnalis* under salinity stress. Molecules 23(1):1–13

Salgado LT, Tomazetto R, Cinelli LP, Farina M, Filho GMA (2007) The influence of brown algae alginates on phenolic compounds' capability for ultraviolet radiation absorption in vitro. Braz J Oceanogr 55(2):145–154

Sallam N, AbdElfatah H, Dawood MFA, Hassan E, Mohamed M, Bagy H (2021) Physiological and histopathological assessments of the susceptibility of different tomato (*Solanum lycopersicum*) cultivars to early blight disease. Eur J Plant Pathol 160(3):755–756

Sano Y (1999) Antiviral activity of alginate against infection by tobacco mosaic virus. Carbohydr Polym 38(3):183–186

Sengha SS, Anderson AJ, Hacking AJ, Dawes EA (1989) The production of alginate by *Pseudomonas mendocina* in batch and continuous culture. Microbiology 135:795–804

Setubal JC, dos Santos P, Goldman BS, Ertesvåg H, Espín G, Rubio LM et al (2009) Genome sequence of *Azotobacter vinelandii*, an obligate aerobe specialized to support diverse anaerobic metabolic processes. J Bacteriol 191:4534–4545. https://doi.org/10.1128/jb.00504-09

Shaw PE (2002) Peptidyl-prolyl isomerases: a new twist to transcription. EMBO Rep 3:521–526
Sheteiwy MS, Ulhassan Z, Qi W, Lu H, AbdElgawad H, Minkina T et al (2022) Association of jasmonic acid priming with multiple defense mechanisms in wheat plants under high salt stress. Front Plant Sci 13:886862
Singh K, Gupta R, Shokat S, Iqbal N, Kocsy G, Pérez-Pérez JM, Riyazuddin R (2024) Ascorbate, plant hormones and their interactions during plant responses to biotic stress. Physiol Plant 176(4):e14388
Skjåk-Bræk G, Grasdalen H, Larsen B (1986) Monomer sequence and acetylation pattern in some bacterial alginates. Carbohydr Res 154:239–250
Smidsrød O, Draget KI (2004) Alginate gelation technologies. In: Food colloids: proteins, lipids and polysaccharides. Woodhead Publishing Limited, Cambridge, pp 279–293
Sofy M, Mohamed H, Dawood M, Abu-Elsaoud A, Soliman M (2022) Integrated usage of Trichoderma harzianum and biochar to ameliorate salt stress on spinach plants. Arch Agron Soil Sci 68(14):2005–2026
Song GC, Choi HK, Ryu CM (2013) The folate precursor Para-aminobenzoic acid elicits induced resistance against cucumber mosaic virus and Xanthomonas axonopodis. Ann Bot 111:925–934
Stengel DB, Connan S (2015) Marine algae: a source of biomass for biotechnological applications. In: Natural products from marine algae. Humana Press, New York, pp 1–37
Surico G (2013) The concepts of plant pathogenicity, virulence/avirulence, and effector proteins by a teacher of plant pathology. Phytopathol Mediterr 52(3):399–417
Szekalska M, Puciłowska A, Szymańska E, Ciosek P, Winnicka K (2016) Alginate: current use and future perspectives in pharmaceutical and biomedical applications. Int J Polym Sci 2016:1–17
Tang J, Zhou Q, Chu H, Nagata S (2011) Characterization of alginase and elicitor-active oligosaccharides from *Gracilibacillus* A7 in alleviating salt stress for *Brassica campestris* L. J Agric Food Chem 59(14):7896–7901
Tavafi H, Ali AA, Ghadam P, Gharavi S (2018) Screening, cloning and expression of a novel alginate lyase gene from P. Aeruginosa tag 48 and its antibiofilm effects on *P. Aeruginosa* biofilm. Microb Pathog 124:356–364
Trouvelot S, Heloir MC, Poinssot B, Gauthier A, Paris F, Adrian M (2014) Carbohydrates in plant immunity and plant protection: roles and potential application as foliar sprays. Front Plant Sci 5:1–14
Urtuvia V, Maturana N, Acevedo F, Peña C, Díaz-Barrera A (2017) Bacterial alginate production: An overview of its biosynthesis and potential industrial production. World J Microbiol Biotechnol 33:198
Valentine ME, Kirby BD, Withers TR, Johnson SL, Long TE, Hao Y et al (2020) Generation of a highly attenuated strain of *Pseudomonas aeruginosa* for commercial production of alginate. Microb Biotechnol 13:162–175
van Aubel G, Buonatesta R, Van Cutsem P (2014) COS-OGA: a novel oligosaccharidic elicitor that protects grapes and cucumbers against powdery mildew. Crop Prot 65:129–137
van Aubel G, Cambier P, Dieu M, Van Cutsem P (2016) Plant immunity induced by COS-OGA elicitor is a cumulative process that involves salicylic acid. Plant Sci 247:60–70
Van Loon L, Rep M, Pieterse C (2006) Significance of inducible defense-related proteins in infected plants. Annu Rev Phytopathol 44:135–162
Varghese S, Hwang NS, Canver AC, Theprungsirikul P, Lin DW, Elisseeff J (2008) Chondroitin sulfate-based niches for chondrogenic differentiation of mesenchymal stem cells. Matrix Biol 27(1):12–21
Wang C, El-Shetehy M, Shine MB, Yu K, Navarre D, Wendehenne D et al (2014) Free radicals mediate systemic acquired resistance. Cell Rep 7(2):348–355
Wang G, Lu M, He Q et al (2024) Low-dose of oligosaccharins boosts antiviral immunity through induction of multiple defense pathways in rice. Phytopathol Res 6:47
Wloka M, Rehage H, Flemming HC, Wingender J (2004) Rheological properties of viscoelastic biofilm extracellular polymeric substances and comparison to the behavior of calcium alginate gels. Colloid Polym Sci 282:1067–1076

Xiang S, Lv X, He L, Shi H, Liao S, Liu C et al (2019) Dual-action pesticide carrier that continuously induces plant resistance, enhances plant anti-tobacco mosaic virus activity, and promotes plant growth. J Agric Food Chem 67(36):10000–10009

Xu X, Iwamoto Y, Kitamura Y, Oda T, Muramatsu T (2003) Root growth-promoting activity of unsaturated oligomeric uronates from alginate on carrot and rice plants. Biosci Biotechnol Biochem 67(9):2022–2025

Yarkent C, Aslanbay Guler B, Gurlek C, Sahin Y, Kose A, Oncel S et al (2022) (2022). Algal alginate in biotechnology: biosynthesis and applications. In: properties and applications of alginates. IntechOpen, London, UK

Yonemoto Y, Tanaka H, Yamashita T, Kitabatake N, Ishida Y, Kimura A, Murata K (1993) Promotion of germination and shoot elongation of some plants by alginate oligomers prepared with bacterial alginate lyase, J. Ferment. Bioeng 75:68–70

Yorgey P, Rahme LG, Tan MW, Ausubel FM (2008) The roles of mucD and alginate in the virulence of *Pseudomonas aeruginosa* in plants, nematodes and mice. Mol Microbiol 41:1063–1076

Younes NA, El-Sherbiny M, Alkharpotly AA, Sayed OA, Dawood AF, Hossain MA et al (2024) Rice-husks synthesized-silica nanoparticles modulate silicon content, ionic homeostasis, and antioxidants defense under limited irrigation regime in eggplants. Plant Stress 11:100330

Yu Y, Gui Y, Li Z, Jiang C, Guo J, Niu D (2022) Induced systemic resistance for improving plant immunity by beneficial microbes. Plants (Basel) 11(3):386. https://doi.org/10.3390/plants11030386

Yue W, Zhang HH, Yang ZN, Xie Y (2021) Preparation of low-molecular-weight sodium alginate by ozonation. Carbohydr Polym 251:117104. https://doi.org/10.1016/J.CARBPOL.2020.117104

Zapata PJ, Guillén F, Martínez-Romero D, Castillo S, Valero D, Serrano M (2008) Use of alginate or zein as edible coatings to delay postharvest ripening process and to maintain tomato (*Solanum lycopersicum*) quality. J Sci Food Agric 88:1287–1293

Zhang Y, Liu H, Yin H, Wang W, Zhao X, Du Y (2013) Nitric oxide mediates alginate oligosaccharides-induced root development in wheat (*Triticum aestivum* L.), Plant Physiol. Biochem 71:49–56

Zhang Y, Xu S, Ding P, Wang D, Cheng YT, He J et al (2010) Control of salicylic acid synthesis and systemic acquired resistance by two members of a plant-specific family of transcription factors. Proc Natl Acad Sci USA 107(42):18220–18225

Zhang S, Tang W, Jiang L, Hou Y, Yang F, Chen W et al (2015) Elicitor activity of algino-oligosaccharide and its potential application in protection of rice plants (*Oryza sativa* L.) against *Magnaporthe grisea*. Biotechnol Biotechnol Equip 29(4):646–652

Zhang C, Howlader P, Liu T, Sun X, Jia X, Yin H (2019) Alginate oligosaccharide (AOS)-induced resistance to Pst DC3000 via salicylic acid-mediated signaling pathway in *Arabidopsis thaliana*. Carbohydr Polym 225:115221

Zhou J, Sun A, Xing D (2013) Modulation of cellular redox status by thiamine-activated NADPH oxidase confers Arabidopsis resistance to *Sclerotinia sclerotiorum*. J Exp Bot 64:3261–3272

Zhu F, Cao MY, Zhang QP, Mohan R, Schar J, Mitchell M et al (2023) Join the green team: inducers of plant immunity in the plant disease sustainable control toolbox. J Adv Res 57:15–42

Zhuo R, Li B, Tian S (2021) Alginate oligosaccharide improves resistance to postharvest decay and quality in kiwifruit (*Actinidia deliciosa* cv. Bruno). Hortic. Plant J 8(1):44–52

Ziane SO, Imehli Z, Talibi ZE, Koraichi SI, Meddich A, El Modafar C (2024) Biocontrol of tomato Verticillium wilt disease by plant growth-promoting bacteria encapsulated in alginate extracted from brown seaweed. Int J Biol Macromol 276:133800

Chapter 10
Proteinaceous Elicitors: Keys to Unlocking Plant Defense Mechanisms

Yachana Jha, Aquino Macwan, Marwa T. El-Mahdy, Mona F. A. Dawood, and Heba I. Mohamed

Abstract Elicitor molecules derived from attenuated virulent strains or non-virulent pathogen strains can be utilized for the stimulation of plant defense response and initiate defense signaling pathways by involving stress hormones such as JA, SA, and ethylene-dependent/independent pathways. Beneficial rhizobacterial species produce many activated defense molecules, and several chemicals, like benzothiadiazole, also have the potential to activate signaling molecules for an active immune response against the pathogenic interactions. By considering all such evidence, numerous microbial cell components are being tested for efficient activation of plant defense response in plants, to overcome the adverse effect of pathogenic plant interactions and to sustain better crop production from the agricultural fields. For this, there is a requirement for broad knowledge about the elicitor's quantitative and qualitative ability to induce the generalized and specific immune responses in the plant, limiting the application of elicitor molecules in agricultural farms. However, reports of studies indicate that the use of elicitors can induce resistance up to 85% in various crop.

Keywords Elicitor · Elicitins · NLPs · Harpins · Virus coat protein · RNA replicates · Biotic stress.

Y. Jha (✉) · A. Macwan
N. V. Patel College of Pure and Applied Sciences, Anand, Gujarat, India

M. T. El-Mahdy
Department of Pomology, Faculty of Agriculture, Assiut University, Assiut, Egypt

M. F. A. Dawood
Botany and Microbiology Department, Faculty of Science, Assiut University, Assiut, Egypt

H. I. Mohamed
Biological and Geological Sciences Department, Faculty of Education, Ain Shams University, Cairo, Egypt

K. A. Abd-Elsalam, H. I. Mohamed (eds.), *Elicitors for Sustainable Crop Production*, Sustainability Sciences in Asia and Africa,
https://doi.org/10.1007/978-981-96-6569-3_10

10.1 Introduction

Elicitors are microbial cell lysis products responsible for the activation of plant defense mechanisms against plant pathogens. These molecules may act as signature pathogenic molecules of a specific pathogen, which are recognized by the host cell receptor. The nature of elicitors primarily highlighted the ability of plant cells to respond to elicitors' signals to induce plant defense response, which includes a range of chemical effectors called "elicitors" (Zehra et al. 2021). Initially, the term "elicitor" was used for effector molecules that stimulate the production of phytoalexins, but now it is normally used for all effector molecules of either abiotic or biotic origin that have the potential to induce any type of plant defense. Elicitors are low-molecular-weight compounds that activate plant immune response by initiating a signal cascade and are classified into two groups, that is, plant-derived elicitors (endogenous elicitor) and pathogen-derived elicitors (exogenous elicitor) (Yang et al. 2022a, b, c, d, e). Endogenous elicitors are commonly derived from the plant cell wall components such as xylan and oligogalacturonic acid, generated by host plants, while exogenous elicitors are derived from pathogens components such as virus coat proteins, lipopolysaccharide, β-glucan, hairpins, chitin, elicitins, RNA replicas, etc. Elicitors activate nonspecific plant defense responses and mainly depend on their chemical nature. Based on their chemical nature, they can be classified as proteinaceous, glycoprotein, oligosaccharides, lipid, chemical-inducing elicitors, and carbohydrate. Activation of the plant defense response during plant-pathogen interactions depends on the specific recognition of elicitors and serves as the basis for pathogen-host specificity (Pršić and Ongena 2020; Jha et al. 2022).

Plants naturally have a continuous struggle to survive in such situations, and if the plants lose the battle, it results in crop and yield loss. Farmers to overcome such loss also develop several strategies like the use of resistant cultivars, crop rotation, the use of biological control measures, and the use of chemical pesticides for the cultivation of disease-free crops. However, uncontrolled uses of such chemical inputs in the crop field have several drawbacks. It results in loss of soil fertility, soil pollution, the evolution of resistant pathogens, and loss of pollinators, as well as being cumbersome and expensive (Münzel et al. 2023). The biochemical study on the nature of elicitors originally showed the ability of plant cells to respond to different biomolecules, and a study in the search of elicitors in different plant-microbe interactions revealed that elicitor-active molecules are essentially a biomolecule. The plant researcher reported that inducible defense responses of the plant can be triggered not only upon the pathogen attack but also upon the elicitor treatment (Jha and Mohamed 2023). So, extensive study is required to uncover the molecular mechanism of defense responses by the elicitors to develop better plant protection measures in an eco-friendly manner.

10.2 Plant Pathogen Resistance and Elicitor Recognition in Plants

The agroecosystem continuously gets affected by many microorganisms, bacteria, fungi, mycoplasmas, and viruses collectively called biotic stress. These microorganisms have both beneficial and detrimental effects on the agroecosystem and are directly related to soil fertility or health and crop productivity (Wang et al. 2022). Pathogenic microorganisms upon interaction with plants alter the metabolic and physiological processes in the plant and result in crop loss and yield. Plants being sessile are forced to face such stress, so plant alters their cell wall composition to create a flaccid structural barrier against pathogen attack. Plant upon pathogen attack generates reactive oxygen species in the infected plant cell to overcome the infection (Jha and Mohamed 2024). This reactive oxygen species causes damage to internal cellular structures and biomolecules due to the photooxidative effect. A plethora of biochemical modifications take place in plants upon pathogenic interaction, which remains associated with the generation of stress signals for the activation of coordinated defense pathways known as induced defense (Jha 2024a, b). The induced defense system includes both enzymatic and nonenzymatic components. Phenolic compounds like flavonoids and lignin are nonenzymatic components (Kiani et al. 2021).

10.3 Mechanism of Elicitation in Plant Cells

Plants being producers serve as a source of nutrients for a wide range of organisms including potential pathogenic organisms which is always around them and constantly attack plants to obtain food (Jha et al. 2024). At the same plants also develop multilayer protective measures to protect themselves from all such attacks of microbes. The struggle to survive is a natural and continuous process for all living organisms on the earth. So, in plants, the production of numerous amounts of varied compounds is not only necessary for growth and survival but also plays an important role in developing resistance to various stresses. The adverse effect of external or internal stress in plants causes the production of specialized chemicals to overcome all such adverse effects, which are generally chemical compounds derived as products of secondary metabolism known as secondary metabolites (Jha and Subramanian 2021). These secondary metabolites, with a broad range of biological activities including phenolics, terpenes, alkaloids, and steroids, help plants overcome the adverse effects of both biotic and abiotic stress. They also have significant uses as flavoring agents, pharmaceuticals, fragrances, agrochemicals, colorants, food additives, and biopesticides. These secondary metabolites also act as elicitors, which induce various types of plant defense responses (Kaur et al. 2022).

The elicitor-induced induction of plant defense response generally initiates after the recognition of elicitor signal by the plasma membrane receptors present on the plant cell surface. Although elicitors have many different chemical structures depending on the source of origin, they have varied effects depending upon the species. There are many elicitor molecular structure receptors in the cytoplasm, which remain coupled with cytosolic membrane and the nuclear membrane. Plant researchers have reported presence of a variety of receptors for different elicitors with diverse structures (Jomova et al. 2023). Elicitors can stimulate different reactions in different species, indicating the presence of general receptors for their perception in plants. The plasma membranes of the plant cell contain several elicitor-binding sites for the recognition of wide range of elicitors with different chemical structures. The transduction of the elicitor signal, perceived by the plasma membranes receptors, activates specific secondary messengers, which further modify the signal for different elicitor-specific reactions to develop resistant against plant pathogens. Signal transduction pathway is not simple; it is highly branched with several secondary messengers at once. The elicitor recognition by the plasma membrane receptor causes reversible phosphorylation and dephosphorylation of plasma membrane proteins as well as cytoplasmic proteins (Giovannoni et al. 2021). This initiates massive modification in the cytoplasm, such as an increase in Ca^{2+} ions; influx of H^+ and efflux of Cl^- and K^+, acidification of cytoplasmic content and alkalinization of extracellular contents; production of reactive nitrogen and oxygen species; accumulation of important bioactive secondary metabolites; and production of stress phytohormones like jasmonate and ABA. Secondary messengers like cytosolic pH, calcium ion, nitrogen oxide, and reactive oxygen species interact simultaneously with different transduction pathways, causing rapid serious effects on cell behavior, which cause remarkable variation in primary metabolic pathways like the pentose phosphate pathway and Krebs cycle. The transduction of the elicitor signal in many cases results in the activation of signal transduction pathways for the deposition of callose for strengthening the cell wall, and the production of pathogenesis-related proteins and phytoalexins, where the specificity of the defense response depending on the critical role of different secondary messengers and targeted proteins at that specific times and places (Jha, 2022). The plant systemic responses used to prevent pathogenic interaction are by the production of antimicrobial bioactive metabolites like pathogenesis-related proteins and phytoalexins, which in combination play a critical role in pathogen rejection (Jha et al. 2024). In plant cellular defensive mechanisms stimulation by biotic elicitors is generally regulated by more than two signaling pathways, resulting in the complex cross-talk between multiple pathways. These work in a highly coordinated manner to induce plant signal transduction networks, enabling plants to modulate different sets of genes both temporally and spatially, including the activation of defense enzymes like NADPH oxidase, mitogen-activated protein kinase, and the expression of early and late defense-responsive gene (Jha 2024a, b).

Elicitor perception can also increase the level of plant resistance against future pathogen attacks. However, a broad range of metabolic responses had been observed in plants due to the complex interaction of the elicitor with its effectors. The mechanisms of action of the elicitors need extensive research to uncover the specific cross-linking and interaction of receptors, secondary messengers, transduction pathways, and responsive genes to develop proper defensive pathogen and crop-specific tools for its application in plant biotechnology (Table 10.1) (Newton et al. 2016).

10.4 Classification of Elicitors

Elicitors are bioactive compounds that upon interaction with plants able to induce several biochemical and physiological changes in the plants, altering plant metabolic pathways and stimulating the production of secondary metabolites similarly, which will take place in the plant in response to pathogenic interaction. So, the application of elicitors has many advantages over other techniques to overcome the adverse effects of different biotic stress. The small amount of elicitors is sufficient to induce long-term protection against a broad range of pathogens as well as it is environmentally friendly, biodegradable, and nontoxic (Thakur and Sohal 2013). The elicitors can be the compounds of pathogen origin are known as exogenous elicitors and compounds released from plants by the action of the pathogen are known as endogenous elicitors. Elicitors are classified as biotic or abiotic, chemical or physical, and complex or simple, depending on their origin and molecular structure (Fig. 10.1). Based on origin, elicitors are classified into two groups: biotic and abiotic elicitors.

10.4.1 Biotic Elicitors

The biotic elicitors have biological origin substances, majorly recognized by the cell-specific receptors bound to the membrane and generate a defense signal, which is transferred by the signal transduction system of the plant to activate plant immune response. Mainly biotic elicitors include microbial cell components responsible for determining pathogenicity and help in the progression of disease signals to stimulate plant immune response. The elicitors were able to induce plant defense by several modes of action that can be used as microbial biocontrol agents to develop disease-resistant crops to sustain yield (Zehra et al. 2021). The fungal elicitor chitooligosaccharide is well recognized as a fungal pathogenicity molecule and the signaling pathway is determined by the length of the chitin chain. It has been reported by Hayafune et al. (2014) that, heptaose, octaose N-acetyl chitooligosaccharides induced a more effective immune response in plants compared to hexamer N-acetyl chitooligosaccharides. The level of plant response also depends on other factors like plant growth state, physiological state, and genetic constitution. Plants have several genes like pathogen avirulence genes, pathogenesis-related enzyme genes, and plant resistance genes for plant defense against pathogenic interaction. The molecular analysis of the elicitors revealed they include fatty acids, oligosaccharides, polysaccharides, proteins, and glycoproteins in nature.

Table 10.1 Mechanism of elicitors in activation of immune responses in plants

Elicitor	Sources	Mechanism	Plants	References
Fungal extracts	Fungus	Antibiosis and induced resistance	Tomato plants,	Barda et al. (2015)
Bacterial-secreted metabolites-pyocyanin	Bacteria	SA accumulation subsequently stimulates defense mechanisms such as phenolic biosynthesis, callose deposition, and stomatal closure	Tomato plants	Jang et al. (2022)
Salicylic acid	Bacteria	Higher transcript levels and the stronger expression of the defense-related genes	*Cannabis sativa*	Sands et al. (2022)
Oligogalacturonide	*Puccinia graminis*, yeast	R gene-mediated defense signaling	Parsley	Montesano et al. (2003)
Harpin	Gram-negative plant-pathogenic bacteria	Causing a hypersensitive response	*Oryza sativa*	Liu et al. (2020)
Flagellin	Bacteria	Effector-triggered susceptibility and causing a hypersensitive response	*Arabidopsis*	Lu et al. (2022)
Chitooligosaccharide	Fungus	Synthesis of ethylene and the synthesis of PR proteins	*Oryza sativa*	Hayafune et al. (2014)
Elicitins	Oomycetes (particularly *Phytophthora* and *Pythium*	Burst of ROS, electrolyte leakage, activation of MAPK cascades, induction of PR genes and phytohormone	Tobacco, tomato, potato, pigeon pea	Janku et al. (2020)
Glucose oxidase peptides (inceptin), and volicitin)	Insect enzyme	Induce defense response in the plant cells	*Oryza sativa*, tomato	War et al. (2018)
Cell wall hydrolytic enzymes	Fungi and bacteria	Dissolve the middle lamella of plant tissues, inducer of membrane depolarization	Soybean, legumes, tomato	Keswani et al. (2019)
Necrosis- and ethylene-inducing peptide 1 (Nep1)-like proteins (NLPs)	Fungi and bacteria	Conidiospore production, formation of appressoria, abiotic stresses in microbes and triggering immune responses in plants	Tomato	Lian et al. (2022)

(continued)

Table 10.1 (continued)

Elicitor	Sources	Mechanism	Plants	References
RNA replicases	RNA virus (*cucumber mosaic virus*)	Causing a hypersensitive response	*Cucumber*	Kachroo et al. (2002)
Coat protein	Turnip crinkle virus	Causing a hypersensitive response, triggering necrosis, and immune responses	Turnip	Garcia and Pallas (2015)

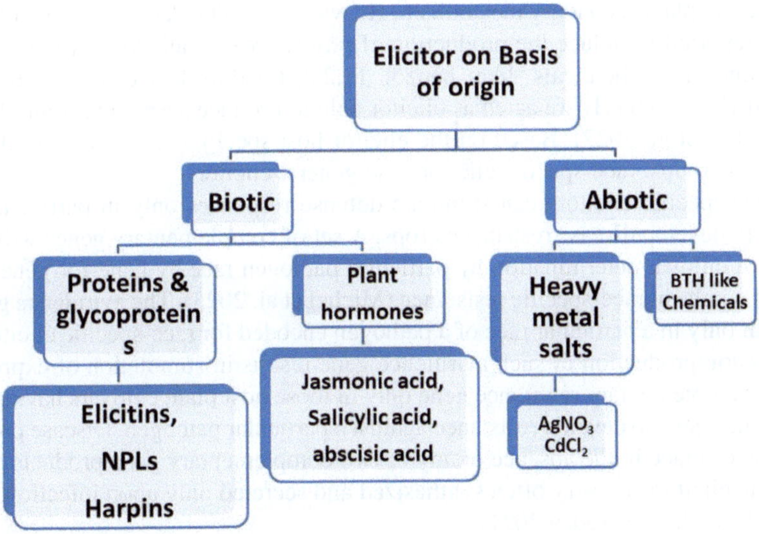

Fig. 10.1 Plant elicitors can be classified in terms of origin, chemical structure, and mechanism of action. Elicitors derived from living sources include biotic elicitors, and elicitors derived from nonliving sources include abiotic elicitors. Chemical structures include polysaccharides, proteins, peptides, and secondary metabolites. Modes of actions include direct and indirect actions, and actions include defensive and growth-promoting elicitors

Proteins and glycoproteins as elicitors-enzymes and proteins are a class of elicitors that induce defense reactions in plant in vivo as well as in vitro. It has been reported by Angelova et al. (2006) that protein elicitors, elicitins secreted by *Phytophtora drechleri,* a plant pathogenic fungus, induce necrosis, while holoproteins are responsible for the phototropic signal transduction and cause concentration-dependent leaf necrosis. Protein elicitors also play a significant role in the alteration of ion channels in plant cell membranes as cell wall-degrading enzyme pectolyase also stimulates depolarization of membrane by chloride effluxes. Glycoproteins are also well known for the induction of phytoalexins in plants.

Plant hormones as elicitors: Plant hormones like jasmonic acid and salicylic acid are well-known for the activation of defense genes. JA is reported to produce various proteins for induction of resistance against herbivores insects, while salicylic acid induces resistance against viral, bacterial, and fungal pathogens. The JA, SA, and karrikin elicitors modulate the biochemical pathways for the induction of various genes and act as the key signals for defense genes (Dawood et al. 2024).

10.4.2 Abiotic Elicitors

Abiotic elicitors have nonbiological origin substances, mainly inorganic compounds as physical factors or salts, which activate the production of bioactive compounds by altering plant secondary metabolism. Heavy metal salts like $AgNO_3$ or $CdCl_2$ are often reported to induce the production of phytoalexin as alkaloids, tropane, hyoscyamine and chemicals like benzol (1,2,3)-thiadiazole-7-carbothermic acid S-methyl ester (BTH) are acted as elicitor able to produce systemic acquired resistance (Liu et al. 2022). Based on the elicitor host specificity, elicitors are divided into two groups: race-specific elicitors and general elicitors.

Race-specific elicitors can stimulate defense responses only in particular host crops to develop disease resistance crops. A set of complementary genes will work for host cultivar determination by particular pathogen race by gene-for-gene interaction results in race-specific resistance (Michel et al. 2023). The avirulence gene is present only in a particular race of a pathogen encoded for race-specific elicitor and expression/production of such avirulence gene results in stimulation of expression of the complementary resistance gene only in those host plant cultivars having such resistance gene to develop resistance against a particular pathogen. Disease progression takes place in the absence of any of two complementary gene products. Race-specific elicitors are very often synthesized and secreted only upon infection of the host plant (Jha and Yadav 2021).

General elicitors are able to stimulate defense responses only in all crops to develop disease resistance in the presence of potential pathogens. General elicitors are nonspecific in nature and the signal is recognized by some plants. Plant scientist reports that there are outstanding similarities between the innate immunity of animals and the defense response stimulated by general elicitors. The recognition of general elicitors by the plant results in the subsequent induction of innate immunity in the plant (Laveti 2018).

10.5 Proteinaceous Elicitors

Elicitors are generally signaling molecules, which can initiate plant response at very low concentrations to develop resistant plants. Toxic compounds always require high concentration, have adverse effects on plants, and do not have the potential to influence active plant metabolism. The primary immune response of the plant is efficiently activated by the signal transduction mechanism of the elicitors (Fig. 10.2). The exogenous proteinaceous elicitors like elicitins, NLPs, and, harpins are involved

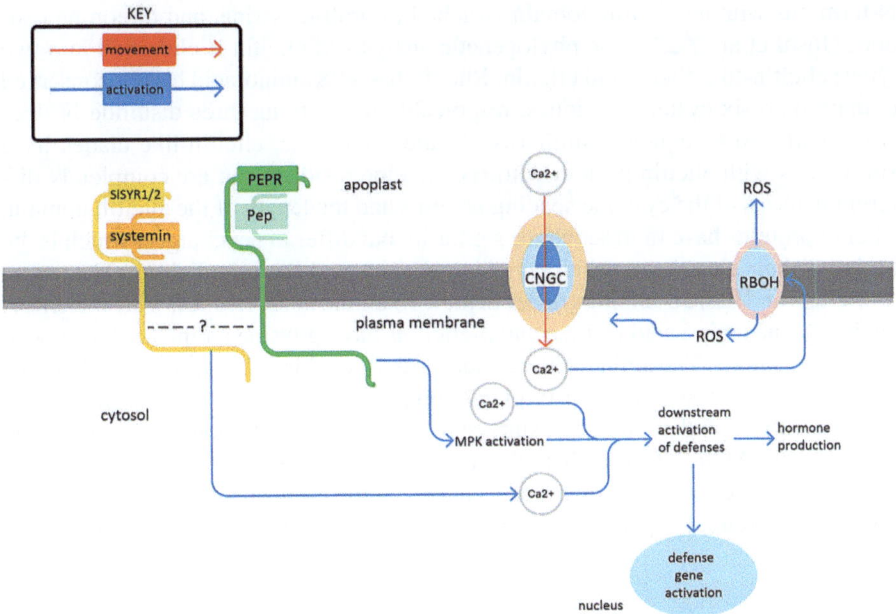

Fig. 10.2 Elicitors in a plant induce a variety of defense processes through a variety of processes, including receptor-mediated perception, calcium signaling, hormonal regulation, reprogramming of gene expression, reactive oxygen species (ROS), systemic acquired resistance (SAR), fortification of the cell wall, metabolite production, and induction of defense enzymes. All these processes make a plant resistant to pathogens and stress, enhance its productivity and health, and induce pathogenesis-related proteins, enzymes, and secondary metabolites. Elicitors also induce secondary metabolites with antimicrobial activity, and stimulate activity of defense enzymes, and lyse pathogen cell walls (Zelman and Berkowitz 2023)

in the signaling mechanisms to activate the immune response of the plants to develop biotic stress tolerance in crop (Li et al. 2023).

10.6 Structure and Mechanism of Proteinaceous Elicitor-Elicitins

10.6.1 Structure of Proteinaceous

Elicitins is a proteinaceous elicitor produced by like Pythium and Phytophthora oomycetes fungus, which can induce immune responses in numerous plant varieties and activate necrotic defense pathways. Elicitins are the first proteinaceous elicitor recognized as capsaicin and cryptogein isolated from *Phytophthora capsici* and *Phytophthora cryptogea*, respectively. It is structures are highly conserved and are only about 10 kDa molecular weights having no structure similarity with plant proteins. Elicitin proteins have variable C-terminal domains, a signal peptide at the

N-terminus with an elicitin domain enriched in proline, serine, and threonine residues (Joshi et al. 2022). The phylogenetic analysis of elicitin is classified into two types: elicitin-like clades and elicitin. Elicitin has a 98 amino-acid highly conserved domain with six cysteine residues, responsible for forming three disulfide bridges with nearly 66% sequence similarity. At the same time, elicitin-like clades have similarities with elicitin protein with six cysteine residues, but are completely different in terms of the cysteine spacing patterns and the length of the elicitin domain. Elicitin proteins have high sequence similarity but differ in net charges, which is the basis for their classification into acidic α-elicitins and basic β-elicitins. The acidic α-elicitins and basic β-elicitins differ in their 13 amino acid residues, with α-elicitins having valine amino acid residue and β-elicitins having lysine amino acid residue at their 13 positions. The amino acid residue at its 13 positions has a critical role in the induction of necrosis response in plants toward pathogen infection, in interaction with receptor/ligand at plant cell surface as well as determination of the isoelectric point (Abdul Malik et al. 2020). The elicitin-like clades and elicitin proteins are encoded by genes, and different species have varying numbers of these genes, which undergo differential expression to induce hypersensitive responses in plants.

10.6.2 Mechanism of Elicitins

The immune and hypersensitive responses induced by the elicitins cause multiple effects such as leakage of electrolytes, the burst of reactive oxygen species, MAPK cascades activation, induction of production of phytohormone, and stimulation of PR genes in many crops like Brassicaceae cultivars, pigeon pea, tobacco, tomato, grapevine, potato, etc. as reported by Yang et al. (2022a, b, c, d, e). The responses elicited in different plants were highly variable. Upon interaction of elicitin with plant cell surface receptor-like protein, stimulates plant immune responses by depolarization of membrane electrochemical potential causes altered membrane permeability by ion flux resulting in the generation of reactive oxygen species resulting in early-phase ROS burst. All these changes activate MAPK signaling cascades and expression of pathogenesis-related genes, collectivity resulting in modified production of phytohormone, enhanced production of phytoalexin, and a higher rate of necrosis of infected cells (Yu et al. 2021).

10.7 Structure and Mechanism of Proteinaceous Elicitor-NLPs

10.7.1 Structure of NLPs

A family of proteins present in numerous bacteria, fungi, and oomycetes is necrosis- and ethylene-inducing peptide 1 (Nep1)-like proteins (NLPs) act as proteinaceous elicitors. Based on sequences of amino acids in NLPs, it is categorized into types I,

II, and III. An extraordinary degree of sequence similarity is present in the protein of this family and has the potential for the induction of hypersensitive-like death response in plants. Type I and type II proteinaceous elicitor NLPs differ from each other based on the number of cysteine residues in it and cysteine amino acid residue is essential for its activity (Xiang et al. 2022). Type I NLPs have one disulfide bridge due to the presence of two cysteine residues, while type II NLPs usually have different disulfide bridges nonessential for the induction of necrosis. In place of that, type II NLPs usually have a calcium-binding motif essential for its cytotoxicity activity. Type III NLPs are conserved for ascomycete fungi having six cysteine residues. The first time it was isolated from the *Fusarium oxysporum* culture filtrates having a molecular weight of about 24-kDa protein responsible for eliciting ethylene production and necrosis. The NLPs are characterized by the presence of NPP1 domains, responsible for the activation of several processes such as the formation of appressoria, virulence, production of conidiospore, in pathogens, and activation of defense responses in plants. The proteinaceous elicitor NLPs are categorized into two classes on the basis of the mechanism of action in the host: noncytotoxic and cytotoxic. When the defense response is without cytolysis, it is non-cytotoxic NLPs, while the defense response with cytolysis is cytotoxic NLPs. The conversion of the pathogen from biotropic to necrotropic is due to the expression of cytotoxic NLPs, which results in the production of toxin for pore formation in plasma membrane having monomeric hexose moieties of glycosyl inositol phosphoryl ceramide at its terminal for cytolysis (Schoonbeek et al. 2022). NLPs consist of a single polypeptide that features signal peptides at the amino terminal and a central conserved domain responsible for binding divalent cations. This binding is crucial for altering plant physiological activity. The most prevalent NLPs, particularly in oomycetes, are classified as type I NLPs.

10.7.2 Mechanisms of NLPs

The cytolytic NLP concentration varies in different plants, which by electrostatic interactions binds to the terminal hexose residues of GIPC sphingolipids, and the binding of GIPC-NLP in the presence of sterols gets strengthened, which causes conformational changes in NLP. The formation of shallow transient pores in the plasma membrane takes place due to the conformational changes NLPs for the generation of functional oligomers, which allow leaking of small molecules from the cytoplasm. Interaction of NLPs with different ligands completely depends on the amino acid residue present in that domain as tryptophan is important for the NLP-GIPC interaction (Schumacher et al. 2020). Noncytotoxic NLPs have hydrophobic amino acid residues at their interaction sites. NLP-ligand binding is necessary for the formation of tripartite complexes for the induction of intracellular innate immune responses.

10.7.3 Plant Response to NLP Promotes Biotic Resistance

As previously stated, NLPs play a variety of roles in both microorganisms and plants. The effects they have on host plants are the main topic of this article (Table 10.2). Rapid cytolysis brought on by treatments with cytotoxic NLPs causes necrosis of plant cells (Lenarcic et al. 2017). Apart from NLPs acting as PFTs, conserved segments and noncytotoxic NLPs also activate innate immunity in plants by acting as PAMPs/MAMPs. The plant immune system responds to exogenous NLP

Table 10.2 Involvement of proteinaceous elicitors in immune responses in plants

Groups	Elicitors/microbes	Plants	Treatments	References
Harpin	RipX(PopA)/*Ralstonia solanacearum*	*N. benthamiana*	Infiltration	Sun et al. (2020)
	HpaXpm/*Xanthomonas phaseoli pv. manihotis*	*N. benthamiana*	Infiltration	Liu et al. (2020)
	HpaXpm/Xanthomonas phaseoli pv. manihotis	*A. thaliana*	Soaking	Liu et al. (2020)
	SSBXoc/*X. oryzae pv. oryzicola*	*N. benthamiana*	Agrobacterium-mediated transformation	Cao et al. (2018)
	HrpZpsta/*P. syringae pv. tabaci*	*Glycine max*	Agrobacterium-mediated transformation	Du et al. (2018)
NLPs	CgNLP1/*Colletotrichumglo eosporioides*	*A. thaliana*	Agrobacterium-mediated transformation	Yang et al. (2022a, b, c, d, e)
	BcNep1/*Botrytis cinerea*	*Allium cepa*	Infiltration	Steentjes et al. (2022)
	PiNPP1.1/*Phytophthora infestans*	*N. benthamiana*	Agroinfiltration and infiltration	Schumacher et al. (2020)
	DserNEP1 and DserNEP2/*Diplodia seriata*	*Vitis vinifera*	Dip and infiltration	Cobos et al. (2019)
	PeNLP1 and PeNLP2/*Penicillium expansum*	*N. benthamiana*	Agroinfiltration	Levin et al. (2019)
	CoNLP1/*Colletotrichum orbiculare*	Several Cucurbitaceae cultivars	Infiltration	Azmi et al. (2018)
Elicitin	INF1/*Phytophthora infestans*	*N. benthamiana*	Agroinfiltration	Turnbull et al. (2019)
	INF1/*Phytophthora infestans*	*N. benthamiana*	Agroinfiltration	Domazakis et al. (2018)
	INF1/*Phytophthora infestans*	*Solanum microdontum*	Agroinfection	Du et al. (2015)
	PoEli8/*Pythium oligandrum*	*N. benthamiana*, tomato, and pepper	Infiltration	Yang et al. (2022a)

or NLP fragment treatment or ectopic expression of NLP or conserved fragments (nlp20/nlp24) in several ways, including necrosis and ethylene-induction, immune-responsive gene expression, MAPK signaling cascade activation, bursts in ROS production, induction of SA, and enhanced plant immunity against pathogens (Yang et al. 2022 a, b; Tian et al. 2021).

10.8 Structure and Mechanism of Proteinaceous Elicitor-Harpins

10.8.1 Structure of Harpins

Harpins is the first cell-free elicitor isolated from Gram-negative plant-pathogenic bacterium *Erwinia amylovora.* It acts as a virulence factor in the host and is responsible for the activation of hypersensitive response. The elicitor harpins are encoded by DNA sequences present on the pathogenicity gene cluster responsible for hypersensitive response having a molecular weight of about 10–60 kDa. The harpins are thermostable, acidic, and hydrophilic proteins, secreted by a type III secretion system (Liu et al. 2020). Harpins contain large amounts of serine and glycine amino acid residue, with few aromatic amino acids and cysteine. Harpins have their characteristic architecture and sequences, which make them chemically sensitive to denaturing proteins. Harpins generally contain 2–9 α-helixes completely different from plant protein, due to which they serve as good elicitors for plants. The hypersensitive response induced by the harpins includes hypersensitive response-related cell death and non-hypersensitive response induced by the harpins including deposition of callose in the cell wall, modified production of phytohormone, bursts in ROS production and activation of enhanced expression of pathogenesis-related genes (Kongala and Mamidala 2023).

10.8.2 Mechanisms of Harpins

Harpin-induced HR initiates firstly with the disruption of the plasma membrane of plant cells, resulting in pores formation in the plasma membrane causing ions conduction through it. Secondly harpins disrupted mitochondrial function by stimulating mitochondrion-dependent programmed cell death in plants inhibiting the mitochondrial electron transport chain, instant release of mitochondrial cytochrome c in the cytoplasm, resulting in the accumulation of reactive oxygen species. Harpins also interact with the F1 subunit of the mitochondrial ATP synthase enzyme and cause reduced synthesis of ATP. Third, harpins induce the activation of MAPK signaling cascades and HR-related gene expression. In plants, harpins interact with oxygen-evolving enhancer-like proteins causing the production of ROS and affecting the photosynthesis process (Daler et al. 2024). Harpins also interact with plasma membrane intrinsic protein on the plasma membrane and alter aquaporin responsible for the translocation of H_2O_2 from apoplasts to the cytosol for the induction of consequent membrane permeability of CO_2 and H_2O as well as immune response, leading to increased photosynthesis to promote plant growth.

10.8.3 Plant Response to Harpins Promotes Biotic Resistance

Harpin treatment helps plants become more resistant to pathogens (Table 10.2), which is an important in agriculture. For instance, the use of HpaXpm, Harpin (Axiom, Rx Green Technologies), and Messenger (a commercial product with Harpin (Ea) as the primary active component) decreased the severity of diseases in tomato, hemp seedlings, and tobacco caused by *Phytophthora infestans* and *Botrytis cinerea, Pythium aphanidermatum,* and tobacco mosaic virus (TMV), respectively (Sands et al. 2022; Liu et al. 2020; Fontanilla et al. 2005a, b).

10.9 Virus and Coat Protein and RNA Replicases Elicitors

Viruses have only two main components: genetic material and protein coat. Among all proteinaceous elicitors for plant defense response, the viral coat protein and RNA replicases of RNA viruses are responsible for both virus infection and the elicitation of plant defense responses. The viral RNA replicase is able to do independent replicase activity and has been reported to induce HR responses in plant. Similarly, viral coat protein is also reported to induce HR responses in plant (Fig. 10.3). Gaulin et al. (2006) reported that the coat protein of turnip crinkle virus elicited HR responses in Arabidopsis and cause cellulose-binding display and lectin-like activity, for the induction of cell necrosis and activation of defense-related gene expression (Shen et al. 2020).

10.10 Applications of Plant Antimicrobial Peptides

The potential and applications of plant antimicrobial peptides to advance agricultural research and food security are enormous. Researchers can use this pathogen elicitor–receptor connection and its specificity to create crops that are more resilient to diseases and pests.

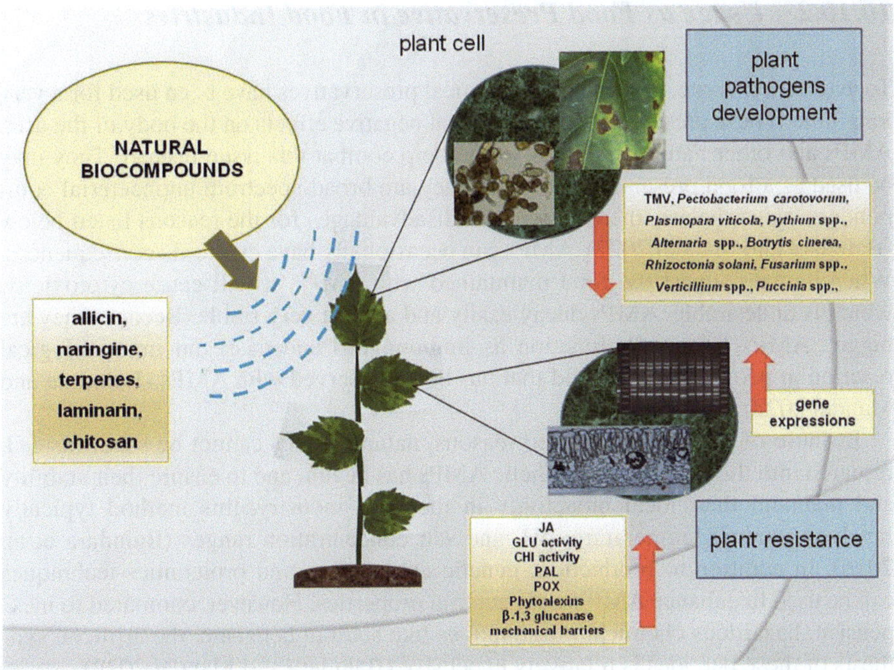

Fig. 10.3 The activation of plant defense processes via viral elicitors entails a sophisticated interrelationship between gene expression, hormonal reprogramming, and recognition and signaling processes. Not only can such processes make a plant resistant to infection with a virus, but its general tolerance toward a range of biotic and abiotic stresses is enhanced (Jamiołkowska 2020)

10.10.1 For Crop Protection

A biotechnological technique called "elicitation" boosts biomass and secondary metabolite synthesis in plant and organ cell cultures. Identifying novel elicitors (natural, chemical, and recombinant elicitors) and a better understanding of their mechanisms of action should make plant defense techniques more effective and environmentally benign (Kaur et al. 2022). It has been demonstrated that introducing the alfalfa antifungal peptide defensin gene into potato plants effectively protects them against *Verticillium dahilae* infection (Tam et al. 2015). It has been demonstrated that potato plants that have the snakin-1 gene included are more resistant to infection from *Erwinia carotovora* and *Rhizoctonia solani* (Ghosh and Roychoudhury 2024).

10.10.2 Usage as Food Preservative in Food Industries

To avoid food waste, artificial and chemical preservatives have been used for a very long time. These preservatives have several negative effects on the body of the user. AMPs and other natural products would help combat this issue instead. They may be used as a food preservative because they are broad-spectrum antibacterial components. Nevertheless, there are certain disadvantages for the reasons listed below (Baindara and Mandal 2022). AMPs can occasionally have cytotoxic consequences. When humans ingest the food maintained with AMPs, it will cause cytotoxicity, which is undesirable. AMPs decay easily and are not very stable. Because they are bigger, AMPs frequently function as immunogens and cause an immunological reaction in people who eat food that has been preserved with AMPs (Baindara and Mandal 2022).

Because of the aforementioned reasons, natural AMPs cannot be used. Instead, research into the creation of synthetic AMPs has begun, and to ensure their stability and maintain their ideal bioactivity in the food industry, this method typically involves extreme temperature, pH, and salt concentration ranges (Baindara et al. 2016). In addition to production, genetic engineering and proteomics techniques can be used to enhance AMP's antibacterial properties. However, compared to inexpensive, hazardous chemical preservatives like sodium benzoate, the artificial synthesis of these peptides is relatively expensive (Baindara and Mandal 2022).

10.10.3 Strategies for Plant Disease Management

Chemically produced fertilizers, agrochemicals, and pesticides pose a serious threat to human health and the environment. In the face of climate change challenges, agrochemicals that use AMPs to strengthen the plant immune system have enormous potential to address global food security and safety issues (Jakupi and Demirbas 2022). By creating extremely effective elite cultivars that can withstand extreme climatic circumstances like excessive temperatures, droughts, and shifting precipitation patterns, this can be utilized to increase crop resilience against a variety of stresses brought on by climate change (Iriti and Vitalini 2020).

Utilizing AMPs can make up for yield losses brought on by plant diseases, such as the 16% yearly loss resulting from pathogen attack (Iriti and Vitalini 2020). Utilizing plant protection products also boosts the immune system of plants, increasing the amounts of bioactive phytochemicals like polyphenols in plant meals. This improves the nutritional value of food and benefits consumer health (Jha, 2019).

These modified techniques encourage safer agriculture practices and support environmental sustainability by lessening reliance on conventional agrochemicals like pesticides, which are harmful to nontarget creatures, the environment, and human health. In order to develop more sustainable and efficient plant protection solutions that not only combat diseases but also promote plant resilience and overall health, additional advancements and studies in this field are required (Jha et al. 2021).

10.10.4 Understanding the Complex Signaling Pathways

Focused strategies to increase plant resilience are made possible by the ability of plant defense responses to differentiate between various elicitor–receptor interactions, such as those linked to induced systemic resistance (ISR) and SAR (Abdul Malik et al. 2020). The search for compounds that can replicate the effects of well-known inducers like acibenzolar-S-methyl (ASM) or benzothiadiazole (BTH) is still ongoing and may result in the development of novel plant defense techniques that could completely transform crop protection against pests and diseases (Kouzai et al. 2018).

According to Ruiz-García and Gómez-Plaza (2013), an incompatible pathogen attack can cause both localized and systemic reactions that lead to the synthesis of secondary metabolites such as phytoalexins, phenolics, terpenoids, and alkaloids. In addition to their potent antibacterial qualities, these secondary metabolites aid in the formation of physical barriers that prevent pathogen entrance and its replication. The idea of "elicitor engineering" aims to increase crop productivity and sustainability by employing elicitors to increase or stimulate the biosynthesis of specific chemicals in plants (Abdul Malik et al. 2020).

Plants' pattern recognition receptors (PRRs) identify pathogen-associated molecular patterns (PAMPs) of extracellular pathogens when they invade. This detection sets off a series of downstream signaling cascades that activate mitogen-activated protein kinases (MAPKs), which in turn causes the nucleus to begin transcription of defense-responsive genes. A baseline level of resistance is typically indicated by these substances. Additionally, plants employ the hypersensitive response (HR), a more potent defense mechanism (Abdul Malik et al. 2020). In an effort to avoid detection by PRRs, pathogens frequently release effector molecules into the cytosol of plant cells, avoiding the cell wall. By blocking the MAPK pathways, these effectors can stop pathogen-triggered immunity (PTI) from being activated. However, certain R genes may recognize these effectors upon entry into the plant cell, causing HR to be induced. It may be possible to use a transgenic method to activate this HR-mediated defense mechanism. The methods entail cloning a pathogen effector gene under a constitutive promoter, for which the plant possesses a corresponding *R* gene. To ensure that the effector gene is consistently produced in response to pathogen infection, any PAMP should activate this promoter. According to Ruiz-García and Gómez-Plaza (2013), effector production causes the relevant *R* gene to become active, which activates HR and confers resistance.

10.11 Conclusion

The global consumption and demand of crops for food, feed, and fuel are rapidly increasing due to a continuous increase in world population at its own pace. To fulfill the growing food demand, increased crop production on current agricultural land and efficient use of underused land for crop production are required. To sustain crop production, the mechanism of the plant immune system needs to be explored for the development of food safety and security measures. A better understanding of the immune system helps plant researchers to develop engineered plants with better resistance to disease. In this regard, a better understanding of the interaction between the cell receptor and elicitor may help plant scientists better understand the present defense system in plants and develop resistant plants against invading organisms.

References

Abdul Malik NA, Kumar IS, Nadarajah K (2020) Elicitor and receptor molecules: orchestrators of plant defense and immunity. Int J Mol Sci 21(3):963

Angelova Z, Georgiev S, Roos W (2006) Elicitation of plants. Biotechnol Biotechnol Equip 20(2):72–83

Azmi NS, Singkaravanit-Ogawa S, Ikeda K, Kitakura S, Inoue Y, Narusaka Y, Shirasu K, Kaido M, Mise K, Takano Y (2018) Inappropriate expression of an NLP effector in Colletotrichum orbiculare impairs infection on Cucurbitaceae cultivars via plant recognition of the C-terminal region. Mol Plant-Microbe Inter 31(1):101–111

Baindara P, Singh N, Ranjan M, Nallabelli N, Chaudhry V, Pathania GL, Sharma N, Kumar A, Patil PB, Korpole S (2016) Laterosporulin10: a novel defensin like class IId bacteriocin from Brevibacillus sp. strain SKDU10 with inhibitory activity against microbial pathogens. Microbiology 162(8):1286–1299

Baindara P, Mandal SM (2022) Plant-derived antimicrobial peptides: novel preservatives for the food industry. Food Secur 11(16):2415

Barda O, Shalev O, Alster S, Buxdorf K, Gafni A, Levy M (2015) Pseudozyma Aphidis induces salicylic-acid-independent resistance to *Clavibacter michiganensis* in tomato plants. Plant Dis 99:621–626

Cao Y, Yang M, Ma W, Sun Y, Chen G (2018) Overexpression of SSBXoc, a single-stranded DNA-binding protein from *Xanthomonas oryzae* pv. Oryzicola, enhances plant growth and disease and salt stress tolerance in transgenic *Nicotiana benthamiana*. Front Plant Sci 9:953

Cobos R, Calvo-Peña C, Álvarez-Pérez JM, Ibáñez A, Diez-Galán A, González-García S, García-Angulo P, Acebes JL, Coque JJ (2019) Necrotic and cytolytic activity on grapevine leaves produced by Nep1-like proteins of *Diplodia seriata*. Front Plant Sci 10:1282

Daler S, Karaca I, Delavar H, Kaya O (2024) Harnessing the potential of Harpin proteins: elicitation strategies for enhanced secondary metabolite accumulation in grapevine callus. Cultures Processes 12(7):1416

Dawood MFA, Mohamed HI, Jha Y, El-Mahdy MT (2024) Karrikin regulation of biosynthesis. In: Hormonal crosstalk, and its role in biotic and abiotic stress adaptation: book plant growth regulators to manage biotic and abiotic stress in agroecosystems. CRC Press

Domazakis E, Wouters D, Visser RG, Kamoun S, Joosten MH, Vleeshouwers VG (2018) The ELR-SOBIR1 complex functions as a two-component receptor-like kinase to mount defense against *Phytophthora infestans*. Mol Plant-Microbe Inter 31(8):795–802

Du Q, Yang X, Zhang J, Zhong X, Kim KS, Yang J, Xing G, Li X, Jiang Z, Li Q et al (2018) Over-expression of the *pseudomonas syringae* harpin-encoding gene hrpZm confers enhanced tolerance to Phytophthora root and stem rot in transgenic soybean. Transgenic Res 27:277–288

Du J, Verzaux E, Chaparro-Garcia A, Bijsterbosch G, Keizer LC, Zhou JI, Liebrand TW, Xie C, Govers F, Robatzek S, Van Der Vossen EA (2015) Elicitin recognition confers enhanced resistance to *Phytophthora infestans* in potato. Nature plants 1(4):1–5

Fontanilla JM, Montes M, De Prado R (2005a) Induction of resistance to the pathogenic agent Botrytis cinerea in the cultivation of the tomato by means of the application of the protein "Harpin"(messenger). Commun Agric Appl Biol Sci 70:35–40

Fontanilla M, Montes M, De Prado R (2005b) Effects of the foliar-applied protein "Harpin(Ea)" (messenger) on tomatoes infected with *Phytophthora infestans*. Commun Agric Appl Biol Sci 70:41–45

Garcia JA, Pallas V (2015) Viral factors involved in plant pathogenesis. Curr Opin Virol 11:21–30

Gaulin E, Drame N, Lafitte C, Torto-Alalibo T, Martinez Y, Ameline-Torregrosa C, Khatib M (2006) Cellulose binding domains of a Phytophthora cell wall protein are novel pathogen-associated molecular patterns. Plant Cell 18:1766–1777

Giovannoni M, Marti L, Ferrari S, Tanaka-Takada N, Maeshima M, Ott T, De Lorenzo G, Mattei B (2021) The plasma membrane-associated Ca^{2+} -binding protein, PCaP1, is required for oligogalacturonide and flagellin-induced priming and immunity. Plant Cell Environ 44(9):3078–3093

Ghosh P, Roychoudhury A (2024) Plant peptides involved in abiotic and biotic stress responses and reactive oxygen species (ROS) signaling. Journal of Plant Growth Regul 43(5):1410–1427

Hayafune M, Berisio R, Marchetti R, Silipo A, Kayama M, Desaki Y, Shibuya N (2014) Chitin-induced activation of immune signaling by the rice receptor CEBiP relies on a unique sandwich-type dimerization. Proc Natl Acad Sci USA 111(3):E404–E413

Iriti M, Vitalini S (2020) Sustainable crop protection, global climate change, food security and safety—plant immunity at the crossroads. Vaccine 8(1):42

Jakupi M, Demirbas S (2022) Production of secondary metabolites through elicitors: their application in agriculture. Tekirdag Namik Kemal Univ Inst Nat Appl Sci 1(3):165–172

Jamiołkowska A (2020) Natural compounds as elicitors of plant resistance against diseases and new biocontrol strategies. Agronomy 10(2):173

Jang H, Kim ST, Sang MK (2022) Suppressive effect of bioactive extracts of bacillus Sp. H8–1 and bacillus Sp. K203 on tomato wilt caused by *Clavibactermichiganensis* subsp. *michiganensis*. Microorganisms 10:403

Janku M, Cincalova L, Luhova L, Lochman J, Petrivalsky M (2020) Biological effects of oomycetes elicitins. Plant Prot Sci 56:1–8

Jha Y (2019) Mineral Mobilizing Bacteria Mediated Regulation of Secondary Metabolites for Proper Photosynthesis in Maize Under Stress. In Photosynthesis, Productivity and Environmental Stress (eds Ahmad P, Abass M, Ahanger M, Nasser A, Alam P).

Jha Y (2022) Enhanced cell viability with induction of pathogenesis related proteins against *aspergillus Niger* in maize by Endo-Rhizospheric bacteria. Jordan J Biol Scie 15(1):139–147

Jha Y (2024a) Role of microbes in altering plant metabolome under stressful environments. In: Abd-Elsalam KA, Mohamed HI (eds) Nanobiotechnology for plant protection. Elsevier, Fungal Secondary Metabolites, pp 427–437

Jha Y, Dehury B, Kumar SPJ (2022) Delineation of molecular interactions of plant growth promoting bacteria induced β-1,3-glucanases and guanosine triphosphate ligand for antifungal response in rice: a molecular dynamics approach. Mol Biol Rep 49:2579–2589

Jha Y, Kulkarni A, Subramanian RB (2021) Psychrotrophic Soil Microbes and Their Role in Alleviation of Cold Stress in Plants. In: Yadav AN (eds) Soil Microbiomes for Sustainable Agriculture. Sustainable Development and Biodiversity, vol 27. Springer, Cham

Jha Y, Mohamed HI (2023) Enhancement of disease resistance, growth potential, and biochemical markers in maize plants by inoculation with plant growth-promoting bacteria under biotic stress. J Plant Pathol 105:729–748

Jha Y, Mohamed HI (2024) Endophytic *pseudomonas pseudoalcaligenes* and arbuscular mycorrhizal fungi mediated anti-autophagy and induction of catalase and antioxidant enzymes in pigeon pea against fungal pathogen. J Plant Pathol 106:225–240

Jha Y, Mohamed HI, El-Mahdy MT, Dawood MFA (2024) Application for growth improvement and protection against biotic stress book plant growth regulators to manage biotic and abiotic stress in agroecosystems. Im print CRC Press, p 23

Jha Y, Subramanian RB (2021) Use of halo-tolerant bacteria to improve the bioactive secondary metabolites in medicinally important plants under saline stress: Book Microbiomes of Extreme Environments. CRC Press, Pages13

Jha Y, Yadav AN (2021) *Piriformo sporaindica*: biodiversity, ecological significances, and biotechnological applications for agriculture and allied sectors. In: Abdel-Azeem AM, Yadav AN, Yadav N, Usmani Z (eds) Industrially important fungi for sustainable development. Fungal Biology. Springer, Cham

Jha Y (2024b) Regulation of photosynthesis under stress. In: Ahanger MA, Bhat JA, Ahmad P, John R (eds) . Improving Stress Resilience in Plants, Academic Press, pp 35–48

Jomova K, Raptova R, Alomar SY (2023) Reactive oxygen species, toxicity, oxidative stress, and antioxidants: chronic diseases and aging. Arch Toxicol 97:2499–2574

Joshi JB, Senthamilselvi D, Maupin-Furlow JA, Uthandi S (2022) Microbial protein elicitors in plant defense. In: Kumar A (ed) Microbial biocontrol: sustainable agriculture and Phytopathogen management. Springer, Cham

Kachroo P, Yoshioka K, Shah J, Dooner HK, Klessig DF (2002) Resistance to turnip crinkle virus in Arabidopsis is regulated by two host genes and is salicylic acid dependent but NPR1, ethylene, and jasmonate independent. Plant Cell 12:677–690

Kaur S, Samota MK, Choudhary M (2022) How do plants defend themselves against pathogens-biochemical mechanisms and genetic interventions. Physiol Mol Biol Plants 28:485–504

Keswani C, Dilnashin H, Birla H, Singh SP (2019) Unravelling efficient applications of agriculturally important microorganisms for alleviation of induced inter-cellular oxidative stress in crops. Acta Agric Slov 114(1):121–130

Kiani R, Arzani A, Mirmohammady Maibody SAM (2021) Polyphenols, flavonoids, and antioxidant activity involved in salt tolerance in wheat, *Aegilops cylindrica* and their amphidiploids. Front Plant Sci 12:646221

Kongala SI, Mamidala P (2023) Harpin-loaded chitosan nanoparticles induced defense responses in tobacco. Carbohydrate Polymer Technologies and Applications 5:100293

Kouzai Y, Noutoshi Y, Inoue K, Shimizu M, Onda Y, Mochida K (2018) Benzothiadiazole, a plant defense inducer, negatively regulates sheath blight resistance in Brachypodium distachyon. Sci Rep 8(1):17358

Laveti G (2018) Role of elicitors in plant defense mechanism. J Pharmacogn Phytochem 7(6):2806–2812

Lenarčič T, Albert I, Böhm H, Hodnik V, Pirc K, Zavec AB, Podobnik M, Pahovnik D, Žagar E, Pruitt R, Greimel P (2017) Eudicot plant-specific sphingolipids determine host selectivity of microbial NLP cytolysins. Science 358(6369):1431–1434

Levin E, Raphael G, Ma J, Ballester AR, Feygenberg O, Norelli J, Aly R, Gonzalez-Candelas L, Wisniewski M, Droby S (2019) Identification and functional analysis of NLP-encoding genes from the postharvest pathogen Penicillium expansum. Microorganisms 7(6):175

Li Z, Liu J, Ma W, Li X (2023) Characteristics, roles and applications of Proteinaceous elicitors from pathogens in plant immunity. Life 13(2):268

Lian J, Han H, Chen X, Chen Q, Zhao J, Li C (2022) *Stemphylium lycopersici* Nep1-like protein (NLP) is a key virulence factor in tomato gray leaf spot disease. J Fungi 8:518

Liu Y, Zhou X, Liu W, Huang J, Liu Q, Sun J, Cai X, Miao W (2020) HpaXpm, a novel harpin of *Xanthomonas phaseoli* pv. *Manihotis*, acts as an elicitor with high thermal stability, reduces disease, and promotes plant growth. BMC Microbiol 20:4

Liu Z, Sun Z, Zeng C, Dong X, Li M, Liu Z, Yan M (2022) The elemental defense effect of cadmium on *Alternaria brassicicola* in *Brassica juncea*. BMC Plant Biol 22(1):17

Lu C, Jiang Y, Yue Y, Sui Y, Hao M, Kang X, Wang Q, Chen D, Liu B, Yin Z (2022) Glutathione and neodiosmin feedback sustain plant immunity. J Exp Bot 74(3):976–990

Michel S, Löschenberger F, Ametz C, Bürstmayr H (2023) Toward combining qualitative race-specific and quantitative race-nonspecific disease resistance by genomic selection. Theor Appl Genet 136(4):79

Montesano M, Brader G, Palva ET (2003) Pathogen derived elicitors: searching for receptors in plants. Mol Plant Pathol 4(1):73–79

Münzel T, Hahad O, Daiber A, Landrigan PJ (2023) Soil and water pollution and human health: what should cardiologists worry about? Cardiovasc Res 119(2):440–449

Newton AC, Bootman MD, Scott JD (2016) Second messengers. Cold Spring HarbPerspect Biol 8(8):a005926

Pršić J, Ongena M (2020) Elicitors of plant immunity triggered by beneficial bacteria. Front Plant Sci 11:594530

Ruiz-García Y, Gómez-Plaza E (2013) Elicitors: a tool for improving fruit phenolic content. Agriculture 3(1):33–52

Sands LB, Cheek T, Reynolds J, Ma Y, Berkowitz GA (2022) Effects of Harpin and Flg22 on growth enhancement and pathogen defense in *Cannabis sativa* seedlings. Plan Theory 11:1178

Schoonbeek HJ, Yalcin HA, Burns R, Taylor RE, Casey A, Holt S, Van den Ackerveken G, Wells R, Ridout CJ (2022) Necrosis and ethylene-inducing-like peptide patterns from crop pathogens induce differential responses within seven brassicaceous species. Plant Pathol 71:2004–2016

Schumacher S, Grosser K, Voegele RT, Kassemeyer H-H, Fuchs R (2020) Identification and characterization of Nep1-like proteins from the grapevine downy mildew pathogen *Plasmopara viticola*. Front Plant Sci 11:65

Shen W, Shi Y, Dai Z, Wang A (2020) The RNA-dependent RNA polymerase NIb of Potyviruses plays multifunctional, contrasting roles during viral infection. *Viruses* 12(1):77

Steentjes MB, Herrera Valderrama AL, Fouillen L, Bahammou D, Leisen T, Albert I, Nürnberger T, Hahn M, Mongrand S, Scholten OE, van Kan JA (2022) Cytotoxic activity of Nep1-like proteins on monocots. New Phytol 235(2):690–700

Sun T, Wu W, Wu H, Rou W, Zhou Y, Zhuo T, Fan X, Hu X, Zou H (2020) *Ralstonia solanacearum* elicitor RipX induces defense reaction by suppressing the mitochondrial atpA gene in host plant. Inter J Mol Sci 21(6):2000

Tam JP, Wang S, Wong KH, Tan W (2015) Antimicrobial peptides from plants. Pharmaceuticals 8(4):711–757

Thakur M, Sohal BS (2013) Role of elicitors in inducing resistance in plants against pathogen infection: a review. ISRN Biochem 2013:762412

Tian H, Wu Z, Chen S, Ao K, Huang W, Yaghmaiean H, Sun T, Xu F, Zhang Y, Wang S, Li X (2021) Activation of TIR signalling boosts pattern-triggered immunity. Nature 598(7881):500–503

Turnbull D, Wang H, Breen S, Malec M, Naqvi S, Yang L, Welsh L, Hemsley P, Zhendong T, Brunner F, Gilroy EM (2019) AVR2 targets BSL family members, which act as susceptibility factors to suppress host immunity. Plant Physiol 180(1):571–581

Wang G, Ren Y, Bai X, Su Y, Han J (2022) Contributions of beneficial microorganisms in soil remediation and quality improvement of medicinal plants. Plan Theory 11(23):3200

War AR, Taggar GK, Hussain B, Taggar MS, Nair RM, Sharma HC (2018) Plant defence against herbivory and insect adaptations. AoB Plants 10(4):ply037

Xiang J, Cheng J, Wei L, Li M, Wu J (2022) Functional analysis of the Nep1-like proteins from *Plasmopara viticola*. Plant Signal Behavior 17(1)

Yang B, Yang S, Zheng W (2022a) Plant immunity inducers: from discovery to agricultural application. Stress Biology 2:5

Yang K, Wang Y, Zhao H, Shen D, Dou D, Jing M (2022d) Novel Elicitin from *Pythium oligandrum* confers disease resistance against *Phytophthora capsici* in *Solanaceae* plants. J Agric Food Chem 70:16135–16145

Yang C, Li W, Huang X, Tang X, Qin L, Liu Y, Xia Y, Peng Z, Xia S (2022e) Ss NEP2 contributes to the virulence of *Sclerotinia sclerotiorum*. Pathogens 11(4):446

Yang K, Chen C, Wang Y, Li J, Dong X, Cheng Y, Zhang H, Zhai Y, Ai G, Song Q, Wang B (2022b) Nep1-like proteins from the biocontrol agent *Pythium oligandrum* enhance plant disease resistance independent of cell death and reactive oxygen species. Front Plant Sci 13:830636

Yang G, Yang J, Zhang Q, Wang W, Feng L, Zhao L, An B, Wang Q, He C, Luo H (2022c) The effector protein CgNLP1 of Colletotrichum gloeosporioides affects invasion and disrupts nuclear localization of necrosis-induced transcription factor HbMYB8-like to suppress plant defense signaling. Front Microbiol 13:911479

Yu WQ, Li P, Yan FC (2021) Protein elicitor EsxA induces resistance to seedling blight and PR genes differential transcription in Rice. Rice 14:91

Zehra A, Raytekar NA, Meena M, Swapnil P (2021) Efficiency of microbial bio-agents as elicitors in plant defense mechanism under biotic stress: a review. Curr Res Microb Sci 2:100054

Zelman AK, Berkowitz GA (2023) Plant elicitor peptide (pep) signaling and pathogen defense in tomato. Plan Theory 12(15):2856

Chapter 11
Innovative Elicitors in Plant Science

Tony Kevork Sajyan, Sana AlHajj Ahmad, Israa El Ghandour, Lourd Abd Al Ahad, and Rania Francis

Abstract The study of plant elicitors, substances that induce or enhance the production of specific compounds in plants, has significantly advanced with the advent of novel technologies and emerging trends. Elicitors are classified into abiotic and biotic categories and interact with plant receptors to stimulate stress responses, leading to increased biosynthesis of secondary metabolites. Recent developments in nanotechnology have introduced innovative elicitors, such as nanoparticles, which offer enhanced efficiency in secondary metabolite production. In agriculture, nanotechnology transforms crop management through nanoformulations, nanosensors, and advanced genetic engineering, aiming to improve productivity and sustainability. Furthermore, metabolomics and proteomics provide deep insights into the molecular mechanisms underlying elicitor responses, revealing how elicitors impact plant defense mechanisms and secondary metabolism. These omics-based approaches are instrumental in understanding the intricate interactions between elicitors and plants, leading to potential applications in enhancing crop resilience and productivity. This chapter explores these advancements and their implications for future research and practical applications in elicitor studies.

Keywords Nanotechnology · Secondary metabolites · Elicitors · Crop production · OMICs

---

T. K. Sajyan (✉)
Faculty of Agricultural and Food Sciences, American University of Beirut, Beirut, Lebanon

Department of Agriculture and Food Engineering, School of Engineering, Holy Spirit University of Kaslik, Jounieh, BP, Lebanon
e-mail: ts71@aub.edu.lb

S. A. Ahmad · I. El Ghandour · L. A. Al Ahad
Faculty of Agricultural and Food Sciences, American University of Beirut, Beirut, Lebanon

R. Francis
Department of Agriculture and Food Engineering, School of Engineering, Holy Spirit University of Kaslik, Jounieh, BP, Lebanon

School of Technology, Al-Kafaàt University, Ain Saadeh, Lebanon

K. A. Abd-Elsalam, H. I. Mohamed (eds.), *Elicitors for Sustainable Crop Production*, Sustainability Sciences in Asia and Africa,
https://doi.org/10.1007/978-981-96-6569-3_11

11.1 Introduction

Nanoparticles (NPs), a vital nanotechnology component, are tiny materials generally classified by their particle size of less than 100 nm (Khan et al. 2019). Due to their nanoscale dimensions, they exhibit physical and chemical properties crucial in understanding how NPs interact biologically and function as elicitors of bioactive compounds in plants. A wide range of these particles, such as titanium oxide (TiO_2), iron oxide (Fe_3O_4), zinc oxide (ZnO), silicon oxide (SiO_2), copper NPs (CuNPs), and selenium NPs, have recently gained considerable attention for their safe applications in agriculture (Hafeez et al. 2015; Alabdallah and Hasan 2021; Hashem et al. 2021; Rasheed et al. 2022). Nanotechnology is crucial in transforming agriculture by introducing beneficial solutions to improve crop production, advanced fertilizers, plant disease diagnostics, pesticides, sustainability practices, post-harvest management, and smart sensors for monitoring soil conditions. Its applications range from crop protection to nutrient delivery.

Plants, being immobile organisms, produce a variety of secondary metabolites to defend against both biotic and abiotic stresses, which have a significant impact on their physiology and ecology. Elicitation refers to the process by which plants are stimulated or enhanced to synthesize higher amounts of secondary metabolites, thus ensuring their survival, persistence, and competitive advantage in their environment (Patel and Krishnamurthy 2013). One intriguing aspect of nanotechnology in agriculture is its role in modulating plant secondary metabolites (PSMs) by using NPs as elicitors.

Elicitors are foreign molecules that induce any form of physiological abnormality in plants. They are categorized into exogenous and endogenous types, both capable of binding to receptor proteins on plant cell membranes. These receptors recognize molecular patterns of elicitors, initiating intracellular defense signaling pathways that enhance the synthesis of metabolites. This process ultimately enhances plant resistance to various stresses. Exogenous elicitors come from pathogens, while endogenous elicitors are released by plants in response to these pathogenic interactions. Elicitors are also used to stimulate the synthesis of secondary metabolites in plants, which is crucial for increasing the production of commercially valuable compounds (Patel and Krishnamurthy 2013). This chapter explores these advancements and their implications for future research and practical applications in elicitor studies.

11.1.1 Nanotechnology and Elicitors' Production

Nanotechnology plays a significant role in agriculture by addressing various challenges, such as enhancing crop resistance to biotic and abiotic stresses and boosting plant productivity (Ahmad et al. 2022). Its advancements have opened up new boundaries for promoting elicitor production, particularly in the agriculture and biotechnological sectors. Elicitors can be classified according to their origin: biotic (biological) or abiotic (nonbiological including physical, chemical, and hormonal elicitors) (Naik and Al-Khayri 2016). Biotic elicitors are biologically derived compounds produced either by pathogens or by the plant itself. Additionally, external elicitors, including bacteria, viruses, herbivore attacks, and fungi, stimulate plants to generate compounds at the site of a pathogen or herbivore invasion (Jain et al. 2024). On the other hand, abiotic elicitors, as mentioned previously, can be categorized into groups based on their mode of action: physical, chemical, and hormonal. Besides, various abiotic elicitors have demonstrated effectiveness in boosting plant resistance to bacterial pathogens by activating multiple resistance mechanisms across different crops, including *Arabidopsis thaliana*, *Brassica napus*, *Capsicum annuum*, *Cicer arietinum*, *Cucumis sativus*, *Nicotiana tabacum*, *Oryza sativa*, and *Solanum lycopersicum* (Khan et al. 2015; Gowtham et al. 2024).

NPs are emerging as promising abiotic elicitors in plant cell and tissue culture, capable of inducing bioactive compounds. Recent studies suggest that they play a role in activating genes for secondary metabolite biosynthesis by inducing reactive oxygen species (ROS) and signaling molecules such as calcium ions, which regulate gene expression in secondary metabolism. The response to NPs, which can trigger oxidative stress and modify primary and secondary metabolite production, varies based on factors such as plant species, tissue type, and type of stressor (Yarizade and Hosseini 2015; Marslin et al. 2017; Tassi et al. 2017; Anjum et al. 2019; Hasanuzzaman et al. 2020; Rivero-Montejo et al. 2021).

11.1.2 Mechanism of Action on Plant Secondary Metabolites

Plant metabolites can be classified into two main types: primary metabolites and secondary metabolites. The primary metabolites are produced by plants as essential compounds for growth and metabolic processes, which are directly involved in primary metabolic functions such as respiration and photosynthesis. Secondary metabolites such as terpenoids, peptides, antibiotics, ergot alkaloids, and growth factors are not directly involved in primary metabolic processes but are required for the plant to survive in the environment. Therefore, secondary metabolites serve crucial roles in plant defense mechanisms. The mechanism by which nanotechnology influences these metabolites involves complex elicitation processes that alter plant metabolic pathways, thereby enhancing the synthesis of these valuable compounds.

PSMs, also known as secondary plant compounds or natural products, are organic chemicals produced by certain plants. They are classified based on their chemical structures into several classes: phenolics, alkaloids, saponins, terpenes, lipids, and carbohydrates. Most of these compounds have an ecological role in regulating the interactions between plants, microorganisms, insects, and animals (Alvarez 2014). These often accumulate in the plant in small quantities, frequently in specialized cells, during specific plant developmental stages, or under stress (Alvarez 2014).

Additionally, they are synthesized at the end of the exponential growth phase (trophophase) or the onset of the stationary phase (idiophase), with their production generally being suppressed during periods of rapid growth. The synthesis of such a rich variety of phytochemicals can be further stimulated with elicitors or by feeding precursors (Guerriero et al. 2018). Such compounds can also function as antimicrobials and can either attract or repel. More than 50,000 secondary metabolites have been identified in the plant kingdom (Teoh 2016).

Plants synthesize a wide range of secondary metabolites with complex chemical compositions, which are produced in response to various biotic and abiotic stresses and to perform essential physiological functions such as attracting pollinators, establishing symbiosis, and providing structural components to lignified cell walls of vascular tissues (Guerriero et al. 2018).

NPs can enter plants in various forms and through different pathways. The primary application methods for NPs in plants include foliar spraying, hydroponic treatment, biolistic gun, injection, and seed incubation. These methods facilitate NP uptake through foliar uptake, root uptake, and seed uptake (Zhang and Su 2024).

The plant root epidermis functions similarly to the leaf surface, but in less developed areas such as root tips and root hairs of primary and secondary roots, NPs can directly penetrate the epidermis. The semi-permeable cell wall of the root epidermis contains small pores that restrict the passage of large particles. Without an exodermis layer, NPs can reach the central root column or xylem. Some of the NPs can disrupt the plasma membrane, creating new pores in the epidermal cell wall and facilitating the entry of larger NPs. Once inside plant tissues, NPs can enter cells through ion channels, endocytosis, binding with cell membrane proteins, or physical damage (Wang et al. 2023).

The other pathway is through foliar uptake. Plant leaves are the main targets for foliar treatment of NPs (Khan et al. 2022). The leaves of higher plants are protected by a waxy cuticle layer, which serves as the first barrier against water loss and NP entry (Khan et al. 2022; Pollard et al. 2008). Solutes typically cross this layer through two pathways: a lipophilic pathway for nonpopular solutes and a hydrophilic pathway for polar solutes (Eichert and Goldbach 2008; Eichert et al. 2008; Khan et al. 2022).

NPs can enter mesophyll cells through the cuticle and stomata. Once inside, they can travel long distances within the plant through either the extracellular pathway or the plastid pathway. The extracellular pathway involves transport through spaces such as the cell wall, channels between cell walls, the intermediate lamella, and the xylem, depending on particle size and surface charge. The plastid pathway primarily uses intercellular channels called plasmodesmata that accumulate in the cytoplasm and can reach the plant's endothelial layer and Casparian strip (Wang et al. 2023).

Moreover, the uptake of NPs through seeds involves treating seeds with NP suspensions via soaking or coating methods. NPs can adhere to the seed coat or penetrate it to enter the seed interior. Studies show that NPs, like carbon nanotubes, can penetrate seed coats and influence seed germination and plant growth. This method, often used for seed priming, enhances seed germination by potentially creating water-permeable pores through the seed coat (Zhang and Su 2024).

Foliar treatments enhance insecticide efficiency, while root treatments boost plant resilience to environmental stresses. In contrast, at the seed stage, NPs are synthesized as nano-priming agents to enhance seed germination and promote plant growth (Zhang and Su 2024). The entry of NPs into plants is affected by several factors. These include the size, chemical, composition, surface charge, and shape of the NPs, as well as the specific plant species and environmental conditions (Wang et al. 2023).

Regarding the movement of NPs in plants, once the NPs penetrate the plant, they can move through tissue via two pathways: apoplastic and symplastic transport. Apoplastic transport occurs outside the plasma membrane, moving through extracellular spaces, cell walls of adjacent cells, and eventually through xylem vessels (Pérez-de-Luque 2017). In contrast, symplastic transport involves movement through the cytoplasm of adjacent cells via specialized structures called plasmodesmata (Pérez-de-Luque 2017). This pathway allows NPs to pass directly from cell to cell, facilitating their transport through plant tissues.

As noted earlier, PSMs including phenolics, flavonoids, and terpenoids, play crucial roles in plant defense mechanisms, growth regulation, and responses to environmental stresses. These metabolites serve diverse functions as phytoalexins, phytoanticides, and phytoncides (defenses against biotic stresses), antioxidants (regulating ROS), chelators (binding free metal ions), ultraviolet (UV) protectants, growth regulators, and mitigators of abiotic stresses (Selvakesavan et al. 2023). When plants are exposed to NPs, these secondary metabolites can be influenced in several ways. For example, several types of NPs, including iron (Fe), cerium (Ce), and SiO_2, have been shown to modify the content of secondary metabolites in lettuce and pepper seedlings (Selvakesavan et al. 2023).

NPs interact with plants, triggering ROS production and oxidative stress, which affects membrane integrity, antioxidant activities, and secondary metabolism. These interactions also alter gene expression related to stress responses and cellular processes. ROS induction by NPs occurs via calcium and MAPK signaling in both apoplast and chloroplast, influencing cell wall dynamics and secondary metabolite synthesis. NP effects on plants are concentration-dependent, with higher concentrations often exhibiting toxicity and lower concentrations potentially benefits (Selvakesavan et al. 2023).

11.1.3 Mechanism of Elicitation

Elicitation, which is the process of stimulating plant stress response to induce the production of specific compounds, harnesses the connection between plant stress and phytochemistry. An elicitor can be defined as a substance that, when applied in small quantities, induces or enhances the biosynthesis of compounds important for plant adaptation to stressful conditions. According to Naik and Al-Khayri (2016), elicitors are chemical substances originating from both non-living (abiotic) and living sources (biotic) that trigger stress responses in plants. This stimulation can result in increased production and accumulation of secondary metabolites or the initiation of entirely new secondary metabolite pathways. Also, based on Khan et al. (2021), this process involves the addition of exogenous elicitors, thus protecting against stress.

Elicitors interact with specific receptors or ion channels in the plasma membrane, triggering gene expression and inducing a stress response in plants. This response can lead to the production of higher amounts or new forms of valuable secondary metabolites (Khan et al. 2021). Among various elicitors, NPs have shown great potential due to their unique properties. Different types of NPs, including metallic NPs, metal oxide NPs, and carbon-related NPs, have been used as novel and effective elicitors of flavonoids in in vitro cultures of various plant species (Fig. 11.1). Notable examples include silver NPs (AgNPs), CuNPs, iron oxide NPs, zinc oxide NPs (ZnONPs), copper oxide NPs (CuONPs), carbon nanotubes, and chitosan-based NPs (Khan et al. 2021). In this section, the focus is to study the role of nanotechnology-based elicitors in enhancing the production of secondary metabolites in plants.

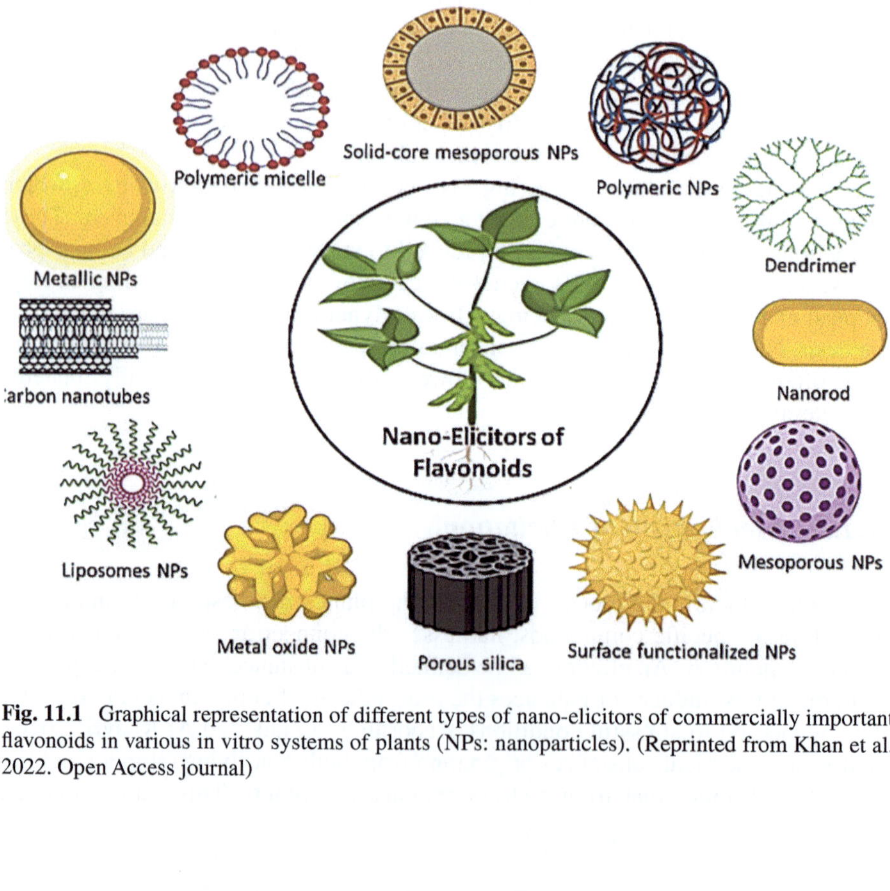

Fig. 11.1 Graphical representation of different types of nano-elicitors of commercially important flavonoids in various in vitro systems of plants (NPs: nanoparticles). (Reprinted from Khan et al. 2022. Open Access journal)

11.2 Application in Agriculture

In the past few years, the focus of agricultural nanotechnology has been on research and practical applications to sustain agricultural and environmental problems sustainably, enhance crop quality, and increase productivity (Neme et al. 2021). Nanotechnology has the potential to stimulate agricultural output significantly. Here are some ways that can be applied: (1) enhanced agrochemicals, using nanoformulations to improve the effectiveness of pesticides and fertilizers, leading to better crop yields; (2) smart crop protection, employing nanosensors to detect plant diseases and monitor agrochemical residues; (3) advanced genetic engineering (GE), utilizing nanodevices for the genetic modification of plants to enhance their traits; (4) disease diagnostics, implementing nanotechnology for more accurate plant disease diagnosis; (5) animal and poultry health, improving animal health, breeding, and poultry production through nanotechnology; and (6) better postharvest management, using nanotech solutions to manage crops after harvest more efficiently (Neme et al. 2021).

11.2.1 Nanoformulations

The development of the applications of NPs, especially agrochemicals, arises widely in its role in minimizing and controlling the use of pesticides and fertilizers. NPs deliver agrochemicals to plants through controlled release systems, reducing the inefficiency of traditional pesticide formulations where over 90% of pesticides are lost to the atmosphere and persist in agricultural products (Yousef et al. 2023). Additionally, nanomaterials are increasingly used in bio-pesticides to offer eco-friendly pest control alternatives that address the health and environmental issues posed by chemical pesticides, and they enhance the stability, effectiveness, and shelf life of microbial, viral, plant-derived, antibiotic, and biochemical pesticides while also improving plant immunity and growth (Table 11.1) (An et al. 2022). According to Liu and Lal, the increase in arable land's carrying capacity from 1.9 to 4.3 persons per hectare between 1908 and 2008, mainly due to synthetic fertilizers, underscores the importance of developing new fertilizer technologies to sustainably feed a projected global population of 9.6 billion by 2050, with a focus on various types of nanofertilizers and plant-growth stimulating nanomaterials (Liu and Lal 2015).

11.2.2 Nanotechnology-Based Genome Editing

NPs have begun to enhance GE through efficient and targeted delivery of plasmids, ribonucleic acid (RNA), and ribonucleoproteins (RNPs). In mammalian cells, NPs are routinely used for direct cytosolic or nuclear delivery of Cas-RNPs, significantly

Table 11.1 Nanomaterials applied to bio-pesticides and their functions (An et al. 2022)

Active ingredients (AIs)	Material(s)	Function	References
Bacillus thuringiensis (*Bt*)	Nanotubular sodium titanate	Effective in controlling cotton leaf-worm	Zaki et al. (2017)
Beauveria bassiana (*Bals.*)	Silica NPs and carbon fibers	Improving mortality to larvae of potato *Spodoptera litura (S. Litura)*	Hersanti et al. (2020)
Essential oils	Chitosan-coated nanosilver	Synergistic effect against a wide range of microorganisms	Gahukar and Das (2020)
Avermectin	Poly (ethylene glycol)-carboxymethyl cellulose (PEG-CMC)	Improved the anti-UV ability and increased the biocompatibility of the Avermectin	Zhu et al. (2020)
Validamycin and thifluzamide	Polylactic acid	The NPs prepared by compounding validamycin with chemical pesticides showed better control of rice sheath blight, which was 4.2 times more effective than the control	Cui et al. (2020)
Gibberellic acid (GA)	Layered double hydroxides (LDH)	Promoted plant growth	Hafez et al. (2018)

reducing off-target effects compared to plasmid-based (CRISPR) clustered regularly interspaced short palindromic repeats systems (Ahmar et al. 2021). However, in plants, the cell wall poses a challenge for passively delivering GE cargo into mature plants. There is potential for developing NP carriers with diverse cargo-loading capabilities and optimal geometry/chemistry to bypass the cell wall and membranes in dense plant tissues. Previous studies have shown that some NP formulations can passively internalize in plants with deoxyribonucleic acid (DNA), RNA, or protein cargo (Ahmar et al. 2021).

11.2.3 Nanotechnology's Role in Post-harvest

A wide range of nanomaterials, including those based on lipids, carbon, inorganic metals, and their oxides, has been developed to control fruit damage in the food and agriculture industry (Khot et al. 2012; Bajpai et al. 2018; Babu et al. 2022). Nanotechnology aids in managing post-harvest diseases, advancing packaging methods with nano-biosensor chips for labeling fresh products. Nanomaterials are also employed as antifungal agents for various fruit infections. For instance, nanopolymers have addressed the challenge of gas impermeability in fruit packaging, while AgNPs, chitosan/silica nanocomposites, and chitosan NPs (CNPs) have been used to protect and extend the shelf-life of fruits such as grapes and bananas. These materials help maintain fruit quality by delaying ripening, reducing moisture loss, and preventing infections, though further research is needed to optimize nanomaterial applications in post-harvest management (Rubentheren et al. 2015; Piña-Barrera et al. 2019; Siddiqi et al. 2018; Babu et al. 2022).

To focus mainly on the role of nano-elicitors, and as mentioned earlier, nano-elicitation in plants enhances its metabolic processes using the NPs as elicitors, thereby stimulating plant yield and productivity, together with the production of PSMs (Jadoon et al. 2024). The latter induces the release of reactive oxidative species (ROS); hence, this highlights the role of NP-mediated ROS in providing protection against both biotic and abiotic stressors. To illustrate, one of the studies by Chandra et al. demonstrates this role. They found that when plants were treated with chitosan (natural biopolymer) and CNPs, the levels of key antioxidant enzymes, SOD and CAT, were significantly higher than in untreated plants. These enzymes help neutralize harmful molecules produced during stress, indicating that chitosan and CNP treatments enhance the plant's ability to handle stressful conditions (Chandra et al. 2015).

Another study by Chung et al. outlined the effects of AgNPs at concentrations of 0.5, 1.0, and 2.0 mg/L on hair root cultures of *Cumis anguria* L. It showed an obvious enhancement in both the biomass accumulation and the total phenolic compounds. Moreover, the AgNPs elicited induced antioxidant, antibacterial, and antifungal activities (Chung et al. 2018a, b). Additionally, a review by Rivero-Montejo et al. reported different experiments that aimed to address the effects of nano-elicitation on PSMs. These studies found that eliciting different types of NPs, such as silver, zinc oxide, and titanium dioxide, can trigger plant defense mechanisms, leading to increased production of bioactive metabolites (Rivero-Montejo et al. 2021). Table 11.2 shows the effect of NP application in food packaging.

Table 11.2 NPs for application in food packaging

Types of NPs	Matrix	Application	References
Silver	Asparagus, orange juice, poultry meat, fresh-cut melon, beef meat exudates	Retards the growth of aerobic psychrotrophics, yeasts, and molds; antimicrobial effect against *Escherichia coli* and *Staphylococcus aureus*	An et al. (2008), Fernández et al. (2009, 2010), Emamifar et al. (2010)
Zinc oxide	Orange juice, liquid egg albumen	Effectively reduces *Lactobacillus plantarum*, *Salmonella,* yeast, and mold counts without changes in quality parameters	Emamifar et al. (2010), Jin and Gurtler (2011)
Titanium oxide	Chinese jujube, strawberry	Reduces browning, slow-down ripening, senescence, and decay	Li et al. (2009)
Silver oxide	Apple slice	Retards microbial spoilage	Zhou et al. (2011)

Reprinted from Bajpai et al. (2018. Open access journal)

11.3 Application in Tissue Culture

Plant tissue culture is defined generally as the growth or regeneration of cells, tissues, organs, or a whole plant in vitro under aseptic conditions. And, due to their special physiochemical properties, NPs can be used as elicitors in plant tissue culture (Kokina et al. 2017; Singh et al. 2023). NPs have shown promising results in plant tissue culture, with studies indicating their effectiveness in improving seed germination, plant growth, and yield, as well as enhancing plant protection, genetic manipulation, and metabolite content (Singh et al. 2023). Research has also demonstrated that NPs can significantly reduce microbial contamination during surface sterilization of explants. Adding NPs to culture media can lead to somaclonal variation, boost morphogenetic potential, and enhance disinfection. Notably, AgNPs have been widely studied for their applications in plant tissue culture (Sarmast and Salehi 2016; Singh et al. 2023). Several researchers have highlighted the role of NPs in various plant biotechnological applications, such as decontamination, callus induction, organogenesis, genetic transformation, and the production of bioactive compounds (Al-Qudah et al. 2022; Kim et al. 2017; Singh et al. 2023). Additionally, the potential of advanced nanomaterials such as quantum dots, carbon nanotubes, and graphene in plant tissue culture is discussed (Singh et al. 2023) (Fig. 11.2).

Plant tissue culture is also used to enhance the production of secondary metabolites, such as alkaloids, terpenoids, quinones, and steroids, which are valuable in pharmaceuticals, cosmetics, fragrances, dyes, and flavors. This technique provides a controlled, sustainable, and economical method for producing these compounds, reducing the need for harvesting wild plants and ensuring consistent yields (Singh et al. 2023). In one of the recent studies by Iqbal et al. that focused on optimizing the production of secondary metabolites in *Vigna radiata* L., researchers have worked on developing a protocol that enhances multiple shooting and callus formation, by using different concentrations of nutrient media and evaluating the role of nano-elicitation in secondary metabolite production (Iqbal et al. 2022). Benzyl amino purine was identified as effective for shoot and leaf development, while indole butyric acid proved superior for root formation. Moreover, callus cultures from the Mung NCM-13 variety using 2,4-D exhibited high phenolic and glycosidic content. Nano-elicitors, such as ZnONPs and CuONPs, further enhanced phytochemical production, with ZnO increasing glycosides and CuO boosting phenolic content (Javed et al. 2018). These findings highlight the potential of combining plant tissue culture techniques with nano-elicitation to optimize secondary metabolite yields.

11.4 Application in Crop Production

In this section, it is essential to discuss the importance of the application of NPs and mainly nano-elicitors in the enhancement of crop production. According to Rivero-Montejo et al., entire studies provided that NPs stimulate the production of plant

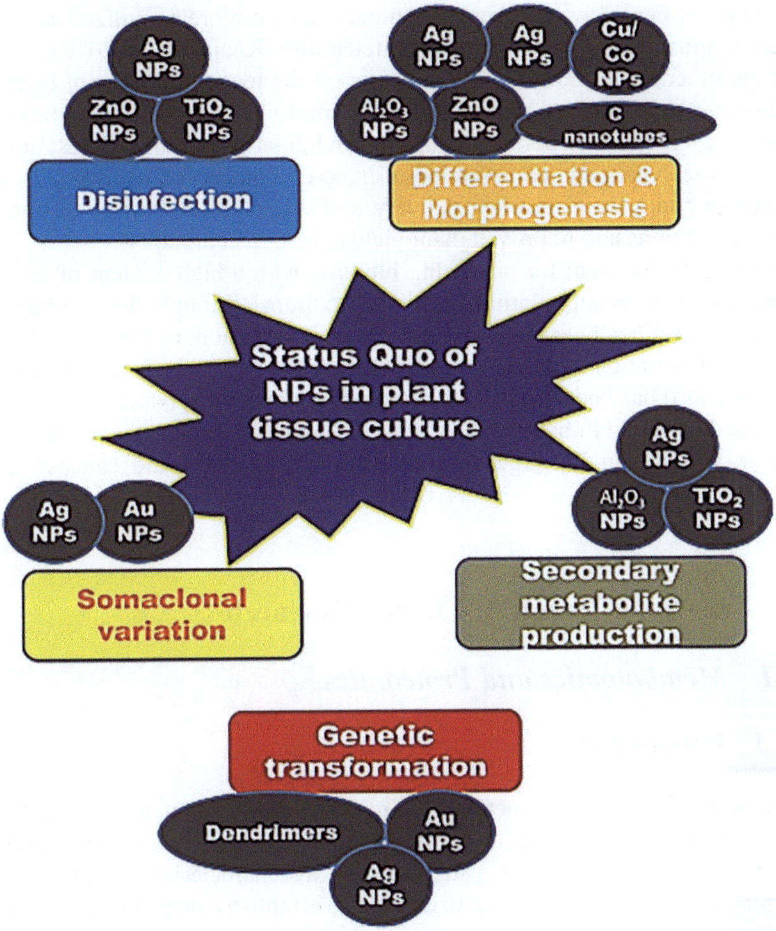

Fig. 11.2 Schematic showing an overview of the various NPs currently involved in plant tissue culture. (Reprinted from Kim et al. 2017. Open Access journal)

bioactive compounds (Rivero-Montejo et al. 2021). Evidence of that is reported by Zhao et al. showing the impacts of CuNPs on cucumber fruits when exposed to different concentrations. This study found an elevation in essential amino acids such as valine (16–32%), leucine (13–41%), isoleucine (12–28%), threonine (0.1–18%), and tyrosine (0.1–32%) (Zhao et al. 2017).

Currently, a new aspect of research is coming ahead, where most researchers focus on the synergistic effect of nanomaterial together with other biotic and abiotic elicitors (Rivero-Montejo et al. 2021). For example, Khajavi et al. examined the effects of titanium dioxide NPs and silica NPs (SiO_2) and salinity stress on Feverfew (*Tanacetum parthenium* L.) as elicitors. The results have shown that both NPs and salinity treatments can be used to stimulate the synthesis of parthenolide in the whole plant of Feverfew, and these treatments can therefore be utilized as effective elicitors to induce the production of this metabolite (Khajavi et al. 2019).

On the other hand, Jadoon et al. also suggest the idea of a fusion of both nano-elicitation and hydroponics that is crucial for minimizing the use of pesticides and fertilizers, escaping soil-borne pathogens, and finally controlling pest invasion. Therefore, the synergistic combination of nano-elicitation and hydroponics allows for accurate control over nutrient delivery and water usage, leading to optimized growing conditions and improved plant yields (Jadoon et al. 2024).

Another advancement for producing biomass with a high content of secondary metabolites involves stimulating hairy root cultures through the mechanism of *Agrobacterium rhizogenes*-mediated genetic transformation, followed by elicitation with NPs. Elicitation with 900 mg/L Ag-SiO_2 core–shell NPs (average size 101.8 ± 8.9 nm) has been reported to increase Artemisinin production from 1.67 to 2.86 mg/g dry weight in hairy root cultures of *Artemisia annua*. This treatment also led to an increase in H_2O_2 generation, lipid peroxidation, and catalase activity (Lala 2021).

11.5 Emerging Trends in Elicitor Research

11.5.1 Metabolomics and Proteomics

11.5.1.1 Introduction

Metabolomics is the study of metabolites in a biological system. It consists of studying the metabolic profile of an organism. In plant sciences, metabolomics refers to analyzing "complex metabolite patterns" that are characteristic of plant extracts (Turumtay et al. 2016). In addition to helping to establish a deeper understanding of the nature of plant metabolic networks and their responses to genetic and environmental change, information on metabolomics can offer a unique perspective into the underlying characteristics of plant phenotypes. Plants can synthesize numerous distinct metabolites. These metabolites serve various purposes, for example, they can offer protection against environmental stress (Kessler and Kalske 2018). Proteomics, on the other hand, is the study of proteomes (a set of proteins produced in an organism), their structure, and how they work together inside the cell (Zhang et al. 2024).

Plant metabolism is a process that reacts to outside stimuli and alterations in the environment have been shown to affect it (Fang et al. 2019). Diseases in plants can be very disastrous because they cause major crop losses, abiotic stress, as well as drought, and high or low temperatures can also affect the plant's health (Pandey et al. 2017). Utilizing proteomic methods helps in identifying pathogenicity, stress-related, and antioxidant proteins that are expressed during interactions between microbes and plants (Jain et al. 2021). When elicitor ToxA was examined for its effect on the wheat proteome, researchers found that photosystem II supercomplexes levels had decreased. This finding suggested that ROS levels would increase

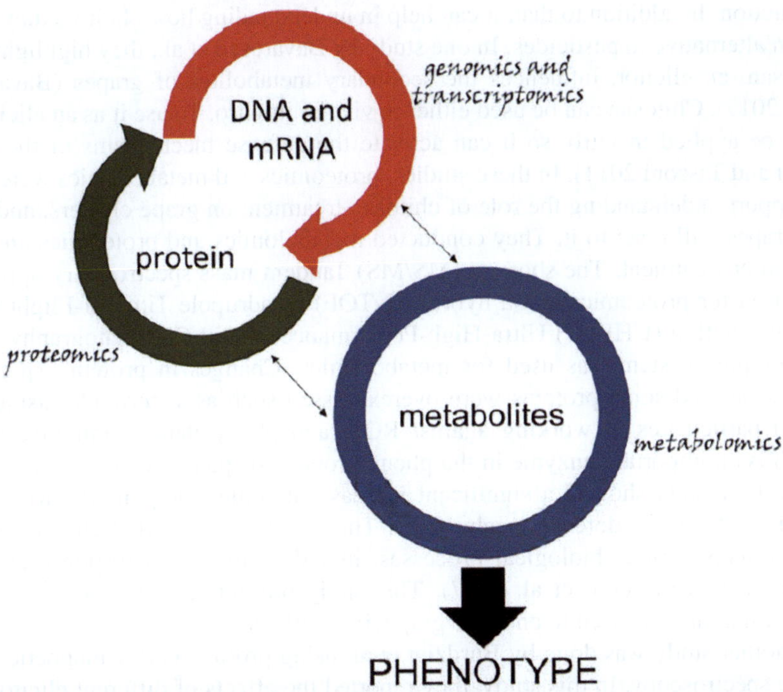

Fig. 11.3 OMICs interaction. DNA and mRNA (red semi-circle) have a direct relationship with proteins (green semi-circle) and can be analyzed by genomics, transcriptomics, and proteomics. Though metabolites (blue circles—which are analyzed by metabolomics) do not have a directly obvious relationship with specific genes and proteins, they are related in important ways that ultimately underlie phenotypes. The smaller arrows symbolize that indirect relationship, while the whole system is connected to phenotypes through the large arrow. (Reprinted from Castro-Moretti et al. 2020. Open Access journal)

and lead to the development of symptoms (Castro-Moretti et al. 2020). Metabolites are particularly important because they perform many distinct roles in plant–pathogen interactions such as checking against a pathogen attack, transduction of signals, regulating enzymes, and so on (Castro-Moretti et al. 2020). Metabolomics may be used to find a group of metabolites related to an infection (Tsuge et al. 2013). An advantage of using metabolomics is that there is no need for genomic resources, so it is easier to study any plant system. Figure 11.3 represents the relationship between the different OMICs.

11.5.1.2 Role of Metabolomics and Proteomics in Elicitor Studies

The use of metabolomic and proteomic approaches in studying elicitors can be of high importance as it can allow us to investigate how the elicitors are affecting the plant's defense mechanisms, how it reacts to stress, and secondary metabolite production. In addition to that, it can help in understanding how elicitors may present an alternative to pesticides. In one study by Bavaresco et al., they highlight how chitosan, an elicitor, influences the secondary metabolites of grapes (Bavaresco et al. 2017). Chitosan can be used either in vivo or in vitro. To use it as an elicitor, it must be applied in vitro, so it can activate the defense mechanisms of the plant (Ferri and Tassoni 2011). In these studies, proteomics and metabolomics were used to support understanding the role of chitosan treatment on grape clusters, and how the grapes will react to it. They conducted metabolomics and proteomics analysis 24 h after treatment. The shotgun (MS/MS) Tandem mass spectrometry approach was used for proteomics and a hybrid (Q-TOF) Quadrupole Time-of-Flight spectrometer with a (UHPLC) Ultra-High-Performance Liquid Chromatography chromatographic system was used for metabolomics. Changes in protein expression were seen, and some proteins were overexpressed such as superoxide dismutase, which participates in working against ROS, and phenylalanine ammonia-lyase, which is an important enzyme in the phenylpropanoids pathway. As for metabolomics, the results showed a significant increase in some triterpenoids, which are important for plant defense mechanisms. The results indicate that chitosan treatment affects various biological processes, including the stress response and the metabolism (Bavaresco et al. 2017). This study presents chitosan as a potential elicitor that may be used to enhance grapevine resilience.

Another study was done by Burdziej et al. using proton nuclear magnetic resonance spectroscopy. In this study, they reported the effects of different elicitors on grapevine leaves' metabolism. They demonstrated that various elicitors, such as methyl jasmonate (MeJA), acibenzolar-S-methyl, and phosphates led to interesting metabolic profiles in grapevine leaves. The study highlighted changes in primary metabolites, such as amino acids and organic acids, which are important for plant defense mechanisms. In sugars, a decrease was noticed among all three treatments, whereas organic acids showed no similar patterns, for example, acetic acid only responded to MeJA in which it increased; similarly, in amino acids, certain amino acids stayed at the same level, but (GABA) gamma-Aminobutyric acid was noticeably reduced by all the used elicitors (Burdziej et al. 2019). These findings suggest that, since different elicitors can trigger unique metabolic responses, they can enhance the grapevine's resistance to stressors. However, more research is still needed to enhance elicitor applications to improve grapevine health and productivity.

Metabolomic studies have also been done on different plant calli and have shown interesting results. One such case presented by Hyeon et al. examined the effects of MeJA being used as an elicitor on *Damnacanthus major* calli as well as using metabolomics to show accumulation of rubiadin. For this study, they used two different concentrations of MeJA, leaves, and unelicited calli. Metabolomic analysis of *Damnacanthus major* calli revealed significant accumulation of rubiadin which is an anthraquinone and is interesting due to its potential therapeutic properties. This study highlighted that MeJA elicitation leads to increased levels of secondary metabolites, including rubiadin, which suggests a mechanism of defense activation in the plant cells. It was noted that, as the MeJA concentration was increased, more rubiadin was seen. This study agreed with other studies that showed anthraquinones accumulation in MeJA elicitation (Hyeon et al. 2024). For example, in a previous study, it was found that *Morinda citrifolia* calli yielded four times more anthraquinones at 100 μM MeJA than the control (Hyeon et al. 2024). MeJA also had effects on the callus in which the color became darker which is correlated with an accumulation in the phenolics. MeJA also increased the activity of compounds that produce secondary metabolites. Furthermore, Zhang et al. discussed how in their study on *Rubia yunnanensis* Diels, anthraquinone levels increased significantly after using the MeJA treatment on the hairy roots (Zhang et al. 2022).

Iula et al. discussed how four molecular elicitors can trigger the metabolomics crosstalk on tomato leaves. By using salicylate, polyamines (PAs) (putrescine, spermidine, and spermine), chitosan, and 2, 1, 3-benzothiadiazole (positive control), they found a modest increase in the biomass in all these treatments, especially PA. Their results suggest that chitosan produces a particular induced systemic resistance, (ISR) as, for the other elicitors, they share a systemic acquired resistance response (Iula et al. 2022). Their results proved that those elicitors could stimulate biochemical responses, which is similar to other studies (Wiesel et al. 2014).

As for proteomics, Pierre et al. conducted a study in which they used natural rhamnolipids (RLs), which are natural defense elicitors that significantly influence the proteomic profiles of *Brassica napus*, particularly in shoots and roots, as they demonstrated through a tandem mass tag (TMT) labeling approach (Pierre et al. 2023). In this research, the impacts of two different RL treatment types—foliar spray and root absorption—on the proteomes of the shoots and roots of rapeseed were assessed using a comparative quantitative proteomic analysis. Their findings indicate that the type of application affects protein accumulation and affects shoot and root protein abundance. They discovered that, when compared to foliar spray, RL treatment by root absorption produced higher differentially accumulated proteins in both shoots and roots. They also discovered that a significantly large over-accumulation of plant defense proteins, particularly in roots, is brought on by RL treatment. Their findings also demonstrated a large over-accumulation of proteins engaged in secondary metabolism. In conclusion, their findings imply that the kind of RL application, whether it is foliar spraying or root absorption has a distinct impact on *Brassica napus* shoots and roots. By applying the TMT labeling approach, they found proteins that are synthesized after several RL treatments, which allowed for the understanding of the molecular responses based on the elicitor-type applications. Since elicitors' level of protection varies depending on the pathogen, this work may aid in the planning of future research into the use of RLs as elicitors against certain *Brassica napus* pathogens, thereby advancing the development of sustainable agriculture and biocontrol tactics (Pierre et al. 2023).

11.5.2 Genomics and Transcriptomics

11.5.2.1 Introduction

Genomics refers to the study of the complete set of genes of an organism. It contains all the biological information needed to maintain a living example of that organism. This biological information is encoded in its DNA which is divided into discrete units called genes. Genes code for proteins that bind to the genome at specific locations and initiate a sequence of events known as gene expression (Keller 2011). The entire set of genes is not viewed as a collection of numerous separate pieces of DNA where each serves as the building block for a single gene. However, it is thought of as the DNA that is a component of the protein X gene and may also function as a component of the protein Y gene (Barnes and Dupres 2008). There are approximately 391,000 land species plants and 8000 species of green algae with remarkably varied genomes (Kersey 2019). The genomes of land plants range from 60 MB (Fleischmann et al. 2014) to 150 GB (Pellicer et al. 2010) which is 50 times larger than the human genome (Kersey 2019), while the smallest algae genomes are only 10 MB. The smallest genomes are made up of mostly coding and regulatory DNA sequences present in low copy, as well as with highly repeated recombinant DNA (rDNA), (ribosomal RNA (rRNA) genes and intergenic spacers), centromeric and telomeric repetitive DNA, and some transposable elements. The larger genomes consist of a comparable number of genes, along with plentiful tandemly repeated sequence motifs, and more than half of the DNA is made up of transposable elements alone (Heslop-Harrison and Schwarzacher 2011).

Transcriptomics refers to the study of the RNA transcripts that the genome produces under specific conditions, including coding and non-coding, in a single or a group of cells. The word transcriptome is a combination of the terms transcript and genome. It first appeared in the domains of life sciences and technology, along with other neologisms with the suffixes -ome and -omics, to signify all studies carried out on a genome-wide scale. Hence, transcriptome and transcriptomics are among the first words that were used with genome and proteome (Jiménez-Chillarón et al. 2014). Transcription consists of three stages: initiation, elongation, and termination (D'Souza and Minczuk 2018). The transcriptome is formed when an organism's DNA is transcribed by RNA polymerase to generate complementary RNA strands, which are then spliced to eliminate introns and produce mature transcripts that contain only exons (Partea 2012). The phenotype of an organism is mostly influenced by the genes that are expressed within it. The transcriptome expresses these genes, and it indicates the identity of every expressed gene and its degree of expression for a defined cell population. Unlike the genome which is essentially fixed and static, the transcriptome is subject to both internal and external factors.

In this section, we will be discovering the role of genomics and transcriptomics in elicitor studies and their applications and implications as well.

11.5.2.2 Role of Genomics in Elicitor Studies

Genomics plays several roles in elicitor studies, including the identification of resistance genes, genome-wide association studies, and comparative genomics. In this section, we will be diving into these roles.

Comparative genomics helps in understanding an organism's function by showing the differences and similarities in the organismal genome and gene structure. There is a wide diversity in the functions among organisms, and comparative genomics can identify a structure/function relationship or an evolutionary relationship between the genes (Sivashankari and Shanmughavel 2007). Comparative genomics analysis was utilized in a study on *Xanthomonas* species which causes bacterial spots in tomatoes and peppers. The aim was to determine the relationship of four species at the genomic level and investigate the molecular causes of their varying host ranges and pathogenicity. The investigations have uncovered a significant degree of variety among the species, offering new perspectives on the differences and similarities that could potentially account for the strains' varied characteristics including pathogen virulence, aggressiveness, and host preferences. Among these findings was a homolog to an elicitor of a resistance gene in rice called XA21, and a functional sulfation system to this homolog (Potnis et al. 2011). Another study showed that a comparative genomics screen identified microbe-associated molecular patterns from *Candidatus Liberibacter* spp. that elicit immune responses in citrus that could prevent or reduce citrus Huanglongbing infection (Chen et al. 2020).

Whole-genome sequencing (WGS) is a laboratory process where the complete DNA sequence of an organism's genome is determined at a single time including all its genes and the non-coding regions of the DNA. This can provide comprehensive information about the genetic makeup of a cell, organism, or individual. A study on the identification of sheath blight (SB) on rice reported that recent advances in WGS allowed the identification of non-synonymous nucleotide polymorphisms and potential genes for resistance to SB which provides the groundwork for further research and marker-assisted breeding of rice for SB resistance (Silva et al. 2012). In another study, a bulked segregate analysis along with WGS was coupled with WGS (BSA-seq) to identify quantitative trait loci (QTLs) governing basal resistance to blast disease in an F2 population originating from two rice varieties whose basal resistance to rice blast varies considerably. A novel haplotype of the durable blast resistance gene was discovered via allelic and genotypic association analysis providing important resources for rice variety breeding. These results provide the foundation for further identification and functional characterization of the remaining QTLs which help us comprehend the molecular mechanisms driving rice basal resistance to blast disease (Liang et al. 2020).

Genome-wide association studies (GWAS) are used to determine genetic variants associated with specific traits or diseases. Genomic selection and the detection of marker–trait associations in response to elicitor treatments involve the analysis of plant materials before and after inoculation. The nature of preparations, elicitor treatment of plants or plant tissues, differences in circumstances between cultivations and stages, and pathogen environmental conditions all contribute to the difficulty of obtaining reliable phenotypic data from both stages. However, despite these challenges, GWAS is widely used in elicitor studies. A study utilized GWAS which revealed that cytokinin is a major component in the root defense system against *Ralstonia solanacearum* where allelic variation at different loci related to cytokinin metabolism, as well as genes responsible for biosynthesis and degradation of cytokinin were identified early during the infection. This is important because the early defense responses that are triggered in the host's root are key to preventing its colonization upon infection (Alonso-Diaz et al. 2021). Another study on the resistance to *sclerotinia* stem rot in *Brassica napus* also utilized GWAS to identify resistance genes and pathogens-related genes (Wei et al. 2016). Moreover, a GWAS using 249 rice accessions found a novel wall-associated kinase (WAK) gene, Pb4, which is responsible for resistance to rice blast, and polygalacturonic acid and chitin were also found to promote its expression (Fan et al. 2024).

11.5.2.3 Role of Transcriptomics in Elicitor Studies

Transcriptomics is particularly useful in elicitor studies for reasons including gene expression profiling, pathway analysis, and temporal dynamics. Transcriptomics has increasingly been used in elicitor studies. Extensive and in-depth gene expressions are designed to explore the efficacy of elicitors on the gene expression changes as a first step of the defensive response. Numerous plants are used for these investigations including crops, ornamentals, trees, ferns, or even grasses that have or do not have reported susceptibility or resistance to respective elicitors. These studies also reveal the response of various plant species to elicitors derived from different sources.

Transcriptomic investigations employ different technologies such as RNA sequencing, sequence-based approaches, and hybridization-based approaches (Imadi et al. 2015). Differential gene expression analysis was used to identify putative genes involved in anthraquinone biosynthesis in *Rubia yunnanensis* which is a highly demanded Chinese medicinal herb used for treating several medical conditions. The biosynthesis of anthraquinone, a type of quinone that is known to induce defensive mechanisms in plants, was largely unknown. Thus, transcriptome analysis was utilized, and 15 putative genes responsible for the biosynthesis of these anthraquinones were identified. Additionally, a positive correlation between MeJA root hair treatment and the expression of 4 out of the 15 putative genes was also found (Zhang et al. 2022). A study integrated transcriptome and targeted metabolome profiling to reveal hormones that are related to genes involved in the growth of *Bletilla striata* which is a medicinal herb with many beneficial effects. The study aimed to determine the molecular and physical mechanisms of *Aspergillus flavus* elicitor, a fungal elicitor, on *B. striata* growth. A fungal elicitor is a unique chemical signal obtained from the extracts or secretions of fungus cells. During the interaction between plants and fungi, it can quickly, extremely specifically, and selectively activate the expression of specific genes in plants, which, in turn, can promote plant growth and the synthesis of secondary metabolites in plants. The results of the study showed that the growth and development of *B. striata* could be significantly promoted by this elicitor which establishes a theoretical framework for the efficient use of endophytic fungi and the enhancement of *B. striata* germplasm resources (Zou et al. 2021). A different study utilized RNA sequencing on tomato (*Solanum lycopersicum*) trichomes, which are the production and storage unit of specialized metabolites like terpenes that act as elicitors playing a role in the plant's defense system, to identify transcription factors (TFs) that activate terpene synthase promoters. The result of high-throughput sequencing of tomato stem trichomes revealed the identification of two TFs that activate several promoters of terpene synthase. This research revealed new features of the transcriptional regulation of tomato terpene biosynthesis in trichomes paving the way for future research and enhancement of tomato-resistant varieties (Spyropoulou et al. 2014). In another subsequent study, TF response to several types of elicitors in *Medicago truncatula* cell suspension culture has been examined. TFs are important components of the signaling pathways of the plant defense responses. They regulate the expression of genes encoding pathogenesis-related proteins and enzymes which have a role in the synthesis of compounds related to defense including various classes of flavonoid and/or triterpene defense molecules when *Medicago truncatula* is exposed to a pathogen or wound. Genes that were over-represented as a response to each type of elicitor under experiment were confirmed to play a significant role in coordinating the metabolic responses to biotic stress, which makes them potentially useful agents for modifying the metabolism to affect the pathogen resistance (Naoumkina et al. 2008).

Overall, these studies reinforce the crucial role of transcriptomics and genomics in understanding the molecular response in elicitor studies, offering more profound insights into plant defense systems.

11.6 Conclusions and Perspectives

The integration of novel technologies and emerging trends in elicitor research promises to revolutionize our understanding and application of plant stress responses. The application of NPs as elicitors demonstrates significant potential for enhancing the production of secondary metabolites in plants, offering a new frontier in elicitor research.

Metabolomics and proteomics have emerged as powerful tools for elucidating the complex interactions between elicitors and plant metabolic pathways. These approaches provide valuable insights into how elicitors trigger stress responses and influence secondary metabolite production, contributing to more effective and sustainable agricultural practices. The use of metabolomics and proteomics also opens new avenues for exploring the role of elicitors in plant defense mechanisms and their potential as eco-friendly alternatives to chemical pesticides. The continued development and application of genomics and transcriptomics will further enhance our understanding of elicitor mechanisms. These technologies will facilitate the identification of key genes and pathways involved in elicitor responses, paving the way for more targeted and efficient elicitor applications. The ongoing advancements in elicitor research hold promise for addressing global agricultural and environmental challenges. By harnessing the potential of novel elicitors and leveraging cutting-edge omics technologies, researchers can drive innovations that lead to more sustainable and effective agricultural practices.

References

Ahmad Z, Tahseen S, Wasi A, Ganie IB, Shahzad A, Emamverdian A, Ramakrishnan M, Ding Y (2022) Nanotechnological interventions in agriculture. Nanomaterials 12:2667

Ahmar S, Mahmood T, Fiaz S, Mora-Poblete F, Shafique MS, Chattha MS, Jung KH (2021) Advantage of nanotechnology-based genome editing system and its application in crop improvement. Front Plant Sci 12:663849

Alabdallah NM, Hasan MM (2021) Plant-based green synthesis of silver nanoparticles and its effective role in abiotic stress tolerance in crop plants. Saudi J Biol Sci 28:5631–5639

Alonso-Díaz A, Satbhai SB, De Pedro-Jové R, Berry HM, Göschl C, Argueso CT, Novak O, Busch W, Valls M, Coll NS (2021) A genome-wide association study reveals cytokinin as a major component in the root defense responses against *Ralstonia solanacearum*. J Exp Bot 72:2727–2740

Al-Qudah T, Mahmood SH, Abu-Zurayk R, Shibli R, Khalaf A, Lambat TL, Chaudhary RG (2022) Nanotechnology applications in plant tissue culture and molecular genetics: a holistic approach. Curr Nanosci 18:442–464

Alvarez MA (2014) Chapter 3 – plant secondary metabolism. In: Plant biotechnology for health, from secondary metabolites to molecular farming, 1st edn. Springer, Cham, pp 15–31

An J, Zhang M, Wang S, Tang J (2008) Physical, chemical and microbiological changes in stored green asparagus spears as affected by coating of silver nanoparticles-PVP. LWT Food Sci Technol 41:1100–1107

An C, Sun C, Li N, Huang B, Jiang J, Shen Y, Wang C, Zhao X, Cui B, Wang C, Li X, Zhan S, Gao F, Zeng Z, Cui H, Wang Y (2022) Nanomaterials and nanotechnology for the delivery of agrochemicals: strategies towards sustainable agriculture. J Nanobiotechnol 20:11

Anjum S, Anjum I, Hano C, Kousar S (2019) Advances in nanomaterials as novel elicitors of pharmacologically active plant specialized metabolites: current status and future outlooks. RSC Adv 9:40404–40423

Babu PJ, Saranya S, Longchar B, Rajasekhar A (2022) Nanobiotechnology-mediated sustainable agriculture and post-harvest management. Curr Res Biotechnol 4:326–336

Bajpai VK, Kamle M, Shukla S, Mahato DK, Chandra P, Hwang SK, Kumar P, Huh YS, Han YK (2018) Prospects of using nanotechnology for food preservation, safety, and security. J Food Drug Anal 26:1201–1214

Barnes B, Dupré J (2008) Chapter 2 – genes, genomes, and molecular genetics. In: Genomes and what to make of them, 1st edn. University of Chicago Press

Bavaresco L, Zamboni M, Squeri C, Xu S, Abramowicz A, Lucini L (2017) Chitosan and grape secondary metabolites: a proteomics and metabolomics approach. BIO Web Conf 9:01004

Burdziej A, Da Costa G, Gougeon L, Le Mao I, Bellée A, Corio-Costet MF, Mérillon JM, Richard T, Szakiel A, Cluzet S (2019) Impact of different elicitors on grapevine leaf metabolism monitored by 1H NMR spectroscopy. Metabolomics 15:67

Castro-Moretti FR, Gentzel IN, Mackey D, Alonso AP (2020) Metabolomics as an emerging tool for the study of plant–pathogen interactions. Meta 10:52

Chandra S, Chakraborty N, Dasgupta A, Sarkar J, Panda K, Acharya K (2015) Chitosan nanoparticles: a positive modulator of innate immune responses in plants. Sci Rep 5:15195

Chen Y, Bendix C, Lewis JD (2020) Comparative genomics screen identifies microbe-associated molecular patterns from '*Candidatus Liberibacter*' spp. that elicit immune responses in plants. Mol Plant-Microbe Interact 33:539–552

Chung IM, Rajakumar G, Thiruvengadam M (2018a) Effect of silver nanoparticles on phenolic compounds production and biological activities in hairy root cultures of *Cucumis anguria*. Acta Biol Hung 69:97–109

Chung IM, Rekha K, Rajakumar G, Thiruvengadam M (2018b) Elicitation of silver nanoparticles enhanced the secondary metabolites and pharmacological activities in cell suspension cultures of bitter gourd. 3 Biotech 8:412

Cui J, Sun C, Wang A, Wang Y, Zhu H, Shen Y, Li N, Zhao X, Cui B, Wang C, Gao F, Zeng Z, Cui H (2020) Dual-functionalized pesticide nanocapsule delivery system with improved spreading behavior and enhanced bioactivity. Nano 10:220

D'Souza AR, Minczuk M (2018) Mitochondrial transcription and translation: overview. Essays Biochem 62:309–320

Eichert T, Goldbach HE (2008) Equivalent pore radii of hydrophilic foliar uptake routes in stomatous and astomatous leaf surfaces–further evidence for a stomatal pathway. Physiol Plant 132:491–502

Eichert T, Kurtz A, Steiner U, Goldbach HE (2008) Size exclusion limits and lateral heterogeneity of the stomatal foliar uptake pathway for aqueous solutes and water-suspended nanoparticles. Physiol Plant 134:151–160

Emamifar A, Kadivar M, Shahedi M, Soleimanian-Zad S (2010) Evaluation of nanocomposite packaging containing Ag and ZnO on shelf life of fresh orange juice. Innov Food Sci Emerg Technol 11:742–748

Fan Y, Ma L, Pan X, Tian P, Wang W, Liu K, Xiong Z, Li C, Wang Z, Wang J, Zhang H, Bao Y (2024) Genome-wide association study identifies rice panicle blast-resistant gene Pb4 encoding a wall-associated kinase. Int J Mol Sci 25:830

Fang C, Fernie AR, Luo J (2019) Exploring the diversity of plant metabolism. Trends Plant Sci 24:83–98

Fernández A, Soriano E, López-Carballo G, Picouet P, Lloret E, Gavara R, Hernández-Muñoz P (2009) Preservation of aseptic conditions in absorbent pads by using silver nanotechnology. Food Res Int 42:1105–1112

Fernández A, Picouet P, Lloret E (2010) Cellulose-silver nanoparticle hybrid materials to control spoilage-related microflora in absorbent pads located in trays of fresh-cut melon. Int J Food Microbiol 142:222–228

Ferri M, Tassoni A (2011) Chapter 22 – chitosan as elicitor of health beneficial secondary metabolites in in vitro plant cell cultures. In: Mackay RG, Tait JMTAIT (eds) Chitosan research and applications. Nova Science Publishers, pp 389–414

Fleischmann A, Michael TP, Rivadavia F, Sousa A, Wang W, Temsch EM, Greilhuber J, Müller KF, Heubl G (2014) Evolution of genome size and chromosome number in the carnivorous plant genus *Genlisea* (*Lentibulariaceae*), with a new estimate of the minimum genome size in angiosperms. Ann Bot 114:1651–1663

Gahukar RT, Das RK (2020) Plant-derived nanopesticides for agricultural pest control: challenges and prospects. Nanotechnol Environ Eng 5:1–9

Gowtham HG, Murali M, Shilpa N, Amruthesh KN, Gafur A, Antonius S, Sayyed RZ (2024) Harnessing abiotic elicitors to Bolster plant's resistance against bacterial pathogens. Plant Stress 11:100371

Guerriero G, Berni R, Muñoz-Sanchez JA, Apone F, Abdel-Salam EM, Qahtan AA, Alatar AA, Cantini C, Cai G, Hausman JF, Siddiqui KS, Hernández-Sotomayor SMT, Faisal M (2018) Production of plant secondary metabolites: examples, tips and suggestions for biotechnologists. Genes 9:309

Hafeez A, Razzaq A, Mahmood T, Jhanzab HM (2015) Potential of copper nanoparticles to increase growth and yield of wheat. J Nanosci Adv Technol 1:6–11
Hafez IH, Osman AR, Sewedan EA, Berber MR (2018) Tailoring of a potential nanoformulated form of gibberellic acid: synthesis, characterization, and field applications on vegetation and flowering. J Agric Food Chem 66:8237–8245
Hasanuzzaman M, Bhuyan MHMB, Zulfiqar F, Raza A, Mohsin SM, Mahmud JA, Fujita M, Fotopoulos V (2020) Reactive oxygen species and antioxidant defense in plants under abiotic stress: revisiting the crucial role of a universal defense regulator. Antioxidants 9:681
Hashem AH, Abdelaziz AM, Askar AA, Fouda HM, Khalil AM, Abd-Elsalam KA, Khaleil MM (2021) *Bacillus megaterium*-mediated synthesis of selenium nanoparticles and their antifungal activity against *Rhizoctonia solani* in faba bean plants. J Fungi 7:195
Hersanti H, Djaya L, Hidayat Y, Pratama LS, Joni I (2020) The effectiveness of suspension of *Beauveria bassiana* mixed with silica nanoparticles (NPs.) and carbon fiber in controlling *Spodoptera litura*. AIP Conf Proc 2219:080011
Heslop-Harrison JS, Schwarzacher T (2011) Organisation of the plant genome in chromosomes. Plant J 66:18–33
Hyeon H, Jang EB, Kim SC, Yoon SA, Go B, Lee JD, Hyun HB, Ham YM (2024) Metabolomics reveals rubiadin accumulation and the effects of methyl jasmonate elicitation in *Damnacanthus major* calli. Plan Theory 13:167
Imadi SR, Kazi AG, Ahanger MA, Gucel S, Ahmad P (2015) Plant transcriptomics and responses to environmental stress: an overview. J Genet 94:525–537
Iqbal Z, Javad S, Naz S, Shah AA, Shah AN, Paray BA, Gulnaz A, Abdelsalam NR (2022) Elicitation of the in vitro cultures of selected varieties of *Vigna radiata* L. with zinc oxide and copper oxide nanoparticles for enhanced phytochemicals production. Front Plant Sci 13:908532
Iula G, Miras-Moreno B, Rouphael Y, Lucini L, Trevisan M (2022) The complex metabolomics crosstalk triggered by four molecular elicitors in tomato. Plan Theory 11:678
Jadoon L, Gul A, Fatima H, Babar MM (2024) Nano-elicitation and hydroponics: a synergism to enhance plant productivity and secondary metabolism. Planta 259:80
Jain A, Singh HB, Das S (2021) Deciphering plant-microbe crosstalk through proteomics studies. Microbiol Res 242:126590
Jain D, Bisht S, Parvez A, Singh K, Bhaskar P, Koubouris G (2024) Effective biotic elicitors for augmentation of secondary metabolite production in medicinal plants. Agriculture 14:796
Javed R, Yucesan B, Zia M, Gurel E (2018) Elicitation of secondary metabolites in callus cultures of *Stevia rebaudiana* Bertoni grown under ZnO and CuO nanoparticles stress. Sugar Tech 20:194–201
Jiménez-Chillarón JC, Díaz R, Ramón-Krauel M (2014) Chapter 4 – omics tools for the genome-wide analysis of methylation and histone modifications. In: Simó C, Cifuentes A, García-Cañas V (eds) Fundamentals of advanced omics technologies: from genes to metabolites, comprehensive analytical chemistry, 1st edn. Elsevier, pp 81–110
Jin T, Gurtler JB (2011) Inactivation of *salmonella* in liquid egg albumen by antimicrobial bottle coatings infused with allyl isothiocyanate, nisin and zinc oxide nanoparticles. J Appl Microbiol 110:704–712
Keller EF (2011) Genes, genomes, and genomics. Biol Theory 6:132–140
Kersey PJ (2019) Plant genome sequences: past, present, future. Curr Opin Plant Biol 48:1–8
Kessler A, Kalske A (2018) Plant secondary metabolite diversity and species interactions. Annu Rev Ecol Evol Syst 49:115–138
Khajavi M, Rahaie M, Ebrahimi A (2019) The effect of TiO2 and SiO2 nanoparticles and salinity stress on expression of genes involved in parthenolide biosynthesis in feverfew (*Tanacetum parthenium* L.). Caryologia 72:3–14
Khan MIR, Fatma M, Per TS, Anjum NA, Khan NA (2015) Salicylic acid-induced abiotic stress tolerance and underlying mechanisms in plants. Front Plant Sci 6:462

Khan I, Saeed K, Khan I (2019) Nanoparticles: properties, applications and toxicities. Arab J Chem 12:908–931

Khan AK, Kousar S, Tungmunnithum D, Hano C, Abbasi BH, Anjum S (2021) Nano- elicitation as an effective and emerging strategy for in vitro production of industrially important flavonoids. Appl Sci 11:1694

Khan I, Awan SA, Rizwan M, Hassan ZU, Akram MA, Tariq R, Brestic M, Xie W (2022) Nanoparticle's uptake and translocation mechanisms in plants via seed priming, foliar treatment, and root exposure: a review. Environ Sci Pollut Res 29:89823–89833

Khot LR, Sankaran S, Maja JM, Ehsani R, Schuster EW (2012) Applications of nanomaterials in agricultural production and crop protection: a review. Crop Prot 35:64–70

Kim DH, Gopal J, Sivanesan I (2017) Nanomaterials in plant tissue culture: the disclosed and undisclosed. RSC Adv 7:36492–36505

Kokina I, Mickeviča I, Jahundoviča I, Ogurcovs A, Krasovska M, Jermaļonoka M, Mihailova I, Tamanis E, Gerbreders V (2017) Plant explants grown on medium supplemented with Fe3O4 nanoparticles have a significant increase in embryogenesis. J Nanomater 2017:4587147

Lala S (2021) Nanoparticles as elicitors and harvesters of economically important secondary metabolites in higher plants: a review. IET Nanobiotechnol 15:28–57

Li H, Li F, Wang L, Sheng J, Xin Z, Zhao L, Xiao H, Zheng Y, Hu Q (2009) Effect of nano-packing on preservation quality of Chinese jujube (*Ziziphus jujuba Mill.* var. *inermis* (Bunge) Rehd). Food Chem 114:547–552

Liang T, Chi W, Huang L, Qu M, Zhang S, Chen ZQ, Chen ZJ, Tian D, Gui Y, Chen X, Wang Z, Tang W, Chen S (2020) Bulked segregant analysis coupled with whole-genome sequencing (BSA-Seq) mapping identifies a novel pi21 haplotype conferring basal resistance to rice blast disease. Int J Mol Sci 21:2162

Liu R, Lal R (2015) Potentials of engineered nanoparticles as fertilizers for increasing agronomic productions. Sci Total Environ 514:131–139

Marslin G, Sheeba CJ, Franklin G (2017) Nanoparticles alter secondary metabolism in plants via ROS burst. Front Plant Sci 8:832

Naik PM, Al-Khayri JM (2016) Chapter 10 – abiotic and biotic elicitors – role in secondary metabolites production through *in vitro* culture of medicinal plants. In: Shanker AK, Shanker C (eds) Abiotic and biotic stress in plants – recent advances and future perspectives, 1st edn. InTech, pp 247–277

Naoumkina MA, He X, Dixon RA (2008) Elicitor-induced transcription factors for metabolic reprogramming of secondary metabolism in *Medicago truncatula*. BMC Plant Biol 8:1–14

Neme K, Nafady A, Uddin S, Tola YB (2021) Application of nanotechnology in agriculture, postharvest loss reduction and food processing: food security implication and challenges. Heliyon 7:e08539

Pandey P, Irulappan V, Bagavathiannan MV, Senthil-Kumar M (2017) Impact of combined abiotic and biotic stresses on plant growth and avenues for crop improvement by exploiting physiomorphological traits. Front Plant Sci 8:537

Patel H, Krishnamurthy R (2013) Elicitors in plant tissue culture. J Pharmacogn Phytochem 2:60–65

Pellicer J, Fay MF, Leitch IJ (2010) The largest eukaryotic genome of them all? Bot J Linn 164:10–15

Pérez-de-Luque A (2017) Interaction of nanomaterials with plants: what do we need for real applications in agriculture? Front Environ Sci 5:12

Pertea M (2012) The human transcriptome: an unfinished story. Genes 3:344–360

Pierre E, Marcelo P, Croutte A, Dauvé M, Bouton S, Rippa S, Pageau K (2023) Impact of Rhamnolipids (RLs), natural defense elicitors, on shoot and root proteomes of *Brassica napus* by a tandem mass tags (TMTs) labeling approach. Int J Mol Sci 24:2390

Piña-Barrera AM, Álvarez-Román R, Báez-González JG, Amaya-Guerra CA, Rivas-Morales C, Gallardo-Rivera CT, Galindo-Rodríguez SA (2019) Application of a multisystem coating based on polymeric nanocapsules containing essential oil of *Thymus vulgaris* L. to increase the shelf life of table grapes (*Vitis vinifera* L.). IEEE Trans NanoBiosci 18:549–557

Pollard M, Beisson F, Li Y, Ohlrogge JB (2008) Building lipid barriers: biosynthesis of cutin and suberin. Trends Plant Sci 13:236–246

Potnis N, Krasileva K, Chow V, Almeida NF, Patil PB, Ryan RP, Sharlach M, Behlau F, Dow JM, Momol M, White FF, Preston JF, Vinatzer BA, Koebnik R, Setubal JC, Norman DJ, Staskawicz BJ, Jones JB (2011) Comparative genomics reveals diversity among xanthomonads infecting tomato and pepper. BMC Genomics 12:1–23

Rasheed A, Li H, Tahir MM, Mahmood A, Nawaz M, Shah AN, Aslam MT, Negm S, Moustafa M, Hassan MU, Wu Z (2022) The role of nanoparticles in plant biochemical, physiological, and molecular responses under drought stress: a review. Front Plant Sci 13:976179

Rivero-Montejo SDJ, Vargas-Hernandez M, Torres-Pacheco I (2021) Nanoparticles as novel elicitors to improve bioactive compounds in plants. Agriculture 11:134

Rubentheren V, Ward TA, Chee CY, Tang CK (2015) Processing and analysis of chitosan nanocomposites reinforced with chitin whiskers and tannic acid as a crosslinker. Carbohydr Polym 115:379–387

Sarmast MK, Salehi H (2016) Silver nanoparticles: an influential element in plant nanobiotechnology. Mol Biotechnol 58:441–449

Selvakesavan RK, Kruszka D, Shakya P, Mondal D, Franklin G (2023) Chapter 6 – impact of nanomaterials on plant secondary metabolism. In: Al-Khayri JM, Alnaddaf LM, Jain SM (eds) Nanomaterial interactions with plant cellular mechanisms and macromolecules and agricultural implications, 1st edn. Springer, Cham, pp 133–170

Siddiqi KS, Husen A, Rao RA (2018) A review on biosynthesis of silver nanoparticles and their biocidal properties. J Nanobiotechnol 16:1–28

Silva J, Scheffler B, Sanabria Y, De Guzman C, Galam D, Farmer A, Woodward J, May G, Oard J (2012) Identification of candidate genes in rice for resistance to sheath blight disease by whole genome sequencing. Theor Appl Genet 124:63–74

Singh Y, Kumar U, Panigrahi S, Balyan P, Mehla S, Sihag P, Sagwal V, Singh KP, White JC, Dhankher OP (2023) Nanoparticles as novel elicitors in plant tissue culture applications: current status and future outlook. Plant Physiol Biochem 203:108004

Sivashankari S, Shanmughavel P (2007) Comparative genomics-a perspective. Bioinformation 1:376

Spyropoulou EA, Haring MA, Schuurink RC (2014) RNA sequencing on *Solanum lycopersicum* trichomes identifies transcription factors that activate terpene synthase promoters. BMC Genomics 15:1–16

Tassi E, Giorgetti L, Morelli E, Peralta-Videa JR, Gardea-Torresdey JL, Barbafieri M (2017) Physiological and biochemical responses of sunflower (*Helianthus annuus* L.) exposed to nano-CeO2 and excess boron: modulation of boron phytotoxicity. Plant Physiol Biochem 110:50–58

Teoh ES (2016) Secondary metabolites of plants. Med Orchids Asia 5:59–73

Tsuge T, Harimoto Y, Akimitsu K, Ohtani K, Kodama M, Akagi Y, Egusa M, Yamamoto M, Otani H (2013) Host-selective toxins produced by the plant pathogenic fungus *Alternaria alternata*. FEMS Microbiol Rev 37:44–66

Turumtay H, Sandalli C, Turumtay EA (2016) Chapter 17 – plant metabolomics and strategies. In: Hakeem K, Tombuloğlu H, Tombuloğlu G (eds) Plant omics: trends and applications, 1st edn. Springer, Cham, pp 399–406

Wang X, Xie H, Wang P, Yin H (2023) Nanoparticles in plants: uptake, transport and physiological activity in leaf and root. Materials 16:3097

Wei L, Jian H, Lu K, Filardo F, Yin N, Liu L, Qu C, Li W, Du H, Li J (2016) Genome-wide association analysis and differential expression analysis of resistance to Sclerotinia stem rot in *Brassica napus*. Plant Biotechnol J 14:1368–1380

Wiesel L, Newton AC, Elliott I, Booty D, Gilroy EM, Birch PR, Hein I (2014) Molecular effects of resistance elicitors from biological origin and their potential for crop protection. Front Plant Sci 5:655

Yarizade K, Hosseini R (2015) Expression analysis of ADS, DBR2, ALDH1 and SQS genes in Artemisia vulgaris hairy root culture under nano cobalt and nano zinc elicitation. Ext J Appl Sci 3:69–76

Yousef HA, Fahmy HM, Arafa FN, Abd Allah MY, Tawfik YM, El Halwany KK, El-Ashmanty BA, Al-anany FS, Mohamed MA, Bassily ME (2023) Nanotechnology in pest management: advantages, applications, and challenges. Int J Trop Insect Sci 43:1387–1399

Zaki AM, Zaki AH, Farghali AA, Abdel-Rahim EF (2017) Sodium titanate-bacillus as a new nanopesticide for cotton leaf-worm. J Pure Appl Microbiol 11:725–732

Zhang HY, Su WH (2024) Classification, uptake, translocation, and detection methods of nanoparticles in crop plants: a review. Environ Sci Nano 11:1847–1870

Zhang R, Miao Y, Chen L, Yi S, Tan N (2022) De novo transcriptome analysis reveals putative genes involved in anthraquinone biosynthesis in *Rubia yunnanensis*. Genes 13:521

Zhang L, Xiao D, Cheng K (2024) Chapter 93 – proteomic analysis of microbial infections. In: Tang YW, Liu D, Spearman P, Hindiyeh MY, Sails A, Zhang JR (eds) Molecular medical microbiology, 3rd edn. Academic Press, pp 1951–1963

Zhao L, Hu J, Huang Y, Wang H, Adeleye A, Ortiz C, Keller AA (2017) 1H NMR and GC–MS based metabolomics reveal nano-cu altered cucumber (*Cucumis sativus*) fruit nutritional supply. Plant Physiol Biochem 110:138–146

Zhou L, Lv S, He G, He Q, Shi BI (2011) Effect of pe/ag2o nano-packaging on the quality of apple slices. J Food Qual 34:171–176

Zhu H, Shen Y, Cui J, Li N, Wang C, Cui B, Sun C, Zhao X, Wang C, Gao F, Zhan S, Guo L, Zhang L, Zeng Z, Wang Y, Cui H (2020) Avermectin loaded carboxymethyl cellulose nanoparticles with stimuli-responsive and controlled release properties. Ind Crop Prod 152:112497

Zou H, Jiang H, Li L, Huang R (2021) Integration of transcriptome and targeted metabolome profiling reveals hormone related genes involved in the growth of *Bletilla striata*. Sci Rep 11:21950

Chapter 12
Utilizing Arbuscular Mycorrhizae to Suppress Agricultural Pathogens and Boost Plant Immunity

Moh Sajid Ansari, Adnan Khan, Abrar Ahmad Khan, and Heba I. Mohamed

Abstract Arbuscular mycorrhizal fungi (AMFs) are vital soil microorganisms that establish symbiotic associations with approximately 80% of terrestrial higher plants, significantly enhancing plant growth, soil fertility, and ecosystem sustainability. Their role as biocontrol agents against soil-borne pathogens has garnered increasing attention, providing an environmentally friendly alternative to synthetic pesticides. AMF-mediated pathogen suppression occurs through multiple mechanisms, including direct competition with pathogens for root colonization sites, modulation of soil physical and chemical properties, and promotion of beneficial rhizosphere microorganisms. Additionally, AMFs enhance plant defense systems by inducing the production of secondary metabolites, regulating phytohormones, and activating defense-related gene expression. This chapter discusses the diverse strategies employed by AMFs to boost plant immunity and suppress agricultural pathogens, including fungi, bacteria, and nematodes. Key aspects include root morphological alterations, AMF-induced systemic resistance, and the regulation of nutrient distribution within the rhizosphere. The potential of AMFs in mitigating disease severity, particularly under low-input or organic farming systems, underscores their significance as cost-effective tools for sustainable agriculture. The synergistic interactions between AMFs and other beneficial rhizosphere organisms further highlight their role in creating a resilient agricultural system. By unraveling the molecular pathways and ecological functions of AMFs, this chapter aims to enhance their application in crop protection and contribute to sustainable farming practices.

M. S. Ansari (✉) · A. Khan · A. A. Khan
Section of Environmental Botany and Plant Pathology, Department of Botany, Faculty of Life Sciences, Aligarh Muslim University, Aligarh, UP, India

H. I. Mohamed
Biological and Geological Sciences Department, Faculty of Education, Ain Shams University, Cairo, Egypt

K. A. Abd-Elsalam, H. I. Mohamed (eds.), *Elicitors for Sustainable Crop Production*, Sustainability Sciences in Asia and Africa,
https://doi.org/10.1007/978-981-96-6569-3_12

Keywords Arbuscular mycorrhizal fungi (AMFs) · Biocontrol agents · Pathogen suppression · Plant immunity · Sustainable agriculture

12.1 Introduction

Arbuscular mycorrhizal fungi (AMFs) constitute an essential group of soil fungi that establish symbiotic associations with nearly 85% of terrestrial plant species (Khaliq et al. 2022). These fungi are widely recognized for their ecological and agricultural significance, pivotal in forestry production, environmental restoration, and sustainable agriculture (Qu et al. 2022). AMF's ability to enhance plant health and productivity through diverse mechanisms has garnered increasing attention for its potential to address modern agricultural challenges (Wahab et al. 2023a, b). One of the most notable attributes of AMFs is their capacity to mitigate biotic stresses caused by various pathogens and pests (Kumar et al. 2024). AMFs play a vital role in enhancing plant immunity by modifying root morphology, outcompeting pathogens for space and resources, stimulating disease-related defense enzymes, and regulating the synthesis of secondary metabolites (Wang et al. 2023; Devi et al. 2022). These fungi have been documented to suppress diseases caused by fungi, viruses, bacteria, and nematodes, making them a vital component of sustainable crop protection strategies (Abarca et al. 2024). The global agricultural sector faces severe challenges from plant pathogens, which contribute to an estimated 30% loss in crop yields annually (Savary et al. 2019). Traditional methods to control diseases, such as crop rotation, sanitation, chemical pesticides, and bioprotectant agents, often fall short of addressing these challenges sustainably (Buitenhuis et al. 2023). Intensive agricultural practices and excessive reliance on chemical pesticides have contributed to pathogen resistance, disrupted soil biodiversity, led to environmental pollution, and posed serious threats to human health (Meena et al. 2020; Huang et al. 2011). These issues highlight the urgent need for eco-friendly, integrated approaches to crop protection (Fig. 12.1).

In this regard, AMFs have gained recognition as a viable alternative. Naturally present in the soil, AMFs act as a bio-protector and bio-stimulator, enhancing plant productivity and resilience while minimizing dependence on harmful chemical pesticides (Bagyaraj et al. 2022; Yang et al. 2012). By improving the rhizosphere environment and activating host defense mechanisms, AMFs effectively suppress soil-borne pathogens and promote sustainable agriculture without negatively impacting beneficial microorganisms or the environment (da Silva Campos, 2020; Koffi et al. 2013). The widespread use of chemical pesticides has led to increased pathogen resistance and the destruction of natural pest enemies, aggravating pest and disease outbreaks in farmland ecosystems (Huang et al. 2011; Dhuldhaj et al. 2023). Additionally, pesticide residues in food and the environment pose significant health and ecological risks, fueling the demand for green and environmentally

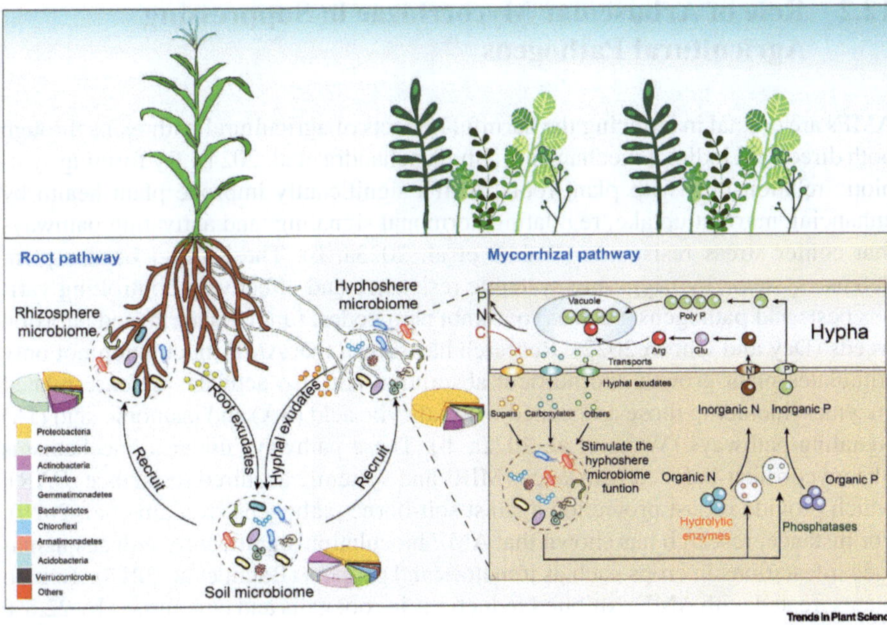

Fig. 12.1 Mycorrhizal is associated with plant roots and the nutrient transport pathway. (Adapted from Zhang et al. 2022)

friendly crop protection methods (Razak and Gange 2021). Biological control has emerged as a sustainable and efficient alternative, with AMFs at the forefront due to its diverse ecological functions and high efficacy in suppressing pathogens (Van Driesche et al. 2010; Pandit et al. 2022).

Research has shown that AMFs are highly effective in mitigating damage caused by pathogens in key crops such as cucumber, tomato, citrus, olive, maize, and potato (Razak and Gange 2021; Begum et al. 2019). These fungi strengthen plant resilience by enhancing root structure, altering rhizosphere conditions, and triggering systemic defense responses (Aseel et al. 2019). Over 30 AMF species have been recognized for their ability to combat soil-borne plant diseases (Jung et al. 2012; Weng et al. 2022), underscoring their significance in sustainable farming practices. This chapter delves into the mechanisms and applications of AMFs in biological control, focusing on their capacity to suppress agricultural pathogens and bolster plant immunity (Bhupenchandra et al. 2024). By unraveling the complex relationships between AMFs and host plants, scientists can devise novel approaches to enhance crop resilience, minimize dependency on chemical treatments, and promote sustainable food systems. The chapter also examines the various strategies AMFs employ to enhance plant immunity and control pathogens, including fungi, bacteria, and nematodes.

12.2 Role of Arbuscular Mycorrhizae in Suppressing Agricultural Pathogens

AMFs are crucial in reducing the harmful impacts of agricultural pathogens through both direct and indirect mechanisms (Bhupenchandra et al. 2024). By forming symbiotic relationships with plant roots, AMFs significantly improve plant health by enhancing nutrient uptake, regulating hormonal signaling, and activating pathways that confer stress resistance (Wahab et al. 2023a, b). These fungi bolster plant defense systems by triggering systemic resistance and effectively controlling various pests and pathogens, such as root-knot nematodes, fungi, bacteria, and parasitic weeds (Dey and Ghosh 2022). Research has shown that AMF inoculation not only stimulates plant growth and nutrient absorption but also activates defense-related enzymes, including those associated with salicylic acid (SA) and jasmonic acid (JA) signaling pathways (Wang et al. 2022a, b). These pathways underlie mechanisms like mycorrhiza-induced resistance (MIR) and systemic acquired resistance (SAR), which provide robust protection against soil-borne pathogens (Kadam et al. 2020). For instance, research has shown that AMF inoculation significantly reduces nematode infestations in crops such as tomatoes and peppers (Baum et al. 2015). Tomato plants treated with AMF exhibited reductions in root galls and egg masses by 22.8% and 23.5%, respectively, while pepper plants showed enhanced resilience against *Nacobbus aberrans* due to decreased nematode populations and physiological stress markers (Vos et al. 2012a, b; Bernardo et al. 2021). The suppressive effect of AMFs extends to bacterial and fungal pathogens as well. AMF symbiosis is associated with increased production of antimicrobial compounds, upregulation of disease resistance genes, and priming of the plant's immune system (Fiorilli et al. 2024). This symbiosis also enhances soil quality by promoting microbial diversity and decomposition processes, creating a protective barrier against harmful microbes (Fig. 12.2). Coffee plants treated with AMF strains such as Huall-pache and Mo-cat demonstrated a 38.3–52.5% reduction in nematode infection severity, further illustrating AMF's efficacy in disease management (Vallejos-Torres et al. 2021).

In addition to pathogen suppression, AMFs improve the photosynthetic rate and biomass accumulation, thereby increasing their resilience to biotic stresses (Wahab et al. 2023a, b). The symbiotic relationship bolsters the accumulation of vital nutrients, such as nitrogen (N) and phosphorus (P), which play a key role in fortifying the plant's defense systems (Bhupenchandra et al. 2024). Furthermore, AMF colonization triggers the production of antioxidant enzymes and defense-related proteins, ensuring comprehensive protection against both localized and systemic threats (Sharma et al. 2021; Mitra et al. 2021). The potential of AMFs as biocontrol agents has been recognized in recent studies, highlighting their ability to mitigate crop losses due to soil-borne pathogens and promote sustainable agricultural practices (Koskey et al. 2021). AMF-mediated resistance reduces the reliance on chemical pesticides, alleviating their adverse impacts on soil health, biodiversity, and the environment (Nie et al. 2024). By facilitating beneficial microbial colonization and

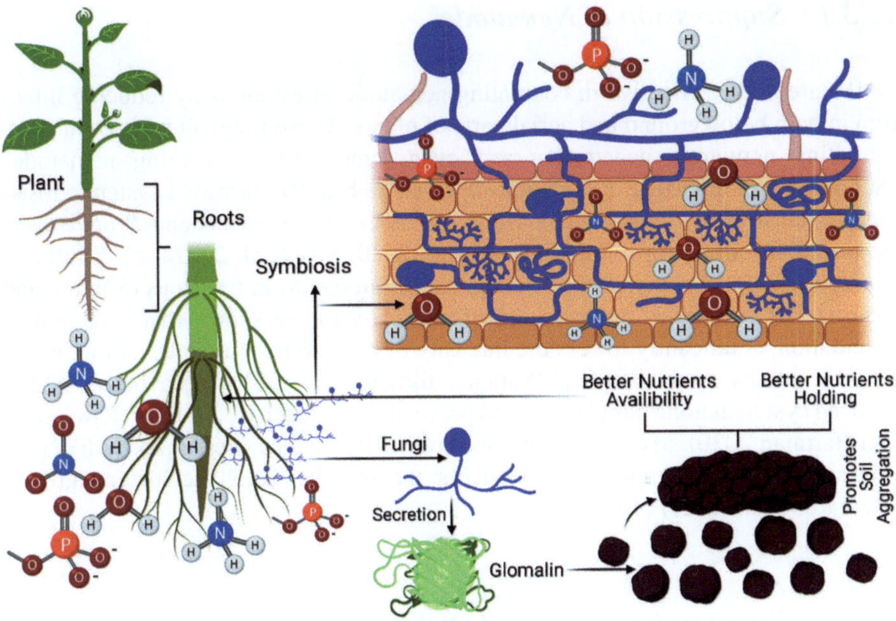

Fig. 12.2 Role of mycorrhizal-root symbiosis in enhancing water and nutrient uptake. (Adapted from Wahid et al. 2022)

inducing tolerance mechanisms, AMF paves the way for eco-friendly and efficient crop management strategies (Delaeter et al. 2024; Weng et al. 2022).

Overall, AMFs are indispensable allies in sustainable agriculture, offering a multifaceted defense system against agricultural pathogens (Wahab et al. 2023a, b). Their ability to enhance nutrient uptake, activate systemic resistance, and suppress diverse pests and diseases underscores their role as a cornerstone of integrated pest management practices. Through these mechanisms, AMFs contribute to improved crop productivity and resilience while promoting environmental health and sustainability.

12.3 Arbuscular Mycorrhizal Fungi in Suppressing Diseases Caused by Pathogens

AMFs are integral to sustainable crop protection strategies, offering suppression of diseases caused by fungi, viruses, bacteria, and nematodes (Singh and Giri 2017; Wahab et al. 2023a, b). Their biocontrol properties enhance plant resilience and reduce pathogen impact through various mechanisms, as discussed below.

12.3.1 Suppression of Nematodes

AMFs are highly effective in combating nematode infestations by reducing infection in both belowground and aerial parts of plants. These fungi alter root exudates, inhibiting nematode penetration and even temporarily paralyzing nematodes (Schouteden et al. 2015). *Funneliformis mosseae* has demonstrated systemic resistance against *Meloidogyne incognita* in tomatoes and soybeans and *Pratylenchus penetrans* in other crops (Bamigboye et al. 2020; Vos et al. 2012a, b). Similarly, *Rhizophagus irregularis* is effective against *M. incognita* in tomatoes (Sharma and Sharma 2017). Studies on coffee (*Coffea arabica*) have shown that mycorrhizal inoculation significantly lowers the intensity of nematode infections, cutting infestation levels by more than half (Vallejos-Torres et al. 2021). In soybeans, AMFs reduced cyst nematode populations and prevented egg hatch by 27–62% (Pawlowski and Hartman 2020). Mycorrhizal colonization also reduces stress markers like proline and phenolic compounds, enhancing the plant's overall health and growth (Bernardo et al. 2021) (Table 12.1).

12.3.2 Suppression of Fungal Pathogens

AMFs exhibit strong biocontrol properties against a range of fungal pathogens. By competing for root space and carbohydrates, AMFs suppress fungal genera such as *Fusarium*, *Alternaria*, and *Colletotrichum* (Song et al. 2015; Singh et al. 2019). Pre-inoculation with *F. mosseae* has been shown to enhance tomato resistance to early blight and prevent *Fusarium wilt* in strawberries and cucumbers (Matsubara 2011; Hu et al. 2010). Combined inoculations, such as *F. mosseae* with *Glomus fasciculatum*, are particularly effective in reducing *Fusarium oxysporum* infections (Devi et al. 2022). In grapevines, *Rhizophagus irregularis* inoculation activates genes linked to stilbenoid biosynthesis, increasing resistance to pathogens such as *Plasmophora viticola* and *Botrytis cinerea* (Bruisson et al. 2016). These fungi also enhance antioxidant enzyme activities, improving photosynthetic performance and mitigating oxidative stress caused by fungal infections (Hashem et al. 2021) (Table 12.2).

12.3.3 Suppression of Bacterial Pathogens

AMFs can suppress bacterial pathogens, such as *Xanthomonas campestris* and *Pseudomonas syringae* (Weng et al. 2022). Colonization with *R. irregularis* in tomatoes and alfalfa reduced necrotic lesions, while *F. mosseae* inoculation

Table 12.1 Effect of AMF species on suppression of nematode-associated plant pathogens

AMF	Nematode phytopathogens	Host plant species	References
Gigaspora albida, Claroideoglomus etunicatum	*Meloidogyne arenaria*	*Alpinia galanga* (Alpinia)	da Silva Campos et al. (2020)
Funneliformis mosseae, Rhizophagus intraradices	*Heterodera glycine*	*Glycine max* (Soybean)	Vallejos-Torres et al. (2021)
Glomus versiforme	*Meloidogyne incognita*	*Vitis vinifera* (Grapevine)	Li et al. (2006)
Glomus aggregatum	*Meloidogyne incognita*	*Mentha arvensis* (Wild mint)	Pandey (2005)
Glomus mosseae	*Meloidogyne incognita*	*S. lycopersicum* (Tomato)	Vos et al. 2013
Gigaspora albida, Claroideoglomus etunicatum	*Meloidogyne enterolobii*	*Psidium guajava* (Guava)	Campos et al. (2013)
Glomus mosseae	*Meloidogyne incognita*	*Vigna mungo* (Black gram)	Sankaranarayanan and Sundarbabu (2010)
Rhizoglomus intraradices	*Pratylenchus coffeae*	*Daucus carota* (Carrot)	Elsen et al. (2008)
Glomus fasciculatum, Funneliformis mosseae	*Pratylenchus coffeae*	*Musa paradisiaca* (Banana)	Elsen et al. (2003)
Funneliformis mosseae	*Pratylenchus penetrans*	*Solanum lycopersicum* (Tomato)	Vos et al. (2012a, b)
Funneliformis mosseae	*Scutellonema bradys*	*Dioscorea* spp. (Yam)	Atti et al. (2016)
Glomus intraradices	*Radophouls similis*	*Musa paradisiaca* (Banana)	Van der Veken et al. (2021)
Glomus sp, Gigaspora sp.	*Meloidogyne incognita*	*Coffea arabica* (Coffee)	Pham et al. (2020)
Funneliformis mossae Rhizophagus irregularis	*Tylenchulus semipenetrans*	*Citrus* spp. (Citrus fruits)	Zoubi et al. (2024)
Nanoglomus plukenetiae Rhizoglomus variabile	*Meloidogyne incognita*	*Plukenetia volubilis*	Corazon-Guivin et al. (2024)

decreased bacterial density in soybeans (Liu et al. 2007; Khrieba 2019). Additionally, AMFs have shown protective effects against actinomycetes causing replant diseases in grapes and apple seedlings (Yang et al. 2012; Reyes-Tena 2017). Notably, AMF inoculation prepares plants' immune systems, enabling a quicker and more robust activation of defense pathways dependent on JA and SA when faced with pathogen attacks (Fujita et al. 2022) (Table 12.3).

Table 12.2 Effect of AMF species in suppression of fungal plant pathogen infections

AMF	Fungal phytopathogens	Host plant species	References
Glomus etunicatum, glomus intraradices	*Rhizoctonia solani*	*Solanum tuberosum* (Potato)	Yao et al. (2002)
Glomus mosseae	*Gaeuomannomyces graminis var tritici*	*Hordeum vulgare* (Barley)	Castellanos-Morales et al. (2011)
Funneliformis mossae	*Alternaria solani*	*S. lycopersicum* (Tomato)	Frey Klett et al. (2007)
Funneliformis mossae	*Fusarium oxysporii*	*Fragaria* × *ananassa* (Strawberry)	Hu et al. (2010)
Glomus etunicatum, Glomus monosporum	*Phytopthora fragariae*	*Fragaria* × *ananassa* (Strawberry)	Norman and Hooker (2000)
Rhizophagus irregularis	*Macrophomina phaseolina*	*Glycine max* (Soybean)	Marcquez et al. (2018)
Rhizophagus fasciculatum	*Plasmophora viticola*	*Vitis vinifera* (Grape)	Cruz-Silva et al. (2021)
Rhizophagus fasciculatum	*Alternaria alternate*	*S. lycopersicum* (Tomato)	Nair et al. (2015)
Glomus intraradices	*Sclerotinia sclerotiorum*	*Fragaria* × *ananassa* (Strawberry)	Delgado et al. (2023)
Glomus mosseae, Glomus intraradices	*Phytophthora parasitica*	*S. lycopersicum* (Tomato)	Pozo et al. (2002)
Glomus intraradices	*Phytophthora nicotianae*	*S. lycopersicum* (Tomato)	Lioussanne et al. (2008)
Glomus intraradices	*Macrophomina phaseolina*	*Cicer arietinum* (Chickpea)	Siddiqui and Akhtar (2006)
Rhizoglomus irregularis	*Botrytis cinerea*	*S. lycopersicum* (Tomato)	Sammartin et al. (2020)
Rhizoglomus intraradices	*Magnaporthe oryzae*	*Oryza sativa* (Rice)	Tian et al. (2019)
Rhizophagus irregularis	*Magnaporthe oryzae*	*Oryza sativa* (Rice)	Campo et al. (2020)
Rhizophagus irregularis	*Verticillium wilt*	*Gossypium herbaceum* (Cotton)	Zhang et al. (2018)
Glomus etunicatum	*Eryspihae graminis*	*Hordeum vulgare* (Barley)	Gerrns et al. (2001)
Glomus mosseae, Glomus etunicatum, Glomus fasciculatum Glomus margarita	*Phytophthora capsici*	*Piper nigrum* (Pepper)	Ozgonen and Erkilic (2007)

(continued)

Table 12.2 (continued)

AMF	Fungal phytopathogens	Host plant species	References
Glomus versiforme	*Fusarium wilt/Fusarium oxysporum*	*Salvia miltiorrhiza* (Chinese sage)	Pu et al. (2022)
Funneliformis mosseae, *Gigaspora margarita*	*Rhizoctonia fragariae*, *Fusarium oxysporum* *Alternaria alternata*	*Fragaria* × *ananassa* (Strawberry)	Demir et al. (2023)
Funneliformis mosseae	*Alternaria solani*	*Solanum lycopersicum* (Tomato)	Boyno et al. (2022)
Funneliformis mosseae, *Glomus fasciculatum*	*Fusarium oxysporum*	*Solanum lycopersicum* (Tomato)	Devi et al. (2022)
An unidentified native AM fungal species	*Fusarium wilt/F. udum*	*Cajanus cajan* (Pigeon pea)	Jha and Mohamed (2024)

Table 12.3 Effect of AMF species on suppression of bacterial plant pathogens

AMF	Bacterial phytopathogens	Host plant species	References
Gigaspora intraradices, *Gigaspora versiforme*	*Xanthomonas comprestis*	*Medicago truncatula* (Barrel clover)	Liu et al. (2007)
Funneliformis mosseae	*Xanthomonas translucens*	*Triticum Aestivum* (Wheat)	Fiorilli et al. (2018)
Glomus versiforme	*Ralstonia solanacearum*	*Solanum lycopersicum* (Tomato)	Zhu and Yao (2004)
Glomus mossae	*Candidatus phytoplasma asteris*	*Catharanthus roseus* (Periwinkle)	Kaminiska et al. (2010)
Glomus macrocarpum	*P. lacrymans*	*Solanum melongena* (Egg plant)	Kapoor et al. (2002)
Rhizophagus irregularis	*Candidatus liberibacter solanacearum*	*Solanum lycopersicum* (Tomato)	Tienebo et al. (2019)
Entrophospora infrequens	*Pseudomonas syringae pv. glycinia*	*Glycine max* (Soybean)	Malik et al. (2016)
Funneliformis mosseae	*Pseudomonas syringae*	*Glycine max* (Soybean)	Khrieba (2019)
Funneliformi mosseae, *Pseudomonas* spp.	*Xanthomonas euvesicatoria*	*Piper nigrum* (Pepper)	Çakar et al. 2024

12.3.4 Suppression of Viral Infections

Although studies on AMF's role in viral disease suppression are limited, evidence indicates that AMF colonization can mitigate viral symptoms and reduce viral titers (Kalra et al. 2023). For instance, *Rhizophagus irregularis* has been shown to mitigate the effects of *Potato Virus Y* (PVY) in potatoes and *Tomato Yellow Leaf Curl Sardinia Virus* (TYLCSV) in tomatoes, promoting plant growth and alleviating symptom severity (Miozzi et al. 2011; Thiem et al. 2014). Similarly, AMFs enhanced resistance to *Cucumber Green Mottle Mosaic Virus* in cucumbers and *Tobacco*

Table 12.4 Impact of AMF species on suppression of viral plant pathogens

AMF	Viral phytopathogens	Host plant species	References
Rhizoglomus irregularis	Tomato Bushy Stunt Virus (TBSV)	*Solanum lycopersicum* (Tomato)	Khoshkhatti et al. (2020)
Funneliformis mosseae	Tomato Spotted Wilt Virus (TSWV)	*Solanum lycopersicum* (Tomato)	Miozzi et al. (2011)
Funneliformis mosseae	TYLCSV	*Solanum lycopersicum* (Tomato)	Maffei et al. (2013)
Funneliformis mosseae	Beet curly top Iran Virus (BCTIV)	*Solanum lycopersicum* (Tomato)	Ebrahimi et al. (2020)
Rhizophagus intraradices	TMV	(*Nicotiana tabacum*) Tobacco	Shaul et al. (1999)
Rhizophagus intraradices	PVY	*Solanum tuberosum* (Potato)	Sipahioglu et al. (2009)
Rhizophagus intraradices	Grapevine fanleaf Virus (GFLV)	*Vitis vinifera* (Grapevine)	Hao et al. (2018)
Funneliformis geosporum	Citrus leaf rugose Virus (CLRV)	Citrus ×paradisi (Duncan grapefruit)	Maffei et al. (2013)
Funneliformis mosseae	Cucumber mosaic Virus (Cucumovirus)	*Cucumis sativus* (Cucumber)	Elsharkawy et al. (2012)
Funneliformis mosseae	Cucumber Mosaic Virus (CMV)	*Cucumis sativus* (Cucumber)	Miozzi et al. (2020)
Funneliformis mosseae, *Azospirillum brasilense*	Tomato brown rugose fruit virus (ToBRFV)	*Solanum lycopersicum* (Tomato)	Ragasová et al. (2022)
Rhizophagus regularis *Funneliformis mosseae*	PVY	*Solanum tuberosum* (Potato)	Deja-Sikora et al. (2023)
Rhizophagus irregularis *Funneliformis mosseae*	PVY	*Solanum tuberosum* (Potato)	Deja-Sikora et al. (2024)

Mosaic Virus (TMV) in tobacco plants (Stolyarchuk et al. 2009). However, interactions between AMF and insect vectors of viruses, such as aphids, may enhance vector survival while simultaneously activating indirect defenses through parasitoid attraction (Gehring and Bennett, 2009). This highlights the complexity of AMF–virus–plant interactions (Table 12.4).

12.4 Plant Innate Immunity: Cellular and Molecular Mechanisms

Understanding the mechanisms behind plant immunity provides essential insights into the role of mycorrhizae in enhancing plant defenses (Fiorilli et al. 2024). Plants rely on innate physical and chemical barriers, as well as inducible biochemical defenses, to resist pathogens (Shittu et al. 2019). These defenses include the production of antimicrobial compounds such as secondary metabolites, terpenes, phenolics, and tannins, which collectively mitigate pathogen attacks (Akbar et al. 2023; Agrios, 2005). When pathogens breach these barriers, plants activate a sophisticated immune system that incorporates systemic signaling and the ability to "remember" prior infections, a phenomenon referred to as induced resistance (Reimer-Michalski and Conrath 2016).

The plant immune system functions through two main pathways: effector-triggered immunity (ETI) and pattern-triggered immunity (PTI) (Naveed et al. 2020). PTI is activated when pattern recognition receptors identify pathogen-associated molecular patterns, microbe-associated molecular patterns, or damage-associated molecular patterns (Bigeard et al. 2015) (Fig. 12.3). This recognition triggers defense mechanisms such as calcium influx, reactive oxygen species (ROS) generation, mitogen-activated protein kinase (MAPK) signaling, and the upregulation of defense-related genes (Saijo et al. 2018; De Lorenzo et al. 2018). If pathogens counteract PTI by releasing effectors, ETI is activated through resistance (R) proteins that detect these effectors. ETI responses are stronger and include hypersensitive reactions, the synthesis of secondary metabolites, and increased expression of defense genes (Poltronieri and Hong 2019). Cellular responses to PTI involve calcium signaling, which activates secondary messengers that regulate defense genes (Jeworutzki et al. 2010; Li et al. 2014), and ROS production, which serves as both a signaling molecule and a direct antimicrobial agent (Jiang et al. 2017). MAPK signaling further coordinates these responses by modulating transcription factors involved in the production of JA, SA, ethylene, and antimicrobial compounds (Bigeard et al. 2015; Jagodzik et al. 2018). In addition to these localized responses, plants develop systemic and long-lasting immunity. SAR enables uninfected tissues to mount defenses against a broad spectrum of pathogens. SAR relies

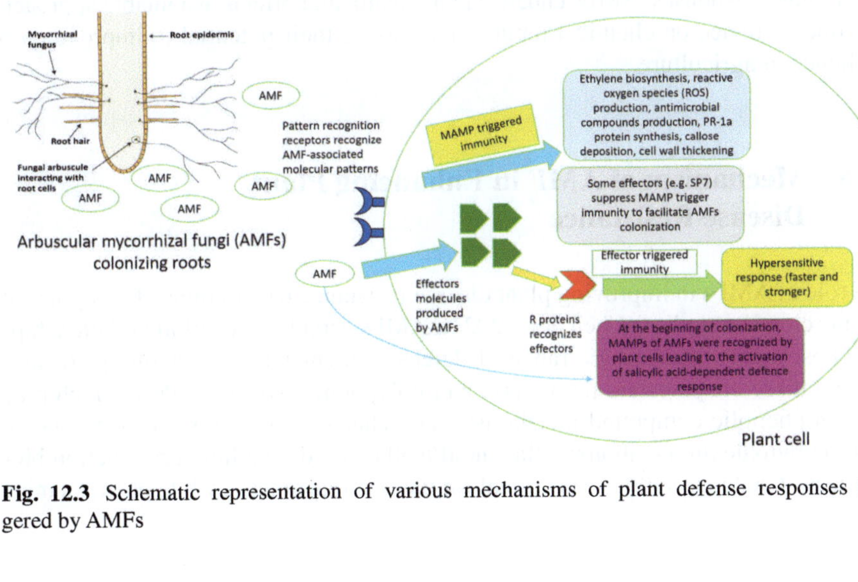

Fig. 12.3 Schematic representation of various mechanisms of plant defense responses triggered by AMFs

on signaling molecules such as JA, SA, and pipecolic acid to activate defense mechanisms in distant tissues, enhancing resilience against future infections (Bernsdorff et al. 2016; Gruner et al. 2013). These interconnected immune strategies provide a comprehensive framework for plants to resist biotic stressors effectively.

12.5 Mycorrhiza-Induced Plant Immune Response

AMFs significantly enhance plant immunity through MIR, providing systemic protection against diverse pathogens such as fungi, nematodes, viruses, and insects (Nguvo and Gao 2019; Abarca et al. 2024). MIR combines aspects of SAR, typically triggered by pathogen infection, and induced systemic resistance, which arises from root colonization by beneficial rhizobacteria (Cameron et al. 2013). This immune response involves the activation of defense genes that produce chitinases, glucanases, pathogenesis-related (PR) proteins, and phytohormones such as JA and SA (Sanmartín et al. 2020; Fiorilli et al. 2018). The coordination between these phytohormones is crucial, as JA-dependent pathways defend against necrotrophic pathogens, while SA-mediated responses target biotrophic pathogens (Pozo et al. 2010; Robert-Seilaniantz et al. 2011). Furthermore, AMF colonization modulates genes involved in JA and abscisic acid biosynthesis, promoting the synthesis of secondary metabolites such as phenolics and lignin, which reinforce physical and chemical defenses against pathogens (Aseel et al. 2019; Lopez-Ráez et al. 2017). AMFs also modify root anatomy, compete with pathogens for resources, and alter root exudate composition, secreting strigolactones that attract beneficial microbes, further bolstering plant defense (Yin et al. 2021; Vos et al. 2014). These interactions create a suppressive soil environment in the mycorrhizosphere, where microbial activity enhances plant resilience (Pršić and Ongena, 2020). By coordinating various immune responses, AMFs enhance plant health and offer a sustainable approach to reduce reliance on chemical inputs, showcasing their potential to improve crop resilience in agriculture.

12.6 Mechanism of AMF in Enhancing Plant Disease Resistance

The role of AMFs in improving plant disease resistance has become a focal point of recent scientific studies (Chen et al. 2021). AMFs establish symbiotic relationships with host plants, activating a range of defense enzymes and metabolic pathways. Among these enzymes are peroxidase and polyphenol oxidase (PPO), which play roles in phenolic compound metabolism, and chalcone isomerase, which contributes to phytoalexin metabolism, flavonoid/isoflavonoid, and lignin production biosynthesis. Chalcone synthase is critical for flavonoid synthesis, while phenylalanine ammonia-lyase (PAL) facilitates phenylpropanoid metabolism (Rashad et al. 2024). Additionally, AMF symbiosis stimulates the expression of PR proteins, which are directly linked to disease resistance (Isayenkov et al. 2005; Fester and Hause 2005).

For instance, PAL activity, a key indicator of plant disease resistance, increases significantly in tomato seedlings colonized by AMF and infected by *Fusarium oxysporum*, leading to reduced disease symptoms (De Román et al. 2011). Similarly, strawberry plants inoculated with *Glomus mosseae* showed elevated superoxide dismutase activity and improved free-radical scavenging capacity, enhancing resistance to *Fusarium oxysporum* and *Glomus anthracis* (Steinkellner et al. 2012; Weng et al. 2022). Tomato plants colonized by *G. mosseae* also exhibited higher levels of glucanase, chitinase, and PR proteins compared to non-mycorrhizal plants (Kloppholz et al. 2011). In date palm trees, inoculation with *G. monosporum*, *G. deserticola*, and *G. clarum* increased PPO activity, reducing chlorosis caused by pathogen attacks (Glazebrook 2005; Kashyap et al. 2024).

A notable example is *Gigaspora rhizogenes*, which secretes a defense protein, SP7, that interacts with the transcription factor ERF19 in the nucleus, alleviating root rot symptoms caused by *Magnaporthe oryzae* (Kloppholz et al. 2011; Aparicio Chacón et al. 2023). The mechanisms through which AMFs enhance disease resistance are complex and often involve both systemic and localized responses (Pozo et al., 2002; Tabin et al. 2009; Ghorui et al. 2025). These mechanisms can function independently or synergistically, with their effectiveness shaped by interactions among AMFs, host plants, pathogens, and abiotic factors such as the timing of inoculation, inoculum size, pH, soil fertility, humidity, and temperature (Liu and Chen 2024). Agricultural practices, including crop rotation, fertilization, and pest management, also influence the biocontrol efficacy of AMF (Chandanie et al. 2009; Ebbisa 2022). These factors must coordinate for AMF to achieve optimal biocontrol effects (Fig. 12.4).

12.7 Conclusion

AMFs are pivotal in modern sustainable agriculture, offering a multifaceted approach to suppress agricultural pathogens and enhance plant immunity. Their ability to provide systemic protection against diverse pathogens, such as fungi, nematodes, viruses, and bacteria while boosting nutrient uptake and promoting plant resilience underscores their utility as natural biocontrol agents. These benefits are facilitated through mechanisms like MIR, which activates defense signaling pathways, including JA and SA, as well as the production of defense-related enzymes and secondary metabolites. Evidence accumulated over the past two decades highlights AMF's role in enhancing plant–microbe interactions, improving rhizosphere health, and reducing dependence on chemical pesticides. AMFs provide physical barriers by colonizing root surfaces and altering root anatomy, while their ability to prime plants for rapid defense responses further strengthens their potential as effective biocontrol agents. However, the efficacy of AMF is highly

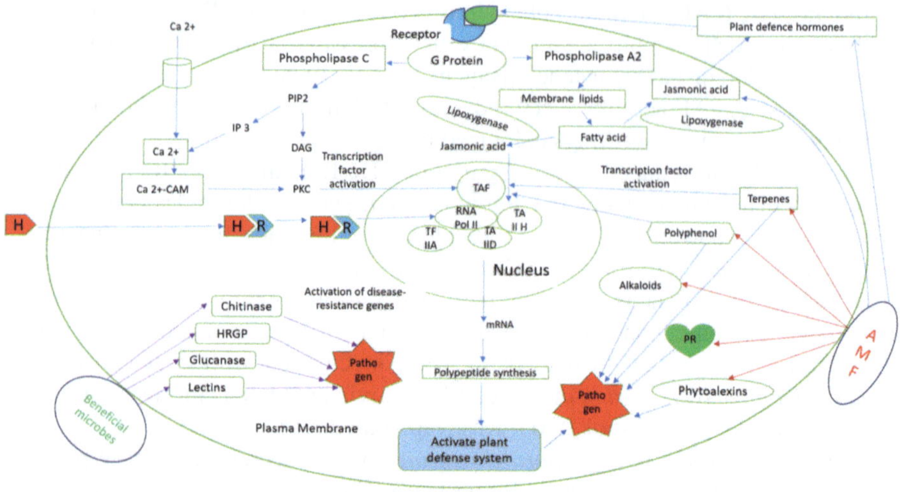

Fig. 12.4 Schematic representation of the mechanisms by which AMFs activate plant defense systems. Note: *PIP2* phosphatidylinositol diphosphate, *IP3* inositol triphosphate, *DAG* diacylglycerol, *PKC* protein kinase C, *HRGP* hydroxyproline-rich glycoprotein, *PR* pathogenesis-related proteins

context-dependent, influenced by environmental factors, host specificity, and interactions with other soil microorganisms. Despite their immense potential, AMF applications remain underutilized in conventional agricultural systems due to reliance on agrochemicals and over-fertilization. Integrating AMFs into sustainable practices such as organic farming, conservation agriculture, and agroecological systems offers a promising solution to mitigate these challenges. AMF's contribution to improving soil health, nutrient cycling, and ecological balance aligns seamlessly with global initiatives aimed at achieving food security and environmental sustainability. Future research should prioritize uncovering the molecular mechanisms behind AMF–host interactions, refining microbial consortia, and addressing performance inconsistencies under shifting agroclimatic conditions. Cutting-edge technologies such as transcriptomics, metabolomics, and genomics can help decode the intricate processes of AMF-mediated defense and nutrient exchange. Additionally, developing approaches to leverage the synergistic effects of AMF and plant growth-promoting rhizobacteria will further amplify their effectiveness. In summary, AMF stands as a foundational element in transforming agriculture by minimizing chemical inputs, boosting crop resilience, and promoting sustainable food production. Collaborative efforts among researchers, policymakers, and practitioners will be crucial to fully unlock their potential in tackling the dual challenges of agricultural productivity and environmental sustainability in the twenty-first century.

References

Abarca C, Fernandez Bidondo L, Bompadre J, Velázquez MS (2024) Arbuscular mycorrhizal fungi in tomato tolerance to pathogens and nematodes: a comprehensive review. Sci Hortic 329:112969

Agrios GN (2005) Plant pathology, 5th edn. Academic Press

Akbar MU, Aqeel M, Shah MS, Jeelani G, Iqbal N, Latif A et al (2023) Molecular regulation of antioxidants and secondary metabolites act in conjunction to defend plants against pathogenic infection. S Afr J Bot 161:247–257

Aparicio Chacón MV, Van Dingenen J, Goormachtig S (2023) Characterization of arbuscular mycorrhizal effector proteins. Int J Mol Sci 24(11):9125

Aseel DG, Rashad YM, Hammad SM (2019) Arbuscular mycorrhizal fungi trigger transcriptional expression of flavonoid and chlorogenic acid biosynthetic pathways genes in tomato against tomato mosaic virus. Sci Rep 9:9692

Atti T, Fabien CC, Bisola O, Louis L, Danny C, Andres W et al (2016) Effect of two species of arbuscular mycorrhizal fungi inoculation on development of micropropagated yam plantlets and suppression of *Scutellonema bradys* (Tylenchidae). J Entomol Nematol 8(1):1–10

Bagyaraj DJ, Sridhar KR, Revanna A (2022) Arbuscular mycorrhizal fungi influence crop productivity, plant diversity, and ecosystem services. In: Rajpal VR, Singh I, Navi SS (eds) Fungal diversity, ecology, and control management. Springer, Singapore, pp 345–362

Bamigboye RA, Adekunle OK, Salami AO (2020) Arbuscular mycorrhiza (AMF) suppressed fungal disease incidence, severity, and population of nematodes on soybean (*Glycine max* L.). Ecologia 10:86–92

Baum C, El-Tohamy W, Gruda N (2015) Increasing the productivity and product quality of vegetable crops using arbuscular mycorrhizal fungi: a review. Sci Hortic 187:131–141

Begum N, Qin C, Ahanger MA, Raza S, Khan MI, Ashraf M et al (2019) Role of arbuscular mycorrhizal fungi in plant growth regulation: implications in abiotic stress tolerance. Front Plant Sci 10:1068

Bernardo VF, Garita SA, Arango MC, Ripodas JI, Saparrat MCN, Ruscitti MF (2021) Arbuscular mycorrhizal fungi against the false root-knot nematode activity in Capsicum annuum: physiological responses in plants. Biocontrol Sci Tech 31(2):119–131

Bernsdorff F, Döring AC, Gruner K, Schuck S, Bräutigam A, Zeier J (2016) Pipecolic acid orchestrates plant systemic acquired resistance and defense priming via salicylic acid-dependent and -independent pathways. Plant Cell 28(1):102–129

Bhupenchandra I, Chongtham SK, Devi AG, Dutta P, Sahoo MR, Mohanty S et al (2024) Unlocking the potential of arbuscular mycorrhizal fungi: exploring role in plant growth promotion, nutrient uptake mechanisms, biotic stress alleviation, and sustaining agricultural production systems. J Plant Growth Regul 1–39

Bigeard J, Colcombet J, Hirt H (2015) Signaling mechanisms in pattern-triggered immunity (PTI). Mol Plant 8(4):521–539

Boyno G, Demir S, Danesh YR (2022) Effects of some biological agents on the growth and biochemical parameters of tomato plants infected with *Alternaria solani* (Ellis & Martin) Sorauer. Eur J Plant Pathol 162(1):19–29

Bruisson S, Maillot P, Schellenbaum P, Walter B, Gindro K, Deglène-Benbrahim L (2016) Arbuscular mycorrhizal symbiosis stimulates key genes of the phenylpropanoid biosynthesis and stilbenoid production in grapevine leaves in response to downy mildew and grey mould infection. Phytochemistry 131:92–99

Buitenhuis R, Cock MJ, Colmenarez YC, De Clercq P, Edgington S, Gadaleta P et al (2023) Sustainable use and conservation of microbial and invertebrate biological control agents and microbial biostimulants. No. 71. FAO

Çakar K, Akköprü A, Uçar CP (2024) Effect of *pseudomonas* spp. and *Funneliformis mosseae* on bacterial spot disease and growth response of pepper. J Plant Pathol:1–13

Cameron DD, Neal AL, Wees SC, Ton J (2013) Mycorrhiza-induced resistance: more than the sum of its parts? Trends Plant Sci 18(10):539–545

Campo S, Martín-Cardoso H, Olivé M, Pla E, Catala-Forner M, Martínez-Eixarch M et al (2020) Effect of root colonization by arbuscular mycorrhizal fungi on growth, productivity, and blast resistance in rice. Rice 13(1):42

Campos M, Silva FSBD, Yano-Melo AM, Melo NFD, Pedrosa EMR, Maia LC (2013) Responses of guava plants to inoculation with arbuscular mycorrhizal fungi in soil infested with Meloidogyne enterolobii. Plant Pathol J 29:242–248. https://doi.org/10.5423/ppj.oa.10.2012.0156

Campos MA (2020) Bioprotection by arbuscular mycorrhizal fungi in plants infected with *Meloidogyne* nematodes: a sustainable alternative. Crop Prot 135:105203

Castellanos-Morales V, Keiser C, Cárdenas-Navarro R, Grausgruber H, Glauninger J, García-Garrido JM et al (2011) The bioprotective effect of AM root colonization against the soil-borne fungal pathogen *Gaeumannomyces graminis* var. tritici in barley depends on the barley variety. Soil Biol Biochem 43(4):831–834

Chandanie WA, Kubota M, Hyakumachi M (2009) Interactions between the arbuscular mycorrhizal fungus *glomus mosseae* and plant growth-promoting fungi and their significance for enhancing plant growth and suppressing damping-off of cucumber (*Cucumis sativus* L.). Appl Soil Ecol 41:336–341

Chen Q, Wu WW, Qi SS, Cheng H, Li Q, Ran Q et al (2021) Arbuscular mycorrhizal fungi improve the growth and disease resistance of the invasive plant *Wedelia trilobata*. J Appl Microbiol 130:582–591

Corazon-Guivin MA, Rengifo del Aguila S, Corrêa RX, Cordova-Sinarahua D, Costa Maia L, Alves da Silva DK et al (2024) Native arbuscular mycorrhizal fungi promote *Plukenetia volubilis* growth and decrease the infection levels of Meloidogyne incognita. J Fungi. 10(7):451

Cruz-Silva A, Figueiredo A, Sebastiana M (2021) First insights into the effect of mycorrhizae on the expression of pathogen effectors during the infection of grapevine with *Plasmopara viticola*. Sustain For 13(3):1226

da Silva Campos MA (2020) Bioprotection by arbuscular mycorrhizal fungi in plants infected with *Meloidogyne* nematodes: a sustainable alternative. Crop Prot 135:105203

De Lorenzo G, Ferrari S, Cervone F, Okun E (2018) Extracellular DAMPs in plants and mammals: immunity, tissue damage, and repair. Trends Immunol 39:937–950

De Román M, Fernández I, Wyatt T, Sahrawy M, Heil M, Pozo MJ (2011) Elicitation of foliar resistance mechanisms transiently impairs root association with arbuscular mycorrhizal fungi. J Ecol 99:36–45

Deja-Sikora E, Werner K, Hrynkiewicz K (2023) AMF species do matter: *Rhizophagus irregularis* and *Funneliformis mosseae* affect healthy and PVY-infected *Solanum tuberosum* L. in a different way. Front Microbiol 14:1127278

Deja-Sikora E, Gołębiewski M, Hrynkiewicz K (2024) Transcriptomic responses of *Solanum tuberosum* cv. Pirol to arbuscular mycorrhiza and potato virus Y (PVY) infection. Plant Mol Biol 114(6):123

Delaeter M, Magnin-Robert M, Randoux B, Lounès-Hadj SA (2024) Arbuscular mycorrhizal fungi as biostimulant and biocontrol agents: a review. Microorganisms. 12(7):1281. https://doi.org/10.3390/microorganisms12071281

Delgado A, Toro M, Memenza-Zegarra M, Zúñiga-Dávila D (2023) Control of white rot caused by *Sclerotinia sclerotiorum* in strawberry using arbuscular mycorrhizae and plant-growth-promoting bacteria. Sustain For 15(4):2901

Demir S, Durak ED, Güneş H, Boyno G, Mulet JM, Rezaee Danesh Y, Porcel R (2023) Biological control of three fungal diseases in strawberry (Fragaria × ananassa) with arbuscular mycorrhizal fungi. Agronomy 13(9):2439

Devi NO, Devi RKT, Debbarma M, Hajong M, Thokchom S (2022) Effect of endophytic *bacillus* and arbuscular mycorrhiza fungi (AMF) against *Fusarium* wilt of tomato caused by *Fusarium oxysporum* f. sp. lycopersici. Egypt J Biol Pest Control 32:1–14

Dey M, Ghosh S (2022) Arbuscular mycorrhizae in plant immunity and crop pathogen control. Rhizosphere. 22:100524

Dhuldhaj UP, Singh R, Singh VK (2023) Pesticide contamination in agro-ecosystems: toxicity, impacts, and bio-based management strategies. Environ Sci Pollut Res 30(4):9243–9270

Ebbisa A (2022) Arbuscular mycorrhizal fungi (AMF) in optimizing nutrient bioavailability and reducing agrochemicals for maintaining sustainable agroecosystems. In: Arbuscular mycorrhizal fungi in agriculture-new insights. IntechOpen

Ebrahimi S, Eini O, Koolivand D (2020) Arbuscular mycorrhizal symbiosis enhances virus accumulation and attenuates resistance-related gene expression in tomato plants infected with beet curly top Iran virus. J Plant Dis Prot. 127(3):341–348

Elsen A, Beeterens R, Swennen R, De Waele D (2003) Effects of an arbuscular mycorrhizal fungus and two plant-parasitic nematodes on Musa genotypes differing in root morphology. Biol Fertil Soils 38:367–376

Elsen A, Gervacio D, Swennen R, De Waele D (2008) AMF-induced biocontrol against plant parasitic nematodes in Musa sp.: a systemic effect. Mycorrhiza 18:251–256

Elsharkawy MM, Shimizu M, Takahashi H, Hyakumachi M (2012) Induction of systemic resistance against cucumber mosaic virus by *Penicillium simplicissimum* GP17-2 in Arabidopsis and tobacco. Plant Pathol 61(5):964–976

Fester T, Hause G (2005) Accumulation of reactive oxygen species in arbuscular mycorrhizal roots. Mycorrhiza 15:373–379

Fiorilli V, Vannini C, Ortolani F, Garcia-Seco D, Chiapello M, Novero M et al (2018) Omics approaches revealed how arbuscular mycorrhizal symbiosis enhances yield and resistance to leaf pathogen in wheat. Sci Rep 8(1):9625

Fiorilli V, Martínez-Medina A, Pozo MJ, Lanfranco L (2024) Plant immunity modulation in arbuscular mycorrhizal symbiosis and its impact on pathogens and pests. Annu Rev Phytopathol 62(1):127–156

Frey Klett P, Garbaye J, Tarkka M (2007) The mycorrhiza helper bacteria revisited. New Phytol 176(1):22–36

Fujita M, Kusajima M, Fukagawa M, Okumura Y, Nakajima M, Akiyama K et al (2022) Response of tomatoes primed by mycorrhizal colonization to virulent and avirulent bacterial pathogens. Sci Rep 12:4686

Gehring C, Bennett A (2009) Mycorrhizal fungal–plant–insect interactions: the importance of a community approach. Environ Entomol 38(1):93–102

Gernns H, Von Alten H, Poehling HM (2001) Arbuscular mycorrhiza increased the activity of a biotrophic leaf pathogen – is a compensation possible? Mycorrhiza 11:237–243

Ghorui M, Chowdhury S, Balu P, Das K, Sunar K (2025) Boosting plant immunity: the functional role and mechanism of arbuscular mycorrhizal fungi in resistance. In: Plant microbiome and biological control: emerging trends and applications. Springer Nature, Cham, pp 195–219

Glazebrook J (2005) Contrasting mechanisms of defense against biotrophic and necrotrophic pathogens. Annu Rev Phytopathol 43:205–227

Gruner K, Griebel T, Návarová H, Attaran E, Zeier J (2013) Reprogramming of plants during systemic acquired resistance. Front Plant Sci 4:252. https://doi.org/10.3389/fpls.2013.00252

Hao Z, van Tuinen D, Fayolle L, Chatagnier O, Li X, Chen B, Gianinazzi S, Gianinazzi-Pearson V (2018) Arbuscular mycorrhiza affects grapevine fanleaf virus transmission by the nematode vector Xiphinema index. Appl Soil Ecol 129:107–111

Hashem A, Akhter A, Alqarawi AA, Singh G, Almutairi KF, Allah EFA (2021) Mycorrhizal fungi-induced activation of tomato defense system mitigates *Fusarium* wilt stress. Saudi J Biol Sci 28(10):5442–5450

Hu J, Lin X, Wang J, Shen W, Wu S, Peng S, Mao T (2010) Arbuscular mycorrhizal fungal inoculation enhances suppression of cucumber *Fusarium* wilt in greenhouse soils. Pedosphere 20(5):586–593

Huang NX, Enkegaard A, Osborne LS, Ramakers PMJ, Messelink GJ, Pijnakker J, Murphy G (2011) The banker plant method in biological control. Crit Rev Plant Sci 30(3):259–278

Isayenkov S, Mrosk C, Stenzel I, Strack D, Hause B (2005) Suppression of allene oxide cyclase in hairy roots of *Medicago truncatula* reduces jasmonate levels and the degree of mycorrhization with *Glomus intraradices*. Plant Physiol 139:1401–1410

Jagodzik P, Tajdel-Zielinska M, Ciesla A, Marczak M, Ludwikow A (2018) Mitogen-activated protein kinase cascades in plant hormone signaling. Front Plant Sci 9:1387

Jeworutzki E, Roelfsema MR, Anschütz U, Krol E, Elzenga JT, Felix G, Becker D (2010) Early signaling through the Arabidopsis pattern recognition receptors FLS2 and EFR involves Ca2+-associated opening of plasma membrane anion channels. Plant J 62(3):367–378

Jha Y, Mohamed HI (2024) Endophytic *Pseudomonas* pseudoalcaligenes and arbuscular mycorrhizal fungi mediated anti-autophagy and induction of catalase and antioxidant enzymes in pigeon pea against fungal pathogen. J Plant Pathol 106(1):225–240

Jiang YN, Wang WX, Xie QJ, Liu N, Liu LX, Wang DP, Zhang X, Yang C, Chen X, Tang D, Wang E (2017) Plants transfer lipids to sustain colonization by mutualistic mycorrhizal and parasitic fungi. Science 356:1172–1175. https://doi.org/10.1126/science.aam9970

Jung SC, Martinez-Medina A, Lopez-Raez JA, Pozo MJ (2012) Mycorrhiza-induced resistance and priming of plant defenses. J Chem Ecol 38(6):651–664

Kadam SB, Pable AA, Barvkar VT (2020) Mycorrhiza-induced resistance (MIR): a defence developed through synergistic engagement of phytohormones, metabolites, and rhizosphere. Funct Plant Biol 47(10):880–890

Kalra C, Bharti S, Bhatnagar S, Giri B, Kumari R (2023) Arbuscular mycorrhizal fungi as potential agents in augmenting growth and stress tolerance in plants. Kavaka 59(2):52–64

Kamińska M, Klamkowski K, Berniak H, Sowik I (2010) Response of mycorrhizal periwinkle plants to aster yellows phytoplasma infection. Mycorrhiza 20:161–166

Kapoor R, Giri B, Mukerji KG (2002) Glomus macrocarpum: a potential bioinoculant to improve essential oil quality and concentration in dill (*Anethum graveolens* L.) and carum (*Trachyspermum ammi* (Linn.) Sprague). World J Microbiol Biotechnol 18(5):459–463

Kashyap P, Sharma I, Kashyap S, Agarwala N (2024) Arbuscular mycorrhizal fungi (AMF)-mediated control of foliar fungal diseases. In: Arbuscular mycorrhizal fungi and higher plants: fundamentals and applications. Springer Nature, Singapore, pp 193–223

Khaliq A, Perveen S, Alamer KH, Haq MZU, Rafique Z, Alsudays IM et al (2022) Arbuscular mycorrhizal fungi symbiosis to enhance plant-soil interaction. Sustain For 14(13):7840

Khoshkhatti N, Eini O, Koolivand D, Pogiatzis A, Klironomos JN, Pakpour S (2020) Differential response of mycorrhizal plants to tomato bushy stunt virus and tomato mosaic virus infection. Microorganisms 8:2038

Khrieba MI (2019) Mycorrhizae's role in plant nutrition and protection from pathogens. Curr Investing Agric Curr Res 8(1)

Kloppholz S, Kuhn H, Requena N (2011) A secreted fungal effector of *Glomus intraradices* promotes symbiotic biotrophy. Curr Biol 21:1204–1209

Koffi MC, Vos C, Draye X, Declerck S (2013) Effects of *Rhizophagus irregularis* MUCL 41833 on the reproduction of *Radopholus similis* in banana plantlets grown under in vitro culture conditions. Mycorrhiza 23(4):279–288

Koskey G, Mburu SW, Awino R, Njeru EM, Maingi JM (2021) Potential use of beneficial microorganisms for soil amelioration, phytopathogen biocontrol, and sustainable crop production in smallholder agroecosystems. Front Sustain Food Syst 5:606308

Kumar A, Dutta J, Chandrawanshi NK, Ekka A, Sethi SK (2024) Arbuscular mycorrhizal fungi in alleviation of biotic stress tolerance in plants: a new direction in sustainable agriculture. In: Entrepreneurship with microorganisms. Academic Press, pp 355–369

Li HY, Yang GD, Shu HR, Yang YT, Ye BX, Nishida I et al (2006) Colonization by the arbuscular mycorrhizal fungus *Glomus versiforme* induces a defense response against the root-knot nematode Meloidogyne incognita in grapevine (Vitis amurensis Rupr.), which includes transcriptional activation of the class III chitinase gene VCH3. Plant Cell Physiol 47:154–163

Li Y, Liu Z, Hou H, Lei H, Zhu X, Li X et al (2013) Arbuscular mycorrhizal fungi-enhanced resistance against *Phytophthora sojae* infection on soybean leaves is mediated by a network involving hydrogen peroxide, jasmonic acid, and the metabolism of carbon and nitrogen. Plant Physiol Biochem 70:1–8

Li L, Eichten SR, Shimizu R, Petsch K, Yeh C, Wu W et al (2014) Genome-wide discovery and characterization of maize long non-coding RNAs. Genome Biol 15(2):R40

Lioussanne L, Jolicoeur M, St-Arnaud M (2008) Mycorrhizal colonization with Glomus intraradices and development stage of transformed tomato roots significantly modify the chemotactic response of zoospores of the pathogen Phytophthora nicotianae. Soil Biol Biochem 40:2217–2224. https://doi.org/10.1016/j.soilbio.2008.04.013

Liu R, Chen Y (2024) Mycorrhizal symbiosis: evolution, opportunities, challenges, and prospects. In: Arbuscular mycorrhizal fungi in sustainable agriculture: inoculum production and application, pp 1–35

Liu J, Maldonado-Mendoza I, Lopez-Meyer M, Cheung F, Town CD, Harrison MJ (2007) Arbuscular mycorrhizal symbiosis is accompanied by local and systemic alterations in gene expression and an increase in disease resistance in the shoots. Plant J 50:529–544

Lopez-Ráez JA, Shirasu K, Foo E (2017) Strigolactones in plant interactions with beneficial and detrimental organisms: the yin and yang. Trends Plant Sci 22(6):527–537

Maffei G, Miozzi L, Fiorilli V, Novero M, Lanfranco L, Accotto GP (2013) The arbuscular mycorrhizal symbiosis attenuates symptom severity and reduces virus concentration in tomato infected by tomato yellow leaf curl Sardinia virus (TYLCSV). Mycorrhiza 24(3):179–186

Malik RJ, Dixon MH, Bever JD (2016) Mycorrhizal composition can predict foliar pathogen colonization in soybean. Biol Control 103:46–53

Marquez N, Giachero ML, Gallou A, Debat HJ, Cranenbrouck S, Di Rienzo JA et al (2018) Transcriptional changes in mycorrhizal and non-mycorrhizal soybean plants upon infection with the fungal pathogen Macrophomina phaseolina. Mol Plant-Microbe Interact 31(8):842–855

Matsubara YI (2011) Tolerance to fusarium wilt and changes in antioxidative ability and free amino acid content in mycorrhizal strawberry plants. In: Genomics, transgenics, molecular breeding and biotechnology of strawberry. Global Science Books, Isleworth, pp 126–131

Meena R, Kumar S, Datta R, Lal R, Vijayakumar V, Brtnicky M et al (2020) Impact of agrochemicals on soil microbiota and management: a review. Land 9(2):34

Miozzi L, Catoni M, Fiorilli V, Mullineaux PM, Accotto GP, Lanfranco L (2011) Arbuscular mycorrhizal symbiosis limits foliar transcriptional responses to viral infection and favors long-term virus accumulation. Mol Plant-Microbe Interact 24(12):1562–1572

Miozzi L, Vaira AM, Brilli F, Casarin V, Berti M, Ferrandino A et al (2020) Arbuscular mycorrhizal symbiosis primes tolerance to cucumber mosaic virus in tomato. Viruses 12(6):675

Mitra D, Djebaili R, Pellegrini M, Mahakur B, Sarker A, Chaudhary P et al (2021) Arbuscular mycorrhizal symbiosis: plant growth improvement and induction of resistance under stressful conditions. J Plant Nutr 44(13):1993–2028

Nair A, Kolet SP, Thulasir AMF, HV, Bhargava S. (2015) Systemic jasmonic acid modulation in mycorrhizal tomato plants and its role in induced resistance against *Alternaria alternata*. Plant Biol 17(3):625–631

Naveed ZA, Wei X, Chen J, Mubeen H, Ali GS (2020) The PTI to ETI continuum in Phytophthora-plant interactions. Front Plant Sci 11:593905

Nguvo KJ, Gao X (2019) Weapons hidden underneath: biocontrol agents and their potentials to activate plant-induced systemic resistance in controlling crop Fusarium diseases. J Plant Dis Prot. 126(3):177–190

Nie W, He Q, Guo H, Zhang W, Ma L, Li J et al (2024) Arbuscular mycorrhizal fungi: boosting crop resilience to environmental stresses. Microorganisms. 12(12):2448

Norman JR, Hooker JE (2000) Sporulation of *Phytophthora fragariae* shows greater stimulation by exudates of non-mycorrhizal than by mycorrhizal strawberry roots. Mycol Res 104:1069–1073

Ozgonen H, Erkilic A (2007) Growth enhancement and Phytophthora blight (*Phytophthora capsica* L.) control by arbuscular mycorrhizal fungal inoculation in pepper. Crop Prot 26(12):1682–1688

Pandey R (2005) Field application of bio-organics in the management of Meloidogyne incognita in Mentha arvensis. Nematol Mediterr 33(1):51–54
Pandit MA, Kumar J, Gulati S, Bhandari N, Mehta P, Katyal R et al (2022) Major biological control strategies for plant pathogens. Pathogens 11(2):273
Pawlowski ML, Hartman GL (2020) Impact of arbuscular mycorrhizal species on Heterodera glycines. Plant Dis 104(9):2406–2410
Pham TT, Giang BL, Nguyen NH, Yen P, Hoang V, Ha B, Le N (2020) Combination of mycorrhizal symbiosis and root grafting effectively controls nematode in replanted coffee soil. Plan Theory 9(5):555
Poltronieri P, Hong Y (eds) (2019) Applied plant biotechnology for improving resistance to biotic stress. Academic Press
Pozo MJ, Cordier C, Dumas-Gaudot E, Gianinazzi S, Barea JM, Azcon-Aguilar C (2002) Localized versus systemic effects of arbuscular mycorrhizal fungi on defense responses to Phytophthora infection in tomato plants. J Exp Bot 53(368):525–534
Pozo MJ, Jung SC, Lopez-Ráez JA, Azcón-Aguilar C (2010) Impact of arbuscular mycorrhizal symbiosis on plant response to biotic stress: the role of plant defense mechanisms. In: Koltai H, Kapulnik Y (eds) Arbuscular mycorrhizas: physiology and function. Springer, pp 193–207
Pršić J, Ongena M (2020) Elicitors of plant immunity triggered by beneficial bacteria. Front Plant Sci 11:594530
Pu C, Ge Y, Yang G, Zheng H, Guan W, Chao Z et al (2022) Arbuscular mycorrhizal fungi enhance disease resistance of salvia miltiorrhiza to *Fusarium* wilt. Front Plant Sci 13:975558
Qu SL, Wu YF, Liu ZK, Wang GL, Chen YJ, Rong YP (2022) Research progress for effects of arbuscular mycorrhizal fungi on growth and development of alfalfa. Acta Agrestia Sin 30:2529–2534
Ragasová L, Hakalová E, Ferby V, Čechová J, Klapcová G, Pokluda R (2022) Effect of plant growth promoting microorganisms on pepper plants infected with tomato brown rugose fruit virus. Diversity 14(8):635
Rashad YM, Bouqellah NA, Hafez M, Abdalla SA, Sleem MM, Madbouly AK (2024) Combined interaction between the diazotrophic *Niallia circulans* strain YRNF1 and arbuscular mycorrhizal fungi in promoting growth of eggplant and mitigating root rot stress caused by *Rhizoctonia solani*. Phytopathol Mediterr 63:25–43
Razak NA, Gange AC (2021) Multitrophic interactions between arbuscular mycorrhizal fungi, foliar endophytic fungi and aphids. Microb Ecol 81:1–11
Reimer-Michalski E, Conrath U (2016) Innate immune memory in plants. Semin Immunol 28(4):319–327
Reyes-Tena A (2017) Effect of mycorrhizae and actinomycetes on growth and bio-protection of *Capsicum annuum* L. against *Phytophthora capsici*. Pak J Agric Sci 54(3):513–522
Robert-Seilaniantz A, Grant M, Jones JD (2011) Hormone crosstalk in plant disease and defense: more than just jasmonate-salicylate antagonism. Annu Rev Phytopathol 49:317–343
Saijo Y, Loo EP, Yasuda S (2018) Pattern recognition receptors and signaling in plant-microbe interactions. Plant J 93(4):592–613
Sankaranarayanan C, Sundarababu R (2010) Influence of application methods of arbuscular mycorrhiza glomus mosseae in the biomanagement of root knot nematode, *Meloidogyne incognita*, on black gram (*Vigna mungo* L.). J Biol Control 24(1):51–57
Sanmartín N, Sánchez-Bel P, Pastor V, Pastor-Fernández J, Mateu D, Pozo MJ et al (2020) Root-to-shoot signalling in mycorrhizal tomato plants upon *Botrytis cinerea* infection. Plant Sci 298:110595
Savary S, Willocquet L, Pethybridge SJ, Esker P, McRoberts N, Nelson A (2019) The global burden of pathogens and pests on major food crops. Nat Ecol Evol 3(3):430–439
Schouteden N, De Waele D, Panis B, Vos CM (2015) Arbuscular mycorrhizal fungi for the biocontrol of plant-parasitic nematodes: a review of the mechanisms involved. Front Microbiol 6:1280
Sharma IP, Sharma AK (2017) Co-inoculation of tomato with an arbuscular mycorrhizal fungus improves plant immunity and reduces root-knot nematode infection. Rhizosphere 4:25–28

Sharma M, Saini I, Kaushik P, Aldawsari MM, Balawi TA, Alam P (2021) Mycorrhizal fungi and Pseudomonas fluorescens application reduces root-knot nematode (Meloidogyne javanica) infestation in eggplant. Saudi J Biol Sci 28:3685–3691. https://doi.org/10.1016/j.sjbs.2021.05.054

Shaul O, Galili S, Volpin H, Ginzberg I, Elad Y, Chet I, Kapulnik Y (1999) Mycorrhiza-induced changes in disease severity and PR protein expression in tobacco leaves. Mol Plant-Microbe Interact 12(11):1000–1007

Shittu HO, Aisagbonhi E, Onyemaechi HO (2019) Plants' innate defence mechanisms against phytopathogens. J Microbiol Biotechnol Food Sci 9(2):314

Siddiqui ZA, Akhtar MS (2006) Biological control of root-rot disease complex of chickpea by AM fungi. Arch Phytopathol Plant Protect 39(5):389–395

Singh I, Giri B (2017) Arbuscular mycorrhiza mediated control of plant pathogens. In: Mycorrhiza-nutrient uptake, biocontrol, eco-restoration, pp 131–160

Singh J, Kitchens C, Ray C, Gerard C, Tharayil N (2019) Arbuscular mycorrhizal fungi affect total phenolic content and antimicrobial activity of *Tamarix gallica* in natural semi-arid Algerian areas. S Afr J Bot 125:39–45

Sipahioglu MH, Demir S, Usta M, Akkopru A (2009) Biological relationship of potato virus Y and arbuscular mycorrhizal fungus *Glomus intraradices* in potato. Pest Technol 3(1):63–66

Song Y, Chen D, Lu K, Sun Z, Zeng R (2015) Enhanced tomato disease resistance primed by arbuscular mycorrhizal fungus. Front Plant Sci 6:786

Steinkellner S, Hage-Ahmed K, García-Garrido JM, Illana A, Ocampo JA, Vierheilig H (2012) A comparison of wild-type, old, and modern tomato cultivars in the interaction with the arbuscular mycorrhizal fungus *Glomus mosseae* and the tomato pathogen *Fusarium oxysporum* f. sp. lycopersici. Mycorrhiza 22:189–194

Stolyarchuk IM, Shevchenko TP, Polischuk VP, Kripka AV (2009) Virus infection course in different plant species under the influence of arbuscular mycorrhiza. Microbial Biotechnol 6:70–75

Tabin T, Arunachalam A, Shrivastava K, Arunachalam K (2009) Effect of arbuscular mycorrhizal fungi on damping-off disease in *Aquilaria agallocha* Roxb. Seedlings. Trop Ecol 50:243–248

Thiem D, Szmidt-Jaworska A, Baum C, Muders K, Niedojadło K, Hrynkiewicz K (2014) Interactive physiological response of potato (Solanum tuberosum L.) plants to fungal colonization and Potato virus Y (PVY) infection. Acta Mycol 1:291–303. https://doi.org/10.5586/AMF.2014.015

Tian L, Chang C, Ma L, Nasir F, Zhang J, Li W, Tran LP, Tian C (2019) Comparative study of the mycorrhizal root transcriptomes of wild and cultivated rice in response to the pathogen *Magnaporthe oryzae*. Rice 12(1):35

Tiénébo EO, Harrison K, Abo K, Brou YC, Pierson LS, Tamborindeguy C, Pierson EA, Levy JG (2019) Mycorrhization mitigates disease caused by "Candidatus Liberibacter solanacearum" in tomato. Plan Theory 8(11):507

Vallejos-Torres G, Espinoza E, Marín-Díaz J et al (2021) The role of arbuscular mycorrhizal fungi against root-knot nematode infections in coffee plants. J Soil Sci Plant Nutr 21:364–373

Van der Veken L, Cabasan MTN, Elsen A, Swennen R, De Waele D (2021) Effect of single or dual inoculation of the arbuscular mycorrhizal fungus *Glomus mosseae* and root-nodulating rhizobacteria on reproduction of the burrowing nematode *Radopholus similis* on non-leguminous and leguminous banana intercrops. J Plant Dis Prot 128(4):961–971

Van Driesche RG, Carruthers RI, Center T, Hoddle MS, Hough-Goldstein J, Morin L et al (2010) Classical biological control for the protection of natural ecosystems. Biol Control 54(Suppl. S1):2–33

Vos C, Claerhout S, Mkandawire R, Panis B, De Waele D, Elsen A (2012a) Arbuscular mycorrhizal fungi reduce root-knot nematode penetration through altered root exudation of their host. Plant Soil 354:335–345

Vos C, Tesfahun A, Panis B, De Waele D, Elsen A (2012b) Arbuscular mycorrhizal fungi induce systemic resistance in tomato against the sedentary nematode *Meloidogyne incognita* and the migratory nematode *Pratylenchus penetrans*. Appl Soil Ecol 61:1–6

Vos C, Schouteden N, Tuinen D, Chatagnier O, Elsen A, De Waele D, Panis B, Gianinazzi-Pearson V (2013) Mycorrhiza-induced resistance against the root-knot nematode *Meloidogyne incognita* involves priming of defense gene responses in tomato. Soil Biol Biochem 60:45–54

Vos CM, Yang Y, De Coninck B, Cammue BPA (2014) Fungal (-like) biocontrol organisms in tomato disease control. Biol Control 74:65–81

Wahab A, Batool F, Muhammad M, Zaman W, Mikhlef RM, Qaddoori SM et al (2023a) Unveiling the complex molecular dynamics of arbuscular mycorrhizae: a comprehensive exploration and future perspectives in harnessing phosphate-solubilizing microorganisms for sustainable progress. Environ Exp Bot:105633

Wahab A, Muhammad M, Munir A, Abdi G, Zaman W, Ayaz A et al (2023b) Role of arbuscular mycorrhizal fungi in regulating growth, enhancing productivity, and potentially influencing ecosystems under abiotic and biotic stresses. Plan Theory 12(17):3102

Wahid F, Sharif M, Shah F, Ali A, Adnan M, Shah S et al (2022) Mycorrhiza and phosphate solubilizing bacteria: potential bioagents for sustainable phosphorus management in agriculture. Phyton 91(2):257

Wang H, Hao Z, Zhang X, Xie W, Chen B (2022a) Arbuscular mycorrhizal fungi induced plant resistance against fusarium wilt in jasmonate biosynthesis defective mutant and wild type of tomato. J Fungi 8(5):422

Wang J, Wang J, He JZ, Jing Z, Xu Y, Ge Y (2022b) Arbuscular mycorrhiza fungi increase soil denitrified abundance relating to vegetation community. Appl Soil Ecol 171:104325

Wang YJ, Li Y, Duan T (2023) Arbuscular mycorrhizal fungus changes alfalfa response to pathogen infection activated by pea aphid infestation. Front Microbiol 13:1074592

Weng W, Yan J, Zhou M, Yao X, Gao A, Ma C et al (2022) Roles of arbuscular mycorrhizal fungi as a biocontrol agent in the control of plant diseases. Microorganisms. 10(7):1266

Yang J, Ruegger PM, Mckenry MV, Becker JO, Borneman J (2012) Correlations between root-associated microorganisms and peach replant disease symptoms in a California soil. PLoS One 7(10)

Yao M, Tweddell R, Desilets H (2002) Effect of two vesicular-arbuscular mycorrhizal fungi on the growth of micro propagated potato plantlets and on the extent of disease caused by *Rhizoctonia solani*. Mycorrhiza 12(5):235–242

Yin C, Casa Vargas JM, Schlatter DC et al (2021) Rhizosphere community selection reveals bacteria associated with reduced root disease. Microbiome 9:86

Zhang F, Zou YN, Wu QS (2018) Quantitative estimation of water uptake by mycorrhizal extraradical hyphae in citrus under drought stress. Sci Hortic 229:132–136

Zhang L, Zhou J, George TS, Limpens E, Feng G (2022) Arbuscular mycorrhizal fungi conducting the hyphosphere bacterial orchestra. Trends Plant Sci 27(4):402–411

Zhu HH, Yao Q (2004) Localized and systemic increase of phenols in tomato roots induced by *Glomus versiforme* inhibits *Ralstonia solanacearum*. J Phytopathol 152(10):537–542

Zoubi B, Mokrini F, Houssayni S, Benkebboura A, Akachoud O, Ghoulam C et al (2024) Effectiveness of the arbuscular mycorrhizal fungi *Funneliformi mossae* and *Rhizophagus irregularis* as biological control agent of the citrus nematode *Tylenchulus semipenetrans*. J Nat Pesticide Res:100104

Chapter 13
Yeast: An Innovative Biological Solution for Plant Health and Productivity

Asmaa M. Mogazy

Abstract Under the concerns about chemical fungicide application on plants affecting the environment, climatic changes, human health, and food safety, there is a growing need for applying biological alternatives that grant sustainable agriculture method implementation, plant pathogen inhibition, and plant production increment for food security. Yeasts as microorganisms could be isolated from various environments; many studies revealed that *Candida, Saccharomyces, Pichia,* and *Rhodotorulastrains* strains belonging to *Saccharomyces* genus could be used as biocontrol agents (BCAs) showing antimicrobial activities against many plant pathogens such as *Alternaria, Botrytis, Colletotrichum, Fusarium, Monilia, Penicillium,* and *Rhizopus.* Antimicrobial activity mechanisms of yeast-based agents could be explained through different strategies. Chemical antimicrobial agents such as essential oils and chlorine-based compounds could be capsulated utilizing dead yeast cells, and forcing chemicals to get into cells using a vacuum is considered an effective technique for pathogen-controlling and protecting antimicrobial chemicals from hydrolysis and degradation serving chemical concentrations and cost reduction. In another strategy, yeast cells trigger plants to develop pathogen resistance reactions depending on yeast cell wall components that mimic pathogen cell wall components. Also, competition between yeast-based agents and plant pathogens on nutrients and space affected pathogen growth and mycelium formation, negatively resulting in decreasing pathogenicity. In other cases, yeast cells as microorganisms produce enzymes, antioxidants, and toxins that affect pathogen growth directly. Yeast showed a beneficial interaction with plants addressing climate-smart agriculture in which plant productivity was enhanced with reducing chemical application. Implementation of such an approach depends on plant growth promoters' selection. Recent research focused on yeast application efficiency in enhancing plant growth and productivity under biotic and abiotic stressors accompanied by minimizing

A. M. Mogazy (✉)
Biological and Geological Sciences Department, Faculty of Education, Ain Shams University, Cairo, Egypt
e-mail: asmaa_mahmoud@edu.asu.edu.eg

K. A. Abd-Elsalam, H. I. Mohamed (eds.), *Elicitors for Sustainable Crop Production*, Sustainability Sciences in Asia and Africa,
https://doi.org/10.1007/978-981-96-6569-3_13

chemical applications to achieve sustainability. This chapter discusses yeast (Saccharomyces and non-Saccharomyces) application as a BCA against plant pathologies causing yield losses and both fruit and vegetable spoilage and the potential of yeast application as plant growth promoters.

Keywords Biocontrol · Biological agent · Yeast-based biocontrol · Pathogenicity · Plant productivity · Plant growth promoters' yeasts · Sustainable agriculture

13.1 Introduction

Sustainable agricultural practices are highly required in the face of global food demand increment and environmental challenges. The intersection of sustainable agriculture and effective plant protection strategies offers a promising avenue to mitigate the adverse effects of traditional farming and pest control depending on pesticides while maintaining crop productivity. The Food and Agriculture Organization of the United Nations 2019 report showed 14% losses in agricultural products before consumption owing to spoilage. Ensuring microbial food safety and controlling food contamination with pathogens is necessary for food security (Muccilli and Restuccia 2015).

Physical techniques including using radiation, thermal treatment, and mechanical methods could affect plant taste, color, and texture (Gonzalez-Gonzalez et al. 2021). The wide application of chemicals such as chitosan, titanium oxide nanoparticles, and sodium hypochlorite as an antimicrobial agent was effective in increasing the shelf life of fruits and vegetables and ensuring microbial food safety (de Matos Fonseca et al. 2021). However, this application resulted in human health risks as well as soil contamination (Dehghan et al. 2018). Consumer awareness increased about food safety and related activities affecting their health; therefore, using alternative methods for plant pathogen control and plant productivity enhancement is urgently needed (Vieira et al. 2022). The relationship between plants and microorganisms has been reported by many previous studies, particularly in plant protection against pathogens and crop productivity enhancement.

Many previous studies have discussed the effectiveness of applying BCAs in the agri-food sector from 1963 to 2021; using biological agents (such as *Bacillus amyloliquefaciens, Pseudomonas synxantha, and Pantoea agglomerans*) illuminated the possible role of such eco-friendly approaches in signaling plant defenses against pathogens and simultaneously enhancing yield potential (Baghban et al. 2019). According to Sui et al. (2020), biological agents also called biocontrol agents (BCAs), such as bacteria (*Bacillus subtilis*), yeasts (*Candida oleophila*), and yeast-like fungi (*Aureobasidium pullulans*), could control the harvested plant diseases. Biological agents should be effective at low concentrations, and for a wide range of pathogens, they survive under environmental stressors, have low nutritional requirements for their growth, are genetically stable, are non-pathogenic to the host, and possess high adaptation ability (Sharma et al. 2009). Among microorganisms, yeasts gained more attention due to their uses in many industries such as wine and bread production, while the potential of application yeasts as antagonistic agents for plant disease biocontrol gained more attention in scientific publications (Zhang et al. 2020a). Yeasts are unicellular fungi that reproduce by forming spores and mostly by budding. Many genera including *Cryptococcus, Candida, Rhodotorula,* and *Brettanomycetes* are known and classified as yeasts. Also, yeasts could be isolated from different habitats such as soil, water, plant tissues, and animal surfaces like insects (Table 13.1). They established relationships with both plants and insects like *Drosophila* (Fenner et al. 2022).

Table 13.1 Yeast habitat and isolation sources

Habitat	Yeast species	References
Animals	*Candida parapsilosis* *Candida albicans* *Meyerozyma guilliermondii* *Rhodotorula mucilaginosa* *Trichosporon coremiiforme* *Prototheca zopfii*	Calderón-Hernández et al. (2024)
Atmosphere	*Cryptococcus* spp. *Moniliella dehoogii* *Moniliella aeria* *Trichosporon asahii* *Candida krusei* *Candida guilliermondii* *Candida albicans* *Rhodotorula* spp.	Lei et al. (2022) and Sanna et al. (2021)
Fresh water	*Candida* spp. *Clavispora* spp. *Cyberlindnera* spp. *Cryptococcus* spp. *Debaryomyces* spp. *Hanseniaspora* spp. *Kluyveromyces* spp. *Metschnikowia* spp. *Meyerozyma* spp. *Pichia* spp. *Rhodotorula* spp. *Saccharomyces* spp.	Monapathi et al. (2020)
Marine water	*Pichia capsulata* *Debaryomyces hansenii* *Geotrichum* spp.	Zaky et al. (2014)
Plants	*Candida blattae* *Pichia kudriavzevii* *Candida glabrata* *Saccharomyces cerevisiae* *Cryptococcus flavus* *Rhodosporidium paludigenum* *Torulaspora globosa*	Koricha et al. (2019) and Grondin et al. (2015)
Soil	*Geotrichum* spp. *Saccharomyces* spp. *Candida* spp.	Oda and Mohammed (2024)

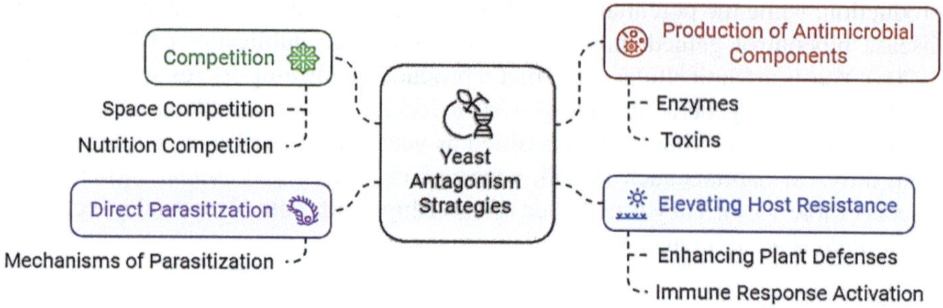

Fig. 13.1 The suggested antagonism strategies of yeasts controlling plant pathogens

Climate change could increase the threat of plant infection with genera such as *Botrytis, Penicillium*, and *Aspergillus* in which huge economic uses could be caused (Di Canito et al. 2021). Biocontrol products based on yeast types such as *Candida oleophila, Aureobasidium pullulans, Metschnikowia fructicola, Cryptococcus albidus*, and *Saccharomyces cerevisiae* have shown their efficiency in controlling plant disease and developed commercially such as Blossom protect that targets *Colletotrichum gloeosporioides* pathogen, Botector that targets *Botrytis cinerea*, and BoniProtect that targets *Penicillium expansum* and *Botrytis cinerea* (Hernandez-Montiel et al. 2021; Kowalska and Krzymariska 2022). The antagonism strategies of yeasts controlling plant pathogens could be divided into four main mechanisms (Fig. 13.1): competition between yeast and pathogens on space and nutrition, production of antimicrobial components such as enzymes and toxins, direct parasitization, and elevating host resistance against the pathogen (Ma et al. 2023) depending on the antagonistic mechanism of yeast a suitable yeast-based BCA that would be used for specific plant infection as shown in Table 13.2.

Besides antagonistic mechanisms, yeast could serve as carriers for antimicrobial agents or capsulated chemical-based antimicrobial compounds such as essential oils, chlorinated N-halamine compounds, and photosensitizers (Tan et al. 2022; Dou et al. 2021; Nakhaee Moghadam et al. 2021).

A combination of antagonistic yeast and physical or chemical treatments showed more efficiency in controlling plant microbial infections. Physical and chemical treatments alone showed high efficiency in controlling plant post and pre-harvested infection. However, it affected food flavor, texture, and shape and did not provide continuous protection for reinfection. The application of antagonistic yeast after physical and/or chemical treatment could provide prolonged plant protection and increased fruit and vegetable shelf life (Zhang et al. 2009). Peach fruits treated with hot air at 45 °C for 4 hours and then inoculated with *Pseudozyma guilliermondii* showed significant inhibitory action to *Penicillium expansum* and recorded a 15% infection rate reduction (Zhao et al. 2019). A combination of ultraviolet irradiation and antagonistic *Candida tropicalis* on pineapple fruits showed effective black rot disease limitation (Ou et al. 2016). Applying chemicals such as antimicrobial agents, metal salts, and plant hormones combined with yeast as a BCA could enhance antimicrobial efficiency (Ons et al. 2020). While antimicrobial chemicals and metal salts must be applied in specific concentrations not to affect antagonistic

Table 13.2 Antagonistic yeasts and target microbes

Yeast species	Yeast strain	Target microbe	References
Candida spp.	*Candida famata*	*Penicillium digitatum*	Arras (1996)
	Candida sake	*Penicillium expansum, Botrytis cinerea, Rhizopus nigricans*	Vinas et al. (1998)
	Candida oleophila	*Penicillium expansum, Penicillium digitatum, Botrytis cinerea*	Wilson et al. (1998)
Cryptococcus spp.	*Cryptococcus laurentii*	*Penicillium expansum*	Lima et al. (1998)
	Cryptococcus flavus	*Penicillium expansum, Helminthosporium oryzae*	Nutaratat et al. (2014)
Pichia spp.	*Pichia caribbica*	*Rhizopus nigricans*	Zhao et al. (2012)
	Pichia anomala	*Botryodiplodia theobromae*	Mohamed and Saad (2009)
	Pichia kluyveri	*Botryodiplodia digitatum*	Liu et al. (2017)
Metschnikowia spp.	*Metschnikowia pulcherrima*	*Botryodiplodia cinerea, Monilinia laxa*	Ruiz-Moyano et al. (2016)
	Metschnikowia pulcherrima	*Botryodiplodia cinerea*	Pawlikowska et al. (2019)
	Mestchnikowia Fructicola	*Penicillium digitatum, Botrytis cinerea, Aspergillus niger*	Droby (2006)
Rhodosporidium spp.	*Rhodosporidium paludigenum*	*Alternaria alternata*	Wang et al. (2008)
	Rhodosporidium paludigenum	*Penicillium expansum*	Zhang et al. (2009)
	Rhodotorula glutinis	Molds and Mucor	Shanmuganathan (1996)
	Rhodotorula mucilaginosa		

yeast, plant hormones could modulate the host plant response to pathogens and increase the yeast flora population (He et al. 2020). Cao et al. (2009) reported that methyl jasmonate increased yeast populations and antimicrobial activities. This chapter aims to lighten the effect of the application of yeast as a plant productivity enhancer as well as BCAs and study the relative mechanism associated with yeast antagonistic activity to microorganisms.

13.2 Role of Yeast in Plant Biotic Stress Resistance and Relevant Mechanisms

Pathogens such as bacteria, fungi, and viruses attack many plants causing disease development that affects plant productivity resulting in food shortage and insecurity. Plants are designed to resist a wide spectrum of pathogens in a phenomenon known as systemic acquired resistance (SAR). This hypersensitive response depends on chemicals such as salicylic acid (SA) that are produced by plants accompanied by primary infection (Addepalli et al. 2006). Fruits, vegetables, and crops safety could be achieved through controlling pathogen infection depending on yeast-based biological treatments. Yeast application not only showing a protective effect on plants from pathogens but also do not produce toxins, secondary metabolites with allergic effects, and mycotoxins in their antipathogenic activity (Liu et al. 2013).

Plant biotic stress also includes insect and weed attacks, depending on pesticides and weed-fighting chemicals leading to toxin accumulation in the soil as well as developing resistance activity in both pests and weed plants. Many reports showed that yeast possesses the ability to produce enzymes that mediate insecticide biodegradation, *Rhodotorula rubra* showed chlorpyrifos biodegradation that is mediated by esterase also. Neonicotinoid insecticides acetamiprid could be degraded by the activity of *R. mucilaginosa* (Bempelou et al. 2018). Genetically modified yeast can affect *Drosophila suzukii* larva survival and can act as biopesticides. *Cryptococcus tephrensis* and *Cydia pomonella* granulovirus affected apple moth larva development and protected crops from larva infestation (Knight and Witzgall 2013). *Solicoccozyma terricola, Clavispora lusitaniae, and Pichia kudriavzevii* soil application showed helpful approaches in glyphosate, pendimethalin, and atrazine herbicide degradation (Han et al. 2019). Housaku Monogatari (HM) is one of the plant activators prepared from yeast showing elevation of *Arabidopsis thaliana* and *Brassica rapa* resistance to bacterial (*Pseudomonas syringae*) and fungal (*Colletotrichum higginsianum*) diseases by the activation of plant defense genes (Narusaka et al. 2015). However, studies are needed to clarify and develop the techniques that are associated with applying biological control agents.

13.2.1 Competition Between Yeast and Pathogens in Space and Nutrition

One of the major antagonistic mechanisms of yeast against plant pathogens is competition on space and nutrients such as carbohydrates, iron, nitrogen, and other elements that are needed for the growth of pathogens (Spadaro and Droby 2016). Iron is an important metal that is required for activating many enzymes related to pathogen growth. A correlation between fungi growth suppression and low iron concentrations was proved by many studies. In addition, iron is an important mineral for pathogen conidia germination and mycelium formation. Yeast cells compete with other fungi on iron sources, which results in iron deficiency affecting the growth and development of pathogens (Sipiczki 2020). Yeast cells produce siderophores that contribute to available iron ion content reduction by forming a stable ferric ion complex (Fe^{3+}). *Metschnikowia Pulcherrima* produces pulcherriminic acid that bonds with ferric ions forming pulcherrimin (red pigment) that inhibits the growth of *B. cinerea* and *P. expansum* on wounded apples (Saravanakumar et al. 2008). Nitrogen is an important mineral in the growth of pathogens as it contributes to protein synthesis. Competition on nitrogenous sources is another antagonistic method that affects pathogen domination in plants. The growth inhibition rate of *Parvularugoglobigerina extensa* on apple plants increased with the application of yeast *Aureobasidium pullulans,* while exogenous application of nitrogen the inhibition rate decreased by almost 30%, which revealed a correlation between nitrogen availability and yeast antagonistic activity (Bencheqroun et al. 2007). That leads us to the fact that enough nutrient supplements are needed for yeast growth and to limit the growth of pathogens at the same time (Spadaro and Droby 2016). Also, competition of yeast and pathogens on wound sites showed a significant effect on restricting infection and limiting colonization site accessibility. However, spatial competition showed less effect than nutrient competition (Kheireddine et al. 2021).

13.2.2 Direct Parasitism and Antimicrobial Component Production

Biocontrol activities of biological agents like yeast-based agents could be figured depending on yeast preying on pathogens affecting hyphae development and cell wall degradation by secreting cell wall degrading enzymes (Lachance et al. 2000). Consuming fungus with other organisms is a known mechanism known as mycoparasitism. Many yeast species such as *Pichia guilliermondii, Pseudozyma aphidis, Saccharomycopsis schoenii,* and *Vishniacozyma tephrensis* are recorded to consume other microbes depending on lytic enzymes (Kowalska amd Krzymariska 2022).

Zhang et al. (2020b) revealed that *Saccharomycopsis capsularis* infected pathogenic fungi in a culture medium resulting in a decrease in its pathogenicity. Other studies on *Aspergillus carbonarius, Penicillium expansum,* and *Botrytis cinerea* pathogens that are responsible for grapes and other fruits infection and human harmful mycotoxin release revealed the ability of the yeasts such as *Trichosporon mycotoxinivorans* and *Yarrowia lipolytica* to limit such pathogen plant infection (Di Canito et al. 2021). Antagonistic activity of *Candida famata* caused the production of *Colletotrichum gloeosporioides* sterile mycelium on papaya plants. *C. famata* produced β-1,3-glucanase that affected pathogen mycelium development (Magallon-Andalon et al. 2012). Previous studies claimed that yeast cells counteract the infection through secreting lytic enzymes (polyphenol oxidase, catalase, and chitinase) and/or biofilm formation (Farbo et al. 2018).

Yeast-based biological agents such as *S. cerevisiae* produce killer toxins 1, 2, and 28 (K1, K2, and K28) which are produced by yeasts, demonstrate toxic activity against various plant pathogens through damaging cell wall, tRNA, and DNA, and prevent Ca^{2+} uptake. Toxins are first produced from yeast cells, bind to receptors present on the pathogen cell wall (chitin, β-1,3, and β-1,6-D-glucan), and then translocated into pathogen cellular plasma membrane, causing pathogenicity inactivation (Crabtree et al. 2019). Antagonistic yeast produces lytic enzymes related to inhibitory effects on pathogens including chitinase, protease, and glucanases. Protease enzymes derive fungi proteins' hydrolysis reaction, decreasing pathogen growth, development, and disease induction ability. Both chitinase and glucanases share the ability to hydrolyze microbial cell wall components (Oztekin and Karbancioglu-Guler 2021). *Wickerhamomyces anomalus* antagonizes the effect of *B. cinerea* pathogen on mango through the production of an exo-β-1,3-glucanase enzyme (Parafati et al. 2017). Volatile organic compounds (VOCs) are the type of products with low molecular weight and produced by microorganisms such as bacteria and fungi. A blend of the volatile compounds known as volatilome yeast shows antimicrobial effects. *Candida nivariensis* produced 1-pentanol that inhibits the growth of *Aspergillus flavus* as well as mycelium and conidia development, resulting in restrictions on maize kernel disease (Jaibangyang et al. 2020). *Hanseniaspora* and *Aureobasidium pullulans* showed inhibition of *Botrytis cinerea* growth by producing VOCs (Ruiz-Moyano et al. 2020; Di Francesco et al. 2020).

13.2.3 Encapsulation

Essential oils, chloride chemicals, and photosensitizers could be used for plant pathogen control. Antimicrobial chemical agents could be encapsulated using inactive yeast-based carriers. This technique used dead yeast cells by immersing them in an antimicrobial solution and applying vacuums that force cell organelles to come out of the cell through generated cell wall pores where the chemical agent solution will diffuse into (Fig. 13.2) (Ren et al. 2021). Studies revealed an enhancement in the activity of encapsulated antimicrobial solutions compared with unencapsulated ones (Young et al. 2017; Ma et al. 2023).

The concentrations, size, and cost of chemicals needed to control pathogens were lowered using the encapsulation technique compared with traditional methods (Dou et al. 2021). Yeast-based carriers encapsulate both hydrophobic and hydrophilic antimicrobial chemical agents, such as essential oils, and chlorine-based antimicrobial agents that have been encapsulated and applied successfully. Encapsulation could protect chemical agents from hydrolysis in a water-rich medium, degradation,

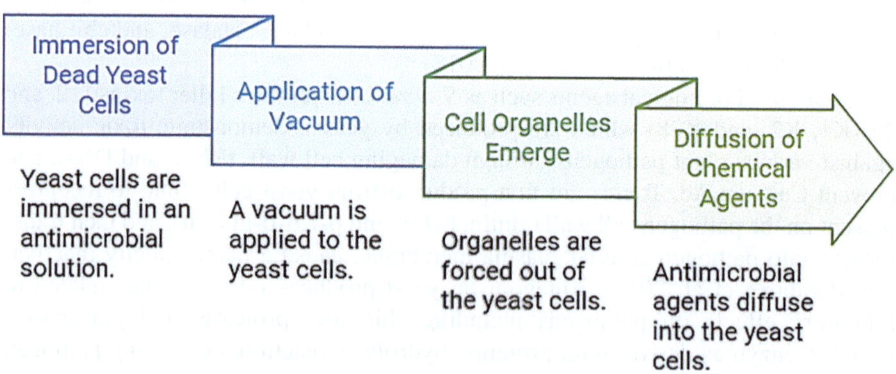

Fig. 13.2 Steps of antimicrobial agent yeast encapsulation process

and oxidation (Liu et al. 2021; Huang et al. 2021). Encapsulated Manuka essential oil (MEO) exhibited antimicrobial efficiency and thermal stability compared to unencapsulated MEO when applied against *B. cereus* (Liu et al. 2021).

13.2.4 Elevating Host Resistance Against the Pathogen

Many plants develop immune responses to pathogens when recognition receptors localized on plant cell walls recognize microbial infection. A pattern of defense responses is triggered to reduce disease severity and attack pathogens (Jones and Dangi 2006). A series of physiological reactions take place within host tissue called hypersensitivity reactions to prevent further infection development (Romanazzi et al. 2016). Plant pathogen resistance depends on the accumulation of resistance markers such as jasmonic acid, azelaic acid, SA, phenylalanine, antioxidant compounds, and stress enzymes (Yaguchi et al. 2017). Plant response to microbes could be estimated by non-pathogenic microbes as cell walls contain polysaccharides and proteins that are responsible for such physiological reactions (Addepalli et al. 2006). The yeast cell wall contains approximately 40% proteins, 60% β-glucan, and chitin that induce plant resistance to bacteria and fungi infection (Klis et al. 2006). Also, other cell wall components like mannan display similar polysaccharides of pathogens eliciting plant defense responses. Plant defense genes are activated parallel to SAR reactions. The expression of these genes is activated by non-pathogenic microbes. Addepalli et al. (2006) reported that plant disease resistance is enhanced by the induction of yeast poly(A)-binding protein gene (PAB1) in plants such as tobacco and Arabidopsis genome.

Yeast such as BCAs could trigger the host to produce substances like pathogen response (PR) proteins that activate host resistance to pathogens. Studies on cherry tomatoes revealed resistance development against *B. cinerea* and *A. alternate* after *C. laurentii* yeast treatment through increasing activities of enzyme antioxidants (Lai et al. 2018). *Arabidopsis* plants that were subjected to *S. cerevisiae* yeasts treatment showed high phytoalexin camalexin content that increased plant resistance to *B. cineria* (Raacke et al. 2006). HM that developed from yeast cell wall extracts induced reduction of plant infections showing resistance of both *Arabidopsis thaliana* and *Brassica rapa* plants to *Pseudomonas syringae* and *Colletotrichum higginsianum*. The resistance was achieved through increasing plant chitinase and glucanase enzyme gene expression (Narusaka et al. 2015; Yaguchi et al. 2017).

13.3 Role of Yeast Application as an Activator of Plant Growth and Productivity

Sustainable and cost-effective farming is a significant challenge in achieving food security. In intensive agriculture, chemicals are widely used to enhance crop growth to meet the population's food demands. However, the accumulation of these

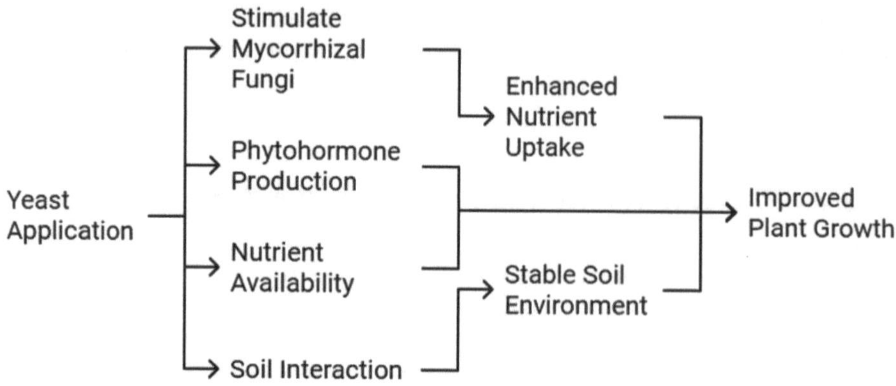

Fig. 13.3 Different mechanisms in which yeast interacts with crops as PGPYs

chemicals in plant tissues poses health risks for humans who consume them (Kowalska and Krzymaariska 2022). Crop productivity stimulation depending on microbes' plant interaction, especially plant growth promoting yeast (PGPs) application, has gained attention recently as a safe alternative to chemical fertilizers to optimize chemical release in soil and enforce sustainable agriculture (Mannino et al. 2020; Poveda et al. 2021). Many studies discussed the positive effect of applying yeast on plants as plant growth promoters and suggested strategies in which yeast enhances plant growth through phytohormone production, controlling plant pathogens, and increasing phosphate solubility for plant needs (Fernández-San Millán et al. 2020). Some of these strategies are shown in Fig. 13.3. Mukherjee et al. (2020) referred to the positive effect of applying yeast as plant growth promoters to phytohormone production such as indole-3-acetic acid (IAA). IAA enhanced plant cell elongation and differentiation that affected plant root growth and plastid development. *Lindera Saturnus* isolated from Mays roots produced IAA and indole-3-pyruvic acid that contribute to enhancing plant growth. The same results were recorded in endophytic yeast (*Rhodotorula*) that were isolated from *Populus* plants (Nassar et al. 2005; Xin et al. 2009). Not only endophytic cells but also epiphytic yeast showed high production of plant growth hormone (IAA) when cultivated in vitro. Yeast produces abscisic acid that is introduced to agriculture to enhance plant seed characteristics. Yeast genera (*Cryptococcus flavus*, *Hannaella coprosmaensis*, *Pseudozyma aphidis*, *Sporisorium reilianum*, and *Ustilago esculenta*) were isolated from *Drosera indica* leaves and showed ameliorative effect on plant growth (Nimsi et al. 2023). Previous studies reported Zeatin production by *Rodotorula mucilaginosa* that controls cellular division (Kudoyarova et al. 2014). Many studies reported a promoting effect of yeast on plant cell division, elongation, increasing leaf area per plant, leaf thickness, and increasing pigment content and carbohydrate content of sugarcane and lettuce plants (Agamy et al. 2013; Ibrahim et al. 2020). The same results were reported by El-Tohamy and Al Raoush (2015) that yeast application on potato plants increased plant growth parameters and productivity. Vargas et al.'s (2024) studies on lettuce and tomato plants subjected to yeast treatment proved an increment in seed germination rate and seedling growth.

In addition, managing PGPY application on plants concerning their interaction with rhizosphere microbes and other soil microorganisms should be focused on by researchers to optimize PGPY uses (Natsheh et al. (2022); Ruspi et al. 2024). The presence of yeasts produces polyamines and stimulates arbuscular mycorrhizal fungi, which enhance nutrient and water uptake and improve plant growth by modifying root structure through developing xylem vessels and cell division stimulation (Khan et al. 2021).

Treating plants with yeast could help in maintaining plant nutrition depending on rocks mineralizing and organic wastes transformation into usable components helping in saving environments and enhancing soil health and water holding capacity (Rayman et al. 2020). Plants require enough mineral nutrients for growth and increased productivity. Yeast produces 1-aminocyclopropane-1-carboxylase (ACC) enzyme that produces a-ketobutyrate and ammonia with decreasing ethylene content in plant tissues, causing large amounts of ammonia releasement helping in nitrogen availability (Kowalska and Krzymariska 2022). Other yeast species such as *Candida*, *Geotrichum*, *Rhodotorula*, *Saccharomyces*, and *Williopsis* can reduce nitrate into nitrite during anaerobic respiration that could be used by plants (Mothapo et al. 2015). Yeast species increase nitrogen availability through enzyme production that plays a role in nitrogen fixation and transferring into a form that could be utilized by plants (Inselsbacher et al. 2007).

Yeast absorbs large amounts of nitrogen and carbon that stabilize the soil carbon–nitrogen ratio, adjusting soil acidity and enhancing plant–soil microorganisms' interaction (Rawichandran et al. 2024). Organic acid formation could be accelerated by the action of PGPYs by solubilizing inorganic soil phosphate that increases the availability of phosphorus for plant cells. *Ascomycetous* genera are able to oxidize sulfur to produce sulfate and tetrathionate (Kuo et al. 2018).

Iron-chelating components like siderophores are produced by microorganisms forming iron ion traps by forming complex compounds with iron ions. Siderophores are taken up with plants supporting plant growth indirectly by providing disease resistance and ensuring iron ion availability to plant cells (Ghosh et al. 2020).

Plants utilize phosphorus in phosphate form that could be present in soil in an insoluble form. It was proved that 79 yeast isolates could solubilize phosphorus, making it usable by plants (Marques et al. 2021). Similarly, microorganisms showed abilities of zinc solubilization to provide plants with their needs from zinc (Nutaratat et al. 2014).

The production of volatile molecules that enhance plant growth performance such as acids, ketones, and aldehydes was recorded by yeast species such as *Geotrichum candidum*. The microorganisms utilize amino acids, carbohydrates, and lipids metabolism to produce such volatile components (Grondin et al. 2015). According to Raklami et al. (2024), alfalfa plants that were subjected to yeast treatment showed growth enhancement in terms of shoot and root lengths, dry and fresh weights, and leaf number. Potassium and nitrogen minerals uptake was enhanced during the experimental period. Vargas et al. (2024) established a comparative study between tomato plants that were fertilized with chemical commercial fertilizer (Hampi) and plants that were treated with *S. cerevisiae*. The results showed a surpass in plant growth parameters that received biological fertilizers.

The advantages of using yeast as biofertilizers as well as plant biocontrol are shown as it is a renewable, soil-safe, eco-friendly, and sustainable approach. However, limitations also appeared as a deficiency of market product and activity period limitation. Among commercialized yeast products, BioGro 2 is a biofertilizer that is used in rice fields, resulting in increased plant resistance to natural pathogens and enhanced plant growth. Also, NutriSmart biofertilizer is based on yeast that adjusts soil mineral balance and changes it into plant-useable forms. Zimevit and Noli or Shemer are yeast and bacteria combined biofertilizers that elevate plant pathogen resistance. However, more studies are needed to optimize biological fertilizer applications including pH, concentration, microorganism cell viability, and temperature (Hernández-Fernández et al. 2021).

13.4 Conclusion

Yeasts are microorganisms that are found in various environments and grow in different environmental conditions showing a wide spectrum of antagonistic activities against many plant pathogens. Biocontrol products depending on the interaction between yeast and plants could serve as sustainable and ecofriendly approaches to plant growth and biotic stress management. Yeast colonizes plants depending on plant nutrients without producing harmful compounds for plants or otherwise plant pathogens. Many mechanisms are involved in yeast interaction with pathogens including toxins and active metabolites production, increasing plant resistance, and generating competition between yeast and pathogens on nutrients and space. Also using yeast cells for transporting and delivering specific components that control pathogen growth is considered a promising low-cost avenue in plant health and growth management. Achieving food security is a worldwide great challenge. PGPYs promote plant growth and increase plant productivity. Yeast produces phytohormones, enhances soil characteristics including phosphorus solubility, and positively affects rhizosphere microbes. Nowadays, yeast products for plant growth are found in markets. However, further studies should be established to optimize its application and reveal mechanisms behind such plant yeast-positive interaction.

References

Addepalli B, Xu R, Dattaroy T, Li B, Bass WT, Li QQ, Hunt AG (2006) Disease resistance in plants that carry a feedback-regulated yeast poly(A) binding protein gene. Plant Mol Biol 61(3):383–397. https://doi.org/10.1007/s11103-006-0019-6

Agamy R, Hashem M, Alamri S (2013) Effect of soil amendment with yeasts as bio-fertilizers on the growth and productivity of sugar beet. Afr J Agric Res 8(1):46–56

Arras G (1996) Mode of action of an isolate of *Candida famata* in biological control of *Penicillium digitatum* in orange fruits. Postharvest Biol Technol 8(3):191–198. https://doi.org/10.1016/0925-5214(95)00071-2

Baghban R, Farajnia S, Rajabibazl M, Ghasemi Y, Mafi A, Hoseinpoor R, Rahbarnia L, Aria M (2019) Yeast expression systems: overview and recent advances. Mol Biotechnol 61:365–384

Bempelou ED, Vontas JG, Liapis KS, Ziogas VN (2018) Biodegradation of chlorpyrifos and 3,5,6-trichloro-2-pyridinol by the epiphytic yeasts *Rhodotorula glutinis and Rhodotorula rubra*. Ecotoxicology 27:1368–1378

Bencheqroun SK, Bajji M, Massart S, Labhilili M, Jaafari SE, & Jijakli MH (2007) In vitro and in situ study of postharvest apple blue mold biocontrol by Aureobasidium pullulans: Evidence for the involvement of competition for nutrients. Postharvest Biol Technol, 46(2), 128–135. https://doi.org/10.1016/j.postharvbio.2007.05.005

Calderón-Hernández A, Castro-Bonilla N, Cob-Delgado M (2024) Chromogenic, biochemical and proteomic identification of yeast and yeast-like microorganisms isolated from clinical samples from animals of Costa Rica. J Fungi 10(3):218. https://doi.org/10.3390/jof10030218

Cao SF, Zheng YH, Wang KT, Jin P, & Rui HJ (2009) Methyl jasmonate reduces chilling injury and enhances antioxidant enzyme activity in postharvest loquat fruit. Food Chem, 115(4), 1458–1463

Crabtree AM, Kizer EA, Hunter SS, Van Leuven JT, New DD, Fagnan MW, & Rowley PA (2019) A rapid method for sequencing double-stranded RNAs purified from yeasts and the identification of a potent K1 killer toxin isolated from Saccharomyces cerevisiae. Viruses, 11(1), 70. https://doi.org/10.3390/v11010070

Dabassa Koricha A, Han DY, Bacha K, Bai FY (2019) Occurrence and molecular identification of wild yeasts from Jimma zone, Southwest Ethiopia. Microorganisms 7(12):633. https://doi.org/10.3390/microorganisms7120633

Di Canito A, Mateo-Vargas MA, Mazzieri M, Cantoral J, Foschino R, Cordero-Bueso G, Vigentini I (2021) The role of yeasts as biocontrol agents for pathogenic fungi on postharvest grapes: a review. Food Secur 10:1650. https://doi.org/10.3390/foods10071650

Di Francesco A, Zajc J, Gunde-Cimerman N, Aprea E, Gasperi F, Placì N, Caruso F, Baraldi E (2020) Bioactivity of volatile organic compounds by Aureobasidium species against gray mold of tomato and table grape. World J Microbiol Biotechnol 36(1):1–11

Droby S (2006) Yeast NRRL Y-30752 for inhibiting deleterious microorganisms on plants US Patent No 6,994,849

Dou F, Huang K, & Nitin N (2021) Targeted photodynamic treatment of bacterial biofilms using curcumin encapsulated in cells and cell wall particles. ACS Appl Bio Mater, 4(1), 514–522. https://doi.org/10.1021/acsabm.0c01051

El-Tohamy AE, Al Raoush AT (2015) The impact of applying total quality management principles on the overall hospital effectiveness: An empirical study on the HCAC accredited governmental hospitals in Jordan. Eur Sci J 11(10)

Farbo MG, Urgeghe PP, Fiori S, Marcello A, Oggiano S, Balmas V, Hassan ZU, Jaoua S, Migheli Q (2018) Effect of yeast volatile organic compounds on ochratoxin A-producing *Aspergillus carbonarius* and *A ochraceus*. Int J Food Microbiol 284:1–10

Fernández-San Millán A, Farran I, Larraya L, Ancin M, Arregui LM, & Veramendi J (2020) Plant growthpromoting traits of yeasts isolated from Spanish vineyards: Benefits for seedling development. Microbiol Res, 237, 126480

Ghosh SK, Bera T, & Chakrabarty AM (2020) Microbial siderophore—a boon to agricultural sciences. Biol Control, 144, 104214

Grondin E, Cheong Sing AS, Caro Y, Raherimandimby M, Randrianierenana AL, James S, Nueno-Palop C, François JM, Petit T (2015) A comparative study on the potential of epiphytic yeasts isolated from tropical fruits to produce flavoring compounds. Int J Food Microbiol 203:101–108. https://doi.org/10.1016/jijfoodmicro201503019

Han Y, Tang Z, Bao H, Wu D, Deng X, Guo G, Ye BC, Dai B (2019) Degradation of pendimethalin by the yeast YC2 and determination of its two main metabolites. RSC Advances 9(1):491–497. https://doi.org/10.1039/C8RA08483A

He F, Zhao L, Zheng X, Abdelhai MH, Boateng NS, Zhang X, & Zhang H (2020) Investigating the effect of methyl jasmonate on the biocontrol activity of Meyerozyma guilliermondii against blue mold decay of apples and the possible mechanisms involved. Physiological and Molecular Plant Pathology, 109, 101454. https://doi.org/10.1016/j.pmpp.2019.101454

Hernández-Fernández M, Cordero-Bueso G, Ruiz-Muñoz M, Cantoral JM (2021) Culturable yeasts as biofertilizers and biopesticides for a sustainable agriculture: a comprehensive review. Plants (Basel, Switzerland) 10(5):822. https://doi.org/10.3390/plants10050822

Huang K, Yang X, Ma Y, Sun G, & Nitin N (2021) Incorporation of antimicrobial bio-based carriers onto poly(vinyl alcohol-co-ethylene) surface for enhanced antimicrobial activity. ACS Appl Mater & Interfaces, 13(30), 36275–36285. https://doi.org/10.1021/acsami.1c07311

Ibrahim N, Sereno PC, Varricchio DJ, Martill DM, Dutheil DB, Unwin DM, Kaoukaya A (2020) Geology and paleontology of the upper cretaceous Kem Kem group of eastern Morocco. ZooKeys 928:1

Inselsbacher E, Cambui CA, Richter A, Stange CF, Mercier H, Wanek W (2007) Microbial activities and foliar uptake of nitrogen in the epiphytic bromeliad *Vriesea gigantean*. New Phytol 175:311–320. https://doi.org/10.1111/j1469-8137200702098x

Jaibangyang S, Nasanit R, Limtong S (2020) Biological control of aflatoxin-producing *Aspergillus flavus* by volatile organic compound-producing antagonistic yeasts. BioControl 65(3):377–386. https://doi.org/10.1007/s10526-020-09996-9

Jones JDG, Dangl JL (2006) The plant immune system. Nature 444:323–329. https://doi.org/10.1038/nature05286

Khan MIR, Palakolanu SR, Chopra P, Rajurkar AB, Gupta R, Iqbal N, & Maheshwari C (2021) Improving drought tolerance in rice: Ensuring food security through multi-dimensional approaches. Physiol Plant, 172(3), 645–668

Kheireddine A, Palmieri D, Vitullo D, Barberio A, Zouaoui M, De Curtis F, & Lima G (2021) Characterization of new yeast isolates collected from different fruits in Tunisia and biocontrol activity against Penicillium expansum on apples. J Plant Pathol, 103(4), 1169–1184. https://doi.org/10.1007/s42161-021-00921-2

Klis FM, Boorsma A, De Groot PJ (2006) Cell wall construction in *Saccharomyces cerevisiae*. Yeast 23:185–202. https://doi.org/10.1002/yea1349

Knight AL, Witzgall P (2013) Combining mutualistic yeast and pathogenic virus – a novel method for codling moth control. J Chem Ecol 39(10):1019–1026. https://doi.org/10.1007/s10886-013-0357-1

Kowalska J, Krzymańska J (2022) Yeasts as a potential biological agent in plant disease protection and yield improvement – a short review. Agriculture 12:1404. https://doi.org/10.3390/agriculture12091404

Kudoyarova GR, Melentiev AI, Martynenko EV et al (2014) Cytokinin producing bacteria stimulate amino acid deposition by wheat roots. Plant Physiol Biochem 83:285–291

Kuo CY, Fu SF, Chou FC, Chen RY, Chou JY (2018) Phosphate-solubilizing characteristics of yeasts. Mycosphere 9:1117–1131. https://doi.org/10.5943/mycosphere/9/6/4

Lachance MA, Pupovac-Velikonja A, Natarajan S, & Schlag-Edler B (2000) Nutrition and phylogeny of predacious yeasts. Can J Microbiol, 46(6), 495–505. https://doi.org/10.1139/w00-021

Lai J, Cao X, Yu T, Wang Q, Zhang Y, Zheng X, & Lu H (2018) Effect of Cryptococcus laurentii on inducing disease resistance in cherry tomato fruit with focus on the expression of defense-related genes. Food Chem, 254, 208–216. https://doi.org/10.1016/j.foodchem.2018.01.100

Lei X, Zheng J, Zhao D, Qiao Z, An M, Zhang X (2022) *Moniliella aeria* sp nov, a novel yeast isolated from the air of a Wuliangye baijiu-making workshop. Int J Syst Evol Microbiol 72(7):101099/ijsem0005464

Lima G, De Curtis F, Castoria R, De Cicco V (1998) Activity of the yeasts *Cryptococcus laurentii* and *Rhodotorula glutinis* against post-harvest rots on different fruits. Biocontrol Sci Tech 8(2):257–267. https://doi.org/10.1080/09583159830324

Liu J, Sui Y, Wisniewski M, Droby S, Liu Y (2013) Review: utilization of antagonistic yeasts to manage postharvest fungal diseases of fruit. Int J Food Microbiol 167:153–160

Liu S, Tao M, & Huang K (2021) Encapsulation of Mānuka essential oil in yeast microcarriers for enhanced thermal stability and antimicrobial activity. Food Bioprocess Technology, 14(12), 2195–2206. https://doi.org/10.1007/s11947-021-02714-y

Liu Y, Wang W, Zhou Y, Yao S, Deng L, Zeng K (2017) Isolation, identification and in vitro screening of Chongqing orangery yeasts for the biocontrol of *Penicillium digitatum* on citrus fruit. Biol Control 110:18–24. https://doi.org/10.1016/jbiocontrol201704002

Ma Y, Wu M, Qin X, Dong Q, & Li Z (2023) Antimicrobial function of yeast against pathogenic and spoilage microorganisms via either antagonism or encapsulation: A review. Food Microbiol, 112, 104242. https://doi.org/10.1016/j.fm.2023.104242

Magallon-Andalon C, Luna-Solano G, Ragazzo-Sanchez J, Calderon-Santoyo M (2012) Parasitism and substrate competitions effect of antagonistic yeasts for biocontrol of *Colletotrichum gloeosporioides* in papaya (*Carica papaya* L) var Maradol *Mexican*. J Sci Res 1:2–9

Mannino G, Campobenedetto C, Vigliante I, Contartese V, Gentile CM, Bertea C (2020) The application of a plant biostimulant based on seaweed and yeast extract improved tomato fruit development and quality. Biomol Ther 10. https://doi.org/10.3390/biom10121662

Marques AR, Resende AA, Gomes FCO, Santos ARO, Rosa CA, Duarte AA, de Lemos-Filho JP, Dos Santos VL (2021) Plant growth-promoting traits of yeasts isolated from the tank bromeliad *Vriesea minarum* LB Smith and the effectiveness of *Carlosrosaea vrieseae* for promoting bromeliad growth. Braz J Microbiol 52(3):1417–1429. https://doi.org/10.1007/s42770-021-00496-1

Mohamed H, Saad A (2009) The biocontrol of postharvest disease (*Botryodiplodia theobromae*) of guava (*Psidium guajava*) by the application of yeast strains. Postharvest Biol Technol 53(3):123–130. https://doi.org/10.1016/jpostharvbio200904001

Monapathi ME, Bezuidenhout CC, Rhode OH (2020) Aquatic yeasts: diversity, characteristics and potential health implications. J Water Health 18(2):91–105. https://doi.org/10.2166/wh2020270

Mothapo N, Chen H, Cubeta MA, Grossman JM, Fuller F, Shi W (2015) Phylogenetic, taxonomic and functional diversity of fungal denitrifiers and associated N_2O production efficacy. Soil Biol Biochem 83:160–175. https://doi.org/10.1016/jsoilbio201502022

Mukherjee A, Verma JP, Gaurav AK, Chouhan GK, Patel JS, Hesham AE-L (2020) Yeast: a potential bio-agent for plant growth and postharvest disease management for sustainable agriculture. Appl Microbiol Biotechnol 104:1497–1510. https://doi.org/10.1007/s00253-019-10321-3

Nakhaee Moghadam M, Jamshidi A, Fazly Bazzaz B S, Azizzadeh M, & Movaffagh J (2021) Saccharomyces cerevisiae as a delivery system of Zataria multiflora Boiss. Essential oil as a natural preservative for food applications. J Sci Food Agric, 101(5), 2006–2013. https://doi.org/10.1002/jsfa.10818

Narusaka M, Minami T, Iwabuchi C, Hamasaki T, Takasaki S, Kawamura K et al (2015) Yeast cell wall extract induces disease resistance against bacterial and fungal pathogens in *Arabidopsis thaliana* and brassica crops. PLoS One 10(1):e0115864. https://doi.org/10.1371/journalpone0115864

Nassar AH, El-Tarabily KA, Sivasithamparam K (2005) Promotion of plant growth by an auxin-producing isolate of the yeast *Williopsis saturnus* endophytic in maize (*Zea mays* L) roots. Biol Fertil Soils 42:97e108

Natsheh B, Masoud W, Abu-Khalaf N (2022) Effect of yeast bio-fertilizer on lettuce growth and productivity *Indian*. J Agric Res 1018805/IJAReAF-700

Nimsi KA, Manjusha K, Kathiresan K, Arya H (2023) Plant growth-promoting yeasts (PGPY), the latest entrant for use in sustainable agriculture: a review. J Appl Microbiol 134(2):lxac088. https://doi.org/10.1093/jambio/lxac088

Nutaratat P, Srisuk N, Arunrattiyakorn P, Limtong S (2014) Plant growth-promoting traits of epiphytic and endophytic yeasts isolated from rice and sugar cane leaves in Thailand. Fungal Biol 118(8):683–694. https://doi.org/10.1016/jfunbio201404010

Oda O, Mohammed A (2024) Isolation and identification of yeasts from soil in Caleb University, Lagos, Nigeria. Dutse J Pure Appl Sci 10(2a)

Ou C, Liu Y, Wang W, & Dong D (2016) Integration of UV-C with antagonistic yeast treatment for controlling postharvest disease and maintaining fruit quality of Ananas comosus. BioControl, 61(5), 591–603. https://doi.org/10.1007/s10526-016-9740-5

Ons L, Bylemans D, Thevissen K, & Cammue BP (2020) Combining biocontrol agents with chemical fungicides for integrated plant fungal disease control. Microorg, 8(12), 1930.https://doi.org/10.3390/microorganisms8121930

Oztekin S, Karbancioglu-Guler F (2021) Bioprospection of *Metschnikowia* sp isolates as biocontrol agents against postharvest fungal decays on lemons with their potential modes of action. Postharvest Biol Technol 181:111634. https://doi.org/10.1016/jpostharvbio2021111634

Parafati L, Cirvilleri G, Restuccia C, Wisniewski M (2017) Potential role of exoglucanase genes (WaEXG1 and WaEXG2) in the biocontrol activity of postharvest mango fruit. J Integr Agric 18(5):1159–1169. https://doi.org/10.1016/S2095-3119(18)62128-8

Pawlikowska E, James SA, Breierova E, Antolak H, Kregiel D (2019) Biocontrol capability of local *Metschnikowia* sp isolates. Antonie Van Leeuwenhoek 112(10):1425–1445. https://doi.org/10.1007/s10482-019-01272-w

Poveda J, Eugui D, Abril-Urías P, Velasco P (2021) Endophytic fungi as direct plant growth promoters for sustainable agricultural production. Symbiosis 85:1–19. https://doi.org/10.1007/s13199-021-00789-x

Raacke IC, von Rad U, Mueller MJ, & Berger S (2006) Yeast increases resistance in Arabidopsis against Pseudomonas syringae and Botrytis cinerea by salicylic acid-dependent as well as -independent mechanisms. Mol Plant-Microbe Interact, 19(10), 1138–1146. https://doi.org/10.1094/MPMI-19-1138

Raklami A, Babalola OO, Jemo M, Nafis A (2024) Unlocking the plant growth-promoting potential of yeast spp: exploring species from the Moroccan extremophilic environment for enhanced plant growth and sustainable farming. FEMS Microbiol Lett 371:fnae015. https://doi.org/10.1093/femsle/fnae015

Rawichandran D, Kasim S, Kee Zuan AT, Effendi MI, Raguraj S (2024) Utilization of yeasts in promoting plant growth in acidic soil – a review *AGRIVITA*. J Agric Sci 46(1):196–210. https://doi.org/10.17503/agrivitav46i14241

Ren T, Zheng X, Bai R, Yang Y, & Jian L (2021) Utilization of PLGA nanoparticles in yeast cell wall particle system for oral targeted delivery of exenatide to improve its hypoglycemic efficacy. International Journal of Pharmaceutics, 601, 120583. https://doi.org/10.1016/j.ijpharm.2021.120583

Ruiz-Moyano S, Martín A, Villalobos M, Calle A, Serradilla M, Córdoba M, Hernández A (2016) Yeasts isolated from figs (*Ficus carica* L) as biocontrol agents of postharvest fruit diseases. Food Microbiol 57:45–53. https://doi.org/10.1016/jfm201601003

Ruiz-Moyano S, Hernández A, Galván AI, Córdoba MG, Casquete R, Serradilla MJ, Martín A (2020) Selection and application of antifungal VOCs-producing yeasts as biocontrol agents of grey mould in fruits. Food Microbiol 92:103556

Ruspi C, Casagrande Pierantoni D, Conti A, Favaro L, Antinori ME, Puglisi E, Corte L, Cardinali G (2024) Beneficial effects of plant growth-promoting yeasts (PGPYs) on the early stage of growth of zucchini plants

Romanazzi G, Sanzani SM, Bi Y, Tian S, Gutiérrez Martínez P, & Alkan N (2016) Induced resistance to control postharvest decay of fruit and vegetables. Postharvest Biol Technol, 122, 82–94. https://doi.org/10.1016/j.postharvbio.2016.08.003

Rayman MP (2020) Selenium intake, status, and health: A complex relationship. Hormones: Int J Endocrinol Metablsm, 19(1), 9–14

Sanna C, Marras L, Desogus A, Marras B, Montero N, Bertolino G, Schintu M, Coroneo V (2021) Evaluation of *Rhodotorula* spp contamination in hospital environments. Environ Monit Assess 193(3):152. https://doi.org/10.1007/s10661-021-08908-3

Saravanakumar D, Ciavorella A, Spadaro D, Garibaldi A, & Gullino ML (2008) Metschnikowia pulcherrima strain MACH1 outcompetes Botrytis cinerea, Alternaria alternata and Penicillium expansum in apples through iron depletion. Postharvest Biology and Technology, 49(1), 121–128. https://doi.org/10.1016/j.postharvbio.2007.11.006

Shanmuganathan N (1996) Yeasts as a biocontrol for microbial diseases of fruit US Patent No 5,525,132
Sharma RR, Singh D, Singh R (2009) Biological control of postharvest diseases of fruits and vegetables by microbial antagonists: a review. Biol Control 50:205–221. https://doi.org/10.1016/jbiocontrol200901002
Sipiczki M (2020) *Metschnikowia pulcherrima* and related pulcherrimin-producing yeasts: fuzzy species boundaries and complex antimicrobial antagonism. Microorganisms 8:1029. https://doi.org/10.3390/microorganisms8071029
Spadaro D, Droby S (2016) Development of biocontrol products for postharvest diseases of fruit: the importance of elucidating the mechanisms of action of yeast antagonists. Trends Food Sci Technol 47:39–49. https://doi.org/10.1016/jtifs201511003
Tan Y, Chen L, Li K, Lou B, Liu Y, & Liu Z (2022) Yeast as carrier for drug delivery and vaccine construction. J Controlled Release, 346, 358–379. https://doi.org/10.1016/j.jconrel.2022
Vargas MF, Mestre MV, Vergara C, Maturano P, Petrignani D, Pesce V, Vázquez F (2024) Residual brewer's *Saccharomyces cerevisiae* yeasts as biofertilizers in horticultural seedlings: towards a sustainable industry and agriculture. Front Ind Microbiol 2:1360263. https://doi.org/10.3389/finmi20241360263
Viñas I, Usall J, Teixidó N, Sanchis V (1998) Biological control of major postharvest pathogens on apple with Candida sake. Int J Food Microbiol 40(1):9–16. https://doi.org/10.1016/S0168-1605(98)00009-9
Wang Y, Bao Y, Shen D, Feng W, Yu T, Zhang J, Zheng XD (2008) Biocontrol of *Alternaria alternata* on cherry tomato fruit by use of marine yeast *Rhodosporidium paludigenum* Fell & Tallman. Int J Food Microbiol 123(3):234–239. https://doi.org/10.1016/jijfoodmicro200802002
Wilson CL, Wisniewski ME, & Chalutz E (1998) Biological control of diseases of harvested agricultural commodities using strains of the yeast Candida oleophila. U.S. Patent No. 5,741,699
Xin G, Glawe D, Doty SL (2009) Characterization of three endophytic, indole-3-acetic acid producing yeasts occurring in Populus trees. Mycol Res 3:973e980
Yaguchi T, Kinami T, Ishida T, Yasuhara T, Takahashi K, Matsuura H (2017) Induction of plant disease resistance upon treatment with yeast cell wall extract. Biosci Biotechnol Biochem 81(11):2071–2078. https://doi.org/10.1080/0916845120171379351
Young S, Dea S, & Nitin N (2017) Vacuum facilitated infusion of bioactives into yeast microcarriers: Evaluation of a novel encapsulation approach. Food Res Int, 100, 100–112. https://doi.org/10.1016/j.foodres.2017.07.067
Zaky AS, Tucker GA, Daw ZY, Du C (2014) Marine yeast isolation and industrial application. FEMS Yeast Res 14(6):813–825. https://doi.org/10.1111/1567-136412158
Zhang H, Ma L, Turner M, Xu H, Dong Y, Jiang S (2009) Methyl jasmonate enhances biocontrol efficacy of *Rhodotorula glutinis* to postharvest blue mold decay of pears. Food Chem 117(4):621–626. https://doi.org/10.1016/jfoodchem200904054
Zhang X, Li B, Zhang Z, Chen Y, Tian S (2020a) Antagonistic yeasts: a promising alternative to chemical fungicides for controlling postharvest decay of fruit. J Fungi (Basel) 6(3):158. https://doi.org/10.3390/jof6030158
Zhang X, Wu F, Gu N, Yan X, Wang K, Dhanasekaran S, Gu X, Zhao L, Zhang H (2020b) Postharvest biological control of *Rhizopus* rot and the mechanisms involved in induced disease resistance of peaches by *Pichia membranefaciens*. Postharvest Biol Technol 163:111146. https://doi.org/10.1016/jpostharvbio2020111146
Zhao Z, Zhang H, Li J, Cui J, Zhang X, & Ren X (2012) Enhancement of biocontrol efficacy of Pichia caribbica to postharvest diseases of strawberries by addition of trehalose to the growth medium. Int J Mol Sci, 13(3), 3916–3932. https://doi.org/10.3390/ijms13033916
Zhao Y, Li Y, & Yin J (2019) Effects of hot air treatment in combination with Pichia guilliermondii on postharvest preservation of peach fruit. J Sci Food Agric, 99(2), 647–655

Chapter 14
Natural Compounds as Elicitors of Plant Resistance Against Diseases

Tony Kevork Sajyan, Rania Francis, Amin Jurdi, and Keyanna Radi

Abstract Natural elicitors play a vital role in activating plant defense mechanisms, aiding plants in resisting infections from various pathogens. These elicitors, such as oligosaccharides, proteins, and lipids, initiate immune responses by simulating the presence of a threat. This prompts plants to strengthen their cell walls, generate antimicrobial compounds, and activate signaling pathways. Plant pathogens, including bacteria, fungi, and viruses, continually seek to overcome these defenses. For example, bacterial pathogens such as *Pseudomonas syringae*, *Ralstonia* spp., and *Xanthomonas* spp. significantly impact crop production by releasing virulence factors into plant tissues, leading to widespread yield losses. Similarly, fungal pathogens and viruses also threaten crop health by manipulating plant defenses. Insect herbivores, through close associations with their host plants, utilize herbivore-associated elicitors (HAEs) present in oral secretions, frass, and egg fluids to suppress plant defenses. These HAEs are recognized by pattern-recognition receptors (PRRs) in plants, initiating specific defense responses. Traditional control methods, including chemical pesticides, copper-based bactericides, and antibiotics, are often ineffective and harmful to human health and the environment. As a result, research has focused on environmentally sustainable alternatives such as abiotic and biotic elicitors. Abiotic elicitors, chemicals of non-biological origin, and biotic elicitors, such as HAEs, activate plant defense mechanisms through early signaling events

T. K. Sajyan (✉)
Faculty of Agricultural and Food Sciences, American University of Beirut, Beirut, Lebanon

Department of Agriculture and Food Engineering, School of Engineering, Holy Spirit University of Kaslik, Jounieh, BP, Lebanon
e-mail: ts71@aub.edu.lb

R. Francis
Department of Agriculture and Food Engineering, School of Engineering, Holy Spirit University of Kaslik, Jounieh, BP, Lebanon

School of Technology, Al-Kafaàt University, Ain Saadeh, Lebanon

A. Jurdi · K. Radi
Faculty of Agricultural and Food Sciences, American University of Beirut, Beirut, Lebanon

K. A. Abd-Elsalam, H. I. Mohamed (eds.), *Elicitors for Sustainable Crop Production*, Sustainability Sciences in Asia and Africa,
https://doi.org/10.1007/978-981-96-6569-3_14

involving calcium ions (Ca^{2+}), reactive oxygen species (ROS), and mitogen-activated protein kinase pathways. These signaling cascades induce both local and systemic resistance, reprogramming plant metabolism and transcription to defend against various pathogens and herbivores. This chapter explores the use of natural compounds as elicitors to enhance plant resistance against a wide range of diseases caused by bacterial, fungal, and viral pathogens and insect herbivores. The main aim is to promote sustainable crop protection, enhance plant immunity, and provide long-term solutions to disease and pest management, ultimately contributing to global food security.

Keywords Biotic stress · Pathogens · Elicitors · Secondary metabolites · Plant defense

14.1 Introduction

The interaction between plants and pathogens set them up to be in a permanent coevolutionary cycle often referred to as "Red Queen dynamics." The Red Queen dynamics can be divided into three broad categories (Brockhurst et al. 2014). The first is known as the Fluctuating Red Queen: which signifies rare host genotypes having higher fitness than common genotypes due to the pathogen's preference for attacking common genotypes. The second is known as Escalatory Red Queen: rounds of arms race are established between plants and pathogens. The host is under selection for pathogen elimination, while the pathogen is evolving to evade the host's immune system. The third is known as Chase Red Queen: an evolutionary chase is created between host and pathogen driven by directional selection. The host undergoes selection to reduce their interactions with pathogens, while the pathogens undergo selection to increase their interactions with the host (Han 2018). Plant–pathogen interactions are not defined as a unit but vary in space and over time (Burdon and Thrall 2014). Plant pathogens, such as bacteria, fungi, viruses, and nematodes, cause significant damage by invading plant tissues, disrupting growth, and reducing yields. To defend against these threats, plants use compounds called elicitors to activate their defense mechanisms. Elicitors can originate from plants (damage-associated molecular patterns (DAMPs)) or pathogenic organisms (pathogen-associated surface receptors (PAMPs)). These compounds activate signaling pathways that stimulate the production of defensive compounds, such as secondary metabolites (alkaloids, phenols, terpenes), structural barriers, and other defenses. These compounds are important as they prime the plant's immune system for more effective responses to future attacks. Plant's immune systems rely on two interconnected categories of the innate immune system to detect and respond to pathogenic infections (Han 2018). One category utilizes cell surface pattern recognition receptors (PRRs) to detect microbe-associated molecular patterns (MAMPs)

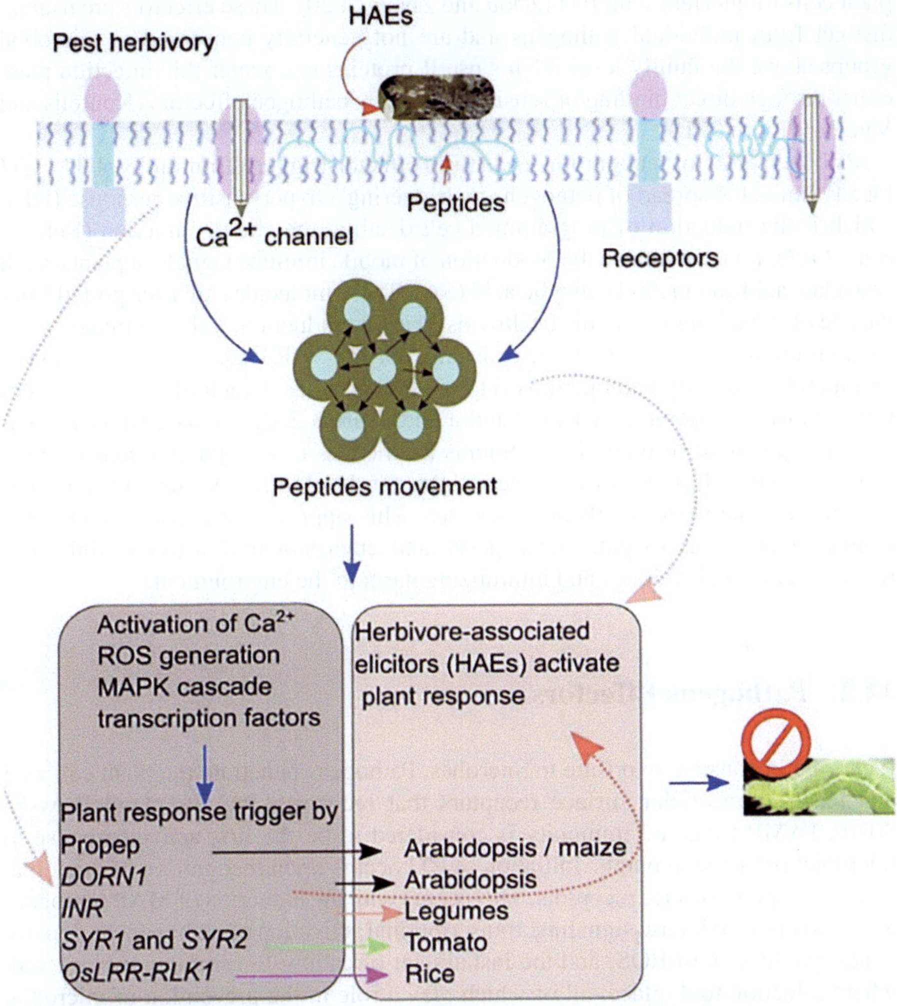

Fig. 14.1 The plant PRRs recognize herbivore-associated elicitors to elicit plant responses against insect herbivory. (Reprinted from Malook et al. (2022). Open Access journal)

found in various groups of microbes and host-derived DAMPs. The other category utilizes resistance (R) proteins deployed against effector molecules secreted by pathogens, aiding infection establishment and plant immunity suppression (Upson et al. 2018). When MAMPs or DAMPs are detected by PRRs, a pattern-triggered immunity (PTI) is activated against invading pathogens (Zipfel 2014) (Fig. 14.1).

An example of PRRs is receptor-like kinases (RLKs) and receptor-like proteins (RLPs), which play a vital role in the development of plants, plant resistance to abiotic stress, and symbiosis (Tang et al. 2017). Pathogens mount a successful infection through the evolution of a varied repertoire of effectors that are carried into plant cells to interfere with PTI (Zhou and Zhang 2020). These effectors are usually distinct from individual pathogens and are not generally conserved in microbial groups above the family level. Plants use R proteins to counter this infection plan, either through direct binding or sensing decoys by pathogen effectors (Kourelis and Van Der Hoorn 2018).

Pathogen effector recognition leads to an effector-triggered immunity (ETI). ETI locally limits the spread of pathogens by triggering a hypersensitive response (HR), which is the induction of programmed cell death at the site of infection (Remick et al. 2023). ETI can induce the production of mobile immune signals in plants, such as azelaic acid and methyl salicylic acid (SA). These molecules are transported from the site of infection to systemic healthy tissues. The induction of this immune mechanism is termed systemic acquired resistance (SAR). SAR leads to the generation of antimicrobial activity pathogenesis-related (PR) proteins, which offer plant protection against pathogenic attacks (Chandra and Sharma 2023). This chapter investigates the potential of natural compounds as elicitors to strengthen plant defenses against diseases. It provides valuable insights for developing a sustainable and eco-friendly alternative to synthetic pesticides. This approach seeks to enhance the resistance of plants to pathogens, pests, and environmental stresses while also reducing chemical residues and minimizing harm to the environment.

14.2 Pathogens Effectors

Plants are in constant exposure to microbes. Pathogens penetrate the plant cell wall and reach extracellular surface receptors that recognize PAMPs (Jamiołkowska 2020). PAMP-triggered immunity is considered to be the first active response to microbial presence in plants. Initiation of PTI occurs upon recognition of microbial patterns by plant surface receptors. This occurs with the induction of MAP (mitogen-activated protein) kinase signaling, transcriptional activation of pathogen-responsive genes, production of ROS, and the installment of callose to reinforce the cell wall where infection took place, all of which play a role in the prevention of microbial growth (Bigeard et al. 2015).

14.2.1 Bacterial-Pathogen Effectors

There has been proof that, during an infection, pathogens actively suppress the host's PAMP-triggered defense. It is thought that all pathogenic microbes encode effectors that act as PAMP suppressors, but the best characterized stems from phytopathogenic bacteria. Gram-negative bacteria rely on a type III secretion system (TTSS) to initiate infections. This needle-like apparatus enables them to inject effector proteins directly into the cytoplasm of host cells. A key example is the ArvE effector family, which is widely conserved across many agriculturally important phytopathogenic bacteria, including species like *Pseudomonas* (Herold et al. 2024). These effectors enhance bacterial virulence by altering the host's immune response and disrupting normal physiological functions. *Pseudomonas syringae* is a plant pathogen that can secrete 20–30 effectors during infection (Popa et al. 2016). The SA pathway provides resistance against *P. syringae*, but the pathogen counters this by injecting effectors like Hopl1 through its TTSS. Hopl1 targets chloroplasts, disrupting the SA biosynthesis pathway and reducing both SA levels and PR gene expression in *Arabidopsis*. It also alters thylakoid membranes and interferes with chloroplast HSP70 protein function, aiding *P. syringae* in host colonization (Kazan and Lyons 2014). Additionally, some pathogenic bacteria utilize effectors that target the Ubiquitin-26S proteasome system (UPS) in plants, a crucial pathway involved in regulating plant immunity. These effectors imitate UPS components and direct plant resistance proteins toward degradation by proteasomes, allowing the pathogens to manipulate host cellular functions and enhance their virulence. *Ralstonia solanacearum*, which is the causative agent of bacterial wilt, has over 80 putative effectors identified in its genome. One notable effector, GALA7, can hijack the host's SCF-type E3 ubiquitin ligases, which are responsible for ubiquitination. This action promotes disease initiation (Ramachandran et al. 2021). Additionally, other bacterial effectors target the host's autophagy pathway. For instance, *Pseudomonas syringae* pv. tomato (PstDC3000) induces autophagy through its effector, HopM1. This leads to the breakdown of proteasomes that would typically restrict immune responses in plants, thereby facilitating bacterial proliferation. The introduction of HopM1 by PstDC3000 enhances both autophagy and autophagic flux (Üstün et al. 2018).

14.2.2 Fungal-Pathogen Effectors

Most biotrophic fungi can form the haustorium, a specialized infection structure. They differ in their delivery ability compared to Gram-negative bacteria as they lack any TTSS and their effectors may be introduced into the plant's apoplast (intracellular space) from haustoria (Wang et al. 2022). Fungal effectors infect hosts using various mechanisms. Some effectors modify hormone and calcium signaling to disrupt immune responses, while others induce cell death and inhibit pathogen recognition (Shao et al. 2021). Effectors that induce cell death contain a necrosis and ethylene-inducing peptide (Nep) domain. *Penicillium expansum* is a postharvest pathogen known for two NLP effectors: PeNLP1 and PeNLP2. Levin et al. showed that these Nep-like protein effectors caused necrosis in *Nicotiana benthamiana* leaves when transiently expressed (Levin et al. 2019). Additionally, plants secrete chitinase, an enzyme that targets pathogenic fungi through fungal cell wall degradation which leads to the release of chitin fragments. These fragments are recognized by plant receptors and trigger an immune response. Some fungi utilize chitin-related effectors that compete with chitin fragments and suppress the chitin-triggered immunity. A recent example is that of the biotrophic fungus *Podosphaera xanthii*, the causal agent of powdery mildew in cucurbits, which utilizes chitin-related effectors (Martinez-Cruz et al. 2021).

14.2.3 Viral-Pathogen Effectors

Plants use a primary defense mechanism against viruses through RNA silencing, which regulates the accumulation of both endogenous and foreign RNA material. The presence of double-stranded RNA is considered a potent trigger. Viral virulence determinants that are considered effectors act as suppressors of host RNA silencing responses. The ability of bacterial and fungal effectors to suppress PAMP-triggered responses limits the host range of those pathogens. Similarly, the success of a viral silencing suppressor may be considered an important factor in determining the host range of viruses (Toruño et al. 2016). Plant viruses and their suppressors have been identified and characterized in detail (Ray and Casteel 2022). Plants have evolved a second defense mechanism against these viruses and their suppressors. This defense mechanism comes in the form of specific R proteins that identify viral components either silencing suppressors or other protein constituents that build up following a successful viral replication and translation. Viral effectors target transcription factors, protein degradation pathways, plant signal transduction, or the detoxification of plant secondary metabolites. The Cucumber mosaic virus (CMV) is transmitted by aphids to over 1200 plant species. It uses its 2b protein to target the host's Jasmonate (JA) hormone pathway, which prevents a JA-dependent defense response (Wu et al. 2017). Similarly, the Tomato yellow leaf virus utilizes its C4 protein in the host's chloroplasts, where it interacts with the calcium-sensing receptor, thereby interfering with SA biosynthesis and hindering the host's immune system (Medina-Puche et al. 2020).

14.3 Mechanism of Elicitor Against Pest

14.3.1 Elicitor Recognition

The pathogen utilizes its entire enzymatic and physical arsenal to invade the intracellular nutrient-rich part of the plant cell. On the other side, the plant detects pathogen signals and prepares itself for defense utilizing established chemical and mechanical barriers as well as inducible defense mechanisms (Zhou and Zhang 2020). In this battle, the winner is determined by either the pathogen's capability to overcome the plant's defense system and establish a disease or the plant's defense system's ability to keep the pathogen at bay. The central ability of a plant to defend itself is due to the recognition of various pathogens. Plants are capable of recognizing conserved microbe patterns that were previously discussed known as MAMPs or PAMPs. The recognition of these conserved patterns leads to the induction of several induced defenses in plants (Remick et al. 2023). They are mostly conserved pathogenic products that vary among diverse groups of microorganisms such as bacterial flagellin, chitin, glucan, lipids, and other membrane elements such as proteins and glycopeptides (Wang et al. 2022). Many PAMPs can be general elicitors of defense that can evade host pattern recognition without compromising their function. When plants recognize pathogens, a myriad of intracellular signaling can occur resulting in events such as rapid ion fluxes, generation of ROS, and activation of kinase cascades (Bigeard et al. 2015). Jasmonic acid (JA), ethylene, and SA are endogenous molecules that play a role in mediating plant defense responses (Putztahelyi et al. 2015). These signaling events also include ROS and nitric oxide leading to the induction of HR, which is a localized form of programmed apoptosis limiting the pathogen's spread. These events occur in both susceptible and resistant interactions but happen at different intensities and speeds (Remick et al. 2023) (Fig. 14.2).

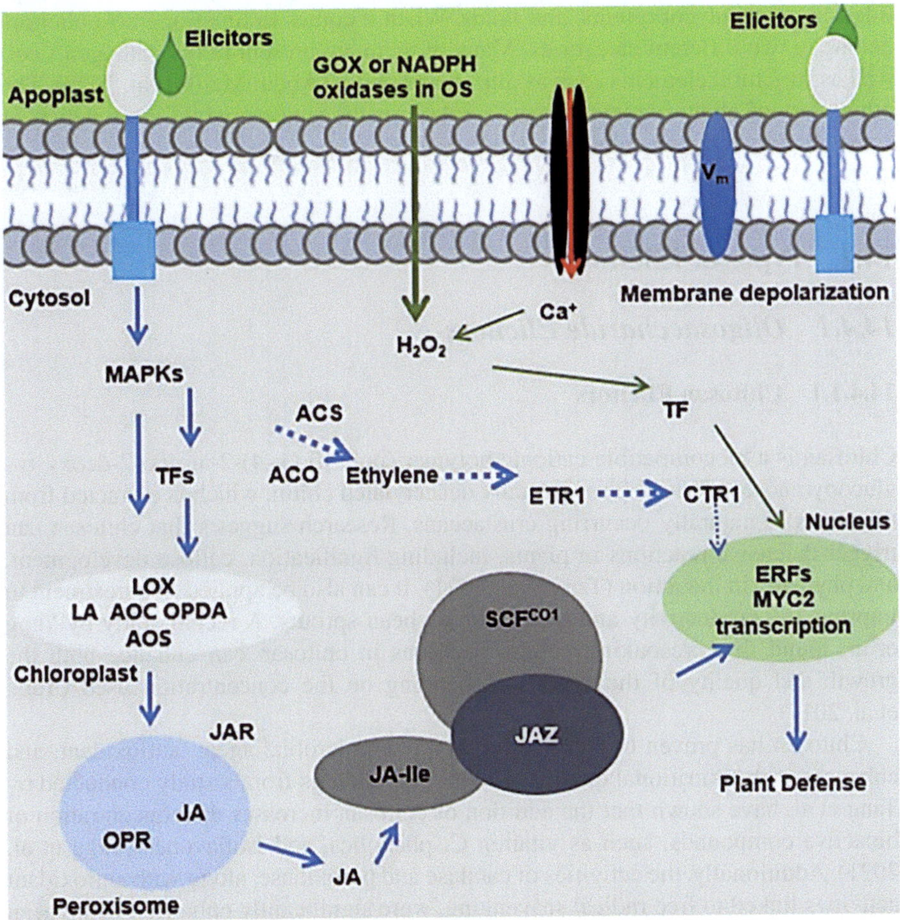

Fig. 14.2 The molecular signaling model of plant response to insect herbivory. (Reprinted from Malook et al. (2022) Open Access journal)

14.3.2 Elicitors and Defense Response

The term "elicitor" was first used to describe molecules that stimulate the production of phytoalexins. Today, the term is commonly applied to compounds that trigger a plant's defense mechanisms by inducing the expression of PR genes. Elicitors can be classified into two categories: endogenous and exogenous. Endogenous elicitors are produced internally by the plants themselves, whereas exogenous elicitors are derived from external sources, typically produced by pathogens such as bacteria, fungi, or viruses (Meena et al. 2022). Microbial elicitors can further be categorized into two groups, general elicitors and race-specific elicitors. General elicitors are elicitors capable of inducing a defense response in host and non-host plants. Race-specific elicitors trigger a defense response, leading to the resistance of disease only in specific host plants. Elicitors do not share a common chemical structure but are a wide range of various classes of compounds consisting of oligosaccharides, peptides, proteins, glycoproteins, and lipids. When it comes to function, most elicitors belong to two different categories. Many of them are present in the pathogen's cell wall as structural elements such as glucan and chitin (Abdul Malik et al. 2020). The other class of elicitors plays an active role in pathogenicity, while the function of some elicitors remains unknown.

14.4 Types of Elicitors

14.4.1 Oligosaccharide Elicitors

14.4.1.1 Chitosan Elicitors

Chitosan is a biocompatible cationic polymer (poly [β-(1–4)-2-amino-2-deoxy-D--glucopyranose]) derived from partially deacetylated chitin, which is extracted from the shells of naturally occurring crustaceans. Research suggests that chitosan can trigger defensive reactions in plants, including lignification, callose development, and phytoalexin induction (Tang et al. 2021). It can also be applied as a treatment to improve the productivity and quality of soybean sprouts. A recent study by Yang et al. found that presoaking yellow soybeans in chitosan can enhance both the growth and quality of the sprouts, depending on the concentration used (Yang et al. 2017).

Chitosan has proven to be effective as an antimicrobial agent, antioxidant, and enhancer of the nutritional quality of food. The findings from a study conducted by Tang et al. have shown that the addition of chitosan increases the concentration of bioactive compounds, such as vitamin C, phenolics, and isoflavones (Tang et al. 2021). Additionally, the activities of catalase and peroxidase, along with antioxidant activities linked to free radical scavenging, were significantly enhanced in soybean sprouts.

14.4.1.2 Glucans

The term "β-glucans" refers to a diverse group of glucose polymers that share a common structure. They consist of a main chain of β-(1,3)- and/or β-(1,4)-glucopyranosyl units, with varying lengths and branching in their side chains. When β-glucans attach to specific receptors on immune cells, they can stimulate immunological responses. However, the structures, physical characteristics, receptor binding affinities, and biological roles of β-glucans can vary significantly depending on their source (Han et al. 2020).

For example, tobacco responds to linear β-1,3-glucans, while rice and soybeans may recognize branched β-glucans. In soybeans and other plants in the Fabaceae family, the heptaglucoside β-1,3 and β-1,6, characterized by a β-1,6 backbone with β-1,3 side chains, leads to the accumulation of phytoalexin. However, this response does not occur in tobacco. Conversely, rice, but not soybeans, produces phytoalexin when exposed to β-1,6 β-1,3 glucans, which feature a β-1,3 backbone with a β-1,6 side chain (Riseh et al. 2023) (Table 14.1).

14.4.2 Protein/Peptide Elicitors

14.4.2.1 Elicitins1

Elicitins are a family of small, highly conserved proteins secreted by certain phytopathogenic oomycetes, particularly those in the genera *Phytophtora* and *Pythium* (Derevnina et al. 2016). These proteins play a critical role in plant–pathogen interactions, particularly in eliciting defense responses in the Solanaceae and Brassicaceae families (Chandra and Sharma 2023). Elicitins function as MAMPs and are known to induce HRs in plants, which involve localized cell death that helps contain pathogen spread (Noman et al. 2020). This is particularly evident in tobacco, where elicitins can trigger robust defense mechanisms even at low concentrations (Kharel et al. 2021). In *Nicotiana* species, specific RLPs known as REL (responsive to elicitins) are recognized by elicitins (Chen et al. 2023). These proteins mediate the plant's response to elicitins. In *Solanum microdontum*, the recognition of elicitins is facilitated by a membrane-localized receptor known as ELR (Elicitin response). Furthermore, when the ELR gene is expressed heterologously in cultivated potatoes (*Solanum tuberosum*), it enhances resistance to *Phytophtora infestans* (Chen et al. 2023).

14.4.2.2 Xylanase Elicitor

One of the most well-known pathogens that cause severe plant diseases and significant agricultural losses worldwide is *Botrytis cinerea*. Among its components, BcXyl1 has been identified as having xylanase activity, which breaks down xylan,

Table 14.1 Some oligosaccharides and their role against plant diseases

Elicitor	Plant species	Disease	Pathogen	Response	References
Chitosan	*Camellia sinensis* (L.) O. Kuntze (tea plant)	Blister blight disease	*Exobasidium vexans Massee*	Upregulation of defense-related enzymes and increased total phenol content	Chandra et al. (2017)
Trichoderma-derived trehalose	*Brassica oleracea* (broccoli)	Broccoli leaf spot disease	*Alternaria brassicicola*	Early hypersensitive reaction in primed seedlings Accelerated deposition of histochemical markers such as callose and phenols Significant enhancement in the activities of defense-related enzymes: Chitinase, catalase, Allene oxide synthase	De Britto and Jogaiah (2022)
Chitosan	*Persea americana* (avocado)	Avocado fruit anthracnose	*Colletotrichum gloeosporioides*	Chitosan induces a specific accumulation of phenylpropanoids and antifungal dienes, contributing to avocado's resistance against fungal pathogens	Xoca-Orozco et al. (2019)
Cello-oligosaccharides Xylo-oligosaccharides (Oligomix)	Tomato (*Solanum lycopersicum*)	Powdery mildew –	*Oidium lycopersicum* –	Upregulation of several defense-related genes: basic secretory protein NtPRp27-like (involved in plant defense), sugar transporter SWEET (potentially linked to defense signaling), calcineurin subunit B-like (involved in stress responses), protease inhibitor I (implicated in insect resistance), **PR** proteins	Pring et al. (2023)

Tramesan (a 23-kDa α-heteropolysaccharide produced by *Trametes versicolor*)	Durum wheat (*Triticum durum*)	Septoria leaf blotch complex	*Zymoseptoria tritici and Parastagonospora nodorum*	Increased levels of JA, a plant defense hormone Enhanced early expression of plant defense genes: PR4 (for SNB), RBOH, PR1, PR9 (for STB)	Scala et al. (2020)
Chitosan and chitosan-derived nanoparticles	Tomato (*Solanum lycopersicum*)	Bacterial wilt	*Ralstonia solanacearum*	Significant increases in lignin, callose, and H_2O_2 deposition, along with superoxide radical accumulation in leaves, indicating a robust defense response Increased activity of key defense-related enzymes: phenylalanine ammonia-lyase, peroxidase, polyphenol oxidase, catalase, β-1,3-glucanase	Narasimhamurthy et al. (2022)
Chitosan	*Solanum lycopersicum* var. *cerasiforme* (cherry tomato)	CMV	CMV	CEVI-1, NPR1, PSY2, and PAL5 genes were upregulated, indicating the activation of SAR and adjustments to the plant's oxidative status.	Rendina et al. (2019)

the second most abundant polysaccharide found in plant cell walls (Yang et al. 2018). When *B. cinerea* infects *Nicotiana benthamiana* and tomato plants, the expression of the BcXyl1 gene is significantly increased. Notably, strains of *B. cinerea* that lack BcXyl1 exhibit greatly reduced virulence.

In various plant species, BcXyl1 induces considerable cell death, but its xylanase activity does not contribute to this cell death. Additionally, purified BcXyl1 has been shown to enhance resistance in tobacco and tomato plants against both *B. cinerea* and *Tobacco Mosaic Virus* (TMV), triggering PAMP-triggered immunity responses. It was also demonstrated that BcXyl1 plays a role in promoting plant defensive responses, thereby enhancing the pathogenicity of *B. cinerea* (Yang et al. 2018).

14.4.2.3 Nep1 Elicitor

Nep1-like proteins (NLPs) are secreted by many significant microbial plant pathogens, including bacteria, fungi, and oomycetes. These proteins play a crucial role in the virulence and spread of these diseases. NLPs bind to specific sphingolipid receptors on plant cell membranes, which disrupts the plasma membrane and makes them cytotoxic (Pirc et al. 2021). Due to their wide taxonomic range and important role in plant infection, NLPs present a promising target for the development of new phytopharmaceuticals. Notably, it has been demonstrated that compounds with the highest affinity for NLPs also inhibited the growth of *Phytophthora infestans* on potato leaves and prevented NLP-mediated necrosis in tobacco leaves. Stabilized binding to the NLP protein resulted in reduced necrotic activity and lower ion leakage from tobacco leaves. This was confirmed through saturation transfer difference nuclear magnetic resonance and molecular modeling of a promising molecule, an anthranilic acid derivative (Pirc et al. 2021).

14.4.2.4 EF-Tu

One crucial aspect of plant innate immunity is the recognition of PAMPs/MAMPs by PRRs. The bacterial PAMP elongation factor Tu (EF-Tu) and its derived peptide elf18 are recognized by the Arabidopsis PRR known as the EF-Tu receptor (EFR) (Schoonbeek et al. 2015). In a study by Schoonbeek et al., a series of bioassays were developed to investigate the activation of PAMP-triggered immunity in wheat. Transgenic wheat plants expressing AtEFR were tested for their response to elf18. When these transgenic EFR wheat plants were inoculated with the cereal bacterial pathogen *Pseudomonas syringae pv. oryzae*, they showed a significant reduction in lesion size and bacterial spread (Schoonbeek et al. 2015).

Saccharide and protein elicitors can be summarized in Fig. 14.3.

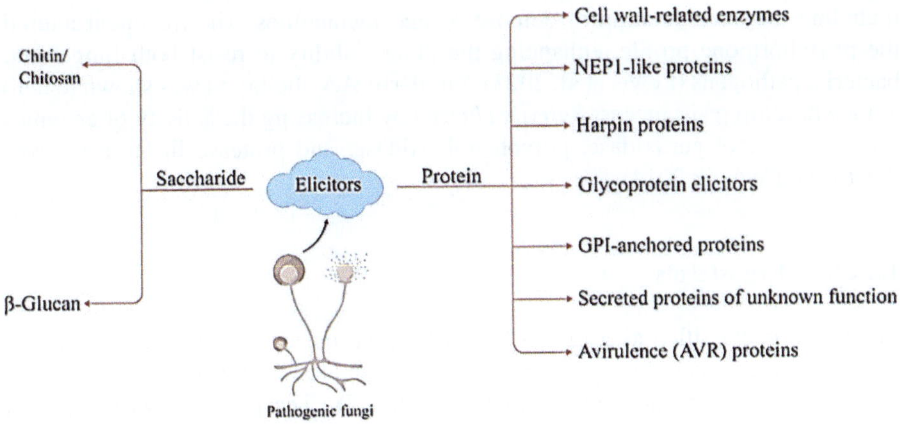

Fig. 14.3 Classification of fungal elicitors. (Reprinted from Guo and Cheng (2022), Open Access journal)

14.4.3 Lipid Elicitors

14.4.3.1 Sphingolipids

Sphingolipids, which contain a sphingoid base backbone, are essential components of eukaryotic membranes and play key roles as signaling molecules. They are crucial for regulating plant immunity, maintaining cellular homeostasis, and adapting to stress (Zeng and Yao 2022). The plant pathogenic oomycete *Phytophthora infestans* produces a ceramide known as *Phytophthora infestans*–ceramide D (Pi-Cer D), which triggers a resistance response in *Arabidopsis*.

An apoplastic ceramidase in *Arabidopsis* called neutral ceramidase 2 (NCER2) cleaves Pi-Cer D, resulting in a 9-methyl-branched sphingoid base. This base is unique to microorganisms and interacts directly with a plasma membrane lectin RLK named Resistant to DFPM Inhibition of Abscisic Acid Signaling 2 (RDA2). The interaction between the 9-methyl-branched sphingoid base and RDA2 initiates plant immune responses. Research by Kato et al. indicates that the ability of *Arabidopsis* to resist *Phytophthora infestans* infection is compromised when either RDA2 or NCER2 is nonfunctional (Kato et al. 2022).

14.4.3.2 Arachidonic Acid

Arachidonic acid (AA) is recognized as a potent elicitor that enhances plant defense mechanisms against various diseases. While plants do not produce AA themselves, the primary producers have been identified among fungi from the genus *Mortierella*, which is part of the class Phycomycete (Dedyukhina et al. 2014). Research has shown that AA can induce systemic resistance in plants. A study involving tomato plants treated with AA revealed extensive local and systemic metabolic changes, including the accumulation of defense-related metabolites. The treatment altered the phytohormone profile, enhancing the plant's ability to resist both fungal and bacterial pathogens (Lewis et al. 2023). Similarly, AA elicitation was shown to help lettuce develop resistance to *Botrytis cinerea* by increasing the activity of enzymes such as guaiacol peroxidase, polyphenol oxidase, and protease in lettuce leaves (Złotek and Wójcik 2014).

14.4.3.3 Ergosterols

Ergosterol is classified as a fungal MAMP, which means that it is recognized by plant cells as a signal of fungal presence. This recognition activates various defense pathways, including those mediated by SA and JA. The pathogen *Zymoseptoria tritici*, which causes Septoria tritici Blotch (STB) disease, poses a persistent threat to wheat harvests. In 2019, Le Mire and colleagues conducted a series of experiments to screen for wheat defense elicitors and evaluate their effectiveness in protecting wheat crops against STB. Foliar treatment with ergosterols, along with four other elicitor molecules, provided 70% protection for wheat against the fungal pathogen under semi-controlled conditions. Additionally, these treatments activated both SA- and JA-dependent signaling pathways in the plants (Le Mire et al. 2019).

14.4.4 Synthetic Elicitors

14.4.4.1 Functional Analogs of Salicylic Acid

14.4.4.1.1 β-Aminobutyric Acid

β-Aminobutyric acid (BABA) and DL-BABA are non-protein amino acids that enhance resistance against a variety of diseases, including those caused by bacteria, fungi, and oomycetes. When applied topically, BABA protects the leaves of tomatoes and potatoes from *P. infestans*, broccoli from *Alternaria*, and lettuce from *Bremia lactucae* (Tripathi et al. 2019).

BABA has been shown to enhance resistance against fungal infections by stimulating the expression of PR1 and promoting SA biosynthesis. It also produces hydrogen peroxide and deposits callose, contributing to defense mechanisms that are independent of SA. Additionally, aminobenzoic acid derivatives, such as para-aminobenzoic acid (PABA), enhance resistance against bacteria and viruses in an SA-dependent manner. Furthermore, the nonprotein amino acid DL-2-amino-N-butanoic acid triggers SAR against *Puccinia helianthi* in sunflowers and *Phytophthora infestans* in tomatoes (Tripathi et al. 2019).

14.4.4.1.2 Laminarin

Laminarin (glucan-polysaccharides with 1, 3-linked β-D-glucose residues) is a natural and exogenous elicitor that activates several defense mechanisms against viruses in potatoes and tomatoes, including alkalization, ROS burst, PAL and LOX activation, SA accumulation, and PR-gene expression (Gauthier et al. 2014). Inorganic salts of phosphoric acid (H_3PO_3) and phosphites (Phosphonates) induce resistance against *Phytophthora* and fungi in many crops by activating defense genes of both SA and JA pathways. Pipecolic acid (Pip) is an endogenous inducer and lysine catabolite that confers SAR by raising levels of free radicals, nitric oxide (NO), and ROS in the Pseudomonas–Arabidopsis pathosystem (Wang et al. 2018).

14.4.4.2 Imprimatins

In a screening of 10,000 small molecules to find plant immune priming agents, three distinct classes of compounds were identified that enhance immunity in *Arabidopsis* cells against both virulent and avirulent *Pseudomonas syringae*. These compounds, named Imprimatins, were categorized based on structural similarities into Imprimatins A, B, and C (Table 14.2). Imprimatin A and B are characterized by their ability to prime plants for enhanced defense responses without directly inducing immune responses (Bektas and Eulgem 2015).

Application of these compounds results in increased levels of endogenous SA and decreased levels of its inactive metabolite SAG, indicating an inhibition of SAGTs. Supporting this, knockout mutants of *Arabidopsis* SAGT genes UGT74F1 and UGT76B1 show increased disease resistance and higher free SA levels, similar to wild-type plants treated with Imprimatins A1, A2, A3, B1, or B2 (Bektas and Eulgem 2015).

Table 14.2 The different classes of Imprimatins

Main type	Common name	Systematic name
Imprimatin A	Imprimatin A1	2-[(E)-2-(2-bromo-4-hydroxy-5-methoxyphenyl)ethenyl] quinolin-8-ol)
	Imprimatin A2	7-Chloro-2-[(E)-2- (4-nitrophenyl) ethenyl]-4H-3,1-benzoxazin-4-one)
	Imprimatin A3	4-[(E)-2-(quinolin-2-yl)ethenyl]phenol)
Imprimatin B	Imprimatin B1	2-(3-(2-Furyl)-3-phenylpropyl) benzo[c]azoline-1,3-dione)
	Imprimatin B2	3-(2-Furyl)-3-phenylpropylamine)
Imprimatin C	Imprimatin C1	[(E)-[1-amino-2-(2-oxopyrrolidin-1-yl)ethylidene]amino] 4-chlorobenzoate)
	Imprimatin C2	[(E)-[1-amino-2-(2-oxopyrrolidin-1- yl)ethylidene] amino]3,4-dichlorobenzoate)

Table 14.3 Effect of biotic elicitors on therapeutic metabolites production in plants (Jeyasri et al. 2023)

Elicitor	Compounds	Plant species	References
Sclerotinia sclerotiorum	Lignan	*Linum album*	Esmaeilzadeh Bahabadi et al. (2014)
Sucrose	Leoligin	*Leontopodium alpinum*	Wawrosch et al. (2014)
Penicillium fellutanum	Glycyrrhizic acid	*Taverniera cuneifolia*	Awad et al. (2014)
Chitin	Trans-resveratrol and viniferins	*Vitis vinifera*	Taurino et al. (2015)
Bacillus cereus	Atropine	*Datura metel*	Shakeran et al. (2015)
Staphylococcus aureus	Atropine	*Datura metel*	Shakeran et al. (2015)
Dextron	Phenylpropanoids and naphtodianthrones	*Hypericum perforatum*	Gadzovska-Simic et al. (2015)
Pectin	Phenylpropanoids and naphtodianthrones	*Hypericum perforatum*	Gadzovska-Simic et al. (2015)
Asperigillus Niger	Oleanolic acid, urosolic acid	*Salvia fruticosa*	Kummritz et al. (2016)
Rhizophagus irregularis	Rosmarinic acid	*Ocimum basilicum*	Srivastava et al. (2016)
Piriformospora indica	Withanolides	*Withania somnifera*	Ahlawat et al. (2017)
Trichoderma harzianum	Ginsenoside and anthocyanin	*Panax sikkimensis*	Biswas et al. (2018)
Bacillus subtilis	Ginsenoside	*Panax quinquefoliusL*	Shaikh et al. (2020)
Chitosan	Withferin	*Withania somnifera (L)*	Shaikh et al. (2020)

14.4.4.3 Jasmonic Acid Analogs

JA and its methyl ester, MeJA, are key members of the JA family, a group of biologically active cyclopentanones derived from fatty acids that are widely found in plants. These compounds are rapidly synthesized through the octadecanoid biosynthetic pathways in response to pathogen or insect attacks, initiating defense mechanisms. JAs are known to manage stress responses to necrotrophic pathogens, herbivores, and wounding, and they also play significant roles in plant development, including leaf aging, growth inhibition, and floral development (Bektas and Euglem 2015) (Table 14.3).

14.5 Secondary Metabolites and Plant Defense

Plants produce a wide variety of organic compounds known as secondary metabolites, secondary products, or natural compounds that are not directly involved in their normal growth, development, or reproduction. The diverse secondary metabolism of plants has evolved through natural selection, with the production of these compounds providing adaptive advantages allowing plants to thrive in their environments (Erb and Kliebenstein 2020). It was previously believed that these compounds were mere byproducts of their metabolic processes with no discernible function. However, it is now understood that they play vital roles as defense mechanisms and communication signals. As analytical techniques like chromatography have advanced, researchers have been able to identify and study over 200,000 secondary metabolites, which can have a wide range of physiological and ecological roles (Teoh 2016). Environmental stressors, such as drought, extreme temperatures, and nutrient deficiencies, can alter the metabolic pathways of plants by affecting the activity of key enzymes. These changes in enzymatic activity can lead to a disruption in the normal physiological processes of the plant, ultimately affecting its growth and productivity. Furthermore, biotic stresses, such as those caused by pathogenic microorganisms such as fungi, viruses, and bacteria, can also have a significant impact on plant metabolism.

14.5.1 Classes of Secondary Metabolites

14.5.1.1 Phenols

Phenolic compounds are a diverse group of naturally occurring aromatic compounds found abundantly in plants. They are not commonly found in bacteria, algae, and fungi, but are extensively documented in higher plants, especially vascular plants. These compounds have gradually evolved to provide specific adaptations to different plant families under stress, aiding their survival in a constantly changing environment (Tak and Kumar 2020). Phenolics are widely distributed throughout the plant kingdom and are among the most prevalent secondary metabolites. They have been reported in diverse types of commodities (Wozniak et al. 2022).

Extensive research has been dedicated to plant phenolics, known for their wide range of health benefits. However, the different classes of plant phenolics also carry diverse functions that are crucial for the survival of all vascular plants. In addition to having roles in the growth and development of plants, these compounds are essential to the plant's physiology, contributing to pigmentation, and pollination, and engaging directly or indirectly in protecting the plant against biotic and abiotic stresses (Tanase et al. 2019).

Phenolic compounds are classified according to their basic carbon skeleton. They can either be simple or polyphenolic compounds based on the number of phenol units in the molecule (Tsimogiannis and Oreopoulou 2019). In general, phenolic compounds are classified into three main categories. Table 14.4 summarizes the major classes of phenolic compounds and their basic carbon skeleton.

Table 14.4 Summary of the major classes of phenolic compounds and their carbon skeleton

Main structure	Phenolic class	Examples
C_6	Simple phenols	Phenol, catechol, resorcinol
$C_6–C_1$	Phenolic acids—Hydroxybenzoic acids	Gallic acid, p-hydroxybenzoic acid, vanillic acid, SA
$C_6–C_2$	Acetophenones	p-Hydroxyphenylacetic acid, tyrozol
$C_6–C_3$	Hydroxycinnamic acid, coumarin	*p*-Coumaric acid, caffeic acid, ferulic acid, sinapic acid
$C_6–C_4$	Naphthoquinones	Juglone, plumbagin
$C_6–C_1–C_6$	Xanthones	Mangiferin, mangostin
$C_6–C_2–C_6$	Stilbenes	Resveratrol
$C_6–C_3–C_6$	Flavonoids	Myricetin, catechins, hesperetin, apigenin
$(C_6–C_3)_2$	Lignans	Pinoresinol, matairesinol, lariciresinol, eusiderin
$(C_6–C_3)_n$	Lignin, suberin, and cutin	
$(C_6–C_3–C_6)_n$	Tannins	Trimeric procyanidin, prodelphinidins, gallotannins

Phenolics and Plant Defense

Plant phenolics play important roles in plant physiology and ecology. They act as antioxidants, protecting plants from oxidative stress and environmental damage. Additionally, these compounds are crucial for pigmentation, influencing the color of flowers and fruits, which helps attract pollinators and seed dispersers. Furthermore, phenolics are involved in maintaining the structural integrity of plants by providing rigidity and strength to cell walls. They also play a key role in various ecological interactions, such as allelopathy, where they can inhibit the growth of neighboring plants. Nevertheless, phenolics are primarily known to contribute to the defense mechanism of plants, protecting them against biotic and abiotic stressors such as extreme temperatures, UV radiation, herbivores, and various pathogens.

Plant phenolics are naturally synthesized during the normal growth and development of plant tissues, which are referred to as "pre-formed." However, when plants are under attack by microbes or pathogens, or when external factors stress them, the production and accumulation of phenolic compounds increase in sub-epidermal tissues. These phenolic compounds are then called "induced" (Pratyusha 2022). Phenolics are protective agents, inhibitors, pesticides, and natural toxins that discourage the growth of living organisms such as herbivores, nematodes, insects, and fungal, viral, and bacterial pathogens. The significance of phenolics in plant defense against diseases is evident in various instances. For example, some pathogens like *Helminthosporium oryzae*, which attack rice plants, produce toxins that specifically target phenolic metabolism. This leads to the suppression of phenolic synthesis (Pandey 2018). Previous studies have demonstrated that certain low-molecular-weight phenols, such as benzoic acids and those produced by the phenylpropanoid pathway, are synthesized at the onset of an infection and accumulate at the infected site of the plant. In response to pathogen attacks, key enzymes of the phenylpropanoid pathway, such as phenylalanine ammonia-lyase and chalcone synthase, are upregulated and significantly increase the production of phenolic compounds (Aqsa and Ambreen 2024). For instance, maize crops accumulate significant amounts of two phenolic caffeic acid esters in response to infections by *Glomerella graminicola* or *C. heterostrophus* (Pusztahelyi et al. 2016). Similarly, certain phenolics with antifungal properties, such as 3-deoxyanthocyanidin phytoalexins, apigeninidin, luteolinidin, and related conjugates, are effective in protecting Sorghum bicolor cereal against *Colletotrichum sublineolum* infection (Tugizimana et al. 2019). Certain families of phenolic compounds found in olive leaves, such as hydroxycinnamic derivatives and flavonol monoglucoside derivatives, have been shown to help the olive tree resist the fungus *Fusicladium oleagineum*, which causes leaf spot disease (Talhaoui et al. 2015). Additionally, the deposition of lignin in plant cell walls serves a crucial defensive function by forming a physical barrier against invading pathogens. For instance, Marques et al. investigated the cell wall defense responses in both smut-resistant and susceptible sugarcane cultivars when infected with *Sporisorium scittamineum*, the causative agent of smut disease. Their findings revealed that resistant cultivars exhibited significant lignin and phenolic compound accumulation at the infection site during the early stages of infection, effectively preventing the initial penetration of the fungus (Marques et al. 2018). Phenolic compounds such as catechin, flavanols, procyanidin B1, dihydrochalcones, epicatechin, and hydroxycinnamic acids are effective against *Venturia inaequalis* (Slatnar et al. 2016).

The diverse antimicrobial, antiviral, and insect-repellent properties of phenolic compounds make them valuable tools in developing eco-friendly, sustainable agricultural practices to combat diseases and pests while promoting plant health and productivity. P-hydroxybenzoic acid and protocatechuic acid from red, brown, and green seaweeds have been used to prevent early blight disease in tomatoes caused by *Alternaria solani* by activating plant defense mechanisms and promoting SA synthesis (Aina et al. 2022). Additionally, an in vitro experiment performed by Rashad et al. suggests that the antifungal activity of "Khella" extract (seed extract of *Ammi visnaga*) against *Rhizoctonia solani* is largely due to its coumarin content, making it a promising candidate for biocontrol strategies in agriculture (Rashad et al. 2018). Phenolic-rich foliar extracts of wild *Capsicum annuum* were shown to be effective in reducing soft rot disease severity on infected cherry tomatoes (Pane et al. 2016).

14.5.1.2 Terpenes

Terpenes represent the largest and most diverse class of organic compounds, primarily produced by plants. They are the aromatic oils responsible for the distinct scents and flavors of many plants, including herbs, fruits, and flowers. To date, over 80,000 terpene compounds have been identified in higher plants such as conifers, citrus, and eucalyptus (Hilgers et al. 2021). They are hydrophobic molecules and are found stored in resin ducts, oil cells, or glandular trichomes (Tiku 2018). However, it has been reported that most of the plant terpenes are emitted in a volatile form and are referred to as volatile terpenes (VTs). These volatile molecules are released in response to stress conditions and play a role in mediating plant–insect interactions, benefitting the plants by repelling insects (Boncan et al. 2020). Terpenes have diverse applications due to their anti-inflammatory, antimicrobial, and analgesic properties (Abdallah and Quanx 2017). Additionally, terpenoids are vital for plant defense, functioning through both direct and indirect mechanisms to protect against a range of threats, while also enhancing plant resilience to environmental stresses.

Terpenes and Plant Defense

Multiple studies have highlighted the crucial role of terpenoidal molecules in plants' defense against biotic stress, such as threats from herbivores, pathogens, and other organisms. Terpenes are naturally synthesized by plants, even in the absence of any biotic or abiotic stressors, and are referred to as "phytoanticipins." When plants face pathogens or herbivores, they release terpenes called "phytoalexins" and are elicitor-induced terpenes. They mainly emit a blend of volatiles which serve as airborne signals that can be detected by other parts of the same plant, triggering defense responses. Terpenes have been reported to have antifungal, antimicrobial, antiparasitic, and antiviral properties. They have been used in the development of terpene-based biopesticides for a safer and more environmentally friendly alternative to pest control, In particular, several diterpenes have been used in the formulation of natural pesticides/herbicides due to their ability to inhibit pathogenic microbes and pests (Ninkuu et al. 2021). For instance, essential oils containing terpenes such as limonene, carvone, and pulegone have demonstrated insecticidal and fungicidal activities against various pests and pathogens (Boncan et al. 2020). Biopesticides such as orange or citrus oils and extracts from *Chenopodium ambrosioides* can effectively reduce aphid pest populations on ornamental crops (Smith et al. 2018). Terpenes are crucial components of plant defense strategies against insect herbivores. Their roles as repellents, toxins, and attractants for natural enemies, combined with their ability to induce defense mechanisms in plants, illustrate their significance in the complex interactions between plants and herbivorous insects. In a study by Cascone et al., the role of the VT β-ocimene as an elicitor of both direct and indirect plant defense in tomatoes was examined. The researchers utilized transgenic tobacco plants enriched in β-ocimene to investigate the direct plant defense of tomatoes against the aphids *Macrosiphum euphorbiae*, as well as the indirect defense mechanism of tomato plants through the attraction of the parasitoid *Aphidius ervi*, which serves as a natural enemy of the aphid. The researchers found that exposure of tomato plants to the volatile led to a decrease in the incidence of aphids on the plant and increased the attractiveness of the plant to the parasitoid (Cascone et al. 2015). In a similar vein, monoterpene esters known as "pyrethroids" have been shown to cause neurotoxic effects in wasps, bees, and beetles. For instance, the spindle twigs of conifers accumulate several terpenes, such as limonene and myrcene, which have toxic effects on beetles and various other pests that affect conifers (Ninkuu et al. 2021).

Certain terpenoidal molecules have been found to have antimicrobial properties. For instance, rice plants contain diterpenes and sesquiterpenes such as momilactone A and B, oryzalexin A-F, and phytocassanes, which have been shown to have antimicrobial effects against *Magnaporthe grisea*, the fungus responsible for causing rice blast diseases (Zhan et al. 2020). Additionally, *Zea mays* produces the terpene epoxydolabranol, which plays a crucial role in defense against two Fusarium species, *F. verticillioides* and *F. graminearum* (Mafu et al. 2018). In *Nicotiana benthamiana*, the sesquiterpene capsidiol 3-acetate was shown to exhibit antiviral activity against the *Potato Virus X* (PVX) (Li et al. 2015). In strawberry fruits, overexpression of the FaTPS1 gene induced by MeJA increases the synthesis of sesquiterpenes such as germacrene D, providing resistance to *Botrytis cinerea* infection (Zhang et al. 2024). Limonene, a monocyclic monoterpene, effectively fights the pathogen R. *solanacearum* by inhibiting its growth, disrupting biofilm formation, and modulating its behavior (Zhang et al. 2024). β-Caryophyllene suppressed the growth of major plant pathogenic fungi such as *Sclerotinia sclerotiorum* and *Fusarium oxysporum* (Hilgers et al. 2021). Numerous studies have highlighted the importance of VTs as immune signals that impact both the same and neighboring plants by activating phytohormone signaling pathways and inducing defense-related gene expression (Rosenkranz et al. 2021). Reidlmeier et al. (2017) reported that the gaseous monoterpenes α/β-pinene and camphene stimulated innate immunity in Arabidopsis through pathways associated with SA.

14.5.1.3 Alkaloids

The term "alkaloid" refers to nitrogen-containing compounds derived from plants. These compounds typically feature one or more nitrogen atoms within a heterocyclic ring, excluding amino acids, peptides, purines, and their derivatives, amino sugars, and antibiotics (Teoh 2016). Over 27,000 distinct alkaloids have been identified, each with unique properties. Some well-known examples include morphine, quinine, and nicotine, which have various applications. Alkaloids are especially prevalent in certain families of flowering plants, with estimates suggesting that around 10–25% of higher plants produce alkaloids. Certain plant families such as *Apocyanaceae*, *Ranunculaceae*, and *Solanaceae* are particularly rich in alkaloids (Gutiérrez-Grijalva et al. 2020). Alkaloids can be found in different plant parts, such as leaves, seeds, roots, and bark, often in varying concentrations depending on the plant species. Usually, the highest concentrations are located in the leaves or seeds. Alkaloids are produced by plants and sequestered in specialized cellular structures. When the plant perceives environmental stressors, the alkaloids are mobilized from the storage sites, such as organelles or glandular tissues, and translocated to the target tissues (Bhambhani et al. 2021). Additionally, some plants may experience an increase in alkaloid levels during specific growth phases such as before seed formation (Pereira et al. 2023).

Alkaloids and Plant Defense
Alkaloids serve various ecological roles, particularly in plant defense against herbivores and pathogens due to their bitter taste (Pereira et al. 2023). This is backed by the fact that *Lupinus* species with higher concentrations of quinolizidine alkaloids

are less palatable to pests and therefore require fewer pesticide applications. Conversely, tomato varieties selected for low or no solanine content, making them suitable for human consumption, necessitate increased pesticide use to compensate for the reduced natural defense provided by alkaloids (Matsuura and Fett-Neto 2015). Tomatine, a key alkaloid found in tomatoes and various potato relatives, exhibits repellent effects. When potato leaves are treated with a tomatine solution at a concentration of 2 mmol/kg, beetle feeding is reduced by 50%. At a higher concentration of 3 mmol.kg, it causes complete mortality in larvae. This indicates that tomatine has a toxic impact on insect pests (Harborne 2014). In a study conducted by Shaheen and Issa, the effects of alkaloid extract from *Peganum harmala* seeds on four phytopathogenic bacteria were investigated. The seed extracts, which are rich in active alkaloids such as quinazoline alkaloids and β-carbolines, proved to be effective against *Ralstonia solanacearum, Erwinia amylovora, Pectobacterium carotovorum* subsp. *carotovorum*, and *Burkholderia gladioli*. These bacteria are responsible for diseases like potato brown rot and pear fire blight, and the extracts demonstrated efficacy at various concentrations (Shaheen and Issa 2020).

Several nitrogen-fixating plants often accumulate high concentrations of alkaloids in their leaves, rendering them more resistant to herbivory. Some of these alkaloids include isoquimoline, quinoline, and indole alkaloids (Bhambhani et al. 2021). Quinolizidine alkaloids extracted from the seed of *Sophora tonkinensis Gagnep* (*Fabaceae* family) have been shown to have insecticidal effects as well as being an anti-TMV substance and are used in the production of biopesticides (Zou et al. 2020). Additionally, several studies have highlighted alkaloids with antiviral properties. For example, 7-deoxytrans-dihydronarciclasine found in plantain lilies was shown to confer resistance to the TMV virus (Anjali et al. 2023). Camalexin (3′-thiazol-2′-yl-indole), which is a characteristic indole alkaloid of *Arabidopsis* and cruciferous plants, has been shown to be integral to plant defense mechanisms as it can trigger an immunity associated with induced systemic resistance (Nguyen et al. 2022). Similarly, the alkaloids canthin-6-one and lycorine have been shown to be effective against *Fusarium graminearum* infections in wheat (Liu et al. 2023). The alkaloid phytoalexin camalexin was reported in high concentrations in plants infected by *Alternaria alternata* and near lesions induced by *Botrytis* species (Pusztahelyi et al. 2015). Several alkaloids such as Sanguinarine, Chelerythrine, Allocryptopine, and Drupacine have been shown to be effective against various plant pathogenic nematodes (Kabdal et al. 2024). Caffeine is an alkaloid known for its antifungal properties. Research has shown that the expression of genes related to caffeine biosynthesis increases in tea plants infected by the fungus *C. fructicola*. Studies indicate that caffeine has antifungal effects by causing damage to the cell walls and plasma membranes of the pathogen (Wang et al. 2016) (Table 14.5).

Table 14.5 Secondary metabolites and plant defense

Plant	Pathogen	Secondary metabolite	Defense mechanism	References
Walnut	*Xanthomonas arboricola* pv. Juglandis	Phenolics	Early defense response	Jiang et al. (2019)
Sorghum bicolor	*Colletotrichum sublineolum*	3-Deoxyanthocyanidin phytoalexins, apigeninidin, luteolinidin, and related conjugates	Protection against infection	Tugizimana et al. (2019)
Olive	*Fusicladium oleagineum*	Hydroxycinnamic derivatives and flavonol monoglucoside derivatives	Resistance to leaf spot disease	Tallaoui et al. (2015)
Sugarcane (resistant)	*Sporisorium scittamineum*	Lignin and phenolic compounds	Physical barrier against fungal penetration	Marques et al. (2018)
Apple	*Venturia inaequalis*	Catechin, flavanols, procyanidin B1, dihydrochalcones, epicatechin, and hydroxycinnamic acids	Effective against the pathogen	Slatnar et al. (2016)
Rice	*Magnaporthe grisea*	Diterpenes and sesquiterpenes such as momilactone A and B, oryzalexin A-F, and phytocassanes	Antimicrobial effects against rice blast diseases	Zhan et al. (2020)
Zea mays	*Fusarium verticillioides* and *F. graminearum*	Terpene epoxydolabranol	Defense against Fusarium species	Mafu et al. (2018)
Nicotiana benthamiana	PVX	Sesquiterpene capsidiol 3-acetate	Antiviral activity	Li et al. (2015)
Strawberry	*Botrytis cinerea*	Sesquiterpenes such as germacrene D	Resistance to Botrytis cinerea infection	Zhang et al. (2024)
Tomato	Insect pests	Tomatine	Repellent and toxic effects	Harborne (2014)
Peganum harmala	*Ralstonia solanacearum*, *Erwinia amylovora*, *Pectobacterium carotovorum* subsp. carotovorum, and *Burkholderia gladioli*	Quinazoline alkaloids and β-carbolines	Inhibition of phytopathogenic bacteria	Shaheen and Issa (2020)
Plantain lilies	TMV virus	7-deoxytrans-dihydronarciclasine	Resistance to TMV virus	Anjali et al. (2023)
Wheat	*Fusarium graminearum*	Canthin-6-one and Lycorine	Antifungal activity	Liu et al. (2023)
Tea	*C. fructicola*	Caffeine	Antifungal effects by damaging cell walls and plasma membranes	Wang et al. (2016)

14.6 Conclusions and Perspectives

Plants use a triggered immunity mechanism as a fundamental defense mechanism against pathogens. This process involves the recognition of conserved microbial patterns, such as flagellin, which activates signaling pathways, including MAP kinase cascades. These pathways lead to the production of ROS, expression of defense genes, and reinforcement of the cell wall. Pathogens, however, deploy effectors to circumvent these defenses.

Future research could include deepening the understanding of the molecular interactions between pathogen effectors and host immune systems. This could reveal new targets for crop protection and guide the development of synthetic elicitors that mimic natural defense signals, such as benzothiadiazole. Additionally, genetic engineering could enhance plant resistance by introducing or overexpressing key immune regulators. Research into the biosynthesis and functions of secondary metabolites—such as phenolics, terpenes, and alkaloids—could uncover new natural defense mechanisms and potential pest control agents.

Studying both pathogen-specific and broad-spectrum immune responses may lead to the development of crops with multi-pathogen resistance, improving agricultural sustainability and resilience.

References

Abdallah II, Quax WJ (2017) A glimpse into the biosynthesis of terpenoids. KnE Life Sci 5:81–98

Abdul Malik NA, Kumar IS, Nadarajah K (2020) Elicitor and receptor molecules: orchestrators of plant defense and immunity. Int J Mol Sci 21:963

Ahlawat S, Saxena P, Ali A, Khan S, Abdin MZ (2017) Comparative study of withanolide production and the related transcriptional responses of biosynthetic genes in fungi elicited cell suspension culture of *Withania somnifera* in shake flask and bioreactor. Plant Physiol Biochem 114:19–28

Aina O, Bakare OO, Daniel AI, Gokul A, Beukes DR, Fadaka AO, Klein A (2022) Seaweed-derived phenolic compounds in growth promotion and stress alleviation in plants. Life 12:1548

Anjali N, Kumar S, Korra T, Thakur R, Arutselvan R, Kashyap AS, Nehela Y, Chaplygin V, Minkina T, Keswani C (2023) Role of plant secondary metabolites in defense and transcriptional regulation in response to biotic stress. Plant Stress 8:100154

Aqsa T, Ambreen A (2024) Plant phenolics production: a strategy for biotic stress management. In: Lone R, Khan S, Mohammed Al-Sadi A (eds) . Springer, Plant phenolics in biotic stress management, pp 441–454

Awad V, Kuvalekar A, Harsulkar A (2014) Microbial elicitation in root cultures of *Taverniera cuneifolia* (Roth) Arn. For elevated glycyrrhizic acid production. Ind Crop Prod 54:13–16

Bektas Y, Eulgem T (2015) Synthetic plant defense elicitors. Front Plant Sci 5:804

Bhambhani S, Kondhare KR, Giri AP (2021) Diversity in chemical structures and biological properties of plant alkaloids. Molecules 26:3374

Bigeard J, Colcombet J, Hirt H (2015) Signaling mechanisms in pattern-triggered immunity (PTI). Mol Plant 8:521–539

Biswas T, Pandey SS, Maji D, Gupta V, Kalra A, Singh M, Mathur A, Mathur AK (2018) Enhanced expression of ginsenoside biosynthetic genes and in vitro ginsenoside production in elicited Panax sikkimensis (Ban) cell suspensions. Protoplasma 255:1147–1160

Boncan DAT, Tsang SS, Li C, Lee IH, Lam HM, Chan TF, Hui JH (2020) Terpenes and terpenoids in plants: interactions with environment and insects. Int J Mol Sci 21:7382

Brockhurst MA, Chapman T, King KC, Mank JE, Paterson S, Hurst GD (2014) Running with the red queen: the role of biotic conflicts in evolution. Proc Biol Sci 281:1797

Burdon JJ, Thrall PH (2014) What have we learned from studies of wild plant-pathogen associations? The dynamic interplay of time, space and life-history. Eur J Plant Pathol 138:417–429

Cascone P, Iodice L, Maffei ME, Bossi S, Arimura GI, Guerrieri E (2015) Tobacco overexpressing β-ocimene induces direct and indirect responses against aphids in receiver tomato plants. J Plant Physiol 173:28–32

Chandra S, Sharma IP (2023) Chapter 4- elicitins as microbe-associated molecular patterns and their role in plant defense. In: Chandra D, Bhatt P (eds) Developments in applied microbiology and biotechnology, unravelling plant-microbe synergy, 1st edn. Academic Press, pp 77–86

Chandra S, Chakraborty N, Panda K, Acharya K (2017) Chitosan-induced immunity in *Camellia sinensis* (L.) O. Kuntze against blister blight disease is mediated by nitric-oxide. Plant Physiol Biochem 115:298–307

Chen Z, Liu F, Zeng M, Wang L, Liu H, Sun Y, Wang L, Zhang Z, Chen Z, Xu Y, Zhang M, Xia Y, Ye W, Dong S, Govers F, Wang Y, Wang Y (2023) Convergent evolution of immune receptors underpins distinct elicitin recognition in closely related Solanaceous plants. Plant Cell 35:1186–1201

De Britto S, Jogaiah S (2022) Priming with fungal elicitor elicits early signaling defense against leaf spot of broccoli underlying cellular, biochemical and gene expression. Microbiol Res 263:127143

Dedyukhina EG, Kamzolova SV, Vainshtein MB (2014) Arachidonic acid as an elicitor of the plant defense response to phytopathogens. Chem Biol Technol Agric 1:18

Derevnina L, Dagdas YF, De La Concepcion JC, Bialas A, Kellner R, Petre B, Domazakis E, Du J, Wu C, Lin X, Aguilera-Galvez C, Cruz-Mireles N, Vleeshouwers VGA, Kamoun S (2016) Nine things to know about elicitins. New Phytol 212:888–895

Erb M, Kliebenstein DJ (2020) Plant secondary metabolites as defenses, regulators, and primary metabolites: the blurred functional trichotomy. Plant Physiol 184:39–52

Esmaeilzadeh-Bahabadi S, Sharifi M, Ahmadian Chashmi N, Murata J, Satake H (2014) Significant enhancement of lignan accumulation in hairy root cultures of *Linum album* using biotic elicitors. Acta Physiol Plant 36:3325–3331

Gadzovska Simic S, Tusevski O, Maury S, Delaunay A, Laine E, Joseph C, Hagege D (2015) Polysaccharide elicitors enhance phenylpropanoid and naphtodianthrone production in cell suspension cultures of *Hypericum perforatum*. Plant Cell Tissue Organ Cult 122:649–663

Gauthier A, Trouvelot S, Kelloniemi J, Frettinger P, Wendehenne D, Daire X, Joubert J, Ferrarini A, Delledonne M, Flors V, Poinssot B (2014) The sulfated laminarin triggers a stress transcriptome before priming the SA-and ROS-dependent defenses during grapevine's induced resistance against *Plasmopara viticola*. PLoS One 9:88145

Guo J, Cheng Y (2022) Advances in fungal elicitor-triggered plant immunity. Int J Mol Sci 23:12003

Gutiérrez-Grijalva EP, López-Martínez LX, Contreras-Angulo LA, Elizalde-Romero CA, Heredia JB (2020) Chapter 5- plant alkaloids: structures and bioactive properties. In: Swamy MK (ed) Plant-derived bioactives, chemistry and mode of action. Springer, Singapore, pp 85–117

Han GZ (2018) Origin and evolution of the plant immune system. New Phytol 222:70–83

Han B, Baruah K, Cox E, Vanrompay D, Bossier P (2020) Structure-functional activity relationship of β-glucans from the perspective of immunomodulation: a mini-review. Front Immunol 11:658

Harborne JB (2014) Chapter 1- the plant and its biochemical adaptation to the environment. In: Introduction to ecological biochemistry, 4th edn. Academic Press, pp 1–35

Herold L, Choi S, He SY, Zipfel C (2024) The conserved AvrE family of bacterial effectors: functions and targets during pathogenesis. Trends Microbiol

Hilgers F, Habash SS, Loeschcke A, Ackermann YS, Neumann S, Heck A, Klaus O, Hage-Hülsmann J, Grundler FMW, Jaeger KE, Schleker ASS, Drepper T (2021) Heterologous production of β-caryophyllene and evaluation of its activity against plant pathogenic fungi. Microorganisms 9:168

Jamiołkowska A (2020) Natural compounds as elicitors of plant resistance against diseases and new biocontrol strategies. Agronomy 10:173

Jeyasri R, Muthuramalingam P, Karthick K, Shin H, Choi SH, Ramesh M (2023) Methyl jasmonate and salicylic acid as powerful elicitors for enhancing the production of secondary metabolites in medicinal plants: an updated review. Plant Cell Tissue Organ Cult 153:447–458

Jiang S, Han S, He D, Cao G, Fang K, Xiao X, Yi J, Wan X (2019) The accumulation of phenolic compounds and increased activities of related enzymes contribute to early defense against walnut blight. Physiol Mol Plant Pathol 108:101433

Kabdal T, Karakoti H, Nagarkoti K, Prakash O, Kumar S (2024) Chapter 13- the rise of plant-based nematicides: a sustainable solution for crop protection. In: Kumar R, de Oliveira M, de Aguiar Andrade M, Suyal D, Soni R (eds) Biorationals and biopesticides: pest management. De Gruyter, pp 261–280

Kato H, Nemoto K, Shimizu M, Abe A, Asai S, Ishihama N, Matsuoka S, Daimon T, Ojika M, Kawakita K, Onai K, Shirasu K, Yoshida M, Ishiura M, Takemoto D, Takano Y, Terauchi R (2022) Recognition of pathogen-derived sphingolipids in *Arabidopsis*. Science 376:857–860

Kazan K, Lyons R (2014) Intervention of phytohormone pathways by pathogen effectors. Plant Cell 26:2285–2309

Kharel A, Islam MT, Rookes J, Cahill D (2021) How to unravel the key functions of cryptic oomycete elicitin proteins and their role in plant disease. Plan Theory 10:1201

Kourelis J, Van Der Hoorn RAL (2018) Defended to the nines: 25 years of resistance gene cloning identifies nine mechanisms for R protein function. Plant Cell 30:285–299

Kummritz S, Louis M, Haas C, Oehmichen F, Gantz S, Delenk H, Steudler S, Bley T, Steingroewer J (2016) Fungal elicitors combined with a sucrose feed significantly enhance triterpene production of a *Salvia fruticosa* cell suspension. Appl Microbiol Biotechnol 100:7071–7082

Le Mire G, Siah A, Marolleau B, Gaucher M, Maumené C, Brostaux Y, Massart S, Brisset MN, Jijakli MH (2019) Evaluation of λ-carrageenan, CpG-ODN, glycine betaine, Spirulina platensis, and ergosterol as elicitors for control of *Zymoseptoria tritici* in wheat. Phytopathology 109:409–417

Levin E, Raphael G, Ma J, Ballester A, Feygenberg O, Norelli J, Aly R, Gonzalez-Candelas L, Wisniewski M, Droby S (2019) Identification and functional analysis of NLP-encoding genes from the postharvest pathogen *Penicillium expansum*. Microorganisms 7:175

Lewis DC, van der Zwan T, Richards A, Little H, Coaker GL, Bostock RM (2023) The oomycete MAMP, arachidonic acid, and an Ascophyllum nodosum-derived plant biostimulant induce defense metabolome remodeling in tomato. Phytopathology 113:1084–1092

Li R, Tee CS, Jiang YL, Jiang XY, Venkatesh PN, Sarojam R, Ye J (2015) A terpenoid phytoalexin plays a role in basal defense of *Nicotiana benthamiana* against potato virus X. Sci Rep 5:9682

Liu Q, Zhang C, Fang H, Yi L, Li M (2023) Indispensable biomolecules for plant defense against pathogens: NBS-LRR and "nitrogen pool" alkaloids. Plant Sci 334:111752

Mafu S, Ding Y, Murphy KM, Yaacoobi O, Addison JB, Wang Q, Shen Z, Briggs SP, Bohlmann J, Castro-Falcon G, Hughes CC, Betsiashvili M, Huffaker A, Schmelz EA, Zerbe P (2018) Discovery, biosynthesis and stress-related accumulation of dolabradiene-derived defenses in maize. Plant Physiol 176:2677–2690

Malook SU, Maqbool S, Hafeez M, Karunarathna SC, Suwannarach N (2022) Molecular and biochemical mechanisms of elicitors in pest resistance. Life 12:844

Marques JPR, Hoy JW, Appezzato-da-Glória B, Viveros AFG, Vieira MLC, Baisakh N (2018) Sugarcane cell wall-associated defense responses to infection by *Sporisorium scitamineum*. Front Plant Sci 9:698

Martínez-Cruz J, Romero D, Hierrezuelo J, Thon M, de Vicente A, Pérez-García A (2021) Effectors with chitinase activity (EWCAs), a family of conserved, secreted fungal chitinases that suppress chitin-triggered immunity. Plant Cell 33:1319–1340

Matsuura HN, Fett-Neto AG (2015) Plant alkaloids: main features, toxicity, and mechanisms of action. Plant Toxins 2:1–15

Medina-Puche L, Tan H, Dogra V, Wu M, Rosas-Diaz T, Wang L, Ding X, Zhang D, Fu X, Kim C, Lozano-Duran R (2020) A defense pathway linking plasma membrane and chloroplasts and co-opted by pathogens. Cell 182:1109–1124

Meena M, Yadav G, Sonigra P, Nagda A, Mehta T, Swapnil P, Harish N, Marwal A (2022) Role of elicitors to initiate the induction of systemic resistance in plants to biotic stress. Plant Stress 5:100103

Narasimhamurthy K, Udayashankar AC, De Britto S, Lavanya SN, Abdelrahman M, Soumya K, Shetty HS, Srinivas C, Jogaiah S (2022) Chitosan and chitosan-derived nanoparticles modulate enhanced immune response in tomato against bacterial wilt disease. Int J Biol Macromol 220:223–237

Nguyen NH, Trotel-Aziz P, Clément C, Jeandet P, Baillieul F, Aziz A (2022) Camalexin accumulation as a component of plant immunity during interactions with pathogens and beneficial microbes. Planta 255:116

Ninkuu V, Zhang L, Yan J, Fu Z, Yang T, Zeng H (2021) Biochemistry of terpenes and recent advances in plant protection. Int J Mol Sci 22:5710

Noman A, Aqeel M, Irshad MK, Qari SH, Hashem M, Alamri S, AbdulMajeed AM, Al-Sadi AM (2020) Elicitins as molecular weapons against pathogens: consolidated biotechnological strategy for enhancing plant growth. Crit Rev Biotechnol 40:821–832

Pandey S (2018) Physiological and biochemical changes associated with the development of brown spot diseases in rice leaves. Int J Adv Agric 5:69–78

Pane C, Fratianni F, Parisi M, Nazzaro F, Zaccardelli M (2016) Control of Alternaria post-harvest infections on cherry tomato fruits by wild pepper phenolic-rich extracts. Crop Prot 84:81–87

Pereira AG, Cassani L, Garcia-Oliveira P, Otero P, Mansoor S, Echave J, Xiao J, Simal-Gándara J, Prieto MA (2023) Chapter – plant alkaloids: production, extraction, and potential therapeutic properties. In: Caracho M, Heleno SA, Barros L (eds) Natural secondary metabolites, 1st edn. Springer, Cham, pp 157–200

Pirc K, Hodnik V, Snoj T, Lenarčič T, Caserman S, Podobnik M, Böhm H, Albert I, Kotar A, Plavec J, Borišek J, Damuzzo M, Magistrato A, Brus B, Sosič I, Gobec S, Nürnberger T, Anderluh G (2021) Nep1-like proteins as a target for plant pathogen control. PLoS Pathog 17:1009477

Popa CM, Tabuchi M, Valls M (2016) Modification of bacterial effector proteins inside eukaryotic host cells. Front Cell Infect Microbiol 6:73

Pratyusha S (2022) Chapter 7 – phenolic compounds in the plant development and defense: an overview. In: Hasanuzzaman M, Nahar K (eds) Plant stress physiology – perspectives in agriculture, 1st edn. IntechOpen, pp 125–140

Pring S, Kato H, Imano S, Camagna M, Tanaka A, Kimoto H, Chen P, Shrotri A, Kobayashi H, Fukuoka A, Saito M, Suzuki T, Terauchi R, Sato I, Chiba S, Takemoto D (2023) Induction of plant disease resistance by mixed oligosaccharide elicitors prepared from plant cell wall and crustacean shells. Physiol Plant 175:14052

Pusztahelyi T, Holb IJ, Pócsi I (2015) Secondary metabolites in fungus-plant interactions. Front Plant Sci 6:573

Pusztahelyi T, Holb IJ, Pócsi I (2016) Chapter 1 – plant-fungal interactions: special secondary metabolites of the biotrophic, necrotrophic, and other specific interactions. In: Merillon JM, Ramawat K (eds) Fungal metabolites, reference series in phytochemistry, 2nd edn. Springer, Cham, pp 1–58

Ramachandran P, Joshi BJ, Maupin-Furlow JA, Uthandi S (2021) Bacterial effectors mimicking ubiquitin-proteasome pathway tweak plant immunity. Microbiol Res 250:126810

Rashad YM, Aseel DG, Hafez EE (2018) Antifungal potential and defense gene induction in maize against Rhizoctonia root rot by seed extract of *Ammi visnaga* (L.) Lam. Phytopathol Mediterr 57:73–88

Ray S, Casteel CL (2022) Effector-mediated plant–virus–vector interactions. Plant Cell 34:1514–1531

Remick BC, Gaidt MM, Vance RE (2023) Effector-triggered immunity. Annu Rev Immunol 41:453–481

Rendina N, Nuzzaci M, Scopa A, Cuypers A, Sofo A (2019) Chitosan-elicited defense responses in cucumber mosaic virus (CMV)-infected tomato plants. J Plant Physiol 234-235:9–17

Riedlmeier M, Ghirardo A, Wenig M, Knappe C, Koch K, Georgii E, Dey S, Parker JE, Schnitzler J, Vlot AC (2017) Monoterpenes support systemic acquired resistance within and between plants. Plant Cell 29:1440–1459

Riseh RS, Vazvani MG, Kennedy JF (2023) Β–glucan-induced disease resistance in plants: a review. Int J Biol Macromol 253:127043

Rosenkranz M, Chen Y, Zhu P, Vlot AC (2021) Volatile terpenes–mediators of plant-to-plant communication. Plant J 108:617–631

Scala V, Pietricola C, Farina V, Beccaccioli M, Zjalic S, Quaranta F, Fornara M, Zaccaria M, Momeni B, Reverberi M, Iori A (2020) Tramesan elicits durum wheat defense against the Septoria disease complex. Biomol Ther 10:608

Schoonbeek H, Wang H, Stefanato FL, Craze M, Bowden S, Wallington E, Zipfel C, Ridout CJ (2015) *Arabidopsis* EF-Tu receptor enhances bacterial disease resistance in transgenic wheat. New Phytol 206:606–613

Shaheen HA, Issa MY (2020) In vitro and in vivo activity of *Peganum harmala* L. alkaloids against phytopathogenic bacteria. Sci Hortic 264:108940

Shaikh S, Shriram V, Khare T, Kumar V (2020) Biotic elicitors enhance diosgenin production in *Helicteres isora* L. suspension cultures via up-regulation of CAS and HMGR genes. Physiol Mol Biol Plants 26:593–604

Shakeran Z, Keyhanfar M, Asghari G, Ghanadian MT (2015) Improvement of atropine production by different biotic and abiotic elicitors in hairy root cultures of *Datura metel*. J Biol 39:111–118

Shao D, Smith DL, Kabbage M, Roth MG (2021) Effectors of plant necrotrophic fungi. Front Plant Sci 12:687713

Slatnar A, Mikulič-Petkovšek M, Veberič R, Štampar F (2016) Research on the involment of phenoloics in the defence of horticultural plants. Acta Agric Slov 107:183–189

Smith GH, Roberts JM, Pope TW (2018) Terpene based biopesticides as potential alternatives to synthetic insecticides for control of aphid pests on protected ornamentals. J Crop Prot 110:125–130

Srivastava S, Conlan XA, Cahill DM, Adholeya A (2016) *Rhizophagus irregularis* as an elicitor of rosmarinic acid and antioxidant production by transformed roots of *Ocimum basilicum* in an in vitro co-culture system. Mycorrhiza 26:919–930

Tak Y, Kumar M (2020) Chapter 13 – phenolics: a key defence secondary metabolite to counter biotic stress. In: Lone R, Shuab R, Kamili A (eds) Plant phenolics in sustainable agriculture, vol 1, 1st edn. Springer, Singapore, pp 309–329

Talhaoui N, Taamalli A, Gómez-Caravaca AM, Fernández-Gutiérrez A, Segura-Carretero A (2015) Phenolic compounds in olive leaves: analytical determination, biotic and abiotic influence, and health benefits. Int Food Res 77:92–108

Tanase C, Bujor OC, Popa VI (2019) Chapter 3 - phenolic natural compounds and their influence on physiological processes in plants. In: Watson RR (ed) Polyphenols in plants, isolation, purification and extract preparation, 2nd edn. Academic Press, pp 45–58

Tang D, Wang G, Zhou JM (2017) Receptor kinases in plant-pathogen interactions: more than pattern recognition. Plant Cell 29:618–637

Tang W, Lei X, Liu X, Yang F (2021) Nutritional improvement of bean sprouts by using chitooligosaccharide as an elicitor in germination of soybean (Glycine max L.). Appl Sci 11:7695

Taurino M, Ingrosso I, D'amico L, Domenico SD, Nicoletti I, Corradini D, Santino A, Giovinazzo G (2015) Jasmonates elicit different sets of stilbenes in *Vitis vinifera* cv. Negramaro cell cultures. Springer Plus 4:49
Teoh ES (2016) Chapter 5- secondary metabolites of plants. In: Medicinal orchids of Asia, 1st edn. Springer, Cham, pp 59–73
Tiku AR (2018) Chapter 13- antimicrobial compounds and their role in plant defense. In: Singh A, Singh IK (eds) Molecular aspects of plant-pathogen interaction, 1st edn. Springer, pp 283–307
Toruño TY, Stergiopoulos I, Coaker G (2016) Plant-pathogen effectors: cellular probes interfering with plant defenses in spatial and temporal manners. Annu Rev Phytopathol 54:419–441
Tripathi D, Raikhy G, Kumar D (2019) Chemical elicitors of systemic acquired resistance-salicylic acid and its functional analogs. Curr Plant Biol 17:48–59
Tsimogiannis D, Oreopoulou V (2019) Chapter 16- classification of phenolic compounds in plants. In: Watson RR (ed) Polyphenols in plants, isolation, purification and extract preparation, 2nd edn. Academic press, pp 263–284
Tugizimana F, Djami-Tchatchou AT, Steenkamp PA, Piater LA, Dubery IA (2019) Metabolomic analysis of defense-related reprogramming in Sorghum bicolor in response to *Colletotrichum sublineolum* infection reveals a functional metabolic web of phenylpropanoid and flavonoid pathways. Front Plant Sci 9:1840
Upson JL, Zess EK, Białas A, Wu CH, Kamoun S (2018) The coming of age of EvoMPMI: evolutionary molecular plant–microbe interactions across multiple timescales. Curr Opin Plant Biol 44:108–116
Üstün S, Hafrén A, Liu Q, Marshall RS, Minina EA, Bozhkov PV, Vierstra RD, Hofius D (2018) Bacteria exploit autophagy for proteasome degradation and enhanced virulence in plants. Plant Cell 30:668–685
Wang YC, Qian WJ, Li NN, Hao XY, Wang L, Xiao B, Wang XC, Yang YJ (2016) Metabolic changes of caffeine in tea plant (*Camellia sinensis* (L.) O. Kuntze) as defense response to *Colletotrichum fructicola*. J Agric Food Chem 64:6685–6693
Wang C, Liu R, Lim G, De Lorenzo L, Yu K, Zhang K, Hunt AG, Kachroo A, Kachroo P (2018) Pipecolic acid confers systemic immunity by regulating free radicals. Sci Adv 4:4509
Wang Y, Pruitt RN, Nürnberger T, Wang Y (2022) Evasion of plant immunity by microbial pathogens. Nat Rev Microbiol 20:449–464
Wawrosch C, Schwaiger S, Stuppner H, Kopp B (2014) Lignan formation in hairy root cultures of edelweiss (*Leontopodium nivale* ssp. alpinum (Cass.) Greuter). Fitoterapia 97:219–223
Woźniak M, Waśkiewicz A, Ratajczak I (2022) The content of phenolic compounds and mineral elements in edible nuts. Molecules 27:4326
Wu D, Qi T, Li W, Tian H, Gao H, Wang J, Ge J, Yao R, Ren C, Wang X, Liu Y, Kang L, Ding S, Xie D (2017) Viral effector protein manipulates host hormone signaling to attract insect vectors. Cell Res 27:402–415
Xoca-Orozco LÁ, Aguilera-Aguirre S, Vega-Arreguín J, Acevedo-Hernández G, Tovar-Pérez E, Stoll A, Herrera-Estrella L, Chacón-López A (2019) Activation of the phenylpropanoid biosynthesis pathway reveals a novel action mechanism of the elicitor effect of chitosan on avocado fruit Epicarp. Food Res Int 121:586–592
Yang F, Luan B, Sun Z, Yang C, Yu Z, Li X (2017) Application of chitooligosaccharides as antioxidants in beer to improve the flavour stability by protecting against beer staling during storage. Biotechnol Lett 39:305–310
Yang Y, Yang X, Dong Y, Qiu D (2018) The *Botrytis cinerea* xylanase BcXyl1 modulates plant immunity. Front Microbiol 9:2535
Zeng H, Yao N (2022) Sphingolipids in plant immunity. Phytopathol Res 4:20
Zhan C, Lei L, Liu Z, Zhou S, Yang C, Zhu X, Guo H, Zhang F, Peng M, Zhang M, Li Y, Yang Z, Sun Y, Shi Y, Li K, Liu L, Shen S, Wang X, Shao J, Jing X, Luo J (2020) Selection of a subspecies-specific diterpene gene cluster implicated in rice disease resistance. Nat Plants 6:1447–1454

Zhang R, Zhang H, Li L, Feng Y, Yang L, Liu J, Gao S, Zhou Z, Yang Y, Xiang H, Li M, Yu J (2024) Limonene's bacteriostatic activity against *Ralstonia solanacearum* by compromised membrane integrity: a transcriptomic and Metabolomic analysis. J Agric Chem Environ 13:312–324
Zhou J, Zhang Y (2020) Plant immunity: danger perception and signaling. Cell 181:978–989
Zipfel C (2014) Plant pattern-recognition receptors. Trends Immunol 35:345–351
Złotek U, Wójcik W (2014) Effect of arachidonic acid elicitation on lettuce resistance towards *Botrytis cinerea*. Sci Hortic 179:16–20
Zou J, Zhao L, Yi P, An Q, He L, Li Y, Lou H, Yuan C, Gu W, Huang L, Hu Z, Hao X (2020) Quinolizidine alkaloids with antiviral and insecticidal activities from the seeds of *Sophora tonkinensis Gagnep*. J Agric Food Chem 68:15015–15026

Chapter 15
Trichoderma and Plant Immunity: A Synergistic Approach to Biotic Stress

Kamel A. Abd-Elsalam, Toka E. Abdelkhalek, Rawan K. Hassan, Ramadan A. Arafa, and Hassan Almoammar

Abstract Various biotic stresses in agroecosystems dramatically impact sustainable agriculture, influencing both the quantity and quality of global crop production. An adequate eco-friendly approach to mitigate the negative effects of biotic stress is indispensable. Trichoderma species promote and enhance plant growth through various specialized mechanisms, such as biologically controlling soil-borne diseases by producing and activating enzymes, generating antibiotics, and invading phytopathogenic fungi. Additionally, as the most prevalent saprophytic fungi in the rhizosphere, Trichoderma acts as a mycoparasite, aiding plants in coping with biotic stresses. This chapter discusses the genetics, mechanisms, and uses of *Trichoderma* species, with particular emphasis on mycoparasitism, antibiosis, and the biosynthesis of volatile organic compounds and secondary metabolites. This study also details the manner in which Trichoderma is induced through systemic resistance by signaling pathways that involve ethylene, salicylic acid, and jasmonic acid. Moreover, the impact on plant growth and yield is also evaluated with respect to the interaction between the soil microbial population, the host plant, and Trichoderma. In this chapter, the potential of *Trichoderma* species in sustainable agriculture is explored, together with the challenges related to their use, including taxonomy, biological properties, and practical application.

Keywords Trichoderma · Biocontrol agent · Biotic stress mitigation · Systemic resistance · Volatile organic compounds · Plant growth promotion · Sustainable agriculture

---

K. A. Abd-Elsalam (✉) · R. A. Arafa
Plant Pathology Research Institute, Agricultural Research Center, Giza, Egypt

T. E. Abdelkhalek · R. K. Hassan
Biotechnology English Program, Faculty of Agriculture, Cairo University, Giza, Egypt

H. Almoammar
Wellness and Preventive Medicine Institute, Health Sector, King Abdulaziz City for Science and Technology (KACST), Riyadh, Saudi Arabia

K. A. Abd-Elsalam, H. I. Mohamed (eds.), *Elicitors for Sustainable Crop Production*, Sustainability Sciences in Asia and Africa,
https://doi.org/10.1007/978-981-96-6569-3_15

15.1 Introduction

Many techniques are being examined to address the challenge of increasing worldwide demand for ecologically friendly agriculture, one of which is the use of Trichoderma species, fungi that benefit plants. As Khan et al. (2021) note, these fungi are currently used to reduce the incidence of soil-borne diseases, promote plant development, and improve soil fertility through mechanisms such as mycoparasitism, antibiosis, and resource competition. By producing signals such as ethylene (ET), salicylic acid (SA), and jasmonic acid (JA), many Trichoderma species have been demonstrated to cause systemic resistance in plants in addition to controlling pathogens (Hermosa et al. 2012). This relationship does enhance not only the immune system of the plant but also the growth and yield. Furthermore, Trichoderma promotes nutrient availability and root system growth by solubilizing minerals and generating phytohormones (Metwally 2020; Fazeli-Nasab et al. 2022).

The application of Trichoderma has recently been extended by the development of technology. The field durability and the performance of Trichoderma strains have been enhanced by recent innovations such as microencapsulation, while the biogenic synthesis of nanoparticles using Trichoderma metabolites presents new options for pest and pathogen control (Rajput et al. 2021). However, these technologies may also have some drawbacks, for instance, the possible adverse effects on other beneficial soil microbes and the environment (Ousley et al. 1993; Abdullah et al. 2021; Asghar et al. 2024). This chapter focuses on the use of Trichoderma species in controlling biotic stress in plants, with emphasis on the biological control mechanisms, relationships with the host plants and soil microbial populations, and their use in practical agriculture.

15.2 Exploring Trichoderma

The first characterization of Trichoderma was done in Germany by Persoon (1794). The first isolation of this fungus in India was done by Thakur and Norris in 1928 in Madras. This fungus was determined to be cellulolytic toward the end of World War II and given the name QM6a *Trichoderma viride* which was later changed to *T. reesei* by Elwyn *T. Reese* because it could decay wood (Simmons 1977). The first mycoparasite, *T. lingorum* (Tode) Harz. (H. virens), was discovered in 1932 as a biocontrol agent against *Rhizoctonia solani*, and gliotoxin, which was the first antimicrobial substance from different Trichoderma species, was discovered in 1934 (Weindling 1934). Gutter (1957) presented a research paper on the effect of light on different conidiation of *T. reesei* in the year 1957. The first time the genus Trichoderma was classified was in 1969 by Rifai (1969) which provided a concept for the characterization of species within the genus of Trichoderma, and by the year 2006, more than 100 species had been recognized (Druzhinina et al. 2006). Pathogen suppression by *T. harzianum* in the field was first recorded in 1972 (Wells et al.

1972). The first cellulase of *T. reesei* was cloned in 1983 through studies of cloning on Trichoderma species, and the first related mycoparasitism gene (prb1) and its induction through cell walls were cloned in 1993 (Geremia et al. 1993). Trichoderma's capacity to promote plant development was initially identified in 1986 (Chang et al. 1986). In particular, the genus' plant immunity was increased by internal colonization of the root systems by Trichoderma (Yedidia et al. 1999) and resistance induction (Bigirimana et al. 1997). 1989 saw the registration of Trichoderma-based BCA product Binab-T, the first commercial formulation used for the biological treatment of different plant diseases (Collinge et al. 2022). *Trichoderma* species are commonly found in soils, as well as on wood and vegetable debris. While many strains are non-pathogenic to growing plants, some aggressive strains have caused severe diseases to commercial mushrooms and soil-borne pathogens. Mitosis only allows Trichoderma species to reproduce asexually (Harman et al. 1998), and meiosis is not required (Harman et al. 1993; Kistler and Miao 1992). The genus is pycnidial with respect to pleiomorphism, the sexual stage of the fungus is Hypocrea, the asexual stage is Trichoderma, and sometimes, the varieties are grouped as Hypocrea/Trichoderma. Though much remains unknown about this genus, recent progress has been significant.

Trichoderma is widely used in agriculture and environmental management because it is useful (Woo et al. 2014). It is an effective biocontrol agent against pathogens such as Fusarium and Phytophthora which cause diseases in soil (Saravanakumar et al. 2017). In addition, Trichoderma enhances root development and phosphorus and micronutrient uptake to enhance nutrient use efficiency and also solubilizes essential minerals. The species can also activate a form of defensing plants, called induced systemic resistance (ISR), to make plants more resistant to pathogens. It improves the soil quality by enhancing the biological processes and movement of nutrients and also decomposes organic pollutants. Though primary infections by Trichoderma are infrequent, they can happen opportunistically in immunocompromised individuals (Woo et al. 2023).

15.2.1 Classification of Trichoderma Fungi

Trichoderma belongs to the kingdom Fungi, phylum Ascomycota, class Euascomycetes, order Hypocreales, and family Hypocreaceae (Esposito and Silva 1998). Fungi from this genus are filamentous, widespread saprobes and quite common in soil, decaying vegetation and plant material. Some Trichoderma species are linked to the teleomorphs of the genus Hypocrea (Doi 1969, 1972). Trichoderma is a genus that includes many species, including *T. harzianum, T. koningii, T. longibrachiatum, T. pseudokoningii*, and *T. viride*, which are well-known (Bissett 1991). Recent molecular studies have revealed that the recognized diversity within this genus has increased from nine species to over 33, reflecting advancements in taxonomic resolution through genetic analysis (Meyer et al. 1992; Muthumeenakshi et al. 1994; Jambhulkar et al. 2024). Recently, about 35 species of Trichoderma have been dependent on the basis of morphological and molecular differences (Kullnig et al. 2000). These species are of importance in the soil and other environments, and an attempt has been made to find out their biodiversity globally. A particular study was made on Trichoderma species in Russia, Siberia, and the Himalayan Mountains. Two regions of cluster from rDNA of ITS 1 and 2 sequences of 75 isolates were compared with ex-typed strains and established isolates dependent on taxonomy. A total of 39 strains were assigned to *T. atroviride, T. virens, T. hamatum, T. asperellum, T. koningii,* and *T. oblongisporum*. This investigation revealed that *T. harzianum* is the most common species genetically within the genus (Kullnig et al. 2000; Baazeem et al. 2021).

The classification of Trichoderma has primarily relied on the morphological features of the fungus. However, the emergence of cryptic species and difficulties in distinguishing the morphological characteristics of certain species have necessitated the use of molecular methods for more accurate classification. Techniques such as DNA sequencing and phylogenetic analysis have been crucial in clarifying taxonomic ambiguities and establishing species identities (Oskiera et al. 2015; Schalamun and Schmoll 2022).

15.2.2 Trichoderma Genetics in Stress Tolerance

Traditionally, genes of interest were cloned using heterologous probes or degenerate PCR. This was succeeded by isolating the full-length genes by screening cDNA, cosmid, or BAC libraries, often requiring some ad hoc strain of manual sequencing work. As a result, getting a full-length gene was a resource-intensive exercise. Gene functions were commonly studied with anti-sense or RNA interference and homologous recombination methods both single- and double-cross-over methods (Mukherjee 2015) either to silence, disrupt, or delete target genes. For Trichoderma biocontrol agents to succeed in natural conditions (Table 15.1), they need to cope with different environmental stresses, including drought, salinity, excessive temperature, and pH variation. As an example, a gene from *T. atroviride, seb1,* that encodes a stress response element binding protein has been implicated in high osmolarity stress tolerance (Peterbauer et al. 2002). Also, *T. virens Hsp23* (a heat shock protein) heterologous overexpression has been demonstrated to increase *T. harzianum* thermo-tolerance as well. Other heat shock proteins such as Hsp70 could also contribute to enhanced thermo-tolerance by *T. harzianum* (Montero-Barrientos et al. 2008; Moran-Diez et al. 2010). Further elucidated role of the *T. harzianum* Dim1 protein pertinence to the TRX fold super-familyin oxidative stress tolerance through over-expression and gene silencing experiments. The nox1 gene from *T. harzianum*, which encodes a NADPH oxidase, has been implicated in biocontrol efficacy against different pathogens such as *Pythium ultimum* (Montero-Barrientos et al. 2011). Hernández-Oñate et al. (2012) explained that NADPH oxidase-dependent reactive oxygen species (ROS) production is crucial for damage response in *T. atroviride*, highlighting the essential roles of the nox1 and noxR genes.

Table 15.1 Various genes of Trichoderma and their role in stress tolerance

Trichoderma species	Genes	Stress type	Pathogen	Disease	References
T. harzianum	*hsp70, Ech42, nag1* (encodes β-N-acetylglucosaminidase), *gla1* (encodes a glycosyl hydrolase), *prb1* (serine protease), *chit33, chit42, Dim1, nox1*	Heat, salt, osmotic, and oxidative tolerances, Mycoparasitism, nutritional stress, exposure to light.	*Rhizoctonia solani, Sclerotium rolfsii, Botrytis cinerea*	Biocontrol of fungal pathogens, inhibits germination of spore and of elongation germ tube	Sharma et al. (2011), Guzmán-Guzmán et al. (2023) and Abbas et al. (2022)
T. Asperellum	*echi42* *chit36Y* 10,541 genes for *T. asperellum* ND-1 and 12,802 genes for *T. asperellum* CBS 433.97	Production of volatile organic compounds (VOCs), expression of genes that are involved in secondary metabolism, degradation of cell wall, and stress tolerance, drought stress	*Rhizoctonia solani, Fusarium oxysporum, Pythium* spp., *Phytophthora* spp.	Biocontrol of soil-borne pathogens, plant growth promotion, ISR, alleviate water deficit stress symptoms	Abbas et al. (2022); Hanaka et al. (2021), and Rawal et al. (2022)
T. atroviride	*Tga1, Tga3, Tmk1, tex1, pga1* (polygalacturonase), *seb18, nox18, noxR8*	Nitrogen shortage	*Botrytis cinerea, Rhizoctonia solani*	Decreased mycoparasitism	Abbas et al. (2022) and Guzmán-Guzmán et al. (2023)
T. hamatum	42 kDa endo chitinase gene tam-ch, B-glucosidase gene, genes responsible for biotic and abiotic stress responses and metabolism of RNA, DNA, and protein, monooxygenase genes, genes involved in the secretion of cysteine-rich proteins, osmotins, and osmotin-like protein	Biotic and abiotic stresses, salt stress, low pH, nutrient poor peat, osmotic stress	*Xanthomonas euvesicatoria, Acidovorax avenae, X. campestris, S. sclerotiorum, Rhizoctonia solani, Alternaria* spp., *Ralstonia solanacearum, Sclerotinia minor, and Sclerotium cepivorum, Phytophthora capsica, Pythium ultimum*	Induces systemic resistance; antibacterial and antifungal activity; mycoparasitic interactions; inhibits growth; promotes growth, damping-off of seedling	Alfano et al. (2007), Lodi et al. (2023), Studholme et al. (2013), Elshahawy & El-Mohamedy (2019) and Bae et al. (2009)

(continued)

Table 15.1 (continued)

Trichoderma species	Genes	Stress type	Pathogen	Disease	References
T. koningii	Chitinase, glucanase, protease genes	Abiotic stresses, mycoparasitism, competition, antibiosis	*Gaeumannomyces graminis var. tritici, Sclerotium rolfsii, Sclerotium cepivorum, Rhizoctonia solani, Fusarium oxysporum*	Induces resistance against fungal pathogens and stimulates growth, take-all disease in wheat, stem rot, inhibits growth of pathogens	Dutta et al. (2023) and Yao et al. (2023); Sharma et al. (2012)
T. virens	TmkA, Tvk1, Gliotoxin biosynthetic genes, Hsp23	Conidiation triggered by light or nutrient stress	*Rhizoctonia solani, Sclerotium rolfsii*	Increased secretion of lytic enzymes, efficient disease control, mycoparasitic coiling, and lysis of host hyphae	Abbas et al. (2022) and Mukherjee et al. (2022); Mukherjee et al. (2003)
T. Oblongisporum	*tef1*	Biotic and abiotic stresses	*Lentinula edodes*	Green mold disease distorts and inhibits mycelia	Cao et al. (2014)

Besides, the advances in molecular genetics have enabled us to better understand the role of categories of genes that are involved in the tolerance to stress in species of Trichoderma. Besides that, the understanding of these ecological functions informs their use as biocontrol agents (Montero-Barrientos et al. 2007; Shahriar et al. 2022) and these genetic mechanisms are being studied in detail and have opened potential avenues to improve the crop resilience against diverse abiotic stresses through biotechnological approaches.

15.3 Trichoderma Used as a Biocontrol Agent

Approximately 100 species of Trichoderma genus are distributed across the globe due to their rapid development, their ability to utilize various surfaces, and their resilience to a broad range of environmental factors and pollutants (Hu et al. 2020). Its application as a biocontrol agent in agriculture and as an enzyme producer in various sectors is its primary present commercial interest (Jangir et al. 2017). Trichoderma primarily interacts with the plant as a root endophyte, occupying only the outermost layers of the root, due to a plant defensive response mediated by SA that prevents the fungus from reaching the vascular bundles and acting as a systemic pathogen (Poveda 2021).

15.3.1 Role in Controlling Plant Pathogens

Its principal habitats include soil, air, plant surfaces, and natural places. Biological control occurs when a biocontrol agent is administered to a host plant in order to inhibit the spread of pathogen-caused plant diseases (Awad et al. 2021). Many studies have extensively established the ability of some strains of Trichoderma to suppress phytopathogens via both direct and indirect methods (Alfiky and Weisskopf 2021; Kamel et al. 2020; Li et al. 2020). This includes metabolite synthesis, competitive exclusion, mycoparasitism, and antibiosis. On the other hand, systemic acquired resistance (SAR) and ISR are indirect pathways that activate the plant immune responses (Prismantoro et al. 2024). By using the antibiosis mechanism, Trichoderma, a biological fungus commonly employed to combat plant pests, may effectively cure a variety of plant diseases (Wang et al. 2022). Trichoderma's ability to inhibit the growth of plant-pathogenic fungi by secreting antagonistic compounds is the primary definition of antibiosis. They are thought to be particularly effective against soil-borne and foliar diseases (Poveda et al. 2020). Trichoderma has shown high resistance to mildew in cucumber and Botrytis cinerea in tomato, cucumber, bean, and strawberry (Maurya et al. 2024; Zhang et al. 2013). A comprehensive analysis is required to create feasible commercial products, such as low-cost biocontrol, Trichoderma strains, and potential plant disease control strategies.

15.3.2 Interaction with Fungal, Bacterial, and Nematode Pathogens

These interactions are generally beneficial for plants; they increase the growth and tolerance to biotic and abiotic stresses of the plants. In this interacting environment, Trichoderma emerges a number of effectors that tune plant development and immunity. It has been found by Ramírez-Valdespino et al. (2019) that Trichoderma interacts with plants by several mechanisms such as production of specific proteins, sRNA, and volatile organic compounds (VOCs). Trichoderma-mediated plant growth regulation could be caused by both direct effects of released molecules on plants and indirect effects of Trichoderma's influence and alteration of the surrounding environment, such as changing the soil microbiome or lowering the soil pH to increase the availability of macro- and micronutrients in plants (Tandon et al. 2020). Oljira et al. (2020) reported that *T. yunnanense and T. afroharzianum* may improve water usage efficiency and net photosynthesis. Members of this genus have a great deal of potential for use in agriculture as biofertilizers, biostimulants, and biocontrol agents. Numerous recent reports have highlighted how *Trichoderma* spp. can modify physiological, biochemical, and molecular mechanisms in a broad range of plants under different conditions of growth (Sun et al. 2020; Lombardi et al. 2020; Guzmán-Guzmán et al. 2019).

15.3.3 Synergistic Effects with Other Biocontrol Agents

In business environments, synthetic chemical fungicides are often more effective than biocontrol agents. As a result, scientists have explored the idea of enhancing the efficacy of biocontrol remedies by combining them with treatments. These additional options include using more biocontrol agents and natural substances and physical treatments, such as heat, ultraviolet C exposure, cold therapy, and other methods (Li et al. 2025). Research has shown that mixing two natural pest control agents in a solution, like pairing yeast with bacteria, offers improved effectiveness, in combating diseases, in apple fruits compared to using one type of combatant (Calvo et al. 2010). Trichoderma strains and pyraclostrobin a specific substance used to treat red pepper anthracnosis outside its environment have displayed potential, in combating anthracnose caused by *C. acutatum* in red pepper fruits preventative way (as stated by Dutta et al. 2023). When two strains at doses were combined with pyraclostrobin fungicide, there was a reduction in anthracnosis severity. This indicates that combining these two Trichoderma strains with fungicides could enhance biocontrol effectiveness. To ensure that Trichoderma works efficiently as a biocontrol agent, combatting illnesses and managing them effectively requires an approach involving mycoparasitism and antibiosis (as stated by Sood et al. 2020). A combination of mycoparasitism alongside exclusion benefits and induced resistance is crucial, for the suppression and control of diseases mediated by Trichoderma.

15.4 Induction of Systemic Resistance in Plants

It is acknowledged that the exchange of microbial and plant elicitors required for interactions between organisms causes the development of opposition, even if the biochemical and molecular processes underpinning Trichoderma-induced plant resistance have not yet been extensively explored (Woo & Pepe 2018). Therefore, a live fungus is not always necessary to stimulate plant defensive systems (Mukherjee et al. 2012). The interaction of these molecules with plant receptors may decide the adherence and recognition of Trichoderma, followed by the induction of resistance in plants, depending on the implicated elicitors secreted by Trichoderma hyphae. The chemicals that Trichoderma releases highlight its capacity to cause resistance (Schuster and Schmoll 2010). It is well known that several protein classes interact with Trichoderma and help combat infections in plants. Fungal hyphae release small extracellular cysteine-rich proteins made by Trichoderma strains that are crucial for mycoparasitism and resistance induction, particularly in the presence of a host plant (Mathys et al. 2012). Among the tiny proteins, are expansin-like proteins associated with cell wall growth and hydrophobins, which originate from the outer layer of the cell wall covering the fungal surface and dictate Trichoderma adhesion to the root surface (Nawrocka and Malolepsza 2013).

15.4.1 Understanding Systemic Acquired Resistance and Induced Systemic Resistance

Effector-triggered immunity (ETI) is a rapid and effective immune response that leads to a hypersensitive response, causing localized cell death to prevent the growth of biotrophic fungal infections. ETI is effective against adapted pathogens, whereas pattern-triggered immunity confers broad and non-host resistance, working against non-adapted infections (Pieterse et al. 2012). Upon recognition of microbe-associated molecular patterns (MAMPs) by pattern recognition receptors on the surface of host cells, host plant immunity is induced. SAR is induced in leaves or the above-ground parts of the plant when pathogens come in contact with it, while the activation of ISR is stimulated by beneficial microorganisms in the plant and is initiated in the roots. The duration of SAR is more than 20 days after induction by necrotrophic infections and is effective against several pathogen species (Hermosa et al. 2013; Yedidia et al. 1999). Likewise, ISR provides a long-lasting systemic resistance against hemi- or biotrophic diseases with the help of beneficial soil microbes. Both SAR and ISR are highly effective against infections and produce a prolonged response. The ISR response, triggered by plant-beneficial microbes, also encourages plant development (Bonanomi et al. 2018). However, artificial SAR inducers can decrease plant growth, likely due to metabolic resource exchange (Yedidia et al. 1999). Phytohormones such as SA, JA, or ET are generally involved

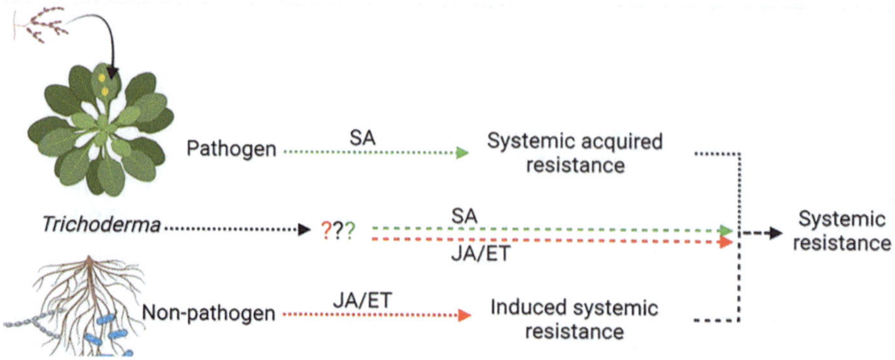

Fig. 15.1 Trichoderma ISR is used by microorganisms to provide protection to plants from many diseases. There is intricate intercommunication of signaling pathways that is activated, and it is triggered by the invasion of plants' rhizosphere by Trichoderma that uses specific chemicals to sense that the fungus is present. There are several signaling pathways such as ET and JA, SA, calcium, ROS, and nitric oxide that contribute to ISR but are not dependent on one of these alone. (Reprinted from Salwan et al. 2022)

in eliciting immune responses after exposure to pathogens (Fig. 15.1). SA plays no part in the ISR-mediated systemic response, which is mediated by JA and ET (Vos et al. 2015; Vlot et al. 2021). Trichoderma species induce both SAR and ISR via SA and JA/ET-mediated pathways, respectively. Trichoderma species boost plant growth and immunity through mutualistic relationships. They enhance plant immunity against phytopathogens through complex signaling events that promote disease resistance and plant growth via specific modulation of phytohormone signaling pathways, increased nutrient bioavailability, and enhanced nutrient uptake[1]. JA, ET, and SA open up pathways needed for SAR and are stimulated by Trichoderma, leading to a chain reaction of signal transduction processes that ultimately aid in defense immunity by regulating the expression of defense-related genes (Gimenez-Ibanez et al. 2009; Hammerschmidt 2009). Trichoderma hyphae can penetrate and insert themselves within the roots of a plant, causing a resistance response. Plants identify Trichoderma through MAMPs such as ergosterol and chitin oligomers, which are recognized by LysM receptor proteins in Arabidopsis. Xylanases and small cysteine-rich proteins secreted by Trichoderma act as elicitors that induce calcium fluxes and the production of ROS, which are part of innate immunity. Mitogen-activated protein kinase (MAPK) cascades are signaling modules that regulate a myriad of plant responses (Khan et al. 2023; Shoresh et al. 2006). The response of hypersensitivity and phytoalexin production depends on the activation of MAPKs such as MPK3 and MPK6. A MAPK induced by *T. asperellum* in cucumber is involved in the JA/ET signaling pathways.

15.4.2 *Role of Trichoderma in Signaling Pathways and as a Plant Growth Regulator*

Trichoderma species are free-living fungi and adaptable plant symbionts that are present in practically every terrestrial habitat. They are well known for facilitating water intake, boosting nutrient acquisition, and promoting plant development (Stewart and Hill 2014). Each isolate's capacity to stimulate plant development is unique, and it depends on a number of variables, including type of crop, growing environment, rate of inoculum, and formulation type. Through mineral chelation and solubilization, Trichoderma promotes greater nutrient absorption, and enhanced isolates can boost root development and nutrient uptake. Trichoderma colonization of plant roots promotes fertilizer and nutrient intake, which increases root growth. Trichoderma species produce gibberellic acid and ET, which change a number of physiological processes in plants, as well as indole-3-acetic acid (IAA), a plant hormone essential for root and plant growth (Sharifah 2018; Asma 2019).

It is possible for the fungus to promote plant development during root colonization or even without direct physical touch. Increases in root and/or shoot biomass are the most typical sign of growth promotion, while modifications to the general morphology and development of plants have also been seen (Doni et al. 2014; Subramaniam et al. 2022). Compounds secreted by Trichoderma can encourage root branching, which enhances the uptake of nutrients and water. Total biomass, height of seedling, length of root, area of root, diameter of root, number of root tips, and number of growth potential of seedlings can all be increased by Trichoderma inoculation (Harman et al. 2004). Trichoderma enhances transpiration, photosynthetic rates, and stomatal conductance, among other physiological functions in plants. They also increase the amount of water, the number of leaves and flowers, the length of the shoots and roots, the amount of chlorophyll, and the rate of photosynthetic activity. Significant biochemical changes in the plant's concentration of lipids, organic acids, amino acids, and carbohydrates can result from association with Trichoderma, potentially improving the health of the plant throughout its life cycle. Trichoderma species are among the greatest ISR agents and have the ability to boost defense responses in host plants (Sharifah 2018; Asma 2019). Some isolates also exhibit a favorable link between activity and germination, where ROS function as signal molecules. Trichoderma isolates can also have a positive impact on root weight, germination energy, and germination in pepper seedlings. Improved *T. harzianum* can boost tolerance against *R. solani* and greatly accelerate the growth of bean roots, increasing their weight and volume. Trichoderma can lessen the severity of illnesses like Fusarium wilt in plants such as cucumbers and bananas. *T. harzianum* can actively participate in plant defense mechanisms by promoting the exudation of enzymes such as peroxidase and polyphenol oxidase for *Phytophthora melonis* exudate. The beneficial impacts of Trichoderma on plants, including defense and growth mechanisms, are probably going to be passed down to future generations. In order to trigger an adaptive physiological response to abiotic stressors like drought, salinity, or environmental pollution, Trichoderma can either rewire and/or activate molecular systems that are frequently affected by IAA, ET, and abscisic acid (Doni et al. 2014; Verma et al. 2021).

15.5 Mechanisms of Biotic Stress Mitigation

To fully leverage the benefits of Trichoderma species as plant bio-stimulants in agriculture, effective application methods such as soil inoculation and seed coating are essential. Such fungi promote plant development through a series of mechanisms (Fig. 15.2), such as biological control of soil-borne pathogens by producing hydrolytic enzymes and antibiotics, and through the parasitization of phytopathogenic fungi (Fazeli-Nasab et al. 2022). They also trigger systemic defense responses in plants through the expression of defense-related genes and the production of bioactive metabolites that interact with plant roots (Khan et al. 2021). The usage of Trichoderma as a biocontrol agent has come into the limelight in the past decades. This fungus secretes a number of metabolites that can inhibit the growth of pathogens, bacteria, and oomycetes. It also has antifungal, antibacterial, and nematicidal

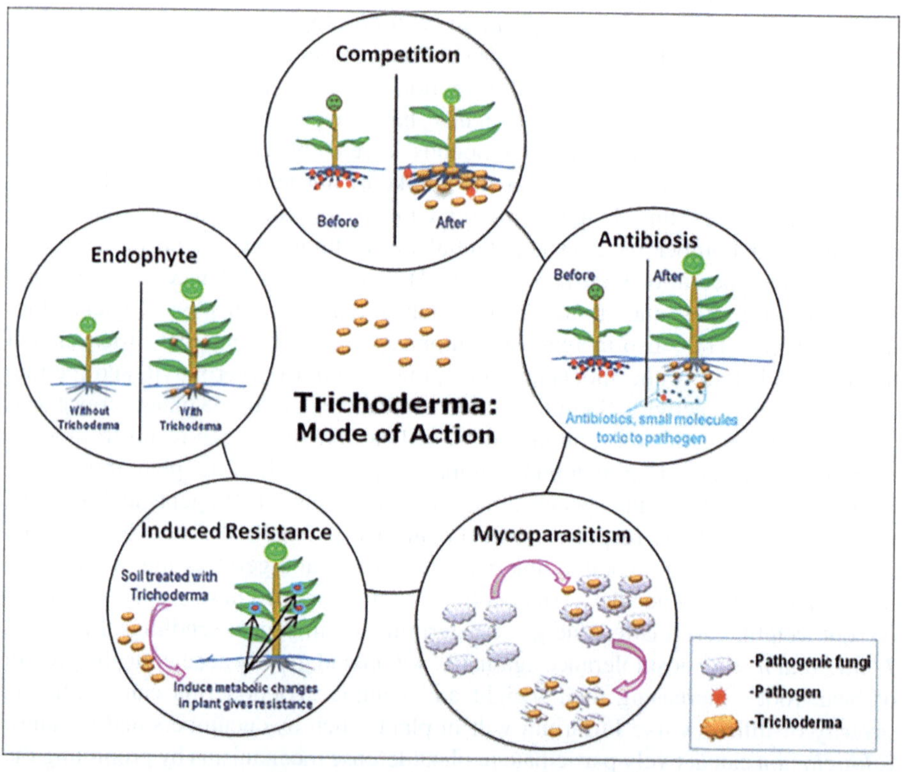

Fig. 15.2 Mode of activity of Trichoderma spp. as biocontrol agents of diseases caused by biotic stresses such as fungi, bacteria, and nematodes. The mode of activity of Trichoderma spp. is multifaceted and complicated in that it involves numerous mechanisms that synergize to repress pathogens. Mechanisms of activity of Trichoderma spp. have been described as broad categories of antibiosis, mycoparasitism, ISR, space and resource competition, and degradation of structures of pathogens by enzymes. (Reprinted from Waghunde et al. 2016)

activity, which makes it a good candidate for integrated pest management strategies (Shah et al. 2020; Zhang et al. 2020, 2021). The application of Trichoderma in sustainable agriculture has been increasing because of the advantages associated with it (Singh et al. 2015). It stimulates the growth of roots and therefore enhances water and other nutrient uptake and increases yield. The use of Trichoderma improves agricultural productivity; however, its continued application may result in adverse effects on soil quality and ecosystem functions. There are, however, some concerns that have been raised by researchers on the effects of heavy metal accumulation and the possibility of pathogenic microorganisms coming up as a result of the treatment of contaminated wastewater with Trichoderma (Khan et al. 2021). Recent studies have highlighted Trichoderma's potential as a long-term solution for increasing agricultural productivity while also addressing environmental concerns. The continued investigation of their processes in stress alleviation is critical for improving their application in sustainable agriculture methods (Gangwar and Singh 2018).

15.5.1 Mode of Action

Trichoderma antagonizes microbes through mechanisms that include nutrients, antibiosis competition, and even mycoparasitism (according to Gomes et al. 2015; Silva et al. 2019; Viterbo et al. 2002). In addition, Trichoderma also produces lytic enzymes that break down the cell tissue of the microbe and also induce host resistance. In addition to the direct antagonism, there are numerous indirect mechanisms and processes that Trichoderma employs, including the induction of biochemical and morphological changes in the host plant to enhance the plant's tolerance to stress. Such methods are effective because alteration of the host plant enhances overall defense responses (Abdelrahman and Jogaiah 2019). Other researchers have noted that, while engaging in mycoparasitism, the antagonist produces several lytic enzymes as well as antibiotics with potent activity that is effective toward the pathogen. These lytic enzymes and mycoparasitism work along with many other toxic metabolites that work together to defeat pathogens (Asad et al. 2015). Various species of Trichoderma are also capable of modifying the rhizosphere, such as soil acidification, as an alternative method to hinder the growth of the pathogen (Table 15.2).

15.5.2 Mycoparasitism

The use of Trichoderma spp. is most widespread for soil-borne pathogens and mycoparasitism is the most effective in limiting the development of the inoculum. The levels of disease suppression by mycoparasitic strains are greater than those mediated by antibiosis (Zeilinger et al. 2016). Mycoparasitism is a vital way by which Trichoderma species control biotic stress that affects plants. This process

Table 15.2 Functional role of Trichoderma enzymes in mitigating biotic stress

Trichoderma spp.	Mode of action	Enzyme(s)	Gene family (example)	References
T. Harzianum	Biocontrol of pathogens; promotes plant growth	Chitinases, β-1,3-glucanases, proteases, toxic volatile metabolites	Chitinases, cellulases, proteases	Harman et al. (2004), Benítez et al. (2004) and Qualhato et al. (2013)
T. Asperellum	Mycoparasitism, production of hydrolytic enzymes	Chitinases, β-1,3-glucanase, NAGAse, acid phosphatase, acid proteases, alginate lyase	Chitinases, cellulases, secondary metabolites	Howell (2003) and Qualhato et al. (2013)
T. Atroviride	Mycoparasitism, production of secondary metabolites, antibiosis, plant defense induction, and priming	Chitinases, cellulases	Chitinases, cellulases, proteases	Hjeljord and Tronsmo (1998), Guzmán-Guzmán et al. (2023); Morán-Diez et al. (2021)
T. Hamatum	Biological control of soilborne pathogens, ISR	Chitinases, Glucanases	Chitinases, lytic enzymes	Sivan et al. (1987), Papavizas (1985) and Manzar et al. (2022)
T. Koningii	Mycoparasitism, production of volatile compounds	Chitinases, proteases	Chitinases, cellulases	Ghisalberti and Sivasithamparam (1991)
T. Virens	Mycoparasitism, production of gliotoxin, antibiosis, terpenoid synthesis induction in cotton roots	Chitinases, gliotoxin synthetases	Chitinases, proteases, secondary metabolites	Woo et al. (2006); Vargas et al. (2014)
T. Oblongisporum	Mycoparasitism	Chitinases, Glucanases, proteases, degradation of cell wall enzymes	Chitinases, cellulases	Yu et al. (2009) and Zhang et al. (2017)

includes the direct parasitism of pathogenic fungi which reduces their population and incidence of plant diseases. Trichoderma secretes hydrolytic enzymes to digest the cell walls of pathogens and also produces antifungal secondary metabolites (Zeilinger et al. 2016).

Thus, Trichoderma is a useful organism in the biocontrol of plant diseases in sustainable agriculture based on its mycoparasitic activities (Mukherjee et al. 2022). However, where mycoparasitic fungi like Trichoderma are concerned, their mode of life is an essential and interesting aspect that leads to the production of cell wall degrading enzymes (CWDEs). These enzymes help in the breakdown of the complex structures of the cell walls of the host fungi and the latter are made of glucans, chitin, and proteins. (Gow et al. 2017; Garcia-Rubio et al. 2020). Chitinases, glucanases, and proteases are the major groups of CWDEs involved in mycoparasitism (Table 15.2). Chitin is the major component of fungal cell walls, and its degradation is very important for mycoparasitism to be effective. With regard to chitinolytic enzymes, Trichoderma is a genus of fungi that is rich in chitinase genes, especially in GH families 18 and 20 (Ihrmark et al. 2010; Kubicek et al. 2011). The GH18 family has also increased in number in other species including *T. virens*, *T. atroviride*, *T. harzianum,* and so on. Endochitinases like ECH42 are important in the metabolism of chitin in the cell wall of the host (Ihrmark et al. 2010; Kubicek et al. 2011). These enzymes react with chitosan, a chitin-deacetylated derivative that can chelate ROS generated by the host fungi. The role of chitosanase genes in *T. atroviride* is well recognized, and removing certain chitosanase genes affects the efficacy of the mycoparasitic process (Kappel et al. 2020). The other important class of fungal cell wall polymers is glucans, which are hydrolyzed by a variety of glucanases. These enzymes, which belong to the GH71 family, are less well-known than other CWDEs. Indeed, *T. harzianum* secretes α-1,3-glucanases that have antifungal properties against certain infections (Ait-Lahsen et al. 2001; Sanz et al. 2004). The enzymes involved in β-glucan breakdown are β-glucanases from families GH16,17, 55, 64, and 81 (Djonović et al. 2006a, b). The increase in the number of genes encoding these enzymes in Trichoderma species is evidence for their importance in mycoparasitism. Thus, the mutants that overexpress the genes encoding β-glucanase displayed improved antagonism toward the pathogen *Globisporangium ultimum* (Djonović et al. 2006a, b). Furthermore, proteolytic enzymes are also important for the cell membrane degradation of parasitic fungi. Parasites have the ability to compromise the structural stability of exposed fungi by hydrolyzing the proteins that constitute the cell wall, as well as neutralizing enzymes alien to the fungus (Yao et al. 2023). Some families, notably the S8 subtilisin-like protease ones, are more abundant in mycoparasitic species of Trichoderma than in their non-mycoparasitic counterpart's classifiers (Kubicek et al. 2011).

Many genes encoding proteases are regulated in a different way during mycoparasitism or when subject to the wall of a fungus. For example, the metalloprotease NMP1 is secreted in response to other fungi and is important in antifungal activity (Zhang et al. 2016).

15.5.3 Production of Secondary Metabolites and Volatile Organic Compounds

The use of Trichoderma species in biocontrol is associated with the production of secondary metabolites that play the role of weapons, signals, and mediators in the relationship with other organisms. Metabolites involved in these interactions include polyketides, terpenes, and volatiles, as well as non-ribosomal peptides that are produced through taxon-specific pathways (Sarma et al. 2014; Zhang et al. 2021). Secondary metabolites are produced in large quantities by these fungi, and for this reason, they are excellent biocontrol agents. A comparative genomics analysis has revealed that these fungi contain a large number of genes encoding the enzymes that catalyze the synthesis of these compounds, polyketide aynthases, 10–25; non-ribosomal peptide synthetases, 12–34; and terpenoid, 6–14 (Kubicek et al. 2019). The produced secondary metabolites could be different from species to strain. For example, the volatile compound 6-pentyl-2H-pyran-2-one (6-PP) is implied in antagonistic interactions in *T. harzianum*. In *T. harzianum*, the pks2 gene that has been induced during the antagonistic encounter with the host fungi has been linked to changes in color and metabolites (Yao et al. 2016). Moreover, in *T. reesei*, the pks4 gene is linked to color and production of inhibiting substances.

Besides the NRPs involved in mycoparasitism, other nonribosomal peptides such as peptaibols, siderophores, and epipolythiodioxopiperazines (ETPs) are also involved in this process. Some fungi produce peptaibols that can disrupt cell membranes. Furthermore, these compounds can work in conjunction with enzymes for cell wall degrading to enhance the antifungal potency. The peptaibol synthetase genes are not expressed during mycoparasitic interactions, curiously. Their expression can be induced by plant roots (Atanasova et al. 2013). Siderophores are also important because they allow Trichoderma to obtain iron and outcompete other microbes in regions where nutrients are scarce. Gliotoxin is another essential compound, which is an ETP produced by some strains of *T. virens* (Crutcher et al. 2013). This compound is notoriously known for its fungistatic properties, and it is thought to employ the "predation by poisoning" strategy against pathogens such as *R. solani* (Zaid et al. 2022; Bulgari et al. 2020). Nevertheless, gliotoxin mutants retain some mycoparasitic activity against other fungi, so it appears that these gliotoxin-derived strains employ other mechanisms of action (Jayalakshmi et al. 2021; Hua et al. 2021). By reducing fungal development and altering the plant response, VOCs aid in mycoparasitism6. 6-PP promotes plant growth by acting as both an antifungal and a Trichoderma self-signaling chemical. Mycotoxins are among the more than 40 distinct metabolites produced by Trichoderma species that have antibacterial properties (Jeleń et al. 2014; Szczepaniak et al. 2015). Two primary categories of Trichoderma antibiotic molecules have been identified through structural analysis: polar metabolites with high molecular weights that may cause direct interactions between Trichoderma2 and volatile metabolites with low molecular weights, such as simple aromatic compounds and certain polyketides, such as pyrones, isocyanates, butenolides, and volatile terpenes (Ait-Lahsen et al. 2001). Numerous volatile secondary metabolites that Trichoderma spp. may create have been reviewed by Reino et al. (2008). Besides secondary metabolites, hydrogen peroxide (H_2O_2) is produced during the interaction with the host fungi, among ROS. H_2O_2 is produced during the course of mycoparasitism in *T. guizhouense* and its production is controlled by genes that are peculiar to the process in this species, for instance, NOX1. The buildup of H_2O_2 leads to the formation of polyketides that are used in the protection against oxidative stress during fungal interactions (Zhang et al. 2019; Pang et al. 2020). Chávez-Avilés and his co-workers determined the ability of *T. asperellum* strains to produce novel VOCs with fungicidal properties against *C. acutatum*, using dual confrontation systems in two culture media. The composition of

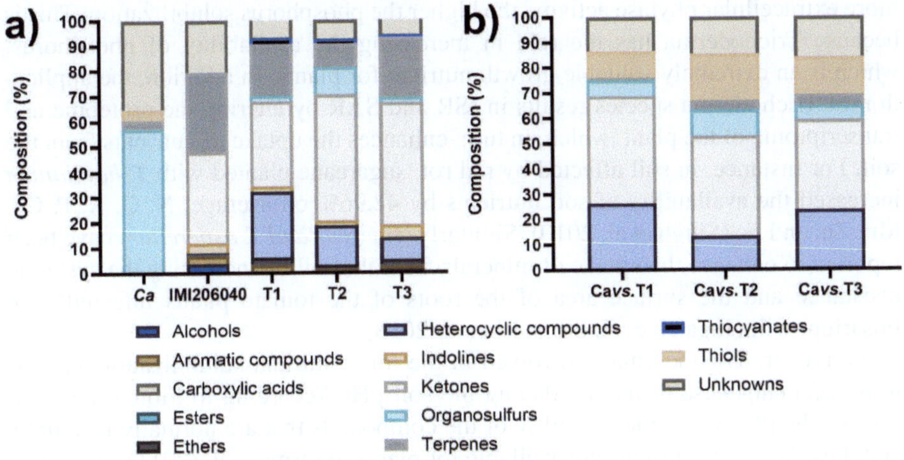

Fig. 15.3 Composition of the VOC profiles generated on Luria-Bertani (LB) broth (LB) medium by the strains under investigation. (**a**) VOC profiles generated by each strain separately. (**b**) VOC profiles generated in systems with dual confrontation

Trichoderma strains' volatile compounds encompasses the variability of chemical compounds produced by members of this genus (Fig. 15.3). This is a set of compounds of potential antifungal activity that need to be explored (Chávez-Avilés et al. 2024).

15.6 Enhance Nutrient Uptake and Root Development

The use of Trichoderma species is already well documented as a biocontrol agent, but these fungi are now also classified as plant growth promoters based on their ability to improve the availability of micro- and nanonutrients from the soil through acidification, chelation, redox reactions, and hydrolysis. Furthermore, these fungi make available essential minerals in the soil, which, in turn, enhance the growth and development of the plant in a better way. The production of organic acids is one of the main mechanisms by which Trichoderma is able to enhance the availability of nutrients. Contreras-Cornejo et al. stated that Trichoderma reduces the pH of the soil and makes available mineral acids. As reported by Atanasova et al. (2013), Trichoderma releases acids that lower the pH of the soil and makes available minerals in forms that can be easily absorbed by the plant. Samples of soluble elements released by *T. harzianum* include iron (III) oxide, zinc phytate, metallic zinc, and copper (II) oxide, and organic acids enhancing plant growth (tartaric acid, succinic acid, lactic acid, and citric acid) (Li et al. 2015). However, it has been observed that some Trichoderma isolates can solubilize phosphorus from calcium phosphate in the rhizosphere of *Avicennia marina*. In a study by Saravanakumar et al. (2013), all isolates were positive for solubilization which agreed with phytase activity; the more extracellular phytase activity, the higher the phosphorus solubilization. This is because Trichoderma has the role in increasing the availability of phosphorus, which is an extremely valuable growth nutrient for plants. In addition, the application of Trichoderma species results in ISR and SAR by altering the proteome and transcriptome of the plant, which, in turn, enhances the uptake of nutrients from the soil. For instance, in soil affected by red rot, sugarcane planted with *T. harzianum* increased the availability of soil nutrients by 42.96% on average: N, C, K, P, Cu, Mn, Zn, and Fe (Singh et al. 2010). Similarly, *T. virens and T. asperellum* have been reported to enhance the uptake of minerals from the soil by increasing the systemic resistance and the surface area of the roots of the tomato plants infected with Fusarium wilt (Jogaiah et al. 2018, Li et al. 2018).

Moreover, Trichoderma is involved in the micronutrient solubilization such as iron and manganese through reducing the soil pH. According to Fiorentino et al. (2018), the pH affects the solubility of the compounds that are normally insoluble, and thus the micronutrients are available for plants. Azarmi et al. (2011) also stated that this can only be achieved when the availability of nutrients in the soil is enhanced. The promotion of root development is another factor that leads to the improvement of nutrient uptake by Trichoderma. According to Lamont (1982), there is a positive relationship between nutrient absorption and root surface area. *T. virens* and *T. atroviride* induced a substantial increase in hairy root and lateral root development by producing IAA in both optimal and adverse conditions. Contreras-Cornejo et al. (2024) also observed that there was a substantial increase in lateral root length in tomato seedlings inoculated with *T. viride*. This increased root architecture improves carbohydrate metabolism and photosynthetic activity so that plants can better manage drought stress. Trichoderma (Hypocreales) is a worldwide genus that contains common species that may be found in a variety of settings. They often occupy a variety of niches, such as soil, dead wood, and organic debris in sediment, above-ground plant tissues (microflora), and other materials (Nascimento et al. 2023). The well-known soil saprophytic fungi, Trichoderma spp., have a number of antifungal characteristics. They use a variety of strategies (mycoparasitism, nutrient competition, etc.) for the compartment of a wide range of nematodes and soil diseases in the rhizosphere (Rebolledo-Prudencio et al. 2020). Plant tolerance to biotic and abiotic stressors is impacted by the symbiotic relationships formed by several organisms that have been identified from plant tissues. Root colonization has been the focus of endophytic Trichoderma for many years (Nascimento et al. 2023). Endophytic Trichoderma species have been isolated from live plant tissues from a broad range of plants, despite the fact that the majority of Trichoderma species have been recovered from various soil types and dead plant tissue. Trichoderma spp. have become more separated as endophytes from woody and herbaceous plants in recent decades, and they often exhibit habitat specialization (Puglielli et al. 2023). As a result, endophyte species are seldom found in soils, and vice versa (Osakabe et al. 2012). The possible effects of Trichoderma in combination with nitrogen fertilizer on microbial richness (Fig. 15.4), diversity, and functional ecology of the soil as well as on rhizosphere microbial community characteristics of maize in Northeast China have been explored by Li et al. (2024).

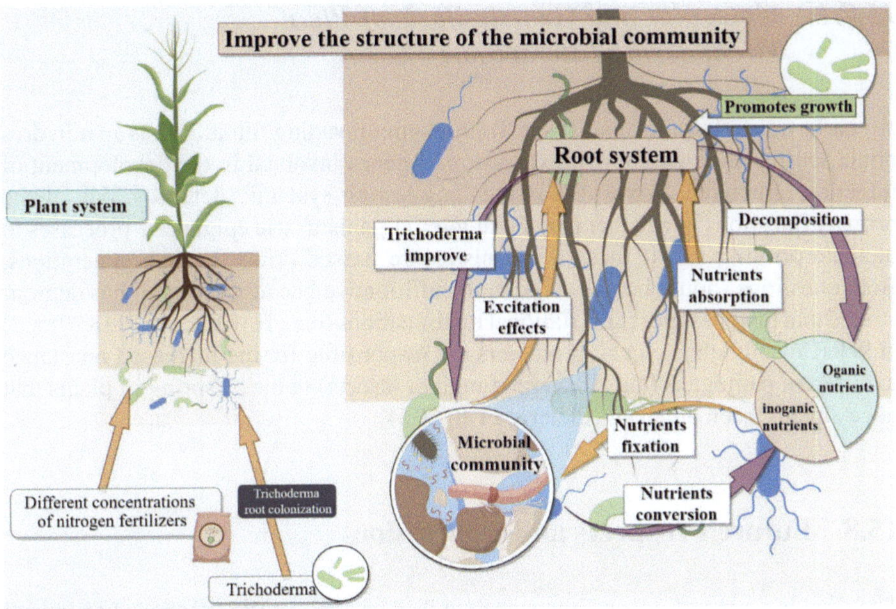

Fig. 15.4 Schematic diagrams elucidate that Trichoderma application boosted beneficial bacteria populations like Penicillium while repressing deleterious bacteria populations like Ophiosphaerella in saline–alkali soil in semi-arid regions (Biradar & Santhosh, 2018). From a microbial functional prediction perspective, Trichoderma would activate the nitrogen cycle in the soil by way of nitrification and nitrogen fixation processes at varying nitrogen application levels while promoting carbon sequestration in the soil by repressing carbon degradation in the soil. (Reprinted from Li et al. 2024)

15.7 Interaction of Trichoderma with Host Plants

Trichoderma (Hypocreales) is a worldwide genus that contains common species that may be found in a variety of settings. They often occupy a variety of niches, such as soil, dead wood, and organic debris in sediment, above-ground plant tissues (microflora), and other materials (Nascimento et al. 2023). The well-known soil saprophytic fungi Trichoderma spp. have a number of antifungal characteristics. They use a variety of strategies (mycoparasitism, nutrient competition, etc.) to combat a wide range of nematodes and diseases in the rhizosphere. Plant tolerance to biotic and abiotic stressors is impacted by the symbiotic relationships formed by several organisms that have been identified from plant tissues. Root colonization has been the focus of endophytic Trichoderma for many years (Nascimento et al. 2023). Endophytic Trichoderma species have been isolated from live plant tissues from a broad range of plants, despite the fact that the majority of Trichoderma species have been recovered from various soil types and dead plant tissue. Trichoderma spp. have become more separated as endophytes from woody and herbaceous plants in recent decades, and they often exhibit habitat specialization. As a result, endophyte species are seldom found in soils, and vice versa (Osakabe et al. 2012).

15.7.1 Molecular and Genetic Basis of Plant–Trichoderma Interactions

An early transcriptome analysis of Trichoderma invading tomato roots in a hydroponic setting revealed the overexpression of genes involved in the development of infection structures. When discussing long-lasting systemic defenses, one of the primary questions is whether beneficial microorganisms use epigenetic processes to mediate priming and if these mechanisms are passed down through generations. Both questions should be answered in the affirmative because research has demonstrated that histones' post-translational modifications (e.g., H3K9ac and H3K27me3) at WRKY and defense gene promoters are responsible for the increased resistance to different pathogens (e.g., bacteria, fungi, or insects) in the offspring of plants that have already been attacked (Pieterse et al. 2014).

15.8 Future Prospects and Conclusion

The growing concern toward the sustainability of agricultural practices has indeed brought into focus the need to find eco-friendly alternatives to the current methods. Therefore, Trichoderma spp. are a possible solution since these fungi have many advantages, which include stimulating growth, improving soil quality, and suppressing diseases. Organic farming is particularly beneficial from their use since they offer effective biocontrol without the use of chemical pesticides. They are also fast-growing, versatile on the substrate, and tolerant to toxins; therefore, they are very useful in preventing soil-borne diseases and in the maintenance of soil fertility. Future agricultural practices should include Trichoderma-based bioorganic fertilizers to maintain soil biodiversity and health. More especially, the research should be directed toward improving the efficacy and shelf life of Trichoderma products for use in different farming systems. Also, the possibility of using Trichoderma in conjunction with other beneficial microorganisms to improve plant diseases and soil control should be investigated. New technologies like microencapsulation have also been introduced to guarantee Trichoderma protection against field conditions, enabling the organism's increased shelf life and enhanced performance. Trichoderma metabolites provide promising avenues for more effective pest and pathogen control through the biogenic synthesis of nanoparticles as well. There is so much more work to be done to ensure that these novel technologies do not inflict harm on non-target organisms or the environment more broadly.

References

Abbas A, Mubeen M, Zheng H, Sohail MA, Shakeel Q, Solanki MK et al (2022) *Trichoderma spp.* genes involved in the biocontrol activity against *Rhizoctonia solani*. Front Microbiol 13:884469

Abdelrahman M, Jogaiah S (2019) Defense mechanism and diverse actions of fungal biological control agents against plant biotic stresses. In: Plant biotechnology: progress in genomic Era, pp 461–478

Abdullah NS, Doni F, Mispan MS, Saiman MZ, Yusuf YM, Oke MA, Suhaimi NSM (2021) Harnessing *Trichoderma* in agriculture for productivity and sustainability. Agronomy 11(12):2559

Ait-Lahsen H, Moya P, Cruz-Magalhães V (2001) Volatile organic compounds produced by *Trichoderma spp.*: potential applications in agriculture

Alfano G, Ivey ML, Cakir C, Bos JIB, Miller SA, Madden LV et al (2007) Systemic modulation of gene expression in tomato by *Trichoderma hamatum* 382. Phytopathology 97(4):429–437

Alfiky A, Weisskopf L (2021) Deciphering *Trichoderma*–plant–pathogen interactions for better development of biocontrol applications. J Fungi 7(1):61

Asad SA, Tabassum A, Hameed A, Afzal A, Khan SA, Ahmed R, Shahzad M (2015) Determination of lytic enzyme activities of indigenous *Trichoderma* isolates from Pakistan. Braz J Microbiol 46:1053–1064

Asghar W, Craven KD, Kataoka R, Mahmood A, Asghar N, Raza T, Iftikhar F (2024) The application of *Trichoderma spp.*, an old but new useful fungus, in sustainable soil health intensification: a comprehensive strategy for addressing challenges. Plant. Stress 100455

Asma A (2019) Characterization of *Fusarium* species associated with wilt disease of cucurbits and effectiveness of *Trichoderma asperellum* as biological enhancer. PhD Thesis, Universiti Putra Malaysia

Atanasova L, et al (2013) Gliotoxin production in *Trichoderma*: implications for mycoparasitism

Awad-Allah EF, Shams AH, Helaly AA (2021) Suppression of bacterial leaf spot by green synthesized silica nanoparticles and antagonistic yeast improves growth, productivity and quality of sweet pepper. Plan Theory 10(8):1689

Azarmi R, Hajieghrari B, Giglou A (2011) Effect of Trichoderma isolates on tomato seedling growth response and nutrient uptake. Afr J Biotechnol 10(31):5850–5

Baazeem A, Almanea A, Manikandan P, Alorabi M, Vijayaraghavan P, Abdel-Hadi A (2021) In vitro antibacterial, antifungal, nematocidal and growth promoting activities of *Trichoderma hamatum* FB10 and its secondary metabolites. J Fungi 7(5):331

Bae H, Sicher RC, Kim MS, Kim SH, Strem MD, Melnick RL, Bailey BA (2009) The beneficial endophyte *Trichoderma hamatum* isolate DIS 219b promotes growth and delays the onset of the drought response in *Theobroma cacao*. J Exp Bot 60(11):3279–3295

Benítez T et al (2004) Biocontrol mechanisms of *Trichoderma* species. In: International microbiology

Bigirimana J, De Mayer G, Poppe J, Elad Y, Hofte M (1997) Induction of systemic resistance on bean (*Phaseolus vulgaris*) by *Trichoderma harzianum*. Med Fac Univ Gent 62:1001–1007

Biradar BP, Santhosh GP (2018) Cell protectants, adjuvants, surfactant and preservative and their role in increasing the shelf life of liquid inoculant formulations of *Pseudomonas fluorescens*. Int J Pure Appl Biosci 6(4):116–122

Bissett J (1991) A revision of the genus *Trichoderma*. II. Infrageneric Classification Canadian J Botany 69:2357–2372

Bonanomi G, Lorito M, Vinale F, Woo SL (2018) Organic amendments, beneficial microbes, and soil microbiota: toward a unified framework for disease suppression. Annu Rev Phytopathol 56(1):1–20

Bulgari D, Fiorini L, Gianoncelli A, Bertuzzi M, Gobbi E (2020) Enlightening Gliotoxin biological system in agriculturally relevant *Trichoderma spp*. Front Microbiol 11:200

Calvo J, Calvente V, Orellano ME, Benuzzi D, Sanz MI (2010) Control of *Penicillium expansum* and *Botrytis cinerea* on apple fruit by mixtures of bacteria and yeast. Food Bioprocess Technol 3:644–650

Cao XT, Bian YB, Xu ZY (2014) First report of *Trichoderma oblongisporum* causing green mold disease on *Lentinula edodes* (shiitake) in China. Plant Dis 98(10):1440–1440

Chang YC, Chang YC, Baker R, Kleifeld O, Chet I (1986) Increased growth of plants in the presence of the biological control agent Trichoderma harzianum. Plant Dis 70:145–148

Chávez-Avilés MN, García-Álvarez M, Ávila-Oviedo JL, Hernández-Hernández I, Bautista-Ortega PI, Macías-Rodríguez LI (2024) Volatile organic compounds produced by Trichoderma asperellum with antifungal properties against Colletotrichum acutatum. Microorganisms 12(10):2007

Collinge DB, Jensen DF, Rabiey M, Sarrocco S, Shaw MW, Shaw RH (2022) Biological control of plant diseases—what has been achieved and what is the direction? Plant Pathol 71(5):1024–1047

Contreras-Cornejo HA, Schmoll M, Esquivel-Ayala BA, González-Esquivel CE, Rocha-Ramírez V, Larsen J (2024) Mechanisms for plant growth promotion activated by Trichoderma in natural and managed terrestrial ecosystems. Microbiol Res 281:127621

Crutcher FK, Parich A, Schuhmacher R, Mukherjee PK, Zeilinger S, Kenerley CM (2013) A putative terpene cyclase, vir4, is responsible for the biosynthesis of volatile terpene compounds in the biocontrol fungus *Trichoderma virens*. Fungal Genet Biol 56:67–77

Djonović S, Pozo MJ, Kenerley CM (2006a) Tvbgn3, a β-1, 6-glucanase from the biocontrol fungus *Trichoderma virens*, is involved in mycoparasitism and control of *Pythium ultimum*. Appl Environ Microbiol 72(12):7661–7670

Djonović S, Pozo MJ, Dangott LJ, Howell CR, Kenerley CM (2006b) Sm1, a proteinaceous elicitor secreted by the biocontrol fungus *Trichoderma virens*, induces plant defense responses and systemic resistance. Mol Plant-Microbe Interact 19(8):838–853

Doi Y (1969) The genus *Hypocrea* and its allies in Japan. Bulletin National Sci Museum Tokyo 12:693–724

Doi Y (1972) Revision of the *Hypocreales* with cultural observations. IV. The genus *Hypocrea* and its allies in Japan. (2) enumeration of the species. Bulletin National Sci Museum Tokyo 15:649–751

Doni F, Isahak A, Zain CRCM, Ariffin SM, Mohamad WNW, Yusoff WMW (2014) Formulation of *Trichoderma* sp. SL2 inoculants using different carriers for soil treatment in rice seedling growth. Springerplus 3(532):3–7

Druzhinina IS, Kopchinskiy AG, Kubicek CP (2006) The first 100 *Trichoderma* species character

Dutta P, Mahanta M, Singh SB, Thakuria D, Deb L, Kumari A et al (2023) Molecular interaction between plants and *Trichoderma* species against soil-borne plant pathogens. Front Plant Sci 14:1145715

Elshahawy IE, El-Mohamedy RS (2019) Biological control of *Pythium* damping-off and root-rot diseases of tomato using *Trichoderma* isolates employed alone or in combination. J Plant Pathol 101:597–608

Esposito E, Silva MD (1998) Systematics and environmental application of the genus *Trichoderma*. Crit Rev Microbiol 24(2):89–98

Fazeli-Nasab B, Shahraki-Mojahed, L, Piri R, & Sobhanizadeh A. (2022). Trichoderma: Improving growth and tolerance to biotic and abiotic stresses in plants. In Trends of Applied Microbiology for Sustainable Economy (pp. 525-564). Academic Press.

Fiorentino N, Ventorino V, Woo SL, Pepe O, De Rosa A, Gioia L et al (2018) Trichoderma-based biostimulants modulate rhizosphere microbial populations and improve N uptake efficiency, yield, and nutritional quality of leafy vegetables. Front Plant Sci 9:743

Gangwar OP, Singh AP (2018) *Trichoderma* as an efficacious bioagent for combating biotic and abiotic stresses of wheat – a review. Agric Rev 39(1):48–54

Garcia-Rubio R, de Oliveira HC, Rivera J, Trevijano-Contador N (2020) The fungal cell wall: *Candida*, *Cryptococcus*, and *aspergillus* species. Front Microbiol 10:2993

Geremia RA, Golgman GH, Jacobs D, Ardiles W, Vila SB, Montagu MV (1993) Molecular characterization of the proteinase-encoding gene, prb1, related to mycoparasitism by *Trichoderma harzianum*. Mol Microbiol 8:603–613. https://doi.org/10.1111/j.1365-2958.1993.tb01604.x

Ghisalberti EL, Sivasithamparam K (1991) Biological control of fungal plant pathogens. In: *Trichoderma koningii* and its metabolites

Gimenez-Ibanez S, Hann DR, Ntoukakis V, Petutschnig E, Lipka V, Rathjen JP (2009) AvrPtoB targets the LysM receptor kinase CERK1 to promote bacterial virulence on plants. Curr Biol 19(5):423–429

Gomes EV, Costa MDN, de Paula RG, Ricci de Azevedo R, da Silva FL, Noronha EF et al (2015) The Cerato-Platanin protein Epl-1 from *Trichoderma harzianum* is involved in mycoparasitism, plant resistance induction, and self cell wall protection. Sci Rep 5(1):17998

Gow NA, Latge JP, Munro CA (2017) The fungal cell wall: structure, biosynthesis, and function. Microbiology Spectrum 5(3):10–1128

Gutter Y (1957) Effect of light on sporulation of *Trichoderma viride* pers. ex. fries. Bulletin Res Council Israel, Section D, Botany 5:273–286

Guzmán-Guzmán P, Kumar A, de Los Santos-Villalobos S, Parra-Cota FI, Orozco-Mosqueda MDC, Fadiji AE et al (2023) *Trichoderma* species: our best fungal allies in the biocontrol of plant diseases—a review. Plan Theory 12(3):432

Guzmán-Guzmán P, Porras-Troncoso MD, Olmedo-Monfil V, Herrera-Estrella A (2019) *Trichoderma* species: versatile plant symbionts. Phytopathology 109(1):6–16

Hammerschmidt R (2009) Systemic acquired resistance. Adv Bot Res 51:173–222

Hanaka A, Ozimek E, Reszczyńska E, Jaroszuk-Ściseł J, Stolarz M (2021) Plant tolerance to drought stress in the presence of supporting bacteria and fungi: an efficient strategy in horticulture. Horticulturae 7(10):390

Harman GE, Howell CR, Viterbo A, Chet I, Lorito M (2004) *Trichoderma* species: opportunistic, avirulent plant symbionts. Nat Rev Microbiol 2:43–56

Harman GE, Hayes CK, Ondik KL (1998) Asexual genetics in *Trichoderma* and *Gliocladium*: mechanisms and implications. In: *Trichoderma* and *Gliocladium*, vol 243

Harman GE, Hayes CK, Lorito M (1993) The genome of biocontrol fungi: modification and genetic components for plant disease management strategies. In: Lumsden RD, Vaughn JL (eds) Pest management biologically based technologies: proceedings of Beltsville symposium XVIII. American Chemical Society, Washington, D.C., pp 347–354

Hermosa R, Rubio MB, Cardoza RE, Nicolás C, Monte E, Gutiérrez S (2013) The contribution of *Trichoderma* to balancing the costs of plant growth and defense. Int Microbiol 16(2):69–80

Hermosa R, Viterbo A, Chet I, Monte E (2012) Plant-beneficial effects of *Trichoderma* and of its genes. Microbiology 158(1):17–25

Hernández-Oñate MA, Esquivel-Naranjo EU, Mendoza-Mendoza A, Stewart A, Herrera-Estrella AH (2012) An injury-response mechanism conserved across kingdoms determines entry of the fungus *Trichoderma atroviride* into development. Proc National Acad Sci USA 109:14918–14923

Hjeljord L, Tronsmo A (1998) Department of Biotechnological Sciences, agricultural University of Norway, Aas, Norway. Enzymes, biological control, and commercial applications, Trichoderma and Gliocladium, p 115

Howell CR (2003) Mechanisms employed by *Trichoderma* species in the biological control of plant pathogens. The Plant Health Instructor

Hu J, Chen K, Li J, Wei Y, Wang Y, Wu Y et al (2020) Large-scale *Trichoderma* diversity was associated with ecosystem, climate, and geographic location. Environ Microbiol 22(3):1011–1024

Hua L, Zeng H, He L, Jiang Q, Ye P, Liu Y et al (2021) Gliotoxin is an important secondary metabolite involved in suppression of *Sclerotium rolfsii* by *Trichoderma virens* T23. Phytopathology 111(10):1720–1725

Ihrmark K, Asmail N, Ubhayasekera W, Melin P, Stenlid J, Karlsson M (2010) Comparative molecular evolution of *Trichoderma* chitinases in response to mycoparasitic interactions. Evol Bioinforma 6:EBO-S4198

Jambhulkar PP, Singh B, Raja M, Ismaiel A, Lakshman DK, Tomar M, Sharma P (2024) Genetic diversity and antagonistic properties of *Trichoderma* strains from the crop rhizospheres in southern Rajasthan. India Sci Rep 14(1):8610

Jangir M, Pathak R, Sharma S (2017) *Trichoderma* and its potential applications. In: Plant-microbe interactions in agro-ecological perspectives: volume 2: microbial interactions and agro-ecological impacts, pp 323–339

Jayalakshmi R, Oviya R, Premalatha K, Mehetre ST, Paramasivam M, Kannan R et al (2021) Production, stability, and degradation of *Trichoderma* gliotoxin in growth medium, irrigation water, and agricultural soil. Sci Rep 11(1):16536

Jeleń H, Błaszczyk L, Chełkowski J, Rogowicz K, Strakowska J (2014) Formation of 6-n-pentyl-2H-pyran-2-one (6-PAP) and other volatiles by different *Trichoderma* species. Mycol Prog 13:589–600

Jogaiah S, Abdelrahman M, Tran LSP, Ito SI (2018) Different mechanisms of *Trichoderma virens*-mediated resistance in tomato against *Fusarium* wilt involve the jasmonic and salicylic acid pathways. Mol Plant Pathol 19(4):870–882

Kamel S, Farag F, Arafa RA, Essa T (2020) Bio-control potentials of *Trichoderma* spp. against *Sclerotium rolfsii*, the causative agent of root and crown rot in tomato, common bean, and cabbage. Egyptian. J Phytopathol 48(1):122–136

Kappel L, Münsterkötter M, Sipos G, Escobar Rodriguez C, Gruber S (2020) Chitin and chitosan remodeling defines vegetative development and *Trichoderma* biocontrol. PLoS Pathog 16(2):e1008320

Khan MR, Parveen G, Zaid A, Wani SH, Jogaiah S (2021) Potential of *Trichoderma* species in alleviating the adverse effects of biotic and abiotic stresses in plants. In: Biocontrol agents and secondary metabolites. Woodhead Publishing, pp 85–112

Khan RAA, Najeeb S, Chen J, Wang R, Zhang J, Hou J, Liu T (2023) Insights into the molecular mechanism of *Trichoderma* stimulating plant growth and immunity against phytopathogens. Physiol Plant 175(6):e14133

Kistler HC, Miao VPW (1992) New modes of genetic change in filamentous fungi. Annu Rev Phytopathol 30:131–152

Kubicek CP, Steindorff AS, Chenthamara K, Manganiello G, Henrissat B, Zhang J et al (2019) Evolution and comparative genomics of the most common Trichoderma species. BMC Genomics 20:1–24

Kubicek CP, Herrera-Estrella A, Seidl-Seiboth V, Martinez DA, Druzhinina IS, Thon M, Grigoriev IV (2011) Comparative genome sequence analysis underscores mycoparasitism as the ancestral lifestyle of *Trichoderma*. Genome Biol 12:1–15

Kullnig C, Szakacs G, Kubicek CP (2000) Molecular identification of *Trichoderma* species from Russia, Siberia, and the Himalaya. Mycol Res 104(9):1117–1125

Lamont BB (1982) Root architecture: a key factor in nutrient uptake

Li J et al (2015) Solubilization of minerals by *Trichoderma* species: implications for plant growth

Li J, Philp J, Li J, Wei Y, Li H, Yang K et al (2020) *Trichoderma harzianum* inoculation reduces the incidence of clubroot disease in Chinese cabbage by regulating the rhizosphere microbial community. Microorganisms 8(9):1325

Li X, Liao Q, Zeng S, Wang Y, Liu J (2025) The use of *Trichoderma* species for the biocontrol of postharvest fungal decay in fruits and vegetables: challenges and opportunities. Postharvest Biol Technol 219:113236

Li Y, Cui J, Kang J, Zhao W, Yang K, Fu J (2024) Trichoderma Rhizosphere soil improvement: regulation of nitrogen fertilizer in Saline–Alkali Soil in semi-arid region and its effect on the microbial community structure of maize roots. Agronomy 14(10):2340

Li YT, Hwang SG, Huang YM, Huang CH (2018) Effects of Trichoderma asperellum on nutrient uptake and Fusarium wilt of tomato. Crop Prot 110:275–282

Lodi RS, Peng C, Dong X, Deng P, Peng L (2023) *Trichoderma hamatum* and its benefits. J Fungi 9(10):994

Lombardi N, Caira S, Troise AD, Scaloni A, Vitaglione P, Vinale F et al (2020) *Trichoderma* applications on strawberry plants modulate the physiological processes positively affecting fruit production and quality. Front Microbiol 11:1364

Manzar N, Kashyap AS, Goutam RS, Rajawat MVS, Sharma PK, Sharma SK, Singh HV (2022) *Trichoderma*: advent of a versatile biocontrol agent, its secrets, and insights into mechanisms of biocontrol potential. Sustain For 14(19):12786

Mathys J, De Cremer K, Timmermans P, Van Kerckhove S, Lievens B, Vanhaecke M et al (2012) Genome-wide characterization of ISR induced in *Arabidopsis thaliana* by *Trichoderma hamatum* T382 against *Botrytis cinerea* infection. Front Plant Sci 3:108

Maurya S, Ntakirutimana R, Debnath B, Rana M, Kaushik D, Srivastava S (2024) *Trichoderma* and their secondary metabolites–a potential approach in plant disease management. Biopesticides International 20(1):21

Metwally RA (2020) Arbuscular mycorrhizal fungi and *Trichoderma viride* cooperative effect on biochemical, mineral content, and protein pattern of onion plants. J Basic Microbiol 60(8):712–721

Meyer W, Morawetz R, Börner T, Kubicek CP (1992) The use of DNA fingerprint analysis in the classification of some species of the *Trichoderma* aggregate. Curr Genet 21:27–30

Montero-Barrientos M, Cardoza RE, Gutiérrez S, Monte E, Hermosa R (2007) The heterologous overexpression of hsp23, a small heat-shock protein gene from *Trichoderma virens*, confers thermotolerance to *T. harzianum*. Curr Genet 52:45–53

Montero-Barrientos M, Hermosa R, Cardoza RE, Gutiérrez S, Monte E (2011) Functional analysis of *Trichoderma harzianum* nox1 gene, encoding a NADPH oxidase, relates ROS production to specific biocontrol activity against *Pythium ultimum*. Appl Environ Microbiol 77:3009–3016

Montero-Barrientos M, Hermosa R, Nicolás C, Cardoza RE, Gutiérrez S, Monte E (2008) Overexpression of Trichoderma HSP70 gene increases fungal resistance to heat and other abiotic stresses. Fungal Genet Biol 45:1506–1513

Morán-Diez ME, Martinez de Alba AE, Rubio MB, Hermosa R, Monte E (2021) *Trichoderma* and the plant heritable priming responses. J Fungi 7(4):318

Morán-Diez ME, Cardoza RE, Gutiérrez S, Monte E, Hermosa R (2010) TvDim1 of *Trichoderma virens* is involved in redox processes and confers resistance to oxidative stresses. Curr Genet 56:63–73

Mukherjee PK (2015) *Trichoderma*–genetics, genomics, and beyond. Indian Phytopathol 66(1):1–7

Mukherjee PK, Horwitz BA, Kenerley CM (2012) Secondary metabolism in *Trichoderma*–a genomic perspective. Microbiology 158(1):35–45

Mukherjee PK, Latha J, Hadar R, Horwitz BA (2003) TmkA, a mitogen-activated protein kinase of *Trichoderma virens*, is involved in biocontrol properties and repression of conidiation in the dark. Eukaryot Cell 2(3):446–455

Mukherjee PK, Mendoza-Mendoza A, Zeilinger S, Horwitz BA (2022) Mycoparasitism as a mechanism of *Trichoderma*-mediated suppression of plant diseases. Fungal Biol Rev 39:15–33

Muthumeenakshi S, Mills PR, Brown AE, Seaby DA (1994) Intraspecific molecular variation among *Trichoderma harzianum* isolates colonizing mushroom compost in the British Isles. Microbiology 140:769–777

Nascimento Brito V, Lana Alves J, Sírio Araújo K, de Souza Leite T, Borges de Queiroz C, Liparini Pereira O, de Queiroz MV (2023) Endophytic *Trichoderma* species from rubber trees native to the Brazilian Amazon, including four new species. Front Microbiol 14:1095199

Nawrocka J, Małolepsza U (2013) Diversity in plant systemic resistance induced by *Trichoderma*. Biol Control 67(2):149–156

Oljira AM, Hussain T, Waghmode TR, Zhao H, Sun H, Liu X et al (2020) *Trichoderma* enhances net photosynthesis, water use efficiency, and growth of wheat (*Triticum aestivum* L.) under salt stress. Microorganisms 8(10):1565

Osakabe Y, Kawaoka A, Nishikubo N, Osakabe K (2012) Responses to environmental stresses in woody plants: key to survival and longevity. J Plant Res 125:1–10

Oskiera M, Szczech M, Bartoszewski G (2015) Molecular identification of strains collected to develop plant growth-promoting and biocontrol agents. J Horticultural Res 23(1):75–86

Ousley MA, Lynch JM, Whipps JM (1993) Effect of *Trichoderma* on plant growth: a balance between inhibition and growth promotion. Microb Ecol 26:277–285

Pang G et al (2020) Azaphilones biosynthesis complements the defense mechanism of *Trichoderma guizhouense* against oxidative stress. Environ Microbiol 22(11):4808–4824

Papavizas GC (1985) *Trichoderma* and *Gliocladium*: biology, ecology, and potential for biocontrol. Annual Rev Phytopathol 23:23

Persoon CH (1794) *Disposita methodica* fungorum. Romer's Neues Mag Bot 1:81–128

Peterbauer CK, Litscher D, Kubicek CP (2002) The *Trichoderma atroviride* seb1 (stress response element binding) gene encodes an AGGGG-binding protein which is involved in the response to high osmolarity stress. Mol Gen Genomics 268:223–231

Pieterse CM, Van der Does D, Zamioudis C, Leon-Reyes A, Van Wees SC (2012) Hormonal modulation of plant immunity. Annu Rev Cell Dev Biol 28(1):489–521

Pieterse CM, Zamioudis C, Berendsen RL, Weller DM, Van Wees SC, Bakker PA (2014) Induced systemic resistance by beneficial microbes. Annu Rev Phytopathol 52(1):347–375

Poveda J (2021) *Trichoderma* as a biocontrol agent against pests: new uses for a mycoparasite. Biol Control 159:104634

Poveda J, Eugui D, Abril-Urias P (2020) Could *Trichoderma* be a plant pathogen? Successful root colonization. In: *Trichoderma:* host pathogen interactions and applications, pp 35–59

Prismantoro D, Akbari SI, Permadi N, Dey U, Anhar A, Miranti M et al (2024) The multifaceted roles of *Trichoderma* in managing rice diseases for enhanced productivity and sustainability. Journal of Agriculture and Food Research 18:101324

Puglielli G, Laanisto L, Gori A, Cardoso AA (2023) Woody plant adaptations to multiple abiotic stressors: where are we? Flora 299:152221

Qualhato TF, Lopes FAC, Steindorff AS, Brandão RS, Jesuino RSA, Ulhoa CJ (2013) Mycoparasitism studies of *Trichoderma* species against three phytopathogenic fungi: evaluation of antagonism and hydrolytic enzyme production. Biotechnol Lett 35:1461–1468

Rajput VD, Minkina T, Feizi M, Kumari A, Khan M, Mandzhieva S, Sushkova S, El-Ramady H, Verma KK, Singh A et al (2021) Effects of silicon and silicon-based nanoparticles on rhizosphere microbiome, plant stress, and growth. Biology 10:791. [Google Scholar] [CrossRef]

Ramírez-Valdespino CA, Casas-Flores S, Olmedo-Monfil V (2019) *Trichoderma* as a model to study effector-like molecules. Front Microbiol 10:1030

Rawal R, Scheerens JC, Fenstemaker SM, Francis DM, Miller SA, Benitez MS (2022) Novel *Trichoderma* isolates alleviate water deficit stress in susceptible tomato genotypes. Front Plant Sci 13:869090

Rebolledo-Prudencio OG, Dautt-Castro M, Estrada-Rivera M, del Carmen González-López M, Jijón-Moreno S, Casas-Flores S (2020) *Trichoderma* in the rhizosphere: an approach toward a long and successful symbiosis with plants. In: New and future developments in microbial biotechnology and bioengineering. Elsevier, pp 3–38

Reino JL, Guerrero RF, Hernandez-Galan R, Collado IG (2008) Secondary metabolites from species of the biocontrol agent *Trichoderma*. Phytochem Rev 7:89–123. [Google Scholar] [CrossRef]

Rifai MA (1969) A revision of the genus *Trichoderma*. Mycol Pap 116:1–56

Salwan R, Sharma A, Kaur R, Sharma R, Sharma V (2022) The riddles of *Trichoderma*-induced plant immunity. Biol Control 174:105037

Sanz L, Montero M, Grondona I, Vizcaíno JA, Llobell A, Hermosa R, Monte E (2004) Cell wall-degrading isoenzyme profiles of *Trichoderma* biocontrol strains show correlation with rDNA taxonomic species. Curr Genet 46:277–286

Saravanakumar K, Jayaraman S, Suresh S (2013) Phosphorus solubilization potential of *Trichoderma* isolates from rhizosphere soils. J Soil Sci Plant Nutr 13(3):559–570. https://doi.org/10.4067/S0718-95162013005000001

Saravanakumar K, Li Y, Yu C, Wang QQ, Wang M, Sun J, Gao JX, Chen J (2017) Effect of *Trichoderma harzianum* on maize rhizosphere microbiome and biocontrol of *Fusarium* stalk rot. Sci Rep 7:1771–1783

Sarma BK, Yadav SK, Patel JS, Singh HB (2014) Molecular mechanisms of interactions of *Trichoderma* with other fungal species. Open Mycol J 8:140–147

Schalamun M, Schmoll M (2022) *Trichoderma*–genomes and genomics as treasure troves for research towards biology, biotechnology, and agriculture. Front Fungal Biol 3:1002161

Schuster A, Schmoll M (2010) Biology and biotechnology of *Trichoderma*. Appl Microbiol Biotechnol 87:787–799

Shah P, Abid M, Abbas A, Basheer N, Sheikh AH, Jabeen N, Sitara U (2020) Screening of Trichoderma viride as a bio-control agent against different soil-borne infectious and phytopathogenic fungi. Int J Biol Biotech 17(2):405–10

Shahriar SA, Islam MN, Chun CNW, Kaur P, Rahim MA, Islam MM et al (2022) Microbial metabolomics interaction and ecological challenges of *Trichoderma* species as biocontrol inoculants in crop rhizosphere. Agronomy 12(4):900

Sharifah SMSAR (2018) Characterization of *Trichoderma* species isolated from soil and efficacy of *Trichoderma asperellum* as a biocontrol agent of Fusarium wilt disease of banana. MSc. Thesis, Universiti Putra Malaysia

Sharma P, Kumar V, Ramesh R, Saravanan K, Deep S, Sharma M et al (2011) Biocontrol genes from *Trichoderma* species: a review. Afr J Biotechnol 10(86):19898–19907

Sharma P, Patel AN, Saini MK, Deep S (2012) Field demonstration of *Trichoderma harzianum* as a plant growth promoter in wheat (*Triticum aestivum* L.). J Agric Sci 4(8):65

Shoresh M, et al (2006) Identification of a MAPK activated by *Trichoderma asperellum* in cucumber

Silva RN, Monteiro VN, Steindorff AS, Gomes EV, Noronha EF, Ulhoa CJ (2019) *Trichoderma*/pathogen/plant interaction in pre-harvest food security. Fungal Biol 123(8):565–583

Simmons EG (1977) Classification of some cellulase-producing *Trichoderma* species. In: Bigelow HE, Simmons EG (eds) International mycological congress, 2nd edn. University of South Florida, Tampa, p 618

Singh BN, Singh A, Singh GS, Dwivedi P (2015) Potential role of *Trichoderma asperellum* T42 strain in growth of pea plant for sustainable agriculture. J Pure Appl Microbiol 9(2):1069–1074

Singh S, Kumar S, Sharma P (2010) Nutrient enhancement in sugarcane infected with *Trichoderma harzianum*. Sugar Tech 12(1):10–15. https://doi.org/10.1007/s12355-010-0040-7

Sivan A, Chet I, Katan J (1987) The role of *Trichoderma* in biological control. Phytopathology 77(9):1416–1420. https://doi.org/10.1094/Phyto-77-1416

Sood M, Kapoor D, Kumar V, Sheteiwy MS, Ramakrishnan M, Landi M et al (2020) *Trichoderma*: the "secrets" of a multitalented biocontrol agent. Plan Theory 9(6):762

Stewart A, Hill R (2014) Applications of *Trichoderma* in plant growth promotion. In: Biotechnology and biology of Trichoderma. Elsevier, pp 415–428

Studholme DJ, Harris B, Le Cocq K, Winsbury R, Perera V, Ryder L et al (2013) Investigating the beneficial traits of *Trichoderma hamatum* GD12 for sustainable agriculture—insights from genomics. Front Plant Sci 4:258

Subramaniam S, Zainudin NAIM, Aris A, Hasan ZAE (2022) Role of *Trichoderma* in plant growth promotion. In: *Advances in Trichoderma biology for agricultural applications*, pp 257–280

Sun X, Sun M, Chao Y, Wang H, Pan H, Yang Q et al (2020) Alleviation of lead toxicity and phytostimulation in perennial ryegrass by the Pb-resistant fungus *Trichoderma asperellum* SD-5. Funct Plant Biol 48(3):333–341

Szczepaniak Z, Cyplik P, Juzwa W, Czarny J, Staninska J, Piotrowska-Cyplik A (2015) Antibacterial effect of the *Trichoderma viride* fungi on soil microbiome during PAH biodegradation. Int Biodeterior Biodegradation 104:170–177. [Google Scholar] [CrossRef]

Tandon A, Fatima T, Shukla D, Tripathi P, Srivastava S, Singh PC (2020) Phosphate solubilization by *Trichoderma koningiopsis* (NBRI-PR5) under abiotic stress conditions. Journal King Saud University-Sci 32(1):791–798

Vargas WA, Mukherjee PK, Laughlin D, Wiest A, Morán-Diez ME, Kenerley CM (2014) Role of gliotoxin in the symbiotic and pathogenic interactions of *Trichoderma virens*. Microbiology 160(10):2319–2330

Verma H, Kumar D, Kumar V, Kumari M, Singh SK, Sharma VK et al (2021) The potential application of endophytes in management of stress from drought and salinity in crop plants. Microorganisms 9(8):1729

Viterbo A, Ramot O, Chernin L, Chet I (2002) Significance of lytic enzymes from Trichoderma spp. in the biocontrol of fungal plant pathogens. Antonie Van Leeuwenhoek 81:549–556

Vlot AC, Sales JH, Lenk M, Bauer K, Brambilla A, Sommer A et al (2021) Systemic propagation of immunity in plants. New Phytol 229(3):1234–1250

Vos CM, De Cremer K, Cammue BP, De Coninck B (2015) The toolbox of T richoderma spp. in the biocontrol of *Botrytis cinerea* disease. Mol Plant Pathol 16(4):400–412

Waghunde RR, Shelake RM, Sabalpara AN (2016) Trichoderma: a significant fungus for agriculture and environment. African J Agric Res 11(22):1952–1965

Wang H, Zhang R, Mao Y, Jiang W, Chen X, Shen X et al (2022) Effects of *Trichoderma asperellum* 6S-2 on apple tree growth and replanted soil microbial environment. J Fungi 8(1):63

Weindling R (1934) Studies on a lethal principle effective in the parasitic action of *Trichoderma*. Phytopathology 24(9):949–950. https://doi.org/10.1094/Phyto-24-949

Wells H, Bell D, Jaworski A (1972) Efficacy of *Trichoderma harzianum* in controlling soilborne pathogens. Plant Disease Reporter 56(5):400–404

Woo SL, Pepe O (2018) Microbial consortia: promising probiotics as plant biostimulants for sustainable agriculture. Front Plant Sci 9:1801

Woo SL et al (2006) The role of Trichoderma in biocontrol. J Plant Pathol

Woo SL, Hermosa R, Lorito M, Monte E (2023) Trichoderma: a multipurpose, plant-beneficial microorganism for eco-sustainable agriculture. Nat Rev Microbiol 21(5):312–326

Woo SL, Ruocco M, Vinale F, Nigro M, Marra R, Lombardi N, Pascale A, Lanzuise S, Manganiello G, Lorito M (2014) Trichoderma-based products and their widespread use in agriculture. Open Mycol J 8:71–126

Yao X, Guo H, Zhang K, Zhao M, Ruan J, Chen J (2023) Trichoderma and its role in biological control of plant fungal and nematode disease. Front Microbiol 14:1160551

Yao Y-J, Wu J, Zhang H, Chen Y (2016) Upregulation of pks2 gene expression during confrontations with host fungi in *Trichoderma harzianum*. Fungal Biol 120(3):377–385. https://doi.org/10.1016/j.funbio.2015.12.002

Yedidia I, Benhamou N, Chet I (1999) The role of root colonization by *Trichoderma* in inducing resistance against pathogens. Phytopathology 89(11):1005–1012. https://doi.org/10.1094/PHYTO.1999.89.11.1005

Yu R, Zhang Y, Wang C (2009) Trichoderma oblongisporum as a biocontrol agent. J Appl Microbiol 107(2):492–499. https://doi.org/10.1111/j.1365-2672.2009.04206.x

Zaid R, Koren R, Kligun E, Gupta R, Leibman-Markus M, Mukherjee PK et al (2022) Gliotoxin, an immunosuppressive fungal metabolite, primes plant immunity: evidence from Trichoderma virens-tomato interaction. MBio 13(4):e00389–e00322

Zeilinger S, Gupta VK, Dahms TE, Silva RN, Singh HB, Upadhyay RS et al (2016) Friends or foes? Emerging insights from fungal interactions with plants. FEMS Microbiol Rev 40(2):182–207

Zhang F, Xu X, Huo Y, Xiao Y (2019) Trichoderma-inoculation and mowing synergistically altered soil available nutrients, rhizosphere chemical compounds and soil microbial community, potentially driving alfalfa growth. Front Microbiol 9:3241

Zhang F, Yuan J, Yang X, Cui Y, Chen L, Ran W, Shen Q (2013) Putative *Trichoderma harzianum* mutant promotes cucumber growth by enhanced production of indole acetic acid and plant colonization. Plant Soil 368:433–444

Zhang JL, Tang WL, Huang QR, Li YZ, Wei ML, Jiang LL et al (2021) Trichoderma: a treasure house of structurally diverse secondary metabolites with medicinal importance. Front Microbiol 12:723828

Zhang J, Bayram Akcapinar G, Atanasova L, Rahimi MJ, Przylucka A, Yang D et al (2016) The neutral metallopeptidase NMP1 of *Trichoderma guizhouense* is required for mycotrophy and self-defence. Environ Microbiol 18(2):580–597

Zhang X, Chen L, Wang Y, Liu Q (2017) Genomic insights into *Trichoderma oblongisporum*. Mycol Prog 16(3):345–357. https://doi.org/10.1007/s11557-017-1334-8

Zhang Y, Tian C, Xiao J, Wei L, Tian Y, Liang Z (2020) Soil inoculation of *Trichoderma asperellum* M45a regulates rhizosphere microbes and triggers watermelon resistance to Fusarium wilt. AMB Express 10:1–13

MIX
Papier aus verantwortungsvollen Quellen
Paper from responsible sources
FSC® C105338

If you have any concerns about our products,
you can contact us on
ProductSafety@springernature.com

In case Publisher is established outside the EU,
the EU authorized representative is:
Springer Nature Customer Service Center GmbH
Europaplatz 3, 69115 Heidelberg, Germany

Printed by Libri Plureos GmbH
in Hamburg, Germany